Concepts and Principles of Rainfed Agriculture & Watershed Management

NIPA GENX ELECTRONIC RESOURCES & SOLUTIONS P. LTD.
New Delhi-110 034

Concepts and Principles of Rainfed Agriculture & Watershed Management

Dr. Mahendra Singh Pal
Ph.D., Post Doc. (P R China), ISA Fellow
Professor
Department of Agronomy
G B Pant University of Agriculture & Technology
Pantnagar-263145, Uttarakhand

NIPA GENX ELECTRONIC RESOURCES & SOLUTIONS P. LTD.
New Delhi-110 034

**NIPA GENX ELECTRONIC
RESOURCES & SOLUTIONS P. LTD.**

101,103, Vikas Surya Plaza, CU Block
L.S.C. Market, Pitam Pura, New Delhi-110 034
Ph. +91 11 27341616, 27341717, 27341718
E-mail: newindiapublishingagency@gmail.com
www: www.nipabooks.com

For customer assistance, please contact
Phone: + 91-11-27 34 17 17
Fax: + 91-11-27 34 16 16
E-Mail: feedbacks@nipabooks.com

ISBN: 978-93-91383-65-7

Composed and Designed by NIPA.

Dedicated to

My Parents

Preface

Agriculture is the backbone of Indian economy as nearly 60% rural population is engaged in agricultural activities. Though the contribution of agriculture in national GDP is declining mainly due to increasing contribution of other activities, it is still the main occupation of our rural masses and its significance is related to food and nutritional security of our country. It is again true that the income of farmers is not increasing as compared to other industrial sectors, so the government has recently launched many farmer's friendly programs for increasing farmer's income. Besides, the Central Government of India has renewed interest in agriculture and related sectors so that it may contribute not only in national GDP but also improve the living standards of our rural India. The Government has also opined that the subject agriculture should be included in new curricula of all secondary and higher secondary education so that the interest of younger generation may be increased in agriculture.

Number of agricultural universities and institutes of Indian Council of Agricultural Research are imparting different master and Ph D degrees but the syllabus differ among the universities/institutes, so to maintain the similar course content and teaching quality of undergraduate and post graduate program in agriculture, ICAR is playing a pivotal role and constituting Dean's Committees time to time to look after the syllabus and content of different courses. In this continuous process, the recommendation of 5th Dean Committee has been implemented in all undergraduate (B Sc Agriculture) degree programs in the country. The course, **'Rainfed Agriculture & Watershed Management'** is one of the important courses of B Sc Agriculture degree program. Therefore the book is written covering all aspects of syllabus as proposed by Dean's Committee. I do believe that the book is written precisely covering the concept and principles of rainfed agriculture and watershed management. The book certainly caters the need of both student and faculty.

Mahendra Singh Pal

Contents

Abbreviation

AICRPDA	:	All India Coordinated Research Project on Dryland Agriculture, Hyderabad
AISSLUP	:	All India Soil Survey and Land Use Planning, New Delhi
CADA	:	Command Area Development Authority
CAZRI	:	Central Arid Zone Research Institute, Jodhpur
CRIDA	:	Central Research Institute for Dryland Agriculture, Hyderabad
CSWCRTI	:	Central Soil & Water Conservation Research & Training Institute, Dehradun
DDP	:	Desert Development Programme
DPAP	:	Drought Prone Area Programme
ICARDA	:	International Centre for Agriculture Research on Dry Areas, Aleppo, Syria
ICRAF	:	International Centre for Research on Agroforestry, Nairobi, Kenya
ICRISAT	:	International Crops Research Institute for the Semi-Arid Tropics, Hyderabad
IFOAM	:	International Federation for Organic Agricultural Movements
IGFRI	:	Indian Grassland and Fodder Research Institute, Jhansi
IPCC	:	Intergovernmental Panel on Climate Change
IREP	:	Integrated Rural Energy Programme
IWDP	:	Integrated Wasteland Development Project
LCC	:	Land Capability Class
NBSS & LUP	:	National Bureau of Soil Survey and Land Use Planning, Nagpur
NRCAF	:	National Research Centre for Agroforestry, Jhansi
NRSA	:	National Remote Sensing Agency, Hyderabad
NWDB	:	National Wasteland Development Board, New Delhi
NWDPRA	:	National Watershed Development Programme for Rainfed Areas
RADAS	:	Reclamation & Development of Alkali and Acid Soil
RVP & FPR	:	River Valley Project & Flood Prone River
UNCCD	:	UN Convention to Combat Desertification
UNCED	:	United Nations Conference on Environment and Development
WALAMTARI	:	Water and Land Management Training Research Institute, Hyderabad
WDF	:	Watershed Development Fund
WDPSCA	:	Watershed Development Project for Shifting Cultivation Area

Glossary of Selected Terms

Aberrant weather: It refers to the deviation from normal weather conditions like early and late arrival as well as retreating of monsoon, long dry spells, flood and snow etc. that affect the crop cultivation adversely.

Accelerated erosion: Erosion much more rapid than normal, natural geological erosion, primarily caused by activities of man or in some cases of animals. So it causes removal of the surface soil from any area and denudes its natural protective cover as a result of human and animal interference. It involves higher rate of soil removal than soil formation. Therefore it is quite serious erosion type that makes soil unproductive and even barren if not controlled timely.

Acid soils: Soils having a pH value of less than 5.5 for most of the year. They are associated with a number of toxicities (Aluminum) as well as deficiencies (Molybdenum) and other plant restricting conditions.

Actual Crop Evapotranspiration (ET_{crop}): It refers to the rate of ET of a particular crop or plant in a specific period under prevailing soil and atmosphere conditions. In other words, it is rate of ET or total ET of a plant or crop in prevailing field conditions, therefore it varies with crop, soil, season and place ($ET_{crop} = K_c \times ET_0$).

Afforestation: Planting of forest trees on bare land or planting of a forest crops on land that has not previously or not recently carried a forest crop.

Agricultural drought : Agricultural drought refers to the conditions that result in adverse plant responses, which can range from reduced crop and forage yields to total crop or forage failure.

Agro-Ecological Cell (AEC: It is defined by a unique combination of land form, soil and climatic characteristics.

Agro-Ecological Zone: It is a land resource mapping unit, having a unique combination of land form, soil and climatic characteristics and/or land cover having a specific range of potentials and constraints for land use (FAO, 1996).

Agro-Ecological Zoning (AEZ): It refers to the division of an area of land into smaller units which have similar characteristics related to land suitability, potential production and environmental impact.

Agro-forestry: Growing of crops and forest trees and shrubs together on the same field and same time. The main objective of Agroforestry is to maximize the utilization of natural resources.

Agroforestry system: Growing of crops and forest trees together, so it is a land-use system in which woody perennials (trees, shrubs, palms, bamboos) are deliberately used on the same land management unit as agricultural crops (woody or not), animals or both, either in some form of spatial arrangement or temporal sequence. In Agro-forestry systems, there are both ecological and economic interactions between the different components.

Agrostology: It refers to study of grasses.

Ahar Pynes: It is a traditional water conservation technique of South Bihar. ***Ahar*** is a catchment basin from which flood water flow in form of rivulet and ***Pynes*** are artificial channel through which rivulet water is directed to field for irrigation.

Alfisols: Also known as red soils. Alfisols form in semi-arid to humid areas, typically under a hardwood forest cover. They have a clay-enriched subsoil and relatively high native fertility.

Alley cropping/ hedge row intercropping: An agro-forestry intercropping system in which crops are grown in alleys (space) of widely spaced shrubs/forest trees.

Altitude: Altitude, like elevation, is the distance above sea level.

Anemometer- Instrument to measure wind speed.

Anicut: An anicut is a masonry check dam that is constructed across a stream to impound water for maintaining and regulating irrigation. The water stored behind an anicut can be used for irrigation of crops or drinking water for humans and livestock.

Apparent drought: The climatic condition of low to medium rainfall mis-matching of the cropping pattern in relation to rainfall and moisture availability for crop production.

Aquaporins: Aquaporins, intrinsic membrane proteins, are synthesized in roots and leaves. They are abundantly found in plasma membrane and vascular membrane. They provide resistance to crop plants to water stress condition.

Aquifer: A saturated, permeable layer of sediment or rock that can transmit significant quantities of water under normal pressure conditions.

Arable land: Land that is capable of being cultivated and supporting agricultural production.

Arboriculture: A general term for the cultivation of trees.

Arid (desert) region: The region having E/PET ratio in range of 0.05-0.20 and rainfall <750mm.

Arid farming: Cultivation of crops in areas receiving <750mm rainfall and drought is fequent.

Aridisols (Desert soils): Aridisols (from the Latin *aridus*, for "dry" and *solum*) form in an arid or semi-arid climate. Aridisols dominate the deserts and xeric shrub lands which occupy about one third of the Earth's land surface.

Aridity Index (AI): An aridity index (AI) is a numerical indicator of the degree of dryness of the climate at a given location. The aridity index defined as the ratio P/PET has been proposed by the UNCCD to identify areas prone to using the following three classes: AI >0.65 (humid). 0.5-0.65 (dry sub humid),0.2-0.5 (semi arid), 0.2 –0.05 (arid) and <0.05 (hyper arid). As the Indian Meteorological Department (IMD), (P/PET)x100, where P - av. annual precipitation & PET-Potential ET. 1).

Available Soil Moisture (ASM): The available soil moisture refers to the soil water between FC and PWP or -0.3 bar to -15 bar. The available soil water is generally expressed in terms of soil depth. The research findings indicate that most of the crops utilize 50-75% of available soil water. The course soils have lower amount of available water than heavy soils.

Bad land: Land devoid of vegetation and broken by an intricate maze of narrow ravines, sharp crests, and pinnacles resulting from serious erosion of soft geological materials.

Baoli: Baoli means a wide mouth well or reservoir with stairs used for reaching to the water level of the baoli. It is traditional rain water harvesting structure and water is used for irrigation purposes.

Barograph*:* measure atmospheric pressure.

Barometer*:* A barometer is the instrument used to measure atmospheric pressure.

Base flow: Stream discharge derived from effluent ground-water seepage.

Basin listing: Small basins or pits are dug along the contour on mild sloppy lands with the help of basin lister. These basins collect rain water and reduce the runoff velocity.

Bawari: It is a unique step well or man made tank built in cities of Rajasthan to regualr supply of drinking water in the cities. The rain water is collected in tanks from hilly outskirts of cities through canals.The series of layered steps were built around the tank/well narrow and deepen the wells.

Bed load: Portion of the stream load that is carried along the stream bed without being permanently suspend in the flowing water.

Bed rock: The solid rock underlying soils and the regolith in depths ranging from zero to several hundred meters.

Bench mark: A point of reference in elevation surveys.

Bench terrace: A shelf-like embankment of earth with a level top and steep or vertical down hill face, constructed along the contour of sloping land to control runoff and erosion.

Blowout: An excavation in areas of loose soil, usually sand produced by wind erosion.

Broad-base terrace: A ridge type terrace 25 to 50 cm high and 4.5 to 9.0m wide with gently sloping sides, a rounded crown and a dish shaped channel along the upper side constructed to control erosion by diverting runoff along the contour at a non-scouring velocity. It may be level or have a grade towards one or both ends.

Buffer strips: Contour strips of grass or other erosion resisting vegetation between or below cultivated strips or fields.

Buffer zone: An area around a forest, national park, or any other conserved place that provides the local community with products that they would otherwise take from the forest, or that provides an opportunity to produce alternative products.

Bund: A barrier on the surface of the soil on sloping land to prevent runoff and soil erosion.

Bunding: Bunding also called a bund wall, is a constructed retaining wall around storage "where potentially polluting substances are handled, processed or stored for the purposes of containing any unintended escape of material from that area until such time as a remedial action can be taken".

Carrying capacity: Amount of animal life, human life or industry that can be supported indefinitely with available resources on a given area. In wildlife management, the optimum population density that a given environment or range is capable of sustaining permanently.

Catchment: It refers to the whole area from where rain water is collected to a common point. However it represents the drainage area of a primary river covering an area of few lakh acres or more.

Catchment area: The land surface on which rain falls. Sometimes it is called a 'water catchment'. When referring to particular streams or rivers, it is the land surface from which rain water flows into them, sometimes through tributaries (feeder streams).

Cavitation: Process of intense erosion due to the surface collapse of air bubbles found in constricted rapid flows of water.

Ceilometer: A ceilometer is a device that uses a laser or other light source to determine the height of a cloud ceiling or a cloud base from the ground. Ceilometers can also be used to measure the aerosol concentration within the atmosphere.

Check dam/Cement Plug/Nala Bund: Check dams are normally constructed across small streams having gentle slope and are feasible both in hard rock as well as alluvial formations. It controls the flow rate of water and checks soil and water erosion. It is usually built of inexpensive and temporary materials where dependence for ultimate protection in placed on vegetative cover.

Chiselling: The breaking or shattering of compacted soil or subsoil layers by use of a chisel.

Clear cutting: Strictly, the removal of an entire standing crop of trees. In practice, may refer to exploitation that leaves much unsaleable material standing. Also termed 'clear felling'. An area from which the entire timber stand has been cut. Removal of the entire stand in one cut.

Chronic drought: It is common in arid areas where rainfall and stored soil moisture are inadequate to meet crop water requirement during most of the years.

Chute Spillways: Chute spillway is an open channel like structure which is constructed on steep slope of the gully face with a suitable inlet and outlet. It usually consists of an inlet, vertical curve section, steep-sloped channel and outlet. The major part of the drop in water surface takes place in a channel.

Climatic constraints: The long-term prevalent weather conditions of an area, determined by latitude, position relative to oceans or continents, altitude, etc.

Cold waves and frost: It refers to the condition of cold winds and frost harmful to crops. The light irrigation, smoking during night, growing seedlings of fruits and vegetables in polyhouses etc. are the protecting measures.

Companion crop: A crop which is grown with another crop.

Concentrated flow: Overland flow is defined as water that flows over the land surface as either diffuse sheet flow (laminar or mixed laminar flow) or concentrated flow (**turbulent flow**) in rills and gullies (Ward and Robinson, 2000), or the flowing of a rather large accumulated body of water over, or one which corresponds to a net rain storm of duration longer than one unit period.

Concentration point: The single geographical location at which all surface drainage from a given area comes together as outflow.

Concentration time: The time required for discharge from the most distant point in a drainage area to reach the concentration point.

Conjunctive use of water: It refers to proper utilization of surface and ground water resources.

Conservation tillage: Conservation tillage is a tillage system that creates a suitable soil environment for growing a crop and that conserves soil, water and energy resources mainly through the reduction in the intensity of tillage and retention of plant residues

Consumptive use of water (CUW): The consumptive use of water is the total quantity of water required for fulfilling the plant ET and metabolic requirements, so it is equal to ET + water required for metabolic activities (Wm).

Contingency crop planning: Contingency planning refers to mitigate any unexpected, unusual, unfavorable and hence unwanted accidental or aberration weather situations occurring without prior knowledge at any time during crop period i.e. before the crop sown or after crop sown. The contingency crop planning therefore is proposed to mitigate such situation through the choice of appropriate crop and varieties, cropping systems or other necessary relevant crop management interventions.

Contingent drought: This involves an abnormal failure of rainfall. It may occur almost anywhere especially in most parts of humid or sub-humid climates.

Contour: An imaginary line connecting points of equal elevation on the surface of the soil.

Contour bunding: Contour bunding (or contour bundling) which involves the placement of lines of stones along the natural rises of a landscape and contour farming. It helps to capture and hold rainfall in the fields.

Contour farming: It means tilling and planting at right angles to the natural slope of the land or on contour lines or terraces.

Contour furrows: Making or digging furrows on the contour lines to prevent soil loss and allow water to infiltrate.

Contour interval: The vertical distance between contour lines.

Contour ploughing: A soil conservation technique involving ploughing parallel to the contour, across a slope rather than up or down it.

Contour strip cropping: The production of crops in comparatively narrow strips planted on the contour and at right angles to the natural direction of slope. Usually strips of grass or close growing crops are alternated with those in cultivated crops.

Cover crop: A close growing crop grown primarily for the purpose of protecting and improving soil between periods of regular crop production, or between trees and vines in orchards and vineyards.

Cover ground: Any vegetation production as a protecting 'mat' on or just above the soil surface. In forestry, low-growing shrubs and herbaceous plants are planted under the trees.

Creep: Slow mass movement of soil and soil material down relatively steep slopes primarily under the influence of gravity, but facilitated by saturation with water and by alternate freezing and thawing.

Crop base constraints: The problems related to crop production like unavailability of improved drought resistant crop varieties, lack of irrigation, poor crop diversification, low fertilizer application, unavailability of farm inputs etc.

Crop rotation: Crop rotation is the practice of planting different crops sequentially on the same plot of land to improve soil health, optimize nutrients in the soil, and combat pest and weed pressure.

Crop rotation: The growing of different crops in recurring succession on the same land.

Cropped area: The total area sown once and/or more than once in a particular year, i.e. the area is counted as many times as there are sowings in year.

Deep ploughing: Deep ploughing is a ploughing to a depth greater than 50cm (20in) as compared to ordinary ploughing which rarely exceeds 20cm (8in). The purpose of deep ploughing is to modify the soil water retention characteristics over the long term.

Deflation: Process where wind erosion creates blowout depressions or deflation hollows by removing and transporting sediment and soil.

Dehydrins: Dehydrins are the late embryogenesis associates (LEA) proteins, synthesized in response to drought and cold and involve in hydrophobic interactions leading to macromolecule stabilization by protecting cellular proteins from denaturation.

Dendrology: The study of trees, that is, a knowledge of the taxonomy, systematic relationships and filed recognition of tree species.

Dendrometer: A device for measuring the diameter of the trunk or branch of tree.

Depression storage: Water in puddles or small ponds depression.

Detention dam: A dam constructed for the purpose of temporary storage of stream flow, or surface runoff, and for releasing the stored water at controlled rates.

Dew: water vapor condenses on the object such as blades of grass when the surface temperature is equal to the dew point.

Dhaam: It refers to water collected from small and big streams in shape of lake. This water is mainly used for domestic animals and irrigation purposes.

Dhara: It is natural springs having a continuous water flow that is collected in stone lined water tanks. The water tanks are covered to avoid pollution and used for drinking purposes in Uttarakhand hills.

Dike: An embankment to confine or control water, especially one built along the banks of a river to prevent overflow of low land.

Disperse: To distribute or suspend fine particles, such as clay, in or throughout a dispersion medium, such as water.

Diversion dam: A structure or barrier built to divert part or all of the water of a stream to a different course.

Diversion: A diversion is a channel with a supporting ridge on the lower side constructed across the slope to intercept runoff and minimize erosion, or to prevent excess runoff from flowing into lower lying areas. In some areas a series of diversions are constructed across the slope similar to terraces but with greater horizontal and vertical spacing. It is also known as diversion terraces.

Drag: The force retarding the flow of water or wind over the surface of the ground.

Drainage area: An area whose runoff is more or less separate from the runoff of adjacent areas, so that it can be considered a hydrologic unit.

Drainage divide: The boundary line, along a topographic ridge or along a sub-surface formation, separating two adjacent drainage basins.

Drainage: The frequency and duration of periods when the soil is free of saturation with water or removing of stagnated water from the field.

Drift: Material of any sort deposited in one place after having been moved from another. The term is most commonly used when referring to glacial drift.

Drizzle: It refers to a light steady rain in fine drops (0.5mm) and intensity is <0.5mm. Drizzle is normally produced by low stratiform clouds and stratocumulus clouds.

Drop Inlet Spillways : A drop inlet or shaft spillway is one in which the water enters through a horizontally positioned circular or rectangular box type riser or inlet and flows to some type of outlet protection through a circular (horizontal or near horizontal) conduit. It consists of an earthen dam and a pipe spillway.

Drop Spillways : The drop spillway is a weir structure. Flow passes through the weir opening, drops to an approximately level apron or stilling basin and then passes into the downstream channel. Drop spillway is one of the most commonly used gully control structures. It is mainly used at the gully bed to create a control point.

Drop structure: A dam constructed in a gully or other water course to lower the grade of the water course and thereby decrease the velocity of flow, promote the deposition of sediment, and prevent further channel erosion.

Drought: It is a condition of insufficient moisture supply to the plants under which they fail to develop and nurture properly. If may be caused by soil, atmosphere or both.

Drought avoidance: 'Drought avoidance' is the ability of plants to maintain (relatively) higher tissue water content despite reduced water content in the soil.

Drought escape: Drought escape is a classical adaptive mechanism which involves rapid plant development to enable the completion of the full life-cycle prior to a coming drought event.

Drought Management: The Drought Management Group was constituted to coordinate the efforts to deal with drought in various states.

Drought resistance: Drought resistance is defined as the mechanism causing minimum loss of yield in a water deficit environment relative to the maximum yield in a water constraint free management of the crop.

Drought tolerance: Drought tolerance is the ability to which a plant maintains its biomass production during arid or drought conditions. Other plants, specifically crops like corn, wheat, and rice, have become increasingly tolerant to drought with new varieties created via genetic engineering.

Dry Farming: Dry farming refers to the cultivation of crops in areas that receives annual rainfall less than 750 mm and crop failures are most common due to prolonged dry spells during crop period.

Dry land: The land or area which receives an annual rainfall of 750 mm or less and there is no irrigation facility for raising crops.

Dry land crops: It refers to all such crops which are drought resistant and can complete their life cycle without irrigation in areas receiving an annual rainfall less than 750 mm.

Dry sowing: Sowing of crops in dry field conditions either after ploughing or without ploughing.

Dry sub humid: The region having 0.52-0.65 P/PET and 600-800mm rainfall.

Drylands farming: Dryland farming refers to the crop cultivation in areas receiving 750- 1150mm annual rainfall.

Dune: A mound or ridge of loose sand piled up by the wind; common where sand is abundant and wind usually strong as along lake and sea shores and in some desert and desert areas.

Dust mulch: A loose, finely granular or powdery condition on the surface of the soil usually produced by shallow cultivation.

Dust mulching: Dust mulches are created by intensively hoeing the soil surface, creating a finely-textured layer of soil.

Early season drought: It occurs due to delay in onset of monsoon or due to long dry spells after early sowing. The early withdrawal of monsoon results in reduced water availability period for crop growth.

Ecological drought: The ecological drought takes place when the productivity of a natural ecosystem fails significantly as a consequence of distress induced environmental damage.

Edaphic: It is related to soil factors.

Effective precipitation: The portion of the total precipitation that becomes available for plant growth.

El-Nino effect: It refers to the large-scale ocean-atmosphere climate interaction linked to a periodic warming in sea surface temperatures across the central and east-central. It is associated with high pressure in the western Pacific. El Nino conditions bring deficient rain, which has in turn resulted in some of the worst droughts in recorded history in India.

Eluviation: The transportation of soil material in suspension, or in solution within the soil, by the downward or lateral movement of water.

Energy Farming: The process of using land to grow crops, woody or otherwise, that provide fuel, for example, close-planted, fast-growing tree species such as poplar (temperate) or Leucaena (tropical). These may be harvested by hand or mechanically. Hydrocarbon fuels can be extracted from other kinds of plants, for example, from some Euphorbia species.

Enriched fallow: A form of Agroforestry in which useful, mainly woody species are sown or planted before cultivation ceases, or at the time it does, so that during the fallow period, or when the land is next cleared for cultivation, products are available for household use or market that would not otherwise have been there (for example, fruits, bamboos, rattans, medicinal plants).

Entisols: It is a new alluvial soils. Entisols are defined as soils that do not show any profile development other than an A horizon. An entisol has no diagnostic horizons, and most are basically unaltered from their parent material, which can be unconsolidated sediment or rock.

Entrainment: One of three distinct processes involved in erosion. More specifically, it is the process of particle lifting by an agent of erosion.

Eri: Eri means tank and it is one of the oldest widely practiced for water conservation in Tamil Nadu, India.

Erode: To wear away or remove the land surface by wind, water, or other agents.

Erodibility: Susceptibility to erosion, erosion proneness or erodability (or erodibility) is an inherent yielding or non resistance of soils and rocks to erosion. A high erodability implies that the same amount of work exerted by the erosion processes leads to a larger removal of material.

Erodible: Susceptible to erosion.

Erosion: (i) The wearing away of the land surface by running water, wind, ice or other geological agents, including such processes as gravitational creep; (ii) Detachment and movement of soil or rock by water, wind, ice, or gravity.

Erosion pavement: A layer of coarse fragments such as sand or gravel, remaining on the surface of the ground after the removal of fine particles by erosion.

Erosive: Tending to cause erosion; the term applies to the eroding agent, such as wind or water.

Erosivity: It is defined as a potential ability of the rain to cause erosion.

Estuary: Somewhat enclosed coastal area at the mouth of a river where nutrient rich fresh water meets with salty ocean water.

Eutrophication: Nutrient enrichment of lakes, ponds and other such waters that stimulates the growth of aquatic organisms, which leads to a deficiency of oxygen in the water body.

Evaporation: Water loss from the soil or water surface due to variation in water potential of water surface and atmosphere.

Evaporimeter: An evaporation gauge or evaporimeter is a pan used in the measurement of the evaporation of water into the atmosphere over specified periods of time.

Fallow: Allowing crop land to lie idle, either tilled or untilled, during the whole or greater portion of a growing season. Tillage is usually practiced to control weeds and encourage the storage of moisture in the soil. Land rested from deliberate cropping, not necessarily without cultivation or grazing but without sowing.

Felling: Cutting tress for their removal.

Field border plantings: Vegetation establishment on field borders to conserve soil and provide cover and food for wildlife.

Field Capacity (FC): The soil moisture content that retains in soil pores after drainage of gravitational water. This stage comes after 48 – 72 hours after either irrigation or heavy rains in the field. It is the upper limit of available water that is held in soil at 0.1 – 0.3 bar or 0.01 – 0.03 MPa or an average of 0.3 bar.

Field strip cropping: A system of strip cropping in which crops are grown in parallel strips laid out across the general slope but which do not follow the contour. Strips of grass or close-growing crops are alternate with those in cultivated crops.

Filter strip: A strip of permanent vegetation of sufficient width and vegetative density above farm ponds, diversion terraces and other structures to retard flow of runoff water, causing it to deposit soil, thereby preventing silting of structure or reservoir below.

Flood plain: The land bordering a stream, built up of sediments from overflow of the stream and subject to inundation when the stream is at flood stage.

Fog: A thin cloud of varying size formed at the surface of the earth by condensation of atmosphere vapour.

Freezing rain: Sleet is simply frozen raindrops and occurs when the layer of freezing air along the surface is thicker. This causes the raindrops to freeze before reaching the ground.

Frost: Dew forms and then it freezes. This commonly occurs during night time. Radiational cooling drops the ground temperature down enough. There are three kinds of frost; delicate crystallized ice on window panes is called ***Hoar frost. Glazed frost*** consists of thick coating of ice on cold surfaces. ***Rime frost*** is formed when super- cooled water droplets freeze on contact with cold surface. It often occurs when freezing fogs or drizzle fall on blanket ground.

Gabions: Small stones and boulders placed in wire nets are placed along the banks particularly to prevent the banks of river from torrents.

Geojute: The jute mates are used to stabilize the landslides mine spoil areas and constructed slopes. These geojutes check the runoff flow and also air and water erosion.

Geological erosion: The normal or natural erosion caused by geological processes acting over long geologic periods and resulting in the wearing a way of mountains, the building up of flood plains, coastal plains, etc.

Geological soil erosion: Natural or geological soil erosion is a very common and continuous process of soil erosion. The physical, chemical and biological weathering of rocks and parent material are pre- requisites for soil formation. Therefore the geologic erosion maintains equilibrium between the removal and formation of soil. If the balance between soil removal and formation is disturbed, the extent of natural erosion in form of soil depth and its characteristics will be changed.

Glacial drift: Rock debris that has been transported by glaciers and deposited, either directly from the ice or from the melt-water. The debris may or may not be heterogeneous.

Glacier: A large long lasting accumulation of snow and ice that develops on land. Most glaciers flow along topographic gradients because of their weight and gravity.

Gley: Soil with impeded drainage. Used in soil classification based on profile leaching.

Graded terrace: A terrace having a constant or variable grade along its length.

Grassed waterway: A natural or constructed waterway, usually broad or shallow, covered with erosion resistant grasses, used to conduct surface water from crop land.

Grassland: Land covered with grasses and other herbaceous species. Woody plants may be present, but if so, they do not cover more than 10% of the ground. There are many different types of grassland designated by ecozone, topography, climate, soil conditions, and so on. Derived grassland is maintained in that condition by regular burning; edaphic grassland arises on particular soil types, for example, those found in or around permanent or seasonal swamps.

Grazing: A method of feeding by animals/herbivores characterized by repeated removal of only a part, generally the leaf of the plant which is most commonly called herbage.

Ground water basin: A ground water storage area more or less separate from neighboring groundwater storage areas, so that it can be considered a hydrologic unit.

Ground water: The general sub-surface water body in the zone of saturation.

Grubbing: Felling trees and shrubs by exposing and cutting the roots.

Gules/Canals: It is cemented channel carrying river water from one place to other for irrigation purposes. The term 'gule' is used in Uttarakhand hills.

Gully: A channel resulting from erosion and caused by the concentrated, but intermittent, flow of water usually during and immediately following heavy rains. Gullies are deep enough to interface with and not to be obliterated by normal tillage operations.

Gully erosion: It is an advanced stage of water erosion in which rills or narrow channels become wider and deeper, known as gullies (usually 30 to 60 cm wide and 22.5 to 30 cm deep) and type of erosion is called gully erosion.

Hail: Water droplets are carried high into the atmosphere by thunderstorm updraft which cause them to freeze. Multiple drops tend freeze together, which is why the diameter of hail can be large.

Hailstorm: A hailstorm is an unusual weather phenomenon in which balls of ice, called hail, fall from the sky. It is widely disastrous to crops.

Haveli system: It is prevalent in Bundelkhand region where rain water is collected in crop fields during rainy season and *Rabi* crops are grown after receding the rain water.

Heat waves: It is a condition of high air temperature in which crop growth and development is affected adversely.

Hedge row: A barrier of bushes, shrubs, or small trees growing close together in a line.

High root/shoot ratio: The high root-shoot ratio is related to efficient root system. It is generally observed that drought resistant plants have lower volume of shoot than root mainly because of reduced size of plant height, number of leaves and branches in short of available water.

Highland climate: Highland climate is the climate of 'high land'. So, this climate is found in high mountain areas.

Humid: Humid is used to describe air that is full of water vapor. Humid is most commonly applied to weather or the general climate of a place, especially when the temperature is hot. Eastern and Western coastal area, West Bengal, Andaman and Nicobar islands and some other regions are considered as humid region.

Humid mesothermal: Humid mesothermal climates are found in the middle latitudes where winters are mild and short.

Humid microthermal: Humid microthermal climates represent the continental climates of North America and Eurasia.

Humidity: *Humidity* is the concentration of water vapor present in the air. Water vapor, the gaseous state of water, is generally invisible to the human eye.

Hydram or Hydraulic dam: It is a water lifting device. It is self powered impulse pump using the energy of falling water for lifting water to higher altitudes. In hilly area, the hydrams may be constructed if the flow is 5 liters / minute and height of the waterfall is minimum one meter from the site of hydram. In this process, the water can be lifted up to 150m otherwise normal lift rarely exceeds 50m.

Hydrological drought: A hydrological drought occurs when a meteorological drought significantly reduces the availability of water resources in rivers, lakes and underground table. Hydrological drought is a natural phenomenon, but it may be exacerbated by human activities.

Hyper arid: As per UNEP classification, hyper arid region is characterized by 0.05 P/PET ratio and 200mm rainfall.

Inceptisols: It is a type of alluvial soils. Inceptisols (from Latin inceptum, "beginning") are soils that exhibit minimal horizon development.

Infiltration: The downward entry of water into the soil.

Inorganic mulch: Polythene, gravel/stone, Al coated plastic and foil etc. are the inorganic mulch materials.

Interpolated cropping: A cropping sequence where two or more species are grown on the same unit of land but at least one species has a later sowing (planting) time and an earlier harvest than the other(s). Refers to situations where the crop duration of species is very different.

Irrigated area: It refers to area under irrigation. The area is assumed to be irrigated for cultivation through such sources as canals (Govt. & Private), tanks, tube-wells, other wells and other sources.

IW/CPE ratio: It is climatological approach of scheduling irrigation.

Jacks: Jacks are wooden posts of 8 cm diameter and 2-3 m length, which are placed in series along the eroded bank and are tied together.

Jetted posts: Jetted posts are green posts of 2-2.5 m. in length and 6-12cm diameter and they are erected along the actively eroded banks in two rows having 1 m distance to protect the banks of streams.

Jhalaras: They are rectangular shaped step wells that have tiered steps on three or four sides. These step wells collect the subterranean seepage of an upstream reservoir or a lake.

Johads: Johad is a small earthen check dam that captures & conserves rainwater, improving percolation & recharging groundwater. So normally the johad is a crescent shaped small check dam built from earth and rock to intercept and conserve rainwater and practiced in Alwar district of Rajasthan.

Khadin: A *khadin*, also called as *dhora*, is an indigenous construction designed to harvest surface runoff water for agriculture. It is an age old rain water harvesting technique in Rajasthan.

Kutta-crate structures: The vacant cement gunnery bags, filled with mine spoil, known as kutta-crates, are used to construct the check dams to stabilize the limestone mine spoil slopes.

La Nina effect: It refers to a decrease in the oceanic temperatures across the equatorial Pacific and so the water of the Pacific Ocean turns colder than usual resulting into strong and sturdy wind flow.

Land capability: The suitability of land for use without damage. Land capability involves consideration of (a) the risks of land damage from erosion or other causes, and (b) the difficulties in land use owing to physical land characteristics including climate.

Land capability classification: The arrangement of land units into various categories based upon the properties of the land or its suitability for some particular purpose. There are 8 land capability class, 1st four class are good for crop cultivation and last 4 for alternate land use systems.

Land degradation: Land degradation is defined as the temporary or permanent decline in the productive capacity of the land and the diminution of the productive potential, including its major land uses (e.g., rainfed arable, irrigation, and forests), its farming systems, and its value as an economic resource.

Landscape: All the natural features such as fields, hills, forests, water etc. which distinguish one part of the earth's surface from another part. Usually that portion of land or territory which the eye can comprehend in a single view, including all its natural characteristics.

Landslide: A mass of material which has slipped down hill under the influence of gravity, frequently assisted by water (when the material is saturated).

Late season or terminal drought: It is the crop conditions in which crop suffer water stress during grain filling stage. It occurs as a result of early cessation of monsoon, mainly during the years with late commencement or weak monsoon activity.

Latitude: Latitude is the measurement of distance north or south of the Equator.

Leaching: The washing out of material from the soil, both in solution and in suspension.

Level terrace: A terrace that follows the absolute contour as contrasted with a graded terrace. Used only on permeable soils where conservation of moisture for crop use is particularly important or where outlet channels are impractical.

Ley farming: Rotation of arable crops with two or more years of sown pasture (crop+pasture combination).

Ley pasture: A temporary pasture grown as a specific phase in a defined crop rotation sequence.

Loess: Deposit of wind transported fine textured material, uniform and unstratified, mostly silt, but may contain some fine sand and clay.

Lopping: Cutting one or more branches of a standing tree, for example, for fuel or fodder.

Lysimeter: Lysimeter is a tank or container filled of soil is installed in the field to grow the crops with similar field conditions. Lysimeters measure crop ET, runoff, deep percolation, evaporation and transpiration. Lysimeter may be weighing or non weighing types. The weighing type lysimeters are useful for short period studies (hourly and daily studies). They are costly and ET measurement is complex. The non weighing type lysimeters are cheap and ET measurement is comparatively simple.

Meadow strip: A sloping field or strip of grassed land which, in addition to yielding a hay crop, acts as a broad shallow water channel during periods of runoff. Often used as a terrace outlet channel and usually much larger in extent than a grassed water way.

Metabolic strain: The metabolic strain refers to the response of drought occurred in plants under drought stress and it is reversible in absence of strain/drought.

Meteorological drought: It occurs when rainfall is less than average over a significant period, often a month. The degree of dryness or rainfall deficit and the length of the dry period are deciding factors to meteorological drought.

Micro irrigation: Micro irrigation is a modern method of irrigation; by this method water is irrigated through drippers, sprinklers and foggers and by other emitters on surface or subsurface of the land. It includes both sprinkler and drip irrigation methods.

Micro watershed: It is a drainage area of a hexad covering an area of less than 1000 acre or 400 ha.

Mid-season drought: Occurs due to inadequate soil moisture availability between two successive rainfall events during crop growth. Its impact depends on crop growth stage, duration and intensity of the drought spell.

Mine spoils: Soil degradation by mining.

Mini watershed: It represents a drainage area of a pentad river/stream with an area of 1,000 to 5,000 acre or 400 to 2000 ha.

Mist - It refers to a very thin fog.

Mixed cropping: Mixed cropping is also known as inter-cropping, polyculture, or co-cultivation. It is a type of agriculture that requires planting two or more plants simultaneously in the same field, interdigitating the plants so that they grow together.

Moisture adequacy index (MAI): Moisture Adequacy Index (MAI) is the ratio of actual evapotranspiration (AET) to the potential evapotranspiration (PET). Agricultural droughts during different seasons (years) were classified into four groups based on average MAI during the season. MAI is also used to quantify the agricultural drought. It is obtained from weekly water balance using soil characteristics, crop growth period and water requirement of major crops. It is calculated as MAI = (AE/PE) x 100, where AE is actual crop ET and PE is potential crop ET.

Moisture deficiency index (MDI): The MDI is usually determined on the basis of annual precipitation and annual potential ET, as adapted by Thornthwaite and Mather in 1955. A regression equation of drainage losses on rainfall was derived: $Dr = O.63R - 8.29R \sim 13.16$ where Dr = drainage and R = rainfall. The mean calculated soil moisture deficit for a month was used to describe the degree of dryness of that month.

Moisture Equivalent (ME): The moisture equivalent refers to the amount of water retained by an initially saturated soil sample against centrifugal force of 1000 times of gravity force for half an hour. It is a laboratory analogue of FC and may be used as equal to FC. The moisture equivalent also denotes the moisture content in the inner capillary or microcapillary pores. The values of FC and ME are same for medium textured soils, slightly less than FC for sandy soils and slightly more than FC for very clayey soils.

Moisture index: The term used instead of 'moisture budget', e.g. by C. W. Thornthwaite (1955), and calculated from the aridity and humidity indices, as $I_m = 100 \times (S - D)/PE$, where I_m is the moisture index, S is the water surplus, D is the water deficit, and PE is the potential ET.

Montane: It is also known as the alpine climate and it is usually found in the northern most tip of India, climatic conditions in this region decrease with elevation.

Mulch: Any material such as straw, sawdust, leaves, plastic film, loose soil etc. that is spread upon the surface of the soil to protect the soil and plant roots from the effects of rain drops, soils crusting, freezing, evaporation etc.

Mulching: Mulching refers to the practice of covering the soil surface with different organic and inorganic materials, known as mulch materials to reduce moisture loss, balance wide variations in diurnal soil temperatures especially in the root zone, control weeds, insect-pests and diseases and finally enhance crop yield.

Mulch farming: Systems of farming in which the organic residues are not ploughed into or otherwise mixed with the soil but are left on the soil surface as mulch.

Multipurpose tree: A woody perennial purposefully grown to provide more than one significant contribution to the production or service functions (food, feed, fodder, fibre, wood, shelter, shade, land sustainability) of the land-use system that it occupies.

Nadi /Pond system: Nadi is a small village pond either excavated or embanked at the lowest point of the village for collection of rain water from adjoining area or catchments.

Natural erosion: Wearing away of the earth's surface by water, ice, or other natural agents under natural environmental conditions of climate, vegetation, etc. undisturbed by man.

Naula: It is a naturally-occurring water aquifer and flowing water is collected in stone lined tanks. The dripping water from aquifers is collected and used mainly for drinking purposes in Uttarakhand hills.

Niche: A place or position adapted to the character, or suited to the merit, of a person or plants. The sum total of adaptations of an organismic unit. In ecology, the total range of conditions under which the individual (or population) lives and replaces itself, or the position or status of an organism within its community and ecosystem resulting from the organism's structural adaptations.

Normal erosion: The gradual erosion of land used by man which does not greatly exceed natural erosion.

Organic mulch: Compost/FYM, crop residues, saw dust, peat, cover cropping, news papers etc. are the organic mulch materials.

Overland flow: Flow of water over the land surface as sheet flow not in channels.

Oxisols: Oxisols (from French oxide, "oxide") are very highly weathered soils that are found primarily in the intertropical regions of the world. These soils contain few weatherable minerals and are often rich in Fe and Al oxide minerals. They are typical laterite soils.

Pan Evaporimeter: It measures the evaporation and crop ET is estimated by multiplying pan evaporation (E_{pan}) to crop factor (K_{crop}).

Parent material: The horizon of weathered rock or partly weathered soil material from which the soil is formed.

Pasture: Land covered with grass and other low plants suitable for grazing animals, especially cattle or sheep. Pastures are properly managed grasslands.

Pasture improvement: Any practice of grazing, mowing, fertilizing, liming, seeding, scattering, droppings, contour furrowing, or other methods of management designed to improve the vegetation for grazing purposes.

Percolation tank: Percolation tanks are the structures for recharging ground water, generally constructed across streams and bigger gullies in order to impound a part of the run-off water. This water, in due course, finds its way into subsoil and recharges the ground water.

Percolation: The downward movement of water through soil.

Permanent drought: It refers to the conditions where rainfall is always scanty and insufficient for crop production, so crop production is possible only after providing irrigation. It occurs in driest climate where sparse vegetation is found. It represents the extremely arid region.

Permanent wilting point (PWP): The soil moisture level or points at which plants are wilted and cannot recoup its original state is called permanent wilting point. At this stage, soil water is not available sufficiently to meet out the ET demand of the field. The force at which plants get wilted is -15 bar or -1.5 MPa. This is also called as *'ultimate wilting point'* as the plants cannot survive but die even if the field is irrigated. It is the lowest limit of available water.

Pitche Atmometer: It is also called 'pitche evaporimeter' and measures evaporation loss.

Plant nutrition: *Plant nutrition* is the study of the chemical elements and compounds necessary for plant growth, plant metabolism and their external supply.

Plastic strain: The plastic strain refers to the response of drought occurred in plants under drought stress and it is irreversible in absence of strain/drought.

Polar climate: A polar climate consists of cool summers and very cold winters, which results in treeless tundra, glaciers or a permanent or semi-permanent layer of ice.

Potential evapotranspiration (PET): The potential evapotranspiration of a plant or crop is the highest rate of ET obtained by a short and actively growing crop or vegetation with abundant foliage completely shading the ground surface with optimum water supply in a given climate. Under specific climatic conditions if any plant or crop achieves maximum ET under optimum growth conditions shading the soil surface, the value of ET is called PET.

Precipitation: In meteorology, *precipitation* is any product of the condensation of atmospheric water vapor that falls under gravitational pull from clouds.

Pressurized irrigation: It includes sprinkler and drip irrigation systems.

Productivity: The capacity of a soil in its normal environment, for producing a specified plant or sequence of plants under a specified system of management. Productivity emphasized the capacity of soil to produce crops and should be expressed in terms of yields.

Pyranometer/ Pyrheliometer: Instrument that measures solar radiation.

Rain: It refers to water vapor that condenses around condensation nuclei (such as dust) and falls when the droplet is heavy enough.

Rainfed farming: Rainfed farming refers to the crop cultivation in the areas receiving >1150 mm annual rainfall. Drainage is a major problem.

Rain forest: Generally, a forest that grows in a region of heavy annual precipitation. There are both tropical and temperate rainforests.

Range: Land that produces primarily native forage suitable for grazing by livestock.

Rangeland: Land that provides or is capable of providing forage for grazing animals.

Range management: The scientific management of range land for the continuous production of forage and livestock, consistent with the use of the land for other purposes.

Ratooning: The practice of growing crop from the crop stems or stubbles left after harvest of main crop and the crop is called ratoon crop as in sugarcane or sorghum.

Ravines: A ravine refers to a small narrow steep-sided valley that is larger than a gully and smaller than a canyon and that is usually worn by runoff water, so ravines are typically classified as larger in scale than gullies, although smaller than valleys.

Reach: A specific portion of the length of a stream channel.

Reference Crop Evapotranspiration (ET_0): The term 'reference crop evapotranspiration' was given by Doorenbos and Pruitt (1973) and it may be defined as "the rate of evapotranspiration from an extended surface of 8 to 15 cm tall green grass cover of uniform height actively growing completely shading the ground and not short of water ($ET_0 = C \times ET_0^*$, where C is a correction factor and ET_0^* is an unadjusted ET).

Reforestation: Planting of forest trees on deforestation or replacing forests after felling.

Relief: The relative differences in elevation between the upland summits and the lowlands or valleys of a given region.

Resilience: The capacity of a soil (or other ecosystem) to return to its original state after a disturbance.

Retaining walls (Gabions): The retaining walls are constructed across the slopes of precipitous hills to check the landslides.

Retards: Retards are special structures used to reduce the stream velocity sufficiently near the bank to prevent bank erosion or scour its toe by inducing deposition. Examples are live hedge, wooden poles (jacks) and jetted posts.

Ridge terrace: A long, low ridge of earth with gently sloping sides, and a shallow channel along the upper side, to control erosion by diverting surface runoff across the slope instead of permitting it to flow uninterrupted down the slope. Types of ridge terraces include the broad base, drainage graded level, narrow base etc.

Rill: A small, intermittent water course with steep sides; usually only a few centimetres to deep and hence no obstacle to tillage operations.

Rill erosion: It is a second stage of water erosion in which small and uniform finger shaped channels or grooves are formed over the entire field at rainfall.

Riparian forest: Vegetation, with trees and shrubs, growing alongside or close to a watercourse, lake, swamp, or the like, and often dependent on its roots reaching the water table.

River wash: Soil material transported and deposited by streams.

Riverine: An environment created along permanent and semi-permanent streams because of the increase in soil moisture.

Rotational grazing: It means dividing the pasture into sub-pastures typically called paddocks that are used for grazing one by one. Rotational grazing allows a producer to better manage forage in a pasture, but requires more labor than continuous grazing systems.

Runoff: Runoff is that part of precipitation, snow or ice melt or irrigation water that flows from the land to the streams or other water surfaces.

Sediment: Fragmented organic or inorganic material derived from the weathering of soil, alluvial and rock materials; removed by erosion and transported by water, wind, ice, and gravity.

Sedimentation: The deposition of sediment from a state of suspension in water or air.

Saltation: The major portion of the soil carried by the wind moves in a series of short bounces or jumps called 'saltations'. Normally the size of soil particles varying from 0.05 to 0.50 mm in diameter is prone to saltation.

Sand dunes: A dune is a mound of sand formed by the wind, usually along the beach or in a desert. Dunes form when wind blows sand into a sheltered area behind an obstacle. Dunes grow as grains of sand accumulate. Every dune has a windward side and a slip face. There are five different shapes of sand dunes i.e. crescentic, linear, dome, parabolic and star.

Scheduling irrigation: It refers to 'when to irrigate, how much to irrigate', how to irrigate' and where to irrigate'. However, scheduling irrigation refers to the proper time of irrigation.

Seasonal drought: It is defined as the condition where drought is experienced seasonally or during the crop growth period, is known as season drought.

Seed priming: Seed priming is a process of regulating the germination process by managing the temperature and seed moisture content, the seed is taken through the first biochemical processes within the initial stages of germination.

Semi arid: the region with 0.2-0.5 P/PET and 400-600mm rainfall.

Sheet erosion: Sheet erosion refers to uniform washing away of top thin layer of soil by run- off water but remains unnoticed.

Shelterbelt: An extended wind break of living trees and shrubs established and maintained for protection of farm lands over an area larger than a single farm.

Silting: The deposition of water-borne sediments in stream channels, lakes, reservoirs, or on flood plains, usually resulting from a decrease in the velocity of the water.

Slash: In forestry, the vegetation (branch and other woody and leafy debris) left on the forest floor after trees have been felled or trimmed.

Sleet: When rain droplets freeze somewhere along its path from the clouds to the ground. Sleet is mixture of snow and rain.

Slope: The degree of deviation of a surface from horizontal, measured in a numerical ratio, percent or degrees.

Social economic constrains: Problems related to socio-economic conditions of the local masses like risk, lack of capital, support price, market intelligence, farm input and supply chain.

Social forestry: The planting and tending of trees or shrubs for the well-being and betterment of local communities.

Soil degradation: The land degradation is the major constraint of dryland regions mainly due to chemical degradation and soil and water erosion. Sequestration of carbon also turns out to be a major problem which further degrades the soil making it less productive.

Soil erosion: Soil erosion is a gradual process that occurs when the impact of water or wind detaches and removes soil particles causing the soil to deteriorate.

Spillway: A conduit in or around a dam for the escape of excess water.

Splash erosion: The spattering of small soil particles caused by the impact of rain drops on very wet soils. The loosened and spattered particles may or may not be subsequently removed by surface runoff.

Spurs: Spurs are the structures constructed at an angle to the stream flow extending from the bank into the stream to divert the speedy water flow and protect the banks of rivers/streams. Spurs are also known as spur dykes or groyne.

Spur terrace: A short terrace used to collect or divert runoff.

Stevenson Screen: Instrument or device to measure both temperature and humidity.

Stream bank erosion: The small streams, rivers and torrents (hill streams) are the main cause of stream bank erosion in which the sides of river banks are cut off and washed away in very unregular shapes with violent flowing water.

Stream gaging: The collection of stream-flow data by direct measurements of discharge and water surface elevation.

Strip cropping: The practice of growing crops alternatively in different strips along contours or across the prevailing direction of wind.

Strip farming: It refers to plant in alternate strips of intensively cultivated crops with strips of sod farming crops. The erosion sediments and runoff from the clean tilled strips are filtered out and retained on sod like crop strips.

Stubble mulch: The stubble of crops or crop residues left essentially in place on the land as a surface cover before and during the preparation of the seedbed and at least partly during the growing of a succeeding crop.

Sub Humid : Regions where the climate is intermediate between semiarid and humid, and the precipitation supports dense growth of tall or short grasses, but not forests or areas with an LGP ranging from 75 to less than 120 days (dry) semiarid and 120 to less than 180 days (moist) semiarid.

Sub Montane soils: These are mountainous soils formed in high rainfall regions of sub- Himalayan under coniferous forests. The soils are acidic in reaction. The organic matter accumulation is high and there is absence of free lime.

Sub soiling: The subsoil hard pan is broken with the help of subsoiler mainly to increase infiltration rate and conserve the soil moisture. The subsoiling also helps in reducing runoff velocity and conserve soil and water.

Sub surface barriers: The sub surface barriers are constructed across the streams particularly to recharge the ground water aquifers.

Sub watershed: It is drainage area of a quaternary river/stream with an area of 5,000 to 10,000 acre or 2,000 to 4000 ha.

Sub-catchment: It is the drainage area of a secondary river having an area of more than a lakh acre or 40,000 ha.

Sunken Screen Pan Evaporimeter: It measures the evaporation loss.

Surangam: Surangam is a traditional water conservation system in which fine tunnels are made and the horizontal wells are dug out at specific distance to collect rainwater for drinking as well as irrigating crops like rice and coconut. It is practiced in Karnataka and Kerala states.

Surface creep: Surface creeping refers to rolling or sliding of large soil particles along the ground surface. These soil particles are too heavy or bigger to be lifted by wind moves by saltation. It causes maximum erosion as the size of soil particles are of 0.5 to 2mm in diameters.

Surface sealing: The orientation and packing of dispersed soil particles in the immediate surface layer of the soil, rendering it relatively impermeable to water.

Suspension: Suspension refers to floating of small size soil particles in the air.

Tanka: Tanka is also known as *Kund* or *Kundi.* Tanka system is an ancient water harvesting technique in dry regions and still very common in eastern Rajasthan.

Technological constrains: Lack of suitable technologies, unavailability of high yielding crop varieties, poor plant protection, little adoption of alternate land resistance, poor extension, poor and alternate farmers.

Terrace: A raised, more or less level or horizontal strip of earth usually constructed on or nearly on a contour and supported on the downslope side by rocks or other similar barrier and designed to make the land suitable for tillage and to prevent accelerated erosion.

Thinning: Thinning is a term used in agricultural sciences to mean the removal of some plants or parts of plants to make room for the growth of others. Selective removal of parts of a plant such as branches, buds or roots is typically known as pruning.

Through fall: The proportion of the gross precipitation that reaches the ground and is not intercepted by or retained in the canopy.

Tie ridging: In contour furrowing or trenching, a narrow transverse strip of ground left unexcavated (or ridge).

Transplantation: It refers to plantation of forest trees and shrubs in uncultivated farm land.

Trenches: Trenches are long, narrow depressions on the sea floor that form at the boundary of tectonic plates where one plate is pushed or subducts beneath another.

Tropical Rainy Climate (Equatorial climate): The Equatorial Climate is characterized by hot average temperature all year round and high monthly precipitation, typically no less than 60 mm a month with annual precipitation tending to be over 2000mm. It is temperature range is just 2°C and its monthly rainfall minimum/maximum is 150 mm/270mm.

Ultisols: Ultisols are Lateritic soils. Ultisols are typical soils with strong acidity and less than 35% base saturation at depth. Ultisols have either an argillic horizon or kandic horizon, while Oxisols have an either oxic horizon or a surface horizon with 40 % clay that overlies a highly weathered kandic horizon with few weatherable minerals remaining.

Virdas: Virdas are shallow wells dug within a natural depression (Lake/Jheel). Since the area around is very saline, when rainwater seeps down the soil, it collects over the saline groundwater due to the difference in density (rainwater being less dense). It is prevalent in Rann of Kutch, Gujarat.

Wasteland: Land not suitable for or capable of producing materials or services of value.

Water Grid: The National Water Grid is a large scale engineering proposal for the management of water resources in India by the interlinking of the Indian rivers through networks of reservoir and canals particularly to control flood and increase water availability for domestic and irrigation purposes in the country.

Water harvesting: It refers to the collection of rainwater and its efficient utilization.

Water Holding Capacity (WHC): It shows the capacity of soil to hold water under fully saturated conditions. The soil pores, macro and micro, are fully filled with water. It is also referred as *'saturation capacity'* or '*maximum water holding capacity' of the soil*'. The soil water is retained with almost zero force or the water remains at free state condition. The WHC of soil varies with soil texture, structure, organic carbon and colloidal matter.

Water spreading: The flood water is deliberately diverted from its natural courses (rivers or larger streams) over adjacent plains through ditches, dikes, small dams and bush fences.

Watershed: A watershed is a natural hydrological unit that covers a specific land surface area from which runoff passes through a well defined outlet or an area that drains water and to a common outlet. It is also knows as drainage-based catchments or contributing area (sub watershed). Normally it has an area of 10,000 to 1,00,000 acres or 4,000 to 40,000 ha and commonly represent a drainage area of a tertiary river/stream.

Watershed Association (WA): This association refers to the local people living in the watershed area. The watershed association plans and executes the program.

Watershed Committee (WC): The watershed association elects the executive committee that has sole power to execute the program.

Watershed Development Team (WDT): It has at least one woman member, one sociologist, three graduate members from agricultural/civil engineering, agriculture, forestry, animal sciences etc. the retired government officers may also be considered for membership.

Waterway: A natural course for the flow of water.

Wave erosion: The coastal area is eroded or lost due to displacement of land or the long term removal of sediment and rocks along the coastal line mainly due to sea waves, currents, tides, wind driven water, water borne ice or impacts of storms.

Weed Control: It refers to removal of weeds from crop field. Weed control is very essential to save the crop from onslaught of drought as weeds rob soil moisture and nutrients as well.

Weir: A dam across a stream for diverting or for measuring the water flow of a stream/river.

Wind erosion: The detachment, transportation and deposition of soil by the action of wind.

Wind Sock: A light, flexible cylinder or cone mounted on a mast to show the direction and strength of the wind.

Wind strip cropping: The production of crops in long, relatively narrow strips, placed cross wise of the direction of the prevailing winds without regard to the contour of the land.

Wind Vane: A device that measures the direction of the wind.

Windbreak: A planting of trees, shrubs or other vegetation, usually perpendicular or nearly so to the principal wind direction to protect soil, crops, homesteads, roads etc. against the effects of winds, such as wind erosion and the drifting of soil and snow. A group of trees or shrubs in any rearrangement that will protect animals or crops or both from high winds. When the arrangement is in a long line, the group is called a shelterbelt. If an associated reason is also to harvest timber at some future date it is sometimes called a 'timber belt'.

Zabo (Ruza) system: Traditional rain water collection on hill tops for irrigation purposed. It is prevalent in Nagaland.

Zero-grazing: A method of keeping animals that involves bringing fodder to them rather than letting the animals graze freely.

1

Rainfed Agriculture Principles and Scope

Outlines

'Rain is essential for cultivation and the cultivation is essential for life, so one should first acquire the knowledge of rainfall carefully'

(Atharveda)

'An increase in supply of milk and butter depends upon the nature of the soil and the quality of fodder and water'

(Arthashashtra)

'A large part of the country depends on rains for agricultural production. It has, therefore, been felt that dry farming practice should be undertaken in such areas'.

(Dr. M S Swaminanthan)

"Intensive irrigated farming is an imperative for survival; but improved dryland agriculture is essential for equity"

(Dr. Gautam, Former DG, ICAR)

Introduction

Rainfed agriculture constitutes 80% of global agriculture and plays a critical role in achieving global food security. However, growing world population, water scarcity, and climate change threaten rainfed farming through increased vulnerability to droughts and other extreme weather events. Out of the total population of 7.3 billion, about 1 billion are food insecure and 60% of these live in South Asia and Sub Saharan Africa. The importance of rainfed agriculture varies regionally, but it produces most food for poor communities in developing countries. The proportion of rainfed agriculture is 95% in Sub Saharan Africa, 90% in Latin America, 60% in South Asia including India, 65% in East Asia and 75% in Near East and North Africa (FAO, 2003).

Food and nutritional security is the major challenge of the country as 21.96% of total population was below poverty line (Census, 2011). India is 2nd in world population sharing 17.74% with highest share of rural population 26.08% that most of them depends on agriculture and allied sectors. Similarly India has 2.44% of total world area with 2nd in arable land with 11.25% share. India also stands 1st in pulses, jute and milk production, 2nd in rice, wheat, sugarcane, tea, cotton, vegetable and fruit production in the world. We are also leader in buffaloes, 2nd in cattle and goat and 3rd in sheep population in the world (Table.1). In recent past, it has been observed decreasing demand of cereals and food grain/capita and increasing of livestock products, fruits and vegetables due to improving annual income. Despite decline in the food grain consumption because of the dietary shift, there is no substitute for cereals and pulses which are the staple foods and the most economic sources of energy and protein and vital for nutrition of poor people, so it is always necessary to produce sufficient food grain and also ensure supply of balanced food to the masses. In 2018-19 the total food production of the country was 284.95 mt and expected 297.63 mt in 2010-21 but the total demand of food grain has been estimated equal to 391.81mt in year 2029-30. It means the food grain production has to be increased by 3.5 mt/year. Similarly, the demand of oilseeds, milk, fruits and vegetables will be met out provided its production is increased at the rate of 5.45, 6.87, 8.02 and 13.05 mt/year, respectively by 2029-30 (Table.2). The future demand of food grain and other essential agricultural items can be fulfilled if both irrigated and rainfed agriculture together are managed properly. The developmental work has been concentrated only in irrigated areas, therefore the productivity of rainfed crops is below half of the irrigated crops, so the rainfed agriculture draws special attention to realize its potential in future.

At present in India, Irrigated area accounts for nearly 48.8 per cent of the 140 million hectare (mha) of agricultural land in India. The remaining 51.2 per cent is rainfed (Table.3). The crop wise rainfed and irrigated area indicates that millets, pulses, oilseeds, cotton are mainly grown under rainfed conditions (Table.4). The mean productivity of rainfed area (71.62 mha) is about 1.1 ton per ha compared to 2.8 ton per hectare of irrigated area. The rainfed area accounts for nearly 40 percent of the total food production. It is also reported that irrigated area cannot be increased more than 50% of the total cropped area. The possibility of area expansion under cultivation is also hugely impossible, so rainfed agriculture is the last option for sustainability of our food security. Though rainfed agriculture is complex, highly diverse and risk prone, it is characterized by low levels of productivity and input usage coupled with vagaries of monsoon emanating from climate change resulting in wide variation and instability in crop yields. In view of the growing demand for food grains in the country, there is a need to develop and enhance the productivity of rainfed areas. If managed properly, these areas have tremendous potential to contribute a larger share in food production and faster agricultural growth compared to the irrigated areas which have reached a plateau.

Besides, the increasing vulnerability of dryland agriculture to climate change may have worst impact on grain production in future mainly due to rainfall variability and reduction in rainy days (Venkateswarlu and Shankar, 2012). Shifts in seasons, change in rainfall pattern and increasing temperatures are already witnessed and so crops may encounter extreme weather events like drought, flood, heat and cold during its life cycle, resulting in substantial yield losses. The yield reductions are likely to be caused by shortening of growing period, negative impacts on reproduction, grain filling and decrease in water availability at critical growth stages. The reduction in rice and wheat yield due to terminal heat due to water stress and number of rainy days have already been reported (Rao and Bapuji Rao, 2013). However the impact climate change may vary with crops as yields of wheat, rice and maize will decrease while it could be neutral or positive with groundnut, soybean and chickpea (Aggarwal, 2008). IPCC (2013) also reported that frequent and more intense extreme events might be the norm of the day for common farming community in near future.

Heath Cote (1983) reported that drylands are characterized not only by low annual precipitation, but also by erratic precipitation that is subject to large temporal and spatial variability. The precipitation in many semi arid lands averages between 200-500 mm/yr which is suitable for some crops but the extreme fluctuations in precipitation from year to year make such averages of little use. He also told that precipitation alone does not define the climatic character of drylands. Aridity describes the normal conditions of drylands that are the result of factors that create low levels of moisture availability which includes not only low annual precipitation amounts, but also high evaporation demands. Branson (1976) stated that ratios of evapotranspiration to precipitation are generally large in drylands exceeding 95% in some areas.

Table 1. India's position in world agriculture in 2017

S.N	Item	India	% share	India's Rank	Next to
1.	Area (mha)				
	Total area	328.73	2.44	7th	Russian Federation, Canada, U.S.A., China, Brazil, Australia
	Land area	297.32	2.29	7th	Russian Federation, China, U.S.A.,Canada, Brazil, Australia
	Arable land	156.46	11.25	2nd	USA
2	Population (millions)				
	Total	1338.68	17.74	2nd	China
	Rural	889.22	26.08	1st	
3	Crop production (million tons)				
a	Total cereals	313.61	10.52	3rd	China, USA
	Wheat	98.51	12.77	2nd	China
	Rice	168.50	21.89	2nd	China
b	Pulses	23.24	24.21	1st	
c	Oilseeds				
	Groundnut (with shell)	9.18	19.49	2nd	China
	Rapeseed	7.92	10.38	3rd	Canada, China
d	Commercial crops				
	Sugarcane	306.07	16.62	2nd	Brazil
	Tea	1.33	21.72	2nd	China
	Jute	1.97	55.69	1st	
	Cotton (lint)*	6.05	24.43	2nd	China
	Tobacco (crude)	0.80	12.30	3rd	China, Brazil
4	Fruits & Vegetables (MT)				
	Vegetables Primary	127.14	11.62	2nd	China
	Fruits Primary	92.30	10.66	2nd	China
	Potatoes	48.61	12.52	2nd	China
	Onion(Dry)	22.43	22.92	2nd	China
5	Livestock (million Heads)				
	Cattle	185.10	12.41	2nd	Brazil
	Buffaloes	113.33	56.39	1st	
	Camels	0.33	0.93	16th	Chad, Somalia, Sudan, Kenya, Niger, Mauritania, Ethiopia, Mali, Pakistan, Saudi Arabia, U.A.E, Yemen, Mongolia, Algeria, Eritrea

S.N	Item	India	% share	India's Rank	Next to
	Sheep	63.07	5.25	3rd	China, Australia
	Goat	133.35	12.89	2nd	China
	Chicken	783.27	3.43	6th	China, Indonesia
6	Dairy products (million tons)				
	Milk Total	176.27	21.29	1ts	China, U.S.A
	Eggs (Primary) Total	4.85	5.57	3rd	
	Meat (Total)	7.30	2.18	6th	China, U.S.A, Brazil, Russian Federation, Germany

Source: FAOSTAT (as on 08.01.2020) and *data relate to year 2013.

Table 2. Production and demand food grains and other major crops in India in 2029-30

S.N	Crops	Production (Million Tons)			
		2018-20	Estimate in 2020-21	Demand in 2029-30	Required increase/ year over 2018-19
1	Rice	116.42	119.76	117.56	0.11
2	Wheat	102.19	107.18	108.62	0.64
3	Coarse cereals	42.95	46.96	60.99	1.80
4	Cereals	261.55	273.90	287.18	2.56
5	Pulses	23.40	23.73	32.64	0.92
6	Foodgrains	284.95	297.63	319.81	3.49
7	Oilseeds	32.26	37.81	86.80	5.45
8	Milk and products	187.7	193.97	256.43	6.87
9	Vegetables	185.88	211.29	316.33	13.05
10	Fruits	98.58	115.25	178.74	8.02

Table 3. Present status of cropped and irrigated area in India (m ha)

S.N	Year	Net sown area	Total cropped area	Area sown more than once	Net irrigated area	Gross irrigated area	Area irrigated more than once
1	1950-51	-	-	-	22	-	-
2	1990-91	143.0	185.74	42.74	48.02	63.20	15.18
3	2000-01	141.34	185.34	44.0	55.20	76.19	20.98
4	2010-11 (P)	141.56	197.68	56.12	63.67	88.94	25.27
5	2010-11(P)	141.56	197.68	56.12	63.67	88.94	25.27
	2014-15(P)	140.13	198.36	58.23	68.38	96.46	28.07

Source: Directorate of Economics & Statistics, DAC&FW P: Provisional

Table 4. Irrigated area under major crops in different states of India (2014-15)

S.N	Crops	Area (%)		Leading States in Irrigated area
		Rainfed	Irrigated	
1	Rice	39.9	60.1	Haryana>Punjab>Sikkim>Telangana>AP UK-70%, UP-86,7%, Lowest in Jharkhand-5% only
2	Jowar	90.1	9.9	Haryana>WB>AP>Guj>Telangana (UK- NA, UP-3.5%)
3	Bajra	90.5	9.5	Punjab>Telangana>WB>HP>Haryana (UK- NA, UP-12.4%)
4	Maize	73.4	26.6	Punjab>AP>Bihar>TN>UP (UK-45.3%, UP-43.4%)
5	Wheat	5.8	94.2	Meghalay/TN/Orissa/Tripura (100%) >Rajasthan >Haryana> Telangana>UP. (UK-98.8%, UP-98.7%)
6	Barley	19.7	80.3	Rajasthan>Punjab>Haryana>AP>UP (UK-77.3%, UP-NA)
7	Total Cereals	39.9	60.1	Punjab>Haryana>AP>UP>Meghalya (UK-87.0%, UP-49.7%)
8	Gram	61.4	38.6	Tripura>Punjab>UK>MP>Rajasthan (UK-80.5%, UP-19.7%)
9	Tur	96.0	4.0	Punjab>Haryana>HP>TN>UP (UK-1.3%, UP-12.4%)
10	Total pulses	80.1	19.9	Goa>Mizoram>Punjab>MP (UK-7.9%, UP-27.4%)
11	Total Food grains	46.9	53.1	Punjab>Haryana>UP>Kerala>Meghalya (UK-44.8%, UP-80.4%)
12	Sugarcane	9.8	90.2	Goa/Haryana/Orissa/TN/Telangana(100%)>Kerala/ MP(99.9%)> Karnataka>Chhattisgarh>UK (UK-98.7%, UP-95.4%)
13	Groundnut	73.3	26.7	Haryana>Telangana>Rajasthan>Goa>Tripura (UK-3.2%, Up-5.3%)
14	Rape & Mustard	23.4	76.6	Telangana>Gujarat>Punjab>Rajasthan>WB (UK-45.4%, UP-82.0 %)
15	Soybean	99.3	0.7	TN>AP>HP>Telangana>Karnataka (UK-2.2%, UP-0.2%)
16	Sunflower	99.9	0.1	Haryana/ Punjab(100%)>Chattisgarh>Bihar>Telangana > UP (UK-0.0%, UP-70.5%)
17	Total Oilseeds	72.6	27.4	Punjab>Haryana>WB>J&K>TN (UK-25.0%, UP-42.5%)
18	Cotton	66.3	33.7	Jharkhand(150%)>Punjab/(27.4%) HP(100%)> Haryana>UP(97.41)>Rajasthan

S.N	Crops	Area (%)		Leading States in Irrigated area
		Rainfed	Irrigated	
19	Tobacco	41.2	58.8	TN/Gujarat/Haryana/UP/ UK(100%)>WB>Chhattisgarh>Rajasthan >Bihar (UK-100%, UP-100%)
20	All Crops	51.4	48.6	Punjab>Haryana>UP>Bihar>WB (UK-49.5%, UP-80.2%)

'Dryland farming' and 'dry farming' encompass specific agricultural techniques for the non-irrigated cultivation of crops. Dry farming refers growing of short duration drought escaping crops on conserved soil moisture regimes. The rainfall is scanty and requires adoption of soil moisture conservation measures for successful farming and these conditions are common in arid region, while dryland farming is associated with areas characterized by a cool wet season (which charges the soil with virtually all the moisture that the crops will receive prior to harvest) followed by a warm dry season. It is associated with semi arid region where water conservation and water harvesting are adopted for life saving irrigation if required. Further dryland agriculture is a special case of rainfed agriculture practiced in arid and semiarid regions in which annual precipitation is about 20–35% of potential evapotranspiration, conditions of moderate to severe moisture stress occur during a substantial part of the year, greatly limiting yield potential and in which farming emphasizes water conservation in all practices throughout the year.

Drylands generally receive < 750mm annual ranfall in arid and semi arid regions and both terms dry farming and dryland farming are used synonymously to indicate almost similar farming situations. 'Rainfed agriculture' specifies the conditions that farming is based on rainfall and region receives sufficient rainfall for good harvest but require safe disposal of runoff water and adoption of soil and water conservation measures to protect fertile upper soil layers. The crop yield can be maximized with high inputs and these conditions normally occur in dry sub humid to humid regions. There are three components of successful dryland agriculture;

i. To retain or store precipitation on the land,

ii. To Reduce or minimize evaporation from the soil surface and increase transpiration, and

iii. To grow drought tolerance crops and fit to the precipitation patterns.

Definition

Some of the important definitions proposed for dryland agriculture are given below for better understanding;

A. Formal definitions

1. 'The profitable production of useful crops without irrigation on lands that receive low annual rainfall of 20" or less and it may be extended to 30"

with unfavorable distribution of rainfall.' (Dr. John A. Widetsoe in his book, 'Dry Farming for Sustainable Agriculture' 1911),

2. The areas with ratio of precipitation (P) and potential evapotranspiration (ETP) (P/ETP) between 0.03-0.02 are arid and areas with P/ETP ratio 0.2-0.5 are semi arid. The dry farming is perfomed in arid and semi arid areas (Steiner *et al.*, 1988).

3. Growing period or growing degree days concept was taken to define dry farming or dryland farming. The growing period or growing degree days is the number of days in a year or season in which precipitation exceeds half of the potential evapotranspiration, plus a period to use an assumed 100mm of water from excess precipitation (or less if not available) stored in the soil profile. According to the growing period concept, dryland agriculture is performed in arid and semi arid areas with 01-74 and 75-119 days growing period, respectively (FAO).

4. 'Dryland farming is adopting the techniques under the conditions of inadequate rainfall using artificial irrigation' (Kutsch).

5. 'Dry farming is using techniques which conserve water for the crop without irrigation' (Geographical Dictionary).

6. 'A type of farming practiced in arid areas without irrigation by planning, drought resistant crops and maintaining a fine surface tilth or mulch that protects the natural moisture of the soil from evaporation. Strategies include mulching, frequent fallowing, working the soil to a fine tilth, and frequent weeding' (Geographical Dictionary).

7. 'Dry farming and Dry land farming have been used synonymously to indicate similar farming situations and strategies including mulching, frequent fallowing, working the soil to a fine tilth and frequent weeding were included' (Free Dictionary).

8. 'Dry farming is farming without irrigation in areas where the moisture supplies is the main limitation to crop production. Efficient dry farming consists of making the best use of a limited water supply by storing in soil as much as the rainfall is possible and by growing suitable crop plants. Usually crops adopted to dry farming are smaller in stature and quicker in maturity than-those grown under more humid conditions' (Encyclopedia Britannica, 1768).

9. 'Dry farming is that system of soil and crop management which is carried out without the help of irrigation in regions of low and uncertain rainfall' (Khan, A.R., 1968).

10. 'Dry farming refers to the type of agricultural operation which involves deep cultivation of the soil to form a sufficient survival for the moisture as

it falls surface cultivation to prevent or reduce, evaporation, selection of drought resistant crops' (Chhatwal Gurdeep, 1989).

11. 'Dry farming areas are those which receive an annual rainfall ranging from 375mm to 1125mm and with very limited irrigation facility' (Planning Commission-4th FYP document-1969-74).

12. Thornthwaite (1955) formula was used for estimating moistures deficiency index (MDI) for defining dryland regions by ICAR for establishing dryland centres in the country in 1970. MDI is the ratio of (P-PE/PE) x 100. P stands for precipitation and PE potential evaporation. According to MDI values, the dryland agriculture is adopted in semi arid region where MDI values ranges from -33.3 to -66.7 (semi arid areas) (ICAR, 1970).

13. 'Dry farming areas are those areas which receive between 40cm and 100cm of annual rainfall and have less than 25 percent of the total sown area under irrigation. They constitute nearly 47 million acres or about 36 percent of the net shown area in the country and account for less than 20 percent of the total food grain production' (Indian Journal of Agricultural Economics, 1971).

14. 'Dry farming consists of making the best use of limited water supply by storing in the soil as much as of the rainfall is possible and by growing suitable crop plants by methods that make the best use of this moisture'.

B. Functional (Operational) definitions

The functional definitions account the availability of natural resources of the area and its better utilization for maximizing the agricultural production and income of the farming communities under specified conditions of dryland areas. Therefore following definitions may be referred as functional or operational as listed below;

a) The term dry farming is applied to farming with seasonal irrigation or without irrigation, under an average rainfall 528.5mm, it is uncertain and unpredictable that creates an atmosphere of high risk and insecurity where farmers tempt not to use costly inputs.

b) The system of dry farming must consider the established facts of science for the better utilization of limited rainfall for better plant growth and development. The knowledge of soil texture and structure being indispensable in relation to water holding capacity of the soil must be utilized for crop production.

c) Dry farming relates to the adoption of package of practice like contour bunding, deep ploughing, surface tillage, bulky organic manures, low seed rate and wider spacing, inter culturing, crop rotations etc. to store soil moisture for improving agricultural production (Bombay Dry Farming System).

d) Dry farming considers protective irrigation for local and hybrid seeds, fertilizer application and plant protection measures, water conservation programs such practices obtain high and sustained productivity in dry land agriculture.

e) Dry farming refers to the diversified farming system including crop production, livestock, poultry, piggery, sericulture, beekeeping, mushroom cultivation and all other possible agro-based entrepreneurs to safeguard the interest of farming communities in dryland areas.

Finally, three words i.e. 'dry farming', 'dryland farming' and 'rainfed farming' or 'dry agriculture', 'dryland agriculture' and 'rainfed agriculture' are used synonymously in the sense that all farm practices or farming is based on natural rainfall only and there is no irrigation facility. It is not true to use these words interchangeably because the amount of precipitation (P) received and potential evapotranspiration (PET) or its ratio (P/PET) are important to divide area in to arid, semi arid and humid regions.

Most of the workers have taken different indicators like annual precipitation, growing period, potential evapotranspiration or ratio of P/PET, soil moisture stress, soil and water conservation measures, disposal of runoff water etc. for defining terms related to dryland agriculture. Following terminologies are important to understand dryland farming as given below;

1. ***Dry farming:*** It is applied to farming without irrigation under an annual precipitation of 25-30 inches or 635-762mm.
2. ***Dry land***: The land or area which receives an annual rainfall of 750 mm or less and there is no irrigation facility for raising crops.
3. ***Dry land Agriculture:*** Basically it refers to scientific management of soil and crops under dry lands without irrigation.
4. ***Dry land crops***: It refers to all such crops which are drought resistant and can complete their life cycle without irrigation in areas receiving an annual rainfall less than 750 mm.
5. ***Drought***: It is a condition of insufficient moisture supply to the plants under which they fail to develop and nurture properly. If may be caused by soil, atmosphere or both.
6. ***Dry farming***: Dry farming is an improved system of cultivation in which maximum amount of moisture is conserved in low and untimely rainfall for

the production of optimum quantities of crop an economic and sustainable basis.

Concept of rainfed farming

Poverty, hunger, water stress and malnutrition are strongly correlated and by chance these factors are dominating in rainfed agriculture where water constitutes a limiting factor to crop growth. The UN Millennium Development Project has identified the 'hot spot' countries located in the semi-arid and dry sub-humid hydro-climates in the world where 850 million undernourished people live in poor, developing countries which predominantly are in tropical regions. The productivity is low in tropical and sub tropical areas including parts of sub-Saharan Africa and South Asia mainly due to degraded soils, high levels of evaporation, droughts, floods and lack of effective water management. International Water Management Institute noted a close correlation between hunger, poverty, and water. However, it concluded that there was much opportunity to raise the productivity of rainfed farming. Managing rainwater and soil moisture more effectively and using supplemental and small-scale irrigation is believed to hold the key to helping the greatest number of poor people.

The semi arid tropics (SAT) is the home to 38% of the developing countries' poor, 75% of whom live in rural areas. Over 45% of the world's hungry and more than 70% of its malnourished children live in the SAT (Raju *et al.*, 2013). The Comprehensive Assessment (ICRISAT) concluded that the crop yield in rainfed areas are two-to five fold lower than achievable potential yields and that current rainwater use efficiency is only 35–45% in most rainfed areas. Water used for food production in rainfed areas is almost three fold higher than that used in irrigated systems. Long-term experiments as well as yield gap analysis using crop simulation models and researchers' managed trials on farmers' fields have demonstrated that crop yields in rainfed areas can go up as high as 5 t/ha under semi-arid tropical Indian conditions. Large yield gaps exist in a number of rainfed crops such as maize, sorghum, pigeonpea, groundnut, soybean, pearl millet, chickpea, wheat and paddy in different countries of Asia and Africa. On the basis of research findings, it was concluded that yields could easily be doubled in rainfed areas of Asia and quadrupled in Africa with adoption of suitable soil, water and crop management options on farmers' fields.

Therefore, the concept of rainfed farming stands for best form of farming in rainfed areas with adoption of all innovative technologies including soil and water conservation and its optimum use for reducing poverty, hunger inequality, unemployment and malnutrition in the region.

Goal of rainfed farming

'The goal of rainfed farming relies on improving the resilience of rainfed and dryland cropping systems through the identification, evaluation, and communication of innovations in production practices as well as soil and water management'.

Principles of rainfed agriculture

The main objective of rainfed agriculture is to adopt suitable site specific technology for maximizing crop and livestock production in the rainfed areas. Therefore the main principles of rainfed agriculture based on efficient utilization of precipitation and soil moisture conservation measures are listed below;

1. Selection of suitable drought resistant crop and varieties,
2. Planning for aberrant weather conditions like drought, early and late arrival of monsoon, mid and terminal drought etc.,
3. Optimizing seed and sowing of crops,
4. 'In situ' conservation of rain water,
5. Adopting site specific soil and water conservation measures,
6. Optimizing evapotranpiration (ET) losses,
7. Application of optimum crop plant nutrition,
8. Reducing evaporation losses,
9. Water harvesting for supplemental life saving irrigation,
10. Safe disposal of excess runoff under heavy rainfall conditions in humid and sub humid regions,
11. Suitable mechanization for tillage, seeding, harvesting, threshing etc. in the region,
12. Maximizing water productivity,
13. Exploring possibility of alternate land use systems like agroforestry, pasture, silviculture, silvi-pasture etc. for higher benefits,
14. Developing market infrastructure as per the need of dryland regions,
15. Selection of livestock and its breeds suitable to arid, semi arid and hilly regions,
16. Adopting suitable information and communication technology for technology transfer,

17. Adopting watershed approach for development , and
18. Improving socio-economic status of the farming communities.

Classification of dryland areas

The important classification of dryland areas is given below;

1. FAO (1993) classification: Dryland were divided into 3 groups based on rainfall as given below;

i. Arid (Desert) : <375 mm rainfall

ii. Dry land : 375-1125 mm rainfall

iii. Rainfed : > 1125 mm rainfall

2. UNEP (1992) classification

World Atlas of Desertification (UNEP, 1992) defined dryland area with <0.65 P/PET ratio so the dryland area of the world was divided into four categories i.e. hyper arid, arid, semi arid and dry sub humid (Table.5).

Table 5. Classification of area/region based on rainfall and P/PET ratio

S.N	Category	P/PET ratio	Rainfall (mm)	Area in World (mha)
1	Hyper arid	<0.05	200	1.0
2	Arid	0.05 – 0.02	< 200 in winter	1.62
			< 400 in summer	
3	Semi- arid	0.20 – 0.50	200-500 in winter	2.37
			400-600 in summer	
4	Dry sub-humid	0.50 – 0.65	500-700 in winter	1.32
			600-800 in summer	

3. The UN Convention to Combat Desertification (UNCCD, 1999) classification

The UNCCD divided drylands of India into 5 groups based on rainfall and growing period as given in table.6.

Table 6. UNCCD classification of dryland region of India

Region	Rainfall (cm)	Growing period (days)	Total land area		Regions in India
			mha	%	
1. Arid	0-10	<90	39.1	1.96	**i. Hot arid** (31.7 mha): Western Plains including Thar desert, Deccan Plateau (parts), **ii. Cold arid** (7.4 mha): Western Himalayas (Leh, Kargil in J & K, some parts of Chambal and Kinanur (HP)

Region	Rainfall (cm)	Growing period (days)	Total land area		Regions in India
			mha	%	
2.Semi arid	10-40	90-150	-	37	Arvalis, Gujarat plains, parts of Deccan Plateau, Eastern Ghats (parts), Central Highlands (parts)
3.Sub humid	40-80	150-180	-	21	Northern Plains, Central Highlands (parts), Eastern Plateau (parts), Western Himalayas (parts)
4. Humid	80-200	180-210	-	22.4	Bengal, Assam, Eastern Himalayas
5. Per humid	200-400	>210	-	-	N-E hills, Eastern Coastal Plains, Western Ghats

4. Miscellaneous classification

Many other workers worked on rainfed agriculture and classified based on irrigated area as well as on annual rainfall without considering the soil type, altitude, slope, temperature, relative humidity, evaporation rate, number of rainy days, rainfall distribution pattern because these factors affect the moisture regime of the soils. Similarly the water holding capacity of soil varied with soil texture and structure, hence the wide variation of average rainfall may hold true for particular region and soil types but not for whole country (Table.7).

Table 7. Alternate criteria to defining rainfed agriculture

S.N	Authors	Criteria used	
		Gross irrigated area	Av annual rainfall (mm)
1	Bapnal *et al.* (1981)	<25	500 -1500
2	Rangaswami (1981)	<30	375 -1125
3	Jodha (1985)	<25	500-1500
4	Subbarao (1985)	<25	<970
5	Shah and Sah (1993)	<25	400 - 750
6	Throat (1993)	<10	375 - 750

5. S. C. Panda (2003) classification

The classification of rainfed agriculture by S. C. Panda (2003) is most accepted in our country. He summarized the definition of rainfed agriculture taking rainfall, evapotranspiration, growing season and moisture conservation measures under consideration as listed below;

i. Dry farming

Dry farming refers to the cultivation of crops in areas that receives annual rainfall less than 750 mm and crop failures are most common due to prolonged dry spells during crop period. The growing season is less than 200 days. It is generally practiced in arid regions of the country with adoption of soil moisture conservation.

The alternate land use system is again very suitable option for sustainable crop production.

ii. Dryland farming

Dryland farming means cultivation of crops in areas receiving annual rainfall above 750 mm but less than 1150cm. The dry spells are not too severe to make crop failure, so crop failures are less frequent than dry farming. The semi-arid regions are included under this category. The potential evapotranspiration is more than total receipt of precipitation and causes moisture deficit in the region. Soil and ware conservation measures are the key factors for successful crop production.

iii. Rainfed farming

Rainfed farming refers to the crop cultivation in the areas receiving 1150 mm annual rainfall. The total rainfall is more than PET, so drainage is a major problem of the region. It is mainly practiced in sub-humid and humid areas. Here chances of crop failure and water stress are very less compared to above both conditions.

6. Singh and Venkateshurulu (2009) classification

Singh and Venkateshurulu (2009) classified Indian climate based on rainfall into 4 regions like arid, semi -arid, sub-humid and humid regions (Table.8). Rainfed agriculture is practicing in all 4 regions mainly without irrigation or deficit / life saving irrigation in emergencies. Therefore, the rainfed agriculture may be grouped into four categories based on above Indian climate classification like arid faring, dry farming, dryland farming and rainfed farming matching with arid, semi arid, sub humid and humid regions, respectively.

Table 8. Singh and Venkateshvarulu (2009) classification of Indian climate

S.N	Region	Rainfall (mm)	Rainfed area (mha)
1	Arid	<500	15.0
2	Semi Arid	500-750	15.0
3	Sub humid	750-1150	42.0
4	Humid	>1150	25.0

1. Arid Farming

Arid farming is practiced in areas where an annual rainfall is < 500 mm. It has following important characteristics;

i. It represents arid region,

ii. Rainfall is <500mm,

iii. Extreme moisture shortage,

iv. Inadequate rainfall with high variability,

v. Crop experiences high temperature,

vi. High potential evapo-transpiration rate,

vii. Poor soil structure and soil fertility,

viii. Highly degradation of natural resources and recurrent drought,

ix. Mono-cropping systems dominate with limited crop choice. However, some grasses, shrubs and trees, only few drought hardy arable cereal, pulses and oilseed crops are grown at the occurrence of rainfall,

x. Partial or complete crop failure is a regular phenomenon, and,

xi. Sustainable livelihood is the major concern in the region,

2. Dry farming

Dry farming refers to the crop cultivation in areas receiving rainfall <750mm. The important features are;

i. It represents the semi arid regions,

ii. Rainfall is <750mm,

iii. Rainfall is always erratic and inadequate,

iv. Crops suffer most often long dry spells,

v. More fragments crop failure,

vi. Wind erosion is predominant,

vii. Conservation of soil moisture or rainwater is a pre-requisite for crop production and

viii. Drought resistance crops and varieties are grown either as a sole, mixed or intercrops.

3. Dryland Farming

Dryland farming refers to the crop cultivation in areas receiving 750 - 1150mm annual rainfall. It may characterized by;

i. It represents the sub humid regions,

ii. Annual rainfall is in between 750 -1150mm,

iii. Double cropping is possible,

iv. Soil & water conservation measures are required,

v. Soils are deficient in essential plant nutrients,

vi. Erratic and Inadequate rainfall,

vii. Dry spells and droughts are common behavior of this region,

viii. Crop contingent plan is required for aberrant weather conditions,

ix. Millets, pulses and oilseeds are important crops and cropping systems mainly grown.

4. Rainfed farming

Rainfed farming is practiced in areas where the annual rainfall is >1150mm. It has following characteristics;

i. It is practiced in humid regions,

ii. Annual rainfall is >1150mm,

iii. Proper drainage is required,

iv. Rainfall is erratic and uneven but sufficient or more,

v. Fewer chances of crop failures due to dry spells,

vi. Most of the crops can be grown in this region in different cropping systems provided the excess rainfall is preserved and utilized efficiently.

7. United Nations Economics and Social commission for Asia and Pacific Region (UNESCAP)

UNESCAP has divided dry level agriculture in to 'dryland' and'rainfed farming' and difference is listed in table.9 below;

Table 9. Difference between dryland farming and rainfed farming

S.N	Parameters	Dryland farming	Rainfed farming
1	Rainfall (mm)	< 750	>750
2	Moisture availability to crop	Shortage	Enough
3	Growing period (days)	Uncertain with inter and intra seasonal variation (< 200)	Almost certain with less inter and intra seasonal variation (>200)
4	Climate	Dry arid/ semi arid	Cold arid/semi arid/sub humid
5	Growing region	Arid and semi arid as well as upland of sub humid and humid region	Humid and sub humid region
6	Cropping systems	Single or intercropping	Intercropping and double cropping
7	Constraints	Wind and water erosion	Water erosion
8	% contribution to arable lands	25	42

Types of rainfed agriculture

a. According to S. C. Panda (2003)

Rainfed agriculture was divided into three categories;

i. Dry farming

ii. Dryland framing

iii. Rainfed farming

b. According to Singh and Venkateshvarulu (2009)

Singh and Venkateshvarulu (2009) divided rainfed agriculture into four categories;

i. Arid farming

ii. Dry farming

iii. Dryland farming

iv. Rainfed farming

c. According to United Nations Economics and Social Commission for Asia and Pacific Region (UNESCAP)

UNESCAP divided rainfed agriculture into two catergories;

i. Dryland farming

ii. Rainfed farming

The detail features of different types of rainfed farming have been described above.

Distribution of drylands areas

The earth's land surface is covered by drylands forming about 41% which is inhabited by 2 billion people (about one third of world population). About 72% of the global dryland are in the developing nations and rest 28 % falls in industrialized nations (www.millenniumassessment.org). India has about 72.21 million hectares of rainfed area which constitutes nearly 51.76% of the net cropped area 139.51 million hectares land in 2015-16. Geographically dryland agriculture area in India includes the north western desert regions of Rajasthan, the plateau region of central India, the alluvial plains of Ganga-Yamuna river basin, the central highlands of Gujarat, Maharashtra and Madhya Pradesh,the rain shadow regions of Deccan in Maharashtra, the Deccan Plateau of Andhra Pradesh and the Tamil Nadu highlands (Rao and Ryan, 2004 and Singh *et al.*, 2004).

Total 128 districts in India have been recognized as dryland farming areas. Out of 128 districts, 25 districts covering 18 mha of net sown area with 10% irrigation receive 375-750mm rainfall annually spread over central Rajasthan, Saurashtra region of Gujarat and rain shadow region of Western Ghats in Maharashtra and Karnataka, 12 districts have irrigation covering 30-50% of the cropped area and do not have serious problem. The remaining 91 districts covering mainly MP, Gujarat, Maharashtra, AP, Karnataka, UP, parts of Haryana, TN etc. represent typical dryland area. The total net sown area in these districts is nearly 42 m ha of which 5 mha is irrigated. The rainfall varied from 75-1150mm (Map.2).

The distribution of rainfed areas can be broadly classified into arid, semiarid, dry-sub humid and humid regions. As per available estimates, 15 million ha of rainfed crop area lies in the arid region which receives less than 500 mm annual rainfall, another 15 million ha is in 500-750 mm annual rainfall zone, and 42 million ha is in the 750-1100 mm rainfall zone. The remaining area lies in annual rainfall zone of greater than 1100 mm (Map.1). Out of the 127 Agro Climatic Zones (ACZs) in India, 73 are predominantly rainfed.

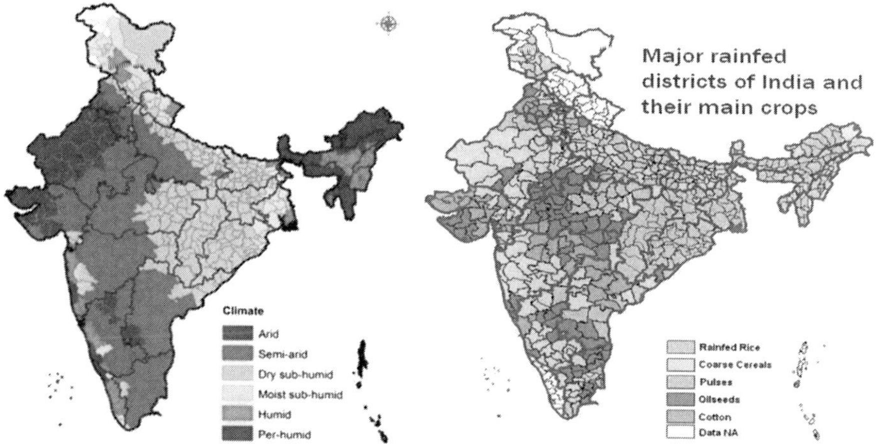

Map 1: Classification of rainfed areas and

Map 2: Rainfed districts and crops

Prospects of rainfed agriculture

1. More than 75% of the peasants involved in dry farming are small and marginal. Therefore, improvement in dry farming would raise the economic status of farmers thus helping in poverty elimination.
2. Dryland farming holds immense significance especially in the context of fluctuating food grain production and expanding population in our country.
3. The biggest employer in our country, the cotton mills are fed by raw cotton grown mostly in dryland areas. Increasing production of cotton subsequently leads to increase in exports of cotton good.
4. The expanding import of oilseeds is a cause of concern to Indian nation. The improvement of production of oilseeds in these regions will save valuable foreign exchange reserves.
5. By enhancing the productivity of crops like *jowar, bajra* and *ragi* which are mainly grown in dryland farming would increase the nutrient consumption levels of our nation.

6. Marginal lands in the semi-arid regions offer potential for fodder production to feed the cattle population which is an integral component of farming practice of this region. Providing importance to these areas can solve the problems of pulses, oilseeds and cotton.

7. The dryland areas have also tremendous potentiality of increased food grain production. Thus enhanced agricultural production in these areas would boost the agriculture dependent economy of India. Moreover it would also be helpful in eliminating the problem of hunger and malnutrition prevailed in below poverty line society of the country.

Significance of Rainfed Agriculture in India

Presently India has nearly 50% area under rainfed agriculture with half or less than half crop productivity compared to irrigated agriculture. Sometimes it is lower than 25% than irrigated area or even complete crop failure mainly because of extreme weather conditions like drought and flood. It is main source of pulses, oilseeds, nutri cereals, cotton as well livestock production. Food security is a major challenge to all developing countries for ever increasing population and present era is more critical due to global warming that will affect the planet with extreme events with temperature, flood, cyclones and more over it will reduce crop productivity drastically unless we check global warming. The rainfed agriculture would be worst affected under global warming particularly dry and dryland farming where high temperature and moisture stress are the main cause of poor crop productivity. With all these natural calamities and constraints, we have to produce enough to our masses and so we have explore all possible ways including vertical and horizontal expansion of crop productivity but possibility of area expansion is almost negligible but even arable land will decrease in years to come due to population and required developmental infrastructure. Therefore, the increase in crop productivity, expansion in irrigated area, development of high yielding climate resilient crop varieties and protective technologies are the options to increase crop production. Rainfed agriculture is so equally important to contribute in crop and livestock production. Some of the following points given below reflect its significance;

1. The acreage under rainfed and irrigated agriculture is 72.21 and 67.3 m ha constituting 51.76 and 48.24%, respectively of net cropped area 139.51 mha (Agriculture Statistics, 2019). It indicates that rainfed agriculture is equally important to boost our agricultural production.

2. India stands 1st in world having largest rural population i.e. 68.9% of total population of India (Census, 2011) and 26% of the world's population. The majority of rural population depends on agriculture and allied sectors. It has been reported that nearly 60% of rural population survive on rainfed agriculture.

3. The rainfed farming contributes substantially towards food grain production i.e.40% rice, 68% cotton, 73% oilseeds, 80% pulses, 83% nutri cereals, 48% food grain area, 40% population and 44% food grain production.

4. It supports two third of livestock including 75% goat, 64% sheep and 78% cattle and are critical to food security, equity and sustainability. Besides, it is a major domain for protein and fat supplements in India diet. The rainfed area faces shortage of water for 7.5 to 10 months, wide spread deficiency of nitrogen, phosphorus and other micronutrients, excessive runoff, mean temperature >18^{0}C and dominance of Alfisols and Aridisols soil orders (Mishra, 2017 –Niti Ayog).

5. Per capita land availability has declined from 0.28 ha in 1990 to 0.19 ha in 2010 and it will further be decreased due to increasing family size. It is also known that 3ha rainfed area is equivalent to 1ha irrigated area for equivalent cereal production. If we manage our available rainfed area to improve its crop productivity by even 11/2 times, India will be self sufficient in food production. It is possible with introduction of new technological innovations and supportive government policies.

6. Indian agricultural/horticultural and processed foods are exported to more than 120 countries, primarily to the Japan, Southeast Asia, SAARC countries, the European Union and the United States.

7. The present scenario of rice-wheat systems and other too indicates that yield level of most of crops have achieved its highest potential and difficult to further increase with available technology. So it is challenge to break the productivity plateau by introducing new strategy in crop production in general but particular in rainfed farming because it is more prone to succumb under natural calamities.

8. Dry land offers good scope for alternate land use systems like agro-forestry, social forestry, horti-silvi-pasture, pasture development, natural park and wild sanctuaries, dairy, goatery, piggery, sericulture, bee keeping, agro based industries. It will not only generate income but also employment for local youth.

9. The biggest employer in our country, the cotton mills are fed by raw cotton grown mostly in dryland areas. Increasing production of cotton subsequently leads to increase in exports of cotton good.

10. The expanding import of oilseeds is a cause of concern to Indian nation. The improvement of production of oilseeds in these regions will save valuable foreign exchange reserves.

11. By enhancing the productivity of crops nutri cereals like ragi, madira,

kakun, etc and also coarse grains *jowar, bajra* and *ragi* which are mainly grown in rainfed conditions would increase the nutrient consumption levels of our nation.

12. Pulses prefer rainfed conditions for higher yield, so rainfed agriculture is good option of higher pulse production in India.

13. Rainfed area is good for tree plantation like medicinal and aromatic plants, fodder trees, timber etc. It will cover forest area and will not only supply different types of valued products but also help in enriching biodiversity and ecology.

Characteristics of the rainfed agriculture

1. ***Scanty, uncertain and uneven rainfall*** - The annual rainfall is very limited, uncertain and also not properly distributed. The total rainfall days are low or even decreasing due to global warming and it makes crop production difficult in the region.

2. ***Aberrant weather or monsoon*** – The monsoon either of south west or north east is quite uncertain in sense of arrival and cessation of monsoon and also total quantity of rainfall. It greatly vary yearly, therefore the crop planning is disturbed. Further long dry spells and crop failure are also common feature of the region, so the timely execution of crop contingent planning is essentially required.

3. ***Occurrence of extensive climatic hazards*** - The climatic hazards like drought and flood are common in the region, in monsoon season, heavy rain causes flood and in other months crop experiences drought conditions.

4. ***Soil Erosion*** - Water and wind cause wide spread soil erosion particularly the high intensity of rainfall during monsoon season and drought in other months, respectively in the region.

5. ***Soil degradation*** - The major causes for land degradation include the chemical degradation of soil, loss of soil structure and texture, loss of natural vegetation leading to soil erosion. Sequestration of carbon also turns out to be a major problem which further degrades the soil making it less productive.

6. ***Undulating soil surface and infertile land*** - The land surface is usually undulated and soils are infertile in rainfed areas mainly due to gully erosion, sand dune formation etc, The soil are normally infertile having deficiencies of many essential plant nutrients.

7. ***Occurrence of extensive and large holdings*** - There are large holdings in dry region and only single or double cropping is practiced in absence of

rainfall. Extensive farming is a common feature of the region with poor management practices.

8. ***Low crop diversification*** - The crop production is based on rainfall, so one or two crops are grown, The crop and varieties may be drought tolerant or resistant and drought evasive in nature so that they may not experience moisture stress conditions, therefore the farmers of the region grow similar type of crops like millets, pulses, oilseeds etc. that suits to region.

9. ***Poor infrastructure of irrigation*** – The dry region does not irrigation facilities, so the adoption of '*in-situ*' moisture conservation and water harvesting measures need to be adopted for life saving irrigation if required.

10. ***Poor market facilities*** – The market infrastructure is very poor so the farmers can not receive the proper cost of their produce. The middle men are enjoyed by cheating the local farmers by purchasing their produce at very low cost from them and earn profit.

11. ***Very low crop yield*** - The crop production is very difficult in the region in absence of irrigation and other inputs like quality seed, fertilizers, pesticides etc, so the crop yield is very low. The farmers don't have suitable crop and varieties and other inputs mainly due to ignorance and sometimes unavailability in the market during crop season.

12. ***Poor technology transfer*** - The extension and communication of modern agro techniques among rural farming communities is very poor, so lack of knowledge and awareness among farmers leave technologies untapped in the region. Central and state governments, agricultural universities and NGOs are working for rural development and technology transfer but the outcome is not very encouraging.

13. ***Poor economy of farmers and insufficient income of farming communities*** – The economy of the region is very poor mainly due to low crop and livestock yield and annual income.

14. ***Poor productivity of livestock*** - The availability of feed and forage is good only during monsoon season. Normally the region has acute shortage of green forage round the year. The productivity of grassland and rangelands are also very low due to lack of management leading to lower livestock productivity.

15. ***Nutrition and health hazards*** - Micronutrient malnutrition or "hidden hunger" has become more conspicuous in the dryland regions. The farmers are poor and cannot afford the cost of balanced nutrition. Normally there is a sheer scarcity of green vegetables, fruits, milks and medical facility in dry areas affecting health of the people.

16. ***Poor living standard of local farming communities*** – The farming communities of the region are devoid of required facilities like education, medical facilities, hygiene, safe drinking water, light, roads etc. making them vulnerable to poor state of life.

Major Constraints of Dry Farming

Agriculture or crop production is very difficult in arid region where the rainfall is very scanty, uneven and erratic, long dry spells occur most often, potential evapo-traspiration is always higher than rainfall, short growing period, late arrival and early cessation of monsoon. Based on these characteristics of the region, following major problems of dry farming may be listed below;

1. ***Insufficient rainfall***: The rainfall is always erratic, uneven and less.
2. ***Soil degradation***: The land degradation is the major constraint of dryland regions mainly due to chemical degradation and soil and water erosion. Sequestration of carbon also turns out to be a major problem which further degrades the soil making it less productive.
3. ***Occurrence of drought***: The dryland are prone to experience extensive climatic hazards and it is further enhanced due to climatic change. Drought is a common scenario in drylands as water availability is less further leading to low productivity.
4. ***Soil moisture stress*** – The soils are very hungry for water because of scarce, erratic, uneven distributed and uncertain rainfall, poor water holding capacity of soils, low organic matter, high mean annual temperature and wind velocity. The crop failure is a common feature of the arid region. Therefore the farmers' faith and belief towards agriculture is not so strong to take agricultural activities seriously, hence the crop yield is very poor and uneconomic. Further it is also a great challenge to farming communities and technologists to conserve water '*in-situ*' or '*ex- situ*' by adopting of suitable technology.
5. ***Extensive agriculture***: Monocropping or single cropping is extensively practiced and makes farm lands lack of nutrients and increased incidence of pest and diseases.
6. ***Production of less remunerative crops***: The farmers have less choice of crops to be grown, so most often millets, nutri cereals, pulses and oilseeds are grown that are less remunerative than rice, wheat and sugarcane.
7. ***Low productivity***: The crops grown in dryland areas are less productive due to water shortage and poor management. The selection of crop to be grown under dry farming is always difficult because there is a limited choice available for dry region crops. Sorghum, pearlmillet, nutricereals, clusterbean, urd, moong, sesame, rape & mustard, etc. can be grown in

limited soil moisture conditions. Sorghum and pearlmillet require more water than other nutri cereals, pulses and oilseeds, so the selection of crop requires to have knowledge of climatic data of the specific area. Sometimes farmers don't care and their crops fail.

8. ***Poor market infrastructure***: The market facility is also poor as the crop production is at low scale, so farmers do not get optimum prices of their products.

9. ***Disposal of produce*** – Most of the crops grown in dry area are similar types, so the demand of the produce is less as compared to supply after crop harvest, Therefore the price is reduced. The market infrastructure in arid area is again very poor for disposal or sale of produce after crop harvest because farmers don't have facility to store for longer period and also need money for other house hold activities and in result the middle men play major role and purchase farm produce at lower rates and get benefitted.

10. ***Lower quality of farm produce*** – The research findings are clear to report that farm produce of dry farming is in poor and inferior quality because of shriveled and unfilled seed due to water stress at different crop growth stages. Mostly number of crops is grown under mixed or intercropping systems that affect the seed purity and infestation of pest, diseases and weeds. The market value of produce is reduced due to mixture and poor quality of seed.

11. ***Poor replacement of crop seeds*** – The replacement of crop seed is very poor i.e. 5-10% only and previous crop seed is stored and used for resowing next year and it causes mixture of crop seed and reduces the quality of final product. Again it also declined the crop yield. The certified or foundation seed should be sown for higher yield. It is considered that crop seed should be replaced after every 2-3 years for cereals and every year for hybrids.

12. ***Poor plant nutrition*** - The farming communities is in general not sure about how much organic and chemical fertilizer to be applied for different crops of the region. High and low both affect the crop yield and economics, therefore judicious use of manures and fertilizers are to be ensured. Besides dose of fertilizer, timing and method of application are also important for higher use efficiencies. There is a little use of manures and fertilizers under dry farming, therefore soils are also hungry and need balanced nutrition.

13. ***Poor economy of farmers***: The economic status of local farming communities is poor mainly due to low production and income.

Major constraints of dryland agriculture may also be categorized in following ways;

1. Climatic constraints
2. Soil based constraints
3. Crop based constraints
4. Technological constraints
5. Socio-economic constraints

1. *Climatic constraints*

i. Erratic, uncertain, insufficient and uneven rainfall that lead to poor crop productivity.

ii. Late arrival, mid season long dry spells and early cessation of rainfall leading to late sowing, mid season and terminal drought, respectively,

iii. Dry soils are poor in nutrients including in nitrogen, phosphorus and micronutirnets, so dry soils are not only thirsty but also hungry too.

iv. Potential evapotranspiration is most often higher than precipitation due to high temperature in early crop growth and chilling and frost injury during grain filling stage leading to forced maturity and poor grain filling and poor quality of produce, respectively.

v. Drought and floods are common features.

2. *Soil based constraints*

i. Soil degradation is very prominent feature due to soil and water erosion,

ii. Soil salinity and alkalinity,

iii. Poor water holding capacity of soil,

iv. Infertile and undulating soil,

v. Poor infiltration rate,

vi. Widespread nutritional deficiencies in soil,

vii. Development of hard pan in soil,

viii. Soil Compaction,

ix. Low soil organic matter,

x. High soil temperature and

xi. Absence of soil conservation measures.

Table 10. Soil related constraints to advancing agricultural production in dryland agriculture of South Asia (Rao and Ryan, 2004)

S.N	Constraint	Aridisols	Alfisols	Inceptisols	3
1	Erosion by water	2	3	2	0
2	Erosion by wind	3	1	3	3
3	Compaction	3	3	3	1
4	Crusting	2	3	3	3
5	Trafficability	0	0	0	3
6	Salinization	2	1	3	2
7	Nutrient depletion	1	2	1	2
8	High soil temperature	3	2	3	2
9	Drought stress	3	3	1	2
10	Soil organic matter depletion	3	3	3	3

Crop based constrains

i. Unavailability of improved drought resistant crop varieties - Green revolution is the evidence of evolving new high yielding varieties but most of the varieties released so far are for irrigated areas. Therefore it is an urgent need to focus on dryland areas and new high yielding climate resilient varieties of should be released for arid and semi arid regions.

ii. Lack of irrigation infrastructure: The irrigation infrastructure should be improved in the regions. In addition, water harvesting and moisture conservation measures should be mandatory in the region so that life saving irrigation may be given to crop in severe drought conditions.

iii. Unavailability of crop varieties resistant to biotic stress like insect-pest and diseases: The arid zone crops are in general hardy in nature but they are not purely resistant to insect-pest and diseases, therefore new high yield varieties should be bred with high level of resistance to biotic stresses.

iv. Crop failure due to mid season long dry spells and terminal drought: It is a common phenomenon in the region that crop suffers mid season or terminal drought due to mid season long dry spells.

v. Reluctance towards application of fertilizers: The farmers are reluctant to apply balanced nutrition due to uncertainty of rain. There is always fear among farmers that crop may be failure due to scanty or absence of sufficient rain during crop period.

vi. Poor crop diversification: The local farmers are reluctant as well as unaware of new crops -varieties and cropping systems, therefore they grow traditional crops of the region. Further they don't have sufficient rainfall to grow more crops. They grow single or rarely double crops in a year, so the cropping intensity is 100 or 200%.

vii. Unavailability of farm inputs timey like herbicides, insecticides, fungicides, nematicides etc.- The demand of inputs is very low in the region because the farmers prefer local seed and use compost or FYM in place of chemical fertilizers. Again the purchasing power of the farmers is very low. Therefore the supply of inputs remains very low during crop period.

4. *Technological constraints*

i. Lack of suitable technologies for dry land areas: The modern technological advances in tillage, leveling, sowing, harvesting, winnowing etc. are flooded in irrigated areas only and dry region farmers hardly adopt these advanced technologies because of poverty as well as low demand in the region.

ii. Unavailability of high yielding crop varieties and its quality seed: The region suffers lack of crop varieties as well as its quality seed timely for timely sowing. The farmers are bound to use local seed resulting in to poor crop yield.

iii. Poor plant protection measures: The farmers hardly use plant protection measures except some progressive farmers. They don't have knowledge about the pesticides and its use.

iv. Little adoption of alternate land use systems: The alternate land use systems like agro-forestry systems, alley cropping, wild life sanctuary, innovation parks etc may be adopted in dryland areas but ignorance of farmers defy then to adopt alternate strategies for higher benefits.

v. Poor extension and communication activities: The Lab to Land programs including other extension activities are very poor in the region, so the new technologies have not been percolated in the dry region areas.

vi. Unaware and inaccessible digital technology: The present era is of digital technology but our farmers are poor and illiterate , so they are unable to handle internet. Again poor light supply hampers the belief of farming communities.

5. *Socio-economic constraints*

i. Lack of capital, support price and market intelligence: The farmers are very poor and don't have much capital to buy technological inputs. The government declared support prices of crops are again not well known by the local farmers. The market intelligence is very poor even in the region.

ii. High risk: The farming in arid region is always very risky, so the farmers bear always high risk.

iii. Input supply chain is again very poor in the region.

iv. Prevailing social systems prevent farmers to adopt new technologies – The farmers are not free to adopt new technologies like piggery, fisheries, dairy etc. as it is decided by the panchyats of the region.

v. Farmers are in general very poor and resource crunch.

vi. The farmers are illiterate so they are unable to adopt new and modern technologies.

vii. The living standard of the local farmers is very low.

Major area of rainfed agriculture for immediate concern

Major area of immediate concern is listed below;

- Need of high yielding climate resilient crops and varieties,
- Timely availability of farm inputs like seed, fertilizers, pesticides etc.
- Proper marketing and price policy for crops and livestock,
- Insurance of crops and livestock due to weather calamities,
- Conservation of soil and water resources,
- Low cost and locally suited agricultural implements,
- Judicious and balance use of manures, fertilizers and chemicals,
- Expansion of acreage under irrigation,
- Training and extension education, and
- Farmers' income and net cash available.

Measures for area development

Our country has fertile cultivable land and receives the highest amount of rainfall per unit area basis in the world but the crop productivity is lower than most of the developed countries. The success story of Israel in field of micro irrigation, water harvesting, off season vegetable and fruit production etc. are enough to understand that we have huge potential to increase our productivity many folds. Similarly the success story of watersheds like 'Ralegaon Siddhi', 'Hivere bajar' in Maharashtra and 'Sukhomajari' in Haryana speak loud to make prosperity in dry regions by adopting site specific measures for area development. Therefore following measures need to be taken to make rainfed agriculture more productive and profitable;

- Selection of climate resilient short duration crop and varieties,
- Optimum seed and sowing,

- Balanced crop nutrition with integration of organic and chemical fertilizers,
- Proper plant protection,
- Timely adoption of contingent crop plan for aberrant weather conditions,
- Adoption of proper soil and water conservation measures,
- In-situ moisture conservation,
- Water harvesting for improving soil moisture regimes and also for life saving irrigation,
- Popularizing micro irrigation ie. Sprinkler and drip irrigation systems in the region,
- Precision mechanization,
- Suitable ICT for technology transfer,
- Accessible market infrastructure,
- Proper disposal of excess run-off,
- Protect soil water by reducing water and wind erosion and
- Adopting alternate land use systems.

Government initiatives to help the dry land farming to make it more profitable and productive

- The National Mission for Sustainable Agriculture (NMSA), which is one of the eight missions under the National Action Plan on Climate Change (NAPCC) seeks to address issues associated with climate change.
- Sub-Mission on Agroforestry under the framework of National Mission for Sustainable Agriculture (NMSA) has been launched during 2016-17 for a period of 4 years (2016-17 to 2019-20).
- National Agroforestry Policy, 2014 has been formulated with the objective to bring coordination, convergence and synergy between various elements of agroforestry scattered in various existing Missions, programmes and schemes of the Government.
- The Pradhan Mantri Krishi Sinchayee Yojana (PMKSY) was launched on 1st July, 2015 with the motto of 'Har Khet Ko Paani' for providing end-to end solutions in irrigation supply chain, viz. water sources, distribution network and farm level applications.
- Agriculture Contingency Plan: CRIDA, ICAR has prepared district level Agriculture Contingency Plans in collaboration with state agricultural

universities using a standard template to tackle aberrant monsoon situations leading to drought and floods, extreme events adversely affecting crops, livestock and fisheries.

- Even GOI has declared 2018 as year of millets and also proposed to declare 2019 as year of millet.

A way forward?

It is not only an urgent need to develop a new paradigm for soil and water management in rainfed systems but also have a holistic approach for proper management of natural resources so that the system's productivity can be enhanced and poverty can be reduced without causing further degradation of the natural resource base. Life saving or supplemental irrigation can play a very important role in reducing the risk of crop failures and in optimizing productivity. Besides the rain water harvesting both *in-situ* and *ex-situ* is essentially required for proper soil moisture regimes and also for recharging ground water.

Key issues in rainfed agriculture include

- Crop productivity can be increased by improved rainwater management in rainfed systems which presently generates excessive runoff and causes soil erosion,
- Investments to maximize rainfall infiltration and the water-holding capacity of soils minimize land degradation while increasing the water available in the soil for crop growth. This infiltration will result in improvements in the quality of natural ecosystems and of water in aquatic ecosystems.
- Policy on water resources management for agriculture remains focused on irrigation, while the framework for integrated water resources management at watershed and basin scales concentrates primarily on allocation and management of blue water in rivers, groundwater, and lakes. What is needed is effective integration that focuses on investment options for water management across the continuum from rainfed to irrigated agriculture and which incorporates a multisectoral approach.
- The rain water is to be conserved to life saving irrigation during long dry spells. Besides it will recharge the ground water table as well as sufficient soil moisture for growing crops with double and often even quadruple yields. In contrast, the rain water is lost causing floods and waterlogging during rainy seasons in most parts of India.
- Apart from water, upgrading rainfed agriculture requires investments in soil, crop, and farm management and improved infrastructure, markets, and better and more equitable access to and security over land and water resources.

- To improve production and thus rural livelihoods in rainfed areas, rainfall-related risks need to be reduced, which means that investments in water management are a entry point to unlock the potential in rainfed agriculture.
- More adoption of modern technology is paramount need in dryalnd areas for sustainable crop production.
- Rationalize subsidies on agricultural inputs so that marginaland sub marginal farmers of dryland areas may be benefitted.
- Cover more crops under minimum support price schemes.
- Cover more households under crop and livestock insurance.
- Address chronic trade deficits in pulses and oilseeds.
- Increase inflow of institutional credit to dryland agriculture.
- Institute measures to cope with globalization and marginalization, and
- Facilitate migration.

References

Aggarwal, P. K. 2008. Global climate change and Indian agriculture: Impacts, adaptation and mitigation. Indian Journal of Agricultural Sciences 78: 10-16.

http://www.indiaenvironmentportal.org.in/content/soil-salinity-threatens-command-areas.

IPCC. 2013. Summary for policy makers. InL Climate change 2013: The physical science basis. Contribution of working group to the 5th Assessment Report of the Intergovernmental Panel on Climate change.

Raju, K. V.; Wani, S. P., eds. (2013). Harnessing dividends from drylands : innovative scaling up *with soil nutrients. ISBN 9780896295537. OCLC 958281358.*

Rao, V.U.M. and Bapuji Rao, B. 2013. Climate change impact on Indian agriculture-adaptaation and mitigation strategies. Journal of Research. Punjab Agricultural University 50:82-91.

Rao, S.C. and Ryan, J. (2004). Crop Science Society of America.and Symposium on Challenges and Strategies for Dryland Agriculture. Challenges and strategies for dryland agriculture. In: Challenges and strategies for dryland agriculture. Crop Science Society of America, Madison, Wis.

Singh, H.P., Sharma, K.P., Reddy, G.S. and Sharma, K.L.(2004). Dryland Agriculture in India. p. 67-92. In: Challenges and Strategies for dryland agriculture. Crop Science Soc of,[S.l.].

Venakeswarlu, B. and Shankar, A.K. 2012. Dryland agriculture: bringing resilience to crop production under changing climate. In: Crop Stress and its Management: perspectives and Strategies, Springer Netherland. Pp. 19-44.

Questions

a. Short and descriptive

1. Write short notes on following;

    a. Dry agriculture

    b. Dryland agriculture

c. Rainfed agriculture

d. Growing period

2. Define rainfed agriculture. Also describe the concept and principles of rainfed agriculture

3. Describe the types of rainfed agriculture. Also enumerate the significance of rainfed agriculture in Indian scenario of food and nutritional security.

4. Differentiate between;

    b. Dry vs. Dryland agriculture

    c. Dryland vs. rainfed farming

    d. Arid vs. dry agriculture

    e. Formal vs. Operational definition of rainfed agriculture

    f. Arid vs. humid region

5. What are the major constraints of rainfed farming?

6. What are the major characteristics of rainfed farming?

7. What are the new initiatives/programs launched for the development of rainfed areas of the country?

8. What are the key issues of rainfed agriculture? Proof with facts and figures that these issues can be tackled by governments for the benefit of rural communities of rainfed areas.

9. Describe the distribution and intensity of rainfed areas in India.

10. What is the future prospect of rainfed agriculture in India

a. Objective

1 Who told about rainfall 'Rain is essential for cultivation and the cultivation is essential for life, so one should first acquire the knowledge rainfall carefully' ?

a	Arthashashtra	b	Atharveda
c	Samveda	d	None above

2 In which of the literature the significance of soil is described as 'An increase in supply of milk and butter depends upon the nature of the soil and the quality of fodder and water' ?

a	Arthashashtra	b	Atharveda
c	Samveda	d	None of these

3 Rainfed agriculture constitutes

a	80% of global agriculture	b	90% of global agriculture
c	70% of global agriculture	d	Nine of these

4 Which of the following is grown mainly under rainfed conditions?

a	Millets	b	Oilseeds
c	Pulses	d	All above

5 Which of the following is grown mainly under rainfed conditions?

a	Cotton	b	Barley
c	Lathyrus	d	All above

6 The mean productivity of rainfed area is

a	1.1 ton per ha	b	1.3 ton per ha
c	2.2 ton per ha	d	None of these

7 India stands 1st in production of

a	Pulses	b	Milk
c	Jute	d	All above

8 India stands 2nd in production of

a	Rice	b	Wheat
c	Sugarcane	d	All above

9 India's position in arable land availability is

a	2nd	b	7th
c	1st	d	All above

10 India stands in tobacco (crude) production;

a	1st	b	3rd
c	2nd	d	All above

11 India stand in fruit production ;

a	2nd	b	3rd
c	1st	d	All above

12 How much percentage of rice comes under rainfed agriculture?

a	39.9 %	b	34.5%
c	36.5%	d	All above

13 How much acreage of maize is under rainfed agriculture?

a	73.4mha	b	66.6mga
c	65.4mha	d	All above

14 What is the net area under irrigation in India?

a 68.38 mha b 98.5 mha

c 94.2 mha d All above

15 What is the total area under irrigation in India?

a 96.5 mha b 99.3 mha

c 96.46 mha d All above

16 The levels of low productivity in Africa and South Asia, are due to

a Degraded soils b High levels of evaporation

c Droughts d All above

17 Dry farming is applied for

a Farming without irrigation b An annual rainfall

c Farming with irrigation d All above

18 Dry land is

a Land or area receiving an annual rainfall of 750 mm or less

b Land or area receiving an annual rainfall of 750 mm or more

c Land or area receiving an annual rainfall of > 1150 mm

d None above

19 Dry land agriculture is

a Soil and crops under dry lands without irrigation b Farming without irrigation

c Both a & b d None above

20 Dry land crops normally receive an annual rainfall around;

a < 750 mm b >750mm

c >730mm d None above

21 According to FAO (1993) classification, rainfall in arid (Desert) area is

a <375 mm rainfall b 375-1125 rainfall

c > 1125mm rainfall d All above

22 According to FAO (1993) classification, rainfall in dry land area is

a <375 mm rainfall b 375-1125 rainfall

c > 1125mm rainfall d All above

23 According to FAO (1993) classification, rainfed area receives annual rainfall equal to

a <375 mm rainfall b 375-1125 rainfall

c > 1125mm rainfall d All above

24 Dryland farming receives annual rainfall approximately

a 750 mm b 650mm

c 689mm d None above

25 Rainfed farming areas receive an annual rainfall approximately

a ≥1150 mm b 750mm

c 650mm d None above

26 Which of the following is an important characteristic of arid farming?

a It represents arid region, b Rainfall is <500mm

c Extreme moisture shortage d All above

27 Which of the following is an important characteristic of Arid farming?

a Inadequate rainfall with high variability b Crop experiences high temperature

c High potential evapo-transpiration rate d All above

28 Which of the following is an important feature for dry farming?

a It represents the semi arid regions b Rainfall is <750mm

c Rainfall is always erratic and inadequate d All above

29 Which of the following is an important feature of dry farming?

a Crops suffer most often long dry spells b More frequently crop failure

c Wind erosion is predominant d All above

30 Which of the followings an important characteristic for dryland farming?

a It represents the sub humid regions b Annual rainfall is in between 750 -1150mm

c Double cropping is possible, d All above

31 Which of the following is an important feature of dryland farming?

a Soil & water conservation measures are required b Erratic and Inadequate rainfall

c Soils are deficient in essential plant nutrients d All above

32 Which of the following is an important characteristic of rainfed farming?

a It is practiced in humid regions, b Proper drainage is required,

c Annual rainfall is >1150mm, d All above

33 Which of the following is an important characteristic of rainfed farming?

a Rainfall is erratic and uneven but sufficient or more
b Fewer chances of crop failures due to dry spells
c Both a & b
d None above

34 Which of the following is a principle of rainfed agriculture ?

a Selection of suitable drought resistant crop and varieties
b Optimizing seed and sowing of crops
c *In situ* conservation of rain water
d All above

35 Which of the following is a principle of rainfed agriculture?

a Adopting site specific soil and water conservation measures
b Optimizing evapotranpiration (ET) losses
c Application of optimum crop plant nutrition
d All above

36 Which of the following is principle of rainfed agriculture?

a Reducing evaporation losses
b Water harvesting for supplemental life saving irrigation
c Both a & b
d All above

37 Dry farming is generally practiced in

a Arid regions
b Semi arid region
c Both a & b
d None above

38 Dryland farming is generally practiced in

a Arid regions
b Semi arid region
c Both a & b
d None above

39 Rainfed farming is generally practiced in

a Sub -humid region
b Humid region
c Both a & b
d None above

40 Which of the following is a main characteristic of the rainfed Agriculture?

a Scanty, uncertain and uneven rainfall
b Aberrant weather or monsoon
c Occurrence of extensive climatic hazards
d All above

41 Rainfed agriculture is characterized mainly by

a Soil Erosion
b Soil degradation
c Poor market facilities
d All above

42 Dryland farming is characterized mainly by

a Very low crop yield | b Poor technology transfer

c Poor productivity of livestock | d All above

43 Which of the following is a major problem of dry farming?

a Insufficient rainfall | b Soil degradation

c Occurrence of drought | d All above

44 Which of the following is a major problem of dry farming ?

a Poor market infrastructure | b Lower quality of farm produce

c Poor replacement of crop seeds | d All above

45 Erratic, uncertain, insufficient and uneven rainfall is related to

a Climatic constraint | b Soil based constraints

c Both a & b | d None above

46 The major soil based constraint is

a Soil salinity and alkalinity | b Soil Compaction

c Poor water holding capacity of soil | d All above

47 Which of the following is a soil based constraint?

a Low soil organic matter | b High soil temperature

c Absence of soil conservation measures. | d All above

48 Which of the following is a crop based constraint?

a Lack of irrigation infrastructure | b Poor crop diversification

c Unavailability of crop varieties resistant to biotic stress like insect | d All above

49 Which of the following is a technological constraint?

a Poor plant protection measures | b Little adoption of alternate land use systems

c Unaware and inaccessible digital technology | d All above

50 What is the P/PET ratio of hyper arid region?

a <0.05 | b 0.05 – 0.02

c 0.20 – 0.50 | d 0.050 – 0.65

51 What is the P/PET ratio of arid region?

a <0.05 | b 0.05 – 0.02

c 0.20 – 0.50 | d 0.050 – 0.65

52 What is the P/PET ratio of semi-arid region?

a <0.05 b 0.05 – 0.02

c 0.20 – 0.50 d 0.050 – 0.65

53 What is the P/PET ratio of dry sub humid region?

a <0.05 b 0.05 – 0.02

c 0.20 – 0.50 d 0.50 – 0.65

54 What is the growing period of arid region?

a <90 days b 90-150 days

c 150-180 days d 180-210 days

55 What is the growing period of semi-arid region?

a <90 days b 90-150 days

c 150-180 days d 180-210 days

56 What is the growing period of sub humid region?

a <90 days b 90-150 days

c 150-180 days d 180-210 days

57 What is the growing period of humid region?

a <90 days b 90-150 days

c 150-180 days d 180-210 days

58 What is the growing period of per-humid region?

a 90-150 days b 150-180 days

c 180-210 days d >210 days

59 Which of the following region comes under hot arid region?

a Western plains including Thar desert b Leh & Kargil

c Northern plains d N-E hills

60 Which of the following region comes under cold arid region?

a Western plains including Thar desert b Leh & Kargil

c Northern plains d N-E hills

61 Which of the following region comes under sub-humid region?

a Western plains including Thar desert b Leh & Kargil

c Northern plains d N-E hills

62 Which of the following region comes under per-humid region?

a	Western plains including Thar desert	b	Leh & Kargil
c	Northern plains	d	N-E hills

63 As per climate classification of Singh and Venkateshvarulu (2009), arid region receives annual rainfall;

a	<500	b	500-750
c	750-1150	d	>1150

64 As per climate classification of Singh and Venkateshvarulu (2009), semi-arid region receives annual rainfall;

a	<500	b	500-750
c	750-1150	d	>1150

65 As per climate classification of Singh and Venkateshvarulu (2009), sub humid region receives annual rainfall;

a	<500	b	500-750
c	750-1150	d	>1150

66 As per climate classification of Singh and Venkateshvarulu (2009), humid region receives annual rainfall;

a	<500	b	500-750
c	750-1150	d	>1150

67 As per climate classification of Singh and Venkateshvarulu (2009), semi-arid region receives annual rainfall;

a	15.0	b	42.0
c	25.0	d	None above

68 As per climate classification of Singh and Venkateshvarulu (2009), Sub humid region receives annual rainfall;

a	15.0	b	42.0
c	25.0	d	None above

69 As per climate classification of Singh and Venkateshvarulu (2009), humid region receives annual rainfall;

a	15.0	b	42.0
c	25.0	d	None above

70 As per climate classification of Singh and Venkateshvarulu (2009), arid region receives annual rainfall;

a	10%	b	15%
c	25%	d	42%

71 How many districts have been identified as dryland farming areas in India?

a	72	B	88
c	128	d	151

72 Which of the following is a watershed?

a	Ralegaon Siddhi	b	Sukhomajri
c	Hivere Bajar	d	All above

73 Which of the following new program has been launched for benefit of dryland agriculture in India?

a	National Mission for Sustainable Agriculture	b	Pradhan Mantri Krishi Sinchayee Yojna
c	Micro Irrigation	d	All above

74 Which of the following measure (s) is helpful for improving the crop yield under rainfed agriculture?

a	Selection of climate resilient short duration crops and varieties	b	In-situ water conservation
c	Mulching	d	All above

75 Water erosion is a major problem in

a	Aridisols	b	Alfisols
c	Inceptisols	d	All above

76 Wind erosion is major problem in

a	Aridisols	b	Alfisols
c	Inceptisols	d	Both a & c

77 Compaction is a major problem in

a	Aridisols	b	Alfisols
c	Inceptisols	d	All above

78 Drought stress is a mjor problem in

a	Aridisols	b	Alfisols
c	Inceptisols	d	Both a & b

79 Salinization is a major problem in

a	Aridisols	b	Alfisols
c	Inceptisols	d	All above

80 Crusting is a major problem in

a	Aridisols	b	Alfisols
c	Inceptisols	d	Both b & c

81 Rainfed rice is mainly grown in

a N-E hills b West Bengal

c Odisha d All above

82 Rainfed cotton is grown in

a Punjab b Telangana & AP

c Maharashtra d All above

83 The book 'Dry Farming for Sustainable Agriculture' was written by

a Dr. John A. Widetsoe b Dr S C Panda

c Dr Rajendra Prasad d None above

84 As per the FAO, the growing degree days in arid area is

a 01-74 days b 75-119 days

c 119-220 days d None above

85 As per the FAO, the growing degree days in semi-arid area is

a 01-74 days b 75-119 days

c 119-220 days d None above

86 Rain water use efficiency is

a 15-20% b 35-45%

c 55-65% d 80-85%

87 The contribution of rainfed agriculture to total food production is

a 20% b 40%

c 60% d 80%

88 Which of the following component is important for successful rainfed cultivation?

a To retain or store precipitation on the land

b To Reduce or minimize evaporation from the soil surface and increase transpiration,

c To grow drought tolerance crops and fit to the precipitation patterns

d All above

89 National Rainfed Area Authority (NRAA) was established in year

a 2006 b 2007

c 2008 d 2010

90 Which of the following institute is working on rainfed agriculture?

a CAZRI, Jodhpur b CRIDA, Hyderabad

c ICRISAT, Hyderabd d All above

91 Which of the following is an international institute working on rainfed crops?

a ICARDA, Lebanon | b ICRISAT, Hyderabad
c Both a & b | d None above

92 Extensive commercial grain cultivation is not practiced in

a Amazon basin | b American and Canadian Prairies
c Pampas of Argentina | d European Steppes

93 Organic matter content in most of the soils under dryland conditions is

a <0.5% | b <1%
c >1% | d None above

94 Planting pattern should always be followed under dryland conditions is

a Rectangular type | b Circular type
c Square type | d Hexagonal

95 ICRISAT was established in year

a 1972 | b 1974
c 1984 | d 1985

96 Abnormal failure of rainfall is known as

a Permanent drought | b Seasonal drought
c Contingent drought | d Meteorological drought

97 Collecting and storing rain water for subsequent use is known as

a Water harvesting | b Water storage
c Water use | d None above

98 National Wasteland Development Board is situated at

a New Delhi | b Hyderabad
c Bengaluru | d Bhopal

99 National Research Centre for Agroforestry is situated at

a New Delhi | b Jhansi
c Hyderabad | d Nagpur

100 International Centre for Research on Agroforestry is situated at

a Nairobi, Kenya | b Hyderabad
c Aleppo, Syria | d New York

Answers

1. **b**	2. **a**	3. **a**	4. **d**	5. **d**	6. **a**	7. **d**	8. **d**	9. **a**	10. **b**
11. **a**	12. **a**	13. **a**	14. **a**	15. **c**	16. **d**	17. **a**	18. **a**	19. **a**	20. **a**
21. **a**	22. **b**	23. **c**	24. **a**	25. **a**	26. **d**	27. **d**	28. **d**	29. **d**	30. **d**
31. **d**	32. **d**	33. **c**	34. **d**	35. **d**	36. **c**	37. **a**	38. **b**	39. **c**	40. **d**
41. **d**	42. **d**	43. **d**	44. **d**	45. **a**	46. **d**	47. **d**	48. **d**	49. **d**	50. **a**
51. **b**	52. **c**	53. **d**	54. **a**	55. **b**	56. **c**	57. **d**	58. **d**	59. **a**	60. **b**
61. **c**	62. **d**	63. **a**	64. **b**	65. **c**	66. **d**	67. **a**	68. **b**	69. **c**	70. **b**
71. **c**	72. **d**	73. **d**	74. **d**	75. **b**	76. **d**	77. **d**	78. **d**	79. **c**	80. **d**
81. **d**	82. **d**	83. **a**	84. **a**	85. **b**	86. **b**	87. **b**	88. **d**	89. **a**	90. **d**
91. **c**	92. **d**	93. **b**	94. **a**	95. **a**	96. **c**	97. **a**	98. **a**	99. **b**	100. **a**

2

Historical Development of Dry Land Agriculture in India and World

Outlines

1. Introduction

Water is essential for life and most of the civilization flourished on bank of rivers like Indus valley, Mesopotamia, May, INCA and also destroyed in lack of water. Scarcities and famines were the common phenomena of the world civilization. Systematic dry cultivation started around 1850 and improved techniques were started in use in USA. The term dry farming was used probably in 1873 in Utah. Later in 1894, Colorado State University initiated experiments on dry farming and established a substation at Cheyyeniwells, USA. Dr. John A Widtsoe, one of the earliest to start dry faming research, defined dry farming as 'the profitable production of useful crops without irrigation on lands that receive annual rainfall of 20" or 50cm'.

Famines and food scarcities are well known in India time immemorial. They were mentioned even in the Mahabharata, Ramayana, Buddhist and Jain records. Some of historical events witness that development of drylands was a big task to food production, so irrigation work was carried out even from Indus Valley Civilization around 4500 BCE, the sophisticated irrigation and water storage systems were developed including artificial reservoirs at Girnar dated to 3000 BCE, and an early canal irrigation system from circa 2600 BCE. There is Jain legend of a 12-year famine in the fourth century A.D. The Mauryan Empire (322–185 BCE) categorized soils and made meteorological observations for agricultural use and also facilitated construction and maintenance of dams. From early Common Era to high middle age (200-1200CE), the water storage systems were designed. Kallanai (1st-2nd century CE), a dam built on river Kaveri in Trichurapalli district of TN during this period, is considered to be one of the oldest water-regulation structures in the world still in use. Chola empire (875-1279) witnessed the distribution of water by tank-and-channel networks to the drier areas. The construction of water works and aspects of water technology in Medieval India (1200-1500 CE) is described in Arabic and Persian works. The diffusion of Indian and Persian irrigation technologies gave rise to irrigation systems which brought about economic growth and growth of material culture. Mughal Era (1526–1757 CE) also worked on development of irrigation work in whole empire. During the reign of Moghul Emperor Shah Jehan in 1630-32, the official historian recorded in 'Badsha Namah' that soup Kitchens were opened, a lakh and half coins were given in charity and one eleventh of land revenue was remitted in the drought effected area. In the remote past, these famines were tided over by saving grains and fodder from years of bumper harvest and by migration of people and cattle to greener pastures. The Bengal famine in 1865 was very famous and later series of famines forced to British Government for setting up of three Famine commissions. The Commissions made several suggestions which formed the basis of the Famine Code promulgated in 1883 onwards and adopted by the then provincial governments. Apart from the relief works to be taken up during famine, the Commission recommended protective irrigation projects in dry and scarcity area and establishment of Department of Agriculture at the center and in the province. Later, the Royal Commission on Agriculture (1928) reiterated in its report the problem of dryland agriculture and recommended a close attention by Departments of Agriculture to these problems.

The first systematic approach was made in 1923 by setting up small plots at Manjri, near Pune. Mr. C. V. Mehta, the then Minster of Agriculture, Dr. H. H. Maan, Director Agriculture and Dr. V. N. Tamhane, soil physicist were responsible for this initiation of small plot studies. In 1926, Dr Tamhane was replaced by Mr. N. V. Kanitkar who continued this dryland research work.

This was followed by a more comprehensive research scheme on dry farming sanctioned by the Imperial (then Indian) Council of Agricultural Research during 1933-35. Under the scheme, research was conducted in low rainfall areas at newly established research centres Solapur (MS) (1933), Bijapur (Karnataka) (1933), Raichur (AP) (1934, Hagari (TN) (1934) and Rohtak (Hisar) (1935) by Imperial Council of Agricultural Research. The recommendation of these centres could increase the grain yield only 5-10% in last 10 years, so the farmers did not adopted the recommendations. Based on the results of this scheme, a series of dry farming practices were evolved which were known as ' Bombay Dry farming', 'the Madras Dry farming ' and ' the Hyderabad Dry farming ' practices. These practices emphasized soil and water conservation as basic to dryland production programs. The Scheme was terminated in 1943, since it was felt that the immediate objectives had been achieved.

Based on these experiences, 8 more soil conservation research centres at Dehradun, Agra, Chandigarh, Kota, Vasad, Bellary, Ibrahimpatnam and Ooty were started during 1954-62 and submitted the report emphasizing the importance of land use classes, rainfall pattern, run off collection and fertilizer use. The crop productivity was not increased significantly due to short span of rainfall period. The first monograph entitled, 'Dry Farming in India' by N V Kanitkar was published in 1944 and later it was modified and updated by S S Sirur and D H Gokhale in 1960 at the suggestion of ICAR.

After independence, nine soil conservation research centers were set up in different parts of the country in 1954. Large-scale action programs in soil and water conservation were taken up which benefited largely the dry farming areas. The rainfed area, which is 75% of the cultivated, contributes 42% to the nation's food. Grain legumes, oilseeds and cotton are almost exclusively grown in drylands. Rainfed rice is also a dominant crop of humid region. Recognizing this, an All India Coordinated Research Project for Dryland Agriculture was initiated in 1970. Financial support came from the Government of Canada. The main goals were to reduce the growing socio-economic gap between the drylands and less water endowed regions and to establish efficient patterns of land utilization and moisture conservation.

2. Brief history of Dryland Agriculture

a. India

i. In 1880: British Govt. of India appointed a 'Royal Famine Commission' after series of famine in India. The commission made following important recommendations;

   a. To establish protective irrigation projects,

   b. To establish a Department of Agriculture and

   c. To establish a Department of Agriculture in every states.

ii. In 1920: The Royal Commission on Agriculture stressed the need for scarcity tract development.

iii. Between 1923 to 1926: First scientific approach was initiated by Dr V. A. Tamhane, a soil physicist in 1923 on a small plot at Manjri farm near Pune and in 1926, this work was passed on to Mr. N V. Kanitkar who made a very comprehensive research project on dry farming and launched with financial support of ICAR.

iv. In 1933-1935: After reviewing the significant achievement of Kanitkar report on dry farming, following dry farming schemes were started;

   a. In 1933: Bombay Dry Farming Research Schemes at Bijapur and Sholapur were started,

b. In 1934: Madras Dry Farming Research scheme at Hagari and Hyderabad Dry Farming Research Scheme at Raichur were started and

c. In 1935: Punjab Dry Farming Scheme satretd at Rohtak.

v. ICAR opened five dry farming centres at Sholapur and Bijapur (Maharashtra) in 1933, Hagari and Raichur (Karnatika) in 1934 and at Rohtak (Haryana) in 1935 and started comprehensive work on dry farming.

vi. 1933-1944: The research work was continued from 1933 to 1944 at different centres on different aspects like rainfall analysis, physio-biochemical properties of soil, physiological studies on millets and other dry crops and its agronomy were studied and its reports were prepared and published in form of ' Bombay Dry Farming Practices', 'Hyderabad Dry Faring Practices' and 'Madras Dry Farming Practices'.

vii. 'Contour bunding', 'deep ploughing', 'application of FYM', 'low seed rate', 'wide spacing', 'mixed cropping' and 'crop rotations' were recommended technologies by above reports for successful dry farming as it was claimed that the yield can be increased by 15-20% over a base yield of 200-400 kg/ha by adoption of above technologies but these recommendations were not sufficient to satisfy the farming communities, so the project was terminated in 1943-44.

viii. In 1942: Bombay Land Development Act was passed.

ix. In 1944: N V Kanitkar published a monograph on 'Dry Farming Practices' based on research under Bombay, Hyderabad and Madras Dry Farming schemes.

x. 1953-1954: By mid 1950s, ICAR realized the importance of soil and water conservation and Central Soil Conservation Board was opened in 1953. Later ICAR opened seven Central Soil and Water Conservation Research and Training Centres at Dehradun (Uttarakhand), Bellary (Karnatika), Kota (Rajasthan), Ooty (Tamil Nadu) and Vasad (Gujrat) in 1954, Agra (UP) and Chandigarh in 1957. The major emphasis was given to contour bunding but the yield enhancement was recorded only up to 15-20%, so the results were not promising.

xi. 1955-1960: Dry land Farming Demonstration Centres were initiated during early 1960s and short duration crops were emphasized to grow so that they can be harvested in the same span of available soil moisture regimes. In mid 1960s, new high yielding and hybrids were available and grown but their performance was very poor because they were nutrient and management responsive.

xii. 1959: Central Arid Zone Research Institute, Jodhpur (Rajasthan) established with following objectives;

  i. Undertaking basic and applied research on sustainable farming systems in the arid ecosystem,

  ii. Act as repository of information on the state of natural resources and desertification processes ,

  iii. Developing livestock-based farming systems and range management practices for the chronically drought-affected areas, and

  iv. Generating and transferring location-specific technologies.

xiii. 1962: Soil Conservation Research Demonstration and Training Center was opned at Inbrahimpatnam (Telangnana).

xiv. 1964-69 (3rd FYP): Three important programs i.e. Intensive Agricultural District Program (IADP), Intensive Agricultural Area Program (IAAP) and High Yielding Variety Program (HYVP) were launched to reduce the socio-economic imbalances between farmers of dryland and irrigated areas.

xv. 1970: All India Coordinated Research Project on Dryland Agriculture (AICRPDA) was started with 23 centres (now 25) with headquarter at Hyderabad (Telangana) in coordination with financial support of Canadian International Development Agency (CIDA). The moisture deficit soil type and rainfall pattern were taken as a criteria for selection of centres in different parts of India. Thornthwaite formula was adopted for moisture deficit or aridity index.

The dryland research project originally proposed in 1967, was approved in 1969 after showing keen interest of Prime Minister Indira Gandhi to eradicate poverty in dryland areas and with this recommendation, All India Research Project on Dryland Agriculture was launched in 1970 at Hyderabad by ICAR in 4th FYP in collaboration with the Government of Canada through Canadian International Development Agency (CIDA). The coordinated project was started with 17 main centres, 02 sub centres and 08 operational research centres.

1. Hisar, Jodhpur, Bellary, Rajkot and Anantpur (05) in moisture index 60-80%,
2. Sholapur, Akola, Kovilpatti, Hyderabd and Varanasi (05) in moisture index 40-60%,
3. Bangalore and Indore (02) in moisture index 20-40%,
4. Bijapur, Rewa, Ludhiana and Samba (04) in moisture index 20-4%,
5. Jhansi, Anand and Agra in moisture index of 20-40%,

6. Special problem areas were Bhubneshwar, Ranchi, Dehradun and discontinued in 1985,
7. Anand was shifted to Sardar Krishinagar, Ludhiana to Hoshiarpur to Ballowal-Saunkhri (Pb), Samba to Rakh Dhiansar and Bhubneshwar to Phulbani and subcentre Udaipur to arjia in 1985,
8. Bijapur and Rewa were upgraded as main centres and
9. Faizabad (now Ayodhya) was added as one of the centres in 1985.

At present, it has now the network of 22 centres (19 main, 3 sub centres) and 8 ORP centres spread across arid, semi-arid, sub-humid and per-humid rainfed agro ecologies. The centres are Akola, Arjia, Anantpur, Bangalore, Bijapur, Hisar, Ballowal-Saunkhri, Indore, Kovilpatti, Phulbani, Ranchi, Rajkot, Rewa, Sardar Krishinagar, Sholapur, Varanasi, Agra, Faizabad (now Ayodhya), Jammu, Bellary, Jhansi and Jodhpur and 08 Operational Research Centres at Anantpur, Arjia, Bangalore, Hisar, Ballowal-Saunkhri, Indore, Ranchi and Sholapur.

In addition, during XII Plan, five new centres were established at ARS, Darsi (ANGRAU, Prakasam Dt, Andhra Pradesh), MRS, Raichur (UAS Raichur, Karnataka), ARS, Aklera (Agricultural University, Kota), Sabour (BAU, Bihar) and, Imphal (CAU Manipur).

Mandate

i. Optimize the use of natural resources, i.e., rainfall, land and water, and minimize soil and water loss and degradation of environment,
ii. Evolve simple technologies to substantially increase crop productivity and profitability,
iii. Increase stability of crop production over years by providing improvements in natural resources management, crop management systems and alternate crop production technologies matching weather aberrations,
iv. Develop alternate and sustainable land use systems and
v. Evaluate and study transferability of improved dryland technologies to farmers' fields.

Significant Achievements

AICRPDA has contributed significantly in dryland research area with following achievements;

i. Location-specific '*in situ*' moisture conservation practices for diverse rainfed agro-ecologies i.e. rainfall (<500 mm to >1500 mm) and soil types (Entisols, Inceptisols, Vertisols, Alfisols, Aridisols),
ii. Rainwater harvesting methods for diverse rainfall and soil types,

iii. Risk resilient and productive intercropping/strip cropping systems for diverse rainfed agro-ecologies i.e. rainfall (<500 mm to >1500 mm) and soil types (Entisols, Inceptisols, Vertisols, Alfisols, Aridisols),

iv. Development and real-time implementation of contingency measures for agricultural drought management,

v. Carbon positive nutrient management practices including residue recycling, green manuring and use of enriched compost for major rainfed cropping systems,

vi. Development, testing and popularization of cost effective and energy efficient tractor / animal drawn and hand-operated implements for timely and precision agricultural operations and drudgery reduction,

vii. Developed dryland horticulture, agri-horti, agri-silvi, silvi-pastoral and agri-silvi-pastoral systems for higher productivity, profitability and risk minimization,

viii. The action/adaptive research and demonstrations through ORPs resulted in technology assessment and refinement, impacted horizontal spread of doable rainfed technologies in convergence with national/state programs,

ix. Documentation & validation of ITKs and their up scaling in farmers' fields,

x. 1972: International Crop Research Institute for Semi arid Tropics (ICRISAT) was established at Hyderabad by Consultative Group on International Agricultural Research (CGIAR). The research emphasis on dry farming was shifted to multidisciplinary approach with coordination of AICRPDA to tackle the dry farming problems,

xi. 1976: Started Operational Research Projects on Dryland farming. The 1st ORP on dryland agriculture was launched at Indore with UK collaboration. It was augmented with addition of 4 more CIDA (Canadian International Development Agency) sponsored ORPs in 1976. In 1984 4 additional ORPs were added to the list. Currently 9 ORPs are in operation at Indore, Bengaluru, Hyderabad, Sholapur, Ranchi, Hoshiyarour, Anantpur, Hisar and Arjia (Rajsathan),

xii. 1977: Establishment of Krishi Vigyan Kendra (KVK) to impart training to farmers in dry regions. The training facilities first created at Hyatnagar Research Farm,

xiii. 1978: Trainer's Training Centre (TTC) was established as a part of AICRPDA and offered its first course to extension personnel,

xiv. 1979: Lab to Land Program was launched to improve technology transfer,

xv. 1980: Integrated Watershed Management in the Catchment of Flood Prone Area was started,

xvi. 1983: The dryland research work was initiated on watershed mode. Initially work was started on 47 model watersheds under ICAR. The average size of each watershed was 700-800 ha. Central Govt. earmarked 25 lakh for each watershed for period of 5 years. The work was started on soil and water conservation measures, crop management and alternate land use systems. Out of 47 watershed, 30 were attached with AICRP on Dryland Agriculture and 17 with CRIDA, Hyderabad,

xvii. 1984: Initiation of World Bank assisted watershed development programs in 04 states and also establishment of 'Dry Land Development Board' in Karnataka,

xviii. 1985: Birth of Central Research Institute for Dryland Agriculture (CRIDA) in Hyderabad.

The Institute was started in 1971 by Government of India as a coordinated unit with various net work centres located in the dryland regions of the country. The mandate of net work centres was to conduct site specific applied research on dryland agriculture. On April 19, 1985, the Directorate of AICRPDA was upgraded as Central Research Institute for Dryland Agriculture (CRIDA). The institute has been concentrating on basic and strategic research in the area of dryland agriculture, technology transfers, national and international collaboration. The institute currently is a nodal centre for coordination of research on dryland agriculture, agrometeorology and climate change. The main research programs of the institute are:

- Rain water management and water productivity enhancement
- Drought mitigation
- Climate resilient cropping systems
- Soil health improvement and enhancing nutrient use efficiency in drylands

xix. 1985: National Wasteland Development Board (NWDB) was established.

xx. 1986: 'National Watershed Development Project on Rainfed Agriculture (NWDPRA) was launched by Govt. of India in 16 states with following objectives;

a. To conserve the basic resources of soil, rain water and vegetation,

b. To achieve higher biomass, production both in arable and non arable areas,

c. To impart stability to crop yields through proper rain water management, restructuring of cropping patterns and ensuring proper land use,

d. To enhance the income of individuals through adoption of alternative enterprises,

e. To restore ecological balance.

xxi. 1989: Integrated Watershed Development Projects (IWDP) was started by National Watershed Development Board (NWDB).

xxii. 1989: Collaborative research project was started on agriculture between India and Govt. of USA. Under this agreement, CAZRI, CRIDA and CSWCRTI (Central Soil and Water Conservation Research and Training Institute, Dehradun) and UD Department of Agriculture started work on following 5 projects;

a. To increase FUE and DLA including crop residue management in dry land crops and cropping systems,

b. Germplasm improvement in castor for drought tolerance and groundnut for higher nitrogen fixation,

c. Crop simulation modeling for mechanization,

d. Evaluation of legume genotypes under drought and higher N fixation, N use efficiency and yield,

e. Development of suitable systems,

f. Implement for efficient use of available power.

xxiii. 2006: National Rainfed Area Authority (NRAA) was established as per decision of Cabinet on 3rd November 2006 as an expert body of Ministry of Agriculture to provide the much needed knowledge inputs regarding systematic up-gradation and management of country's dry land and rainfed agriculture. The mandate of NRAA is listed below;

i. To prepare a perspective plan, outlining the national strategy and road map for holistic and sustainable development of rainfed farming areas.

ii. To evolve common guidelines for all schemes of different Ministries including Externally aided project (EAPs) for development of Rainfed/ Dry land Farming Systems

iii. To coordinate and bring convergence within and among agricultural and wasteland development programs being implemented in rainfed areas of the country.

iv. To identify rainfed areas in different States which need priority attention and prepare watershed development programs for integrated natural resource management in consultation with States, focusing on multi dimensional crop, livestock, horticulture, agri-pasture integrated systems and programs for landless farming communities.

v. To identify gaps in input supply, credit availability, dissemination of appropriate technology and other requirements relevant for development of rainfed areas.

vi. To guide the implementing agencies on priority setting and monitor the specific interventions required.

vii. To develop plans/ programs for capacity building of Centre/ State Government functionaries in rainfed areas.

viii. To suggest modalities to strengthen National and State Level Institutions concerned with Rainfed/Dryland areas and establish institutional linkages with prioritized watersheds.

ix. To monitor disbursement of rural credit/insurance cover/ safety net programs developed for rainfed areas.

x. To set the research agenda including a critical appraisal of on-going programs and promote diffusion of required knowledge for integrated farming in rainfed areas to district and lower level authorities.

xi. To evaluate the effectiveness of completed watersheds and concurrent evaluation of on-going programs.

b. World

Dryland farming is practiced in various parts of the world. The specific practices vary because of differences in local conditions, both physical and social. Development histories of some of the larger dryland regions of the world are presented below:

A. Australia

1. In Australia, dryland-farming systems combine crops, pastures and fallow periods for the purpose of making efficient use of the limited water. Moisture is usually the deciding factor in the success of cereal cropping. Fallowing can be an important strategy to store and conserve water for the establishment and maturation of the crop. Winter crop production often depends on fallowing over the summer.

2. Australian dryland-farming systems have distinctive characteristics (Tow, 1991). These reflect: the importance of livestock products, particularly sheep, and wheat as export commodities; the availability of pasture legumes for incorporation into crop rotations; and the need for efficiency imposed by farm size. However, in general, Australian dryland-farming systems have similarities to three major world farming systems: Mediterranean agriculture; mixed farming of Western Europe and eastern North America; and large-scale grain production.

3. In general, Australian dryland-farming systems have similarities to three major world farming systems: Mediterranean agriculture; mixed farming of Western

Europe and eastern North America; and large-scale grain production. Southern Australia has Mediterranean type climate where crops, pasture, weed and livestock have great significance. Similarly, Australian cereal production systems are similar to United States of America's mixed farming areas where large farm size, mechanization, fertile land and scarcity of labor are greatly featured. Large-scale cereal production is common in parts of Argentina, Australia, Canada, the former Soviet Union and the United States of America, and their early successes came mainly through the exploitation of large areas of initially fertile lands brought under cultivation. Australian examples are the black, grey and brown clays (mainly Vertisols) of northern New South Wales and southern and central Queensland. Because of the initial high fertility, inputs of fertilizers were low for many years. However, these systems were not sustainable because of the lack of flexibility, risk of soil erosion, buildup of disease organisms, and periodic low prices (Tow, 1991). However Australian farmers are finding alternative crop and pasture species overcoming erosion problems by maintaining crop residues on the soil surface through conservation tillage.

4. The crops, pasture and fallow periods are the main characteristics of dryland faming in Australia mainly for efficient use of limited moisture. Fallowing is considered an important strategy to conserve *in-situ* moisture for crop production and fallowing in summer is the pre-requisite for success of winter crop production.

5. Following distinct features of dryland farming in Australia are listed below;

   i. Livestock mainly sheep production and its products,
   ii. Cereal mainly wheat production on fertile lands,
   iii. Large farm size,
   iv. Fallowing in summer for moisture conservation,
   v. Crop rotation,
   vi. Crop residues (pasture and legumes) management for maintaining moisture and fertility and
   vii. Practice of conservation tillage.

B. China

The total area under dryland agriculture in China is nearly 51% of the total land area including 31 and 21% of arid and semi arid areas, respectively (Li Shengxiu and Xiao Ling, 1992). It accounts 30% of the total arable land of the country and its development was rated as most important for crop production. These lands are located between 30 and 50°N from the warm temperate belt in the south to the temperate zone in the north. Rainfall is very variable; annual totals average 300–500 mm and increase gradually from northwest to southeast. Rainfall from June to September accounts for 70–80 percent of the annual total. Based on the research findings over the period of time, following practices are used in China for conservation of soil and water in these areas;

i. Fallowing once in 3 years and weed control,
ii. Deep ploughing (20-30 cm deep) in summer for conserving rain water and reducing run off. Building low earthen banks between fields,
iii. Making ridges and furrows,
iv. Digging ditches in fields,
v. Contour ploughing and planting,
vi. Digging pits for seeding,
vii. Contour plough furrows,
viii. Cultivation in pits or furrows,
ix. Designing soil- and water-conserving cropping systems to increase plant cover,
x. Narrow crop rows,
xi. Intercropping and inter planting,
xii. Dust mulching, and
xiii. Plastic mulching mainly for increasing temperature for higher plant growth.

C. Ethopia

1. Occurrence of famine is a common feature of Ethopian agriculture. Famine of 1984-85 is considered worst affecting 10 million people (Hurni, 1993). The production as well as availability of sufficient food grain is a major challenge of the country.
2. Seventy percent of the people live in the mountainous areas, where 60 percent of the land has slopes greater than 16 percent (FAO, 1995). Annual soil losses can reach 300 tonnes/ha. So soil and water conservation measures are essentially required for sustainable agricultural production.
3. There were 194 000 ha of irrigated land in Ethiopia in 1994, representing about 3.2 percent of the cultivated area (FAO, 1995). The development of new cropping systems in both irrigated and rainfed areas is required for meeting the future demand of the country but requires untiring efforts at national level.
4. Ancient irrigation systems have a long history in Ethiopia, with modern systems dating from the 1960s. The first large irrigated farms, established by private investors, were located in the middle Awash Valley, where there are large sugar estates, and fruit and cotton farms.

5. There are small-scale, medium-scale and large-scale irrigation systems in Ethiopia (FAO, 1995). Small-scale projects are smallholder projects for a single peasant association up to 200 ha in size. Medium-scale systems are between 200 and 3000 ha, extending beyond one peasant association. Large-scale systems are centrally managed state farms for commercial production and cover 3000 ha or more.

D. India

1. Dryland farming in India began centuries earlier than in North America. However, there are some striking similarities between the two regions with respect to the scientific study of dryland farming. Hegde (1995) reported that, in 1917, Aiyer had listed the important farmer practices and found them quite similar to those that Campbell (1907) had proposed for the Great Plains in the United States of America in the early 1900s. Field bunding, fall ploughing, frequent inter cultivations, drill sowing, and growing drought-resistant crops, such as finger millet, grain sorghum and pearl millet, were some of the practices listed.

2. Scientific study of dryland farming was initiated by the Government of India in 1923. Early research focused on improving crop yields. Important practices included: (i) bunding to conserve soil and water; (ii) deep ploughing once in three years for better intake and storage of water; (iii) use of farmyard manure to supply plant nutrients; (iv) use of a low seeding rate; and (v) inter cultivation for weed and evaporation control. These practices gave a 15–20 percent increase over the base yields (Hegde, 1995, Singh, 1995).

3. By the mid-1950s, the emphasis had shifted to soil management. Soil conservation research and training centres were established at eight locations, focusing on contour bunding. However, negative results were often obtained because of water accumulation and runoff problems, particularly on Vertisols. Even where yield increases were observed, they were again not more than 15–20 percent above the base yields.

4. The importance of shorter-duration crops to match the soil-water availability period was recognized in the 1960s. It was also in the mid-1960s that high-yielding hybrids and cultivars became available that were responsive not only to fertilizers but also to management.

5. An All-India Coordinated Research Project for Dryland Agriculture was established, and the research emphasis shifted to a multidisciplinary approach to tackle the problems. Similar efforts were initiated at the International Crops Research Institute for the Semi-Arid Tropics at Hyderabad in 1972.

6. Although many of the recommended practices for dryland farming in India are similar to those for North America, there are differences. A highly recommended practice for water conservation in India is the use of dust mulch (Hedge, 1995), similar to that recommended in the North America in the early 1900s, which is commonly considered to have contributed to the Dust Bowl. These contrasting recommendations illustrate the importance of recognizing and addressing the differences between semi-arid regions. Semi-arid conditions in North America are very different from those in India. For example, summer fallow has played an important role in North America because some precipitation occurs throughout the year, but at no time does monthly precipitation exceed PET. Many dryland regions in the Great Plains in the United States of America do not have any month with precipitation that even reaches half of PET. In contrast, most dryland areas in India have more than seven months with essentially zero precipitation. They then have a monsoon season of varying length when the precipitation greatly exceeds the potential evapotranspiration for at least a portion of the growing season. Therefore, fallow for storing soil water is not a viable alternative because much of the water saved during the rainy season would be lost during the prolonged dry period. More importantly, there is usually more than enough precipitation during the monsoon season to fully wet the soil profile.

E. Mediterranean Regions

1. The worldwide distribution of Mediterranean climates is in between latitudes 32 and 40 ° north and south of the equator on the west coasts of the continents (Boyce, Tow and Koocheki, 1991). This type climate is characterized by the concentration of rainfall in the winter half-year – from November to April in the northern hemisphere, and from May to October in the southern – with drought in summer. A winter precipitation exceeding 65 percent of the annual total has been used to define the climate zone. North of the Mediterranean Sea, they extend into higher latitudes and in Western Australia into lower latitudes.

2. Major dryland-farming areas with a Mediterranean-type climate are located in Morocco, Algeria, Tunisia, the Libyan Arab Jamahiriya, Syrian Arab Republic, Jordan, Iraq, Turkey, the Islamic Republic of Iran, Chile, Australia, and parts of California.

3. Annual precipitation values suitable for agriculture are generally considered to be from 250 to 500 mm. Most areas show a wide fluctuation in total precipitation from year to year; distribution during the season; time of onset of the rainy season; duration of the rainy season; and intensity of rainfall during precipitation events. Except for Australia, where the topography

is moderately flat, countries with Mediterranean-type climates are characterized by rugged mountain ranges. Increasing elevation correlates positively and strongly with increasing precipitation.

4. Increasing elevation correlates positively and strongly with increasing precipitation. Dryland agriculture has been practiced in some of these areas for perhaps 10 000 years or more (Boyce, Tow and Koocheki, 1991).

5. The West Asia – North Africa region constitutes a large part of the dryland farming areas of the Mediterranean regions. There are two major types of dryland-farming systems in these areas: the wheat-based systems in wetter areas; and the barley-based systems in drier areas. Although there are some variations owing to elevation, soil depth and soil type, the transition between the two systems is generally considered to be the 300 mm isohyets (Boyce, Tow and Koocheki, 1991).

6. Yields are often low as a consequence of low soil fertility, exacerbated by erosion.

7. Fallow is also common in these areas, but it is practiced very differently from in other major dryland cereal-cropping areas. Fallows are generally uncultivated weedy fallows. The fallow fields are grazed by animals, with a primary purpose of increasing soil fertility rather than conserving soil water. The main goal of the barley-based farming systems is animal production. Reliance on animals increases with decreasing amount and reliability of rainfall. Animal production in these low-rainfall areas, as in other dryland regions of the world, is an important survival mechanism in subsistence-type agriculture where the risk of crop failure from inadequate rainfall is high.

8. Farming systems in these regions continue to evolve depending on a variety of pressures, particularly those related to increasing population, increasing shortage of land, increasing mechanization, changing market forces, and a variety of social and political factors. In the higher-rainfall wheat-based areas, crop production will probably continue to dominate although there is increasing tree-fruit production. In the lower-rainfall wheat-based systems and the barley-based systems, animal production is assuming greater importance. Barley is replacing both wheat and fallow to satisfy the increasing need for animal-feed production.

9. Although animal production is increasing but an important feature of all of the farming systems is that cropping and livestock-production systems are not generally integrated. Cropland is predominantly owned and worked by farmers who do not own livestock, while livestock owners do not usually crop the land. This has been the general pattern for centuries or millennia in

most of the cropped areas of the Mediterranean basin. This presents both a physical and a social barrier to the use of systems that efficiently integrate cropping and livestock production under one management system, such as the Australian ley farming system.

10. The cropping systems in the other Mediterranean-type climate regions – Chile, Australia, and parts of California and the Pacific North West in the United States of America – usually include cereal production as an important component (Boyce, Tow and Koocheki, 1991). However, the mix of crop and farm animals is somewhat different. The degree of industrialization accounts for many of the differences between the farming systems.

F. North America

1. The Prairies in Canada and the Great Plains, Pacific North-West, parts of the South-West and intermountain areas in the United States of America constitute the major dryland-farming areas of North America. These areas are major contributors to the food and fibre production system.
2. Gras (1946) suggested that, historically, six general types of farming developed in the United States of America. Listed in chronological order, they were: *woodland, prairie, ranch, irrigated, dryland and left-over farming*. This is a natural progression that is typical of other regions in the world in that agriculture expands to increasingly marginal lands. In the United States of America, as settlers pushed west from the sub-humid eastern edge of the Great Plains, they found that crop production became more erratic and precarious.
3. United States Department of Agriculture (USDA) was formed in 1862 and provided a basis for scientific study. This later led to development of a series of dryland experiment stations throughout the Great Plains, some as early as 1903. In 1914, the USDA Division of Dryland Agriculture was established, consisting of 22 field stations to study crop adaptation and cultural practices (map in Burnett, Stewart and Black, 1985). Although a few of these stations remain today as part of the USDA Agricultural Research Service and a few others as units of the Agricultural Experiment Stations of various states, the Division of Dryland Agriculture was terminated in 1938.
4. The expansion in cultivated land took place primarily between 1915 and 1925, but dryland farming had become well established in the early 1900s. One of the early promoters of dryland farming was Campbell (1907), who published a manual of recommended practices that were widely followed. Campbell believed erroneously that a "dust mulch" on the surface not only conserved substantial amounts of soil water but also attracted it from the atmosphere.

5. Summer fallow was an important practice in the development of dryland farming in North America.
6. Fallow continues to be an important component of many 90 cropping systems. However, its length has been shortened and the use of crop residues as surface mulches has been found very beneficial in increasing the proportion of precipitation stored in the soil profile for subsequent crops.
7. Conservation tillage and no-tillage systems have proved extremely important for water conservation and protection against wind erosion.

G. West Africa

1. Sub-Sahelian West Africa lies between 10 and 140N of the equator and is largely semi-arid, with 300–800 mm annual rainfall. Soils are mostly Yermosols (Aridisols), Cambisols (Inceptisols) and Luvisols (Alfisols) (Lal, 1993). These soils are highly susceptible to wind and water erosion.
2. Estimates of annual soil loss range from 10 to 50 ton/ha for wind alone. Much of the dryland farming in West Africa occurs in the sub-Sahelian region.
3. Rains tend to be intense and erratic, droughts are frequent and may persist for several consecutive years. The erratic character of the rains makes dryland cropping very unreliable despite relatively high rainfall during parts of the year.
4. Sorghum and millet are the principal crops in the African summer rainfall zones (Dregne, 1982). Dregne also reported that the dominant soil problems in the sub-Sahelian region are low inherent soil fertility caused by coarse textures, soil acidity, low nutrient holding capacity, and high phosphorus fixation.
5. Other important problems relate to surface sealing as well as compaction and hardening to a depth of several centimeters. Lal (1993) reported that cultivation and accelerated soil erosion leads to a rapid loss of soil organic matter and plant-nutrient reserves, which results in a serious decline in soil productivity.
6. Several traditional techniques are used to increase water-use efficiency in West Africa. For example, the zai system in Burkina Faso and Mali consists of small pits, about 10 cm deep and 10–30 cm in diameter, at intervals of about 1 m. After mixing in some manure and dead weeds, seeds are planted in the pits at the start of the cropping season. Zai are often combined with contour bunds to capture still more of the rainfall for infiltration into the pits.

7. Pieri (1995) summarized long-term experiments in semi-arid francophone Africa and concluded that continuous cropping without any fallow period is technically achievable while increasing yields. However, annual application of mineral fertilizer with periodic liming is required. A combination of mineral fertilizer, periodic liming, and periodic application of cattle manure brings several benefits, including higher and more-stable yields. Yield stability is particularly improved where water-harvesting techniques are used in combination with fertilizer applications.

3. Collaborative Research Work on Dryland Agriculture

i. *Indo-Canadian Dryland Research*

The collaborative research work between India and Canada Govt. on dryland agriculture was initiated on August 25, 1970. The important components of the agreement were;

i. Canada Govt. will provide expertise in priority areas of research,

ii. Canadian Govt will supply lab and field equipment,

iii. Canadian Govt. will provide training to Indian scientist in Canada and 3rd world.

It had three distinct phases as given below;

A. Phase I (1970-75)

B. Phase II (1976-80)

C. Phase III (1981-85)

ii. *ICAR-ICRISAT Collaborative Research Project*

In 1970, the MOU was signed between ICAR and ICRISAT and Project Director cum ADG (Dry Farming) was identified as Member-Secretary of the policy committee meetings held annually. Later CRIDA also joins ICRISAT for number of research initiatives.

iii. *INDO-US Collaborative in Dryland Agricultural Research*

The work was started in 1988 under the agreement of Indo-US Joint Sub Commission on Agriculture. CRIDA-Hyderabad, CAZRI-Jodhpur and CSWCRI-Dehradun were collaborating with US Department of Agriculture in areas of enhancement of fertilizer use efficiency, germplasm enhancement in castor for drought tolerance and in groundnut for higher N_2 fixation, crop simulation modeling and farm mechanization.

4. New Initiatives to Combat Rainfed agriculture

1. ***National Mission for Sustainable Agriculture (NMSA):*** NMSA is envisaged as one of the eight Missions outlined under National Action Plan on Climate

Change (NAPCC), NMSA aim at promoting Sustainable Agriculture through climate change adaptation measures. The major thrust is enhancing agriculture productivity especially in rainfed areas focusing on integrated farming, soil health management, and synergizing resource conservation. Besides, NMSA also a committed target to fulfill National and International commitments on Sustainable Development Goals (SDG) & Intended Nationally Determined Contribution (INDC). All the components of NMSA such as Rainfed Agriculture, Soil Health Management, Organic Farming, etc. have significant role in achieving SDGs and INDC.

NMSA focuses mainly on conservation agriculture to make farm sector more productive, sustainable, and remunerative and climate resilient by promoting location specific integrated/composite farming systems; soil and moisture conservation measures; comprehensive soil health management and mainstreaming rainfed technologies. NMSA has one of the components of Rainfed Area Development being implemented by RFS division.

2. *Rainfed Area Development (RAD)* : RAD focuses on Integrated Farming System (IFS) for enhancing productivity and minimizing risks associated with climatic variabilities. Under this system, crops/cropping system is integrated with activities like horticulture, livestock, fishery, agro-forestry, apiculture etc. to enable farmers not only in maximizing farm returns for sustaining livelihood, but also to mitigate the impacts of drought, flood or other extreme weather events with the income opportunity from allied activities during crop damage. The total area covered up to 2019-20 was 58187 ha but the leading states were TN (13797ha), UP (10043ha), Odisha (6320 ha), AP (5866 ha) and Maharashtra (4749ha). The area was covered in Uttarakhand 3386 ha, HP (2957ha) and J&K (0 ha). Among the NE states, Nagaland was leading followed by Mizoram, Arunachal Pradesh and Tripura.

For the Year 2020-21 (B.E), budget provision of Rs. 202.50 crore has been made for implementation of the program. The major findings of impact evaluation carried out by Agricultural Corporation Finance Ltd in 2017 are as under:

- Integrated Farming System (IFS) is an effective initiative of DAC&FW for climate resilient and sustainable livelihoods of the farmers adopting cluster approach.
- Horticulture and Livestock based farming system are more popular intervention among most of the farmers. There is high demand for assistance for these interventions. The Livestock based farming system not only enhances the livelihoods but also increases the organic residue in the soils which further improves soil health and ecology.
- Percentage of local breed of cow increased in Andhra Pradesh (73%), Jharkhand (50%), Rajasthan (52%) and Tamil Nadu (54%) during RAD

period. Similarly the population of local breed of buffalo increased in Andhra Pradesh, Karnataka, Rajasthan, and Uttar Pradesh.

- The annual income from cows increased and beneficiaries were benefitted by 49% in both Rajasthan and Uttar Pradesh, 36% in Maharashtra and 19% in Karnataka and for beneficiaries in other states the annual income mostly remained unchanged.
- It has been observed that the scheme has helped the farmers in increasing their income by 20% to 30% depending on different interventions under IFS and supplementary activities. This estimate was corroborated by field study as the overall increase in income during RAD was about 38%.

3. *Pradhan Mantri Krishi Sinchai Yojana (PMKSY)*

The Pradhan Mantri Krishi Sinchayee Yojana (PMKSY) was launched during the year 2015-16 with the motto of 'Har Khet Ko Paani' for providing end-to end solutions in irrigation supply chain, viz. water sources, distribution network and farm level applications. The PMKSY not only focuses on creating sources for assured irrigation, but also creating protective irrigation by harnessing rain water at micro level through 'Jal Sanchay' and 'Jal Sinchan'.

PMKSY has the following components

1. **Accelerated Irrigation Benefit Program (AIBP):** It was implemented by Ministry of Jal Shakti.
2. **PMKSY -Har Khet ko Pani:** It was implemented by Ministry of Jal Shakti.
3. **PMKSY -Watershed:** It is implemented by Department of Land Resources.
4. **PMKSY- Per Drop More Crop (PMKSY- PDMC):** Department of Agriculture, Cooperation & Farmers Welfare is implementing PMKSY-PDMC which is operational in the country from 2015-16. The PMKSY-PDMC focuses on enhancing water use efficiency at farm level through Micro Irrigation viz Drip and Sprinkler Irrigation System. Besides promoting micro Irrigation, this component also supports micro level water storage or water conservation/management activities as Other Interventions to supplement source creation for Micro Irrigation.
5. ***Micro Irrigation*:** Micro Irrigation technologies viz. drip and sprinkler irrigation is a proven technology which has gained immense popularity amongst the farmers in India. Micro irrigation not only helps in water saving, but also in reducing fertilizer usage, labour expenses, and other input costs and enhancing farmers' income. With this technology, additional area can be irrigated with the same amount of water compared to conventional method of irrigation. In addition, water deficient, cultivable waste land and

undulating land areas can easily be brought under cultivation due to ease of irrigation.

At present, the area under micro irrigation in the country is 1.17 m ha (2019-20) including 2.63 lakh ha in TN, 2.51 lakh ha in Karnatika, 1022 lakh ha in AP, 1.71 lakh ha in Maharashtra and 1.08 lakh ha in Gujarat. Among the hill states Uttarakhand has 6790ha, HP 1086 ha and J & K only 35ha area under micro irrigation.

An impact evaluation study for Micro Irrigation was carried out in the year 2014 and major findings of the study are:

- Irrigation cost reduced by 20% to 50% with average of 32.3%.
- Electricity consumption reduced by about 31%.
- Saving of fertilizers vary from 7% to 42%.
- Average productivity of fruits and vegetables increased by about 42.3% and 52.8%.
- Overall income enhancement of farmers in the range of 20% to 68% with average of 48.5%.

Summary

Famines and food scarcities are well known in India time immemorial. They were mentioned even in the Mahabharata, Ramayana, Buddhist and Jain records. The history reveals about the dryland crop production was initiated as early as Indus civilization with irrigation and water storage structures at Girnar dated to 3000 BCE and canal irrigation system in 2600 BCE. Mauryan Empire (322–185 BCE) categorized soils and made meteorological observations for agricultural use and also facilitated construction and maintenance of dams. Kallanai (1st-2nd century CE), a dam built on river Kaveri in Trichurapalli district of TN and it is considered one of the oldest water-regulation structures in the world still in use. The tanks and canal system were developed in Chola Empire (875-1279). Mughal Era (1526–1757 CE) also witnessed the development of irrigation work in whole empire. The Bengal famine in 1865 was very famous and later series of famines forced to British Government for setting up of three Famine commission and later the Department of Agriculture and Royal Commision on Agriculture (1928) came into existence. In 1905-06, six Agriculture colleges were opened at Layalpur (Pakistan), Sabour (Bihar), Kanpur (UP), Pune (Maharashtra) and Coimbatore (TN) to support agricultural research and human resources development.

Systematic dry cultivation started around 1850 and improved techniques were started in use in USA. The term dry farming was used probably in 1873 in Utah. Later in 1894, Colorado State University initiated experiments on dry farming and established a substation at Cheyyeniwells, USA. USA. Dr. John A Widtsoe, one of the earliest to start dry faming research, defined dry farming as 'the profitable production of useful crops without irrigation on lands that receive annual rainfall of 20" or 50cm'. In India, systematic research work on dryland agriculture was initiated in 1923 with establishment of Manjri farm near Pune by Dr. V A Tamhane and Mr. N V Kanitkar. Later in between 1933-35, five more research centres on dryland agriculture were opened at Sholapur and Bijapur (Maharashtra) in 1933, Hagari and Raichur (Karnatika) in 1934 and at Rohtak (Haryana) in 1935 for comprehensive work on dry farming. Central Soil Conservation Board was opened in 1953. Later ICAR opened eight Soil Conservation Research and Training Centres in 1954 at Dehradun (Uttarakhand), Agra (UP), Chandigarh, Kota (Rajasthan), Vasad (Gujrat), Bellary (Karnatika), Ibraheempatnam (Telangana) and Ooty (Tamil Nadu). The major emphasis was given to contour bunding but the yield enhancement was recorded only up to 15-20%, so the results were not promising.

The Central Arid Zone Research Institute, Jodhpur (Rajasthan) was established in 1959 and during 1964-69 (3rd FYP, new programs like Intensive Agricultural District Program (IADP), Intensive Agricultural Area Program (IAAP) and High Yielding Variety Program (HYVP) were initiated for intensive research in dryland areas. All India Coordinated Research Project on Dryland Agriculture (AICRPDA) was started in 1970 with 23 centres with headquarter at Hyderabad (Telangna) in coordination with financial support of Canada Govt. In 1972, International Crop Research Institute for Semi arid Tropics (ICRISAT) was established at Hyderabad. In 1976, Operational Research Projetcs on Dryland Farming was started at Indore with financial support of UK. At present, 9 ORPs are in operation at Indore, Bengaluru, Hyderabad, Sholapur, Ranchi, Hoshiyarpur, Anantpur, Hisar and Arjia (Rajsathan). The Lab to Land Program was launched in 1979 to improving technology transfer. In 1983, the dryland research work was initiated on watershed mode. Initially work was started on 47 model watersheds under ICAR. Out of 47 watersheds, 30 were attached with AICRP on Dryland Agriculture and 17 with CRIDA, Hyderabad. In 1985, Central Research Institute for Dryland Agriculture (CRIDA) was established at Hyderabad for strategic planning and research in dryland areas of whole country. National Watershed Development Project on Rainfed Agriculture (NWDPRA) was launched in 1986 by Govt. of India in 16 states. Later National Rainfed Area Authority (NRAA) was established in 2006 as a nodal agency on dryland work in India. Recently the Govt. of India has started new programs like National Mission on Sustainable Agriculture (NMSA), Pradhan Mantri Krishi Sinchai Yojna (PMKSY) and Micro Irrigation for development of rainfed area with due importance of every drop of water with higher water use efficiency.

Questions

a. Short and Descriptive

1. Write short notes on following;

    i. CAZRI

    ii. ORP on Dryland areas

    iii. CRIDA

    iv. NWDPRA

    v. AICRPDA

    vi. NRAA

    vii. NMSA

    viii. PMKSY

    ix. Micro irrigation

2. Write down the brief history on development of dryland agriculture in India.
3. Briefly describe the history of dryland agriculture in China, USA, Australia, Ethopia and West Africa.
4. Describe collaborative research work on dryland agriculture in India.

b. Objective (MCQs)

1 Who defined dry farming as 'the profitable production of useful crops without irrigation on lands that receive annual rainfall of 20" or 50cm'?

a John A Widtsoe   b S S Sirur

c N V Kanitkar   d None above

2 The Kallanai dam is built on river;

a Kaveri   b Krishana

c Ganga   d None above

3 Which of the following is an oldest water-regulation structure in the world and still in use ?

a Bhakra Dam   b Rihand Dam

c Kallanai dam   d None above

4 The soil conservation research centre is found at

a Dehradun   b Chandigarh

c Kota   d All above

5 The first monograph entitled, 'Dry Farming in India' was written by

a N V Kanitkar
b John A Widtsoe
c S S Sirur
d None above

6 Which of the following is exclusively grown on drylands?

a Grain legumes
b Cotton
c Oilseeds
d All above

7 Which of the following is a important recommendation of 'Royal Famine Commission' proposed after series of famine in India?

a To establish protective irrigation projects,
b To establish a Department of Agriculture,
c To establish a Department of Agriculture in every states
d All above

8 Which of the following is a correct recommendation given by 'Bombay Dry Farming Practices'?

a Rainfall analysis
b Physic -chemical properties of soils
c Physiological studies in millets
d All above

9 Which of the following is a recommended technology for successful dry farming?

a Contour bunding
b Deep ploughing
c Wide spacing
d All above

10 Which of the following is a correct recommendation for successful dry farming?

a Low seed rate
b Crop rotations
c Mixed cropping
d All above

11 In which year, the 'Bombay Land Development Act' was passed ?

a 1942
b 1943
c 1944
d None above

12 In which year, N. V. Kanitkar published a monograph on 'Dry Farming Practices?

a 1942
b 1944
c 1943
d None above

13 What is the objective (s) of CAZRI, Jodhpur ?

a Sustainable farming systems in the arid ecosystem
b Generating and transferring location-specific technologies
c Developing livestock-based farming systemsd
d All above

14 Which of the following program (s) was launched to reduce the socio-economic imbalances between farmers of dryland and irrigated areas?

a Intensive Agricultural District Program (IADP)
b Intensive Agricultural Area Program (IAAP))
c High Yielding Variety Program (HYVP)
d All above

15 Which of the following is a correct achievement (s) of AICRPDA program?

a Rainwater harvesting methods for diverse rainfall and soil types
b Risk resilient and productive intercropping/strip cropping systems
c '*in situ*' moisture conservation practices
d All above

16 Which of the following achievements AICRPDA has contributed significantly in dryland research?

a Development and real-time implementation of contingency measures for agricultural drought management
b Rainwater harvesting methods for diverse rainfall and soil types
c '*in situ*' moisture conservation practices
d All above

17 Which of the following achievements AICRPDA has contributed significantly in dryland research?

a Development, testing and popularization of cost effective and energy efficient technologies
b Developed dryland horticulture, agri-horti, agri-silvi, silvi-pastoral systems
c Documentation & validation of ITKs and their up scaling in farmers' fields
d All above

18 International Crop Research Institute for Semi arid Tropics (ICRISAT) was established at

a Hyderabad
b New Delhi
c Indore
d All above

19 NWDPRA, Govt. of India was launched with following objective (s);

a To conserve the basic resources of soil, rain water and vegetation
b To achieve higher biomass, production both in arable and non arable areas
c To restore ecological balance
d All above

20 In which year, the collaborative research project on watershed was started between India and Govt. of USA?

a 1989
b 1986
c 1987
d None above

21 Which of the following is a distinct feature of dryland farming in Australia?

a Livestock mainly sheep production and its products
b Cereal mainly wheat production on fertile lands
c Large farm size
d All above

22 Which of the following is a special feature of dryland farming in Australia?

a Fallowing in summer for moisture conservation,
b Crop rotation
c Practice of conservation tillage
d All above

23 Which of the following practice (s) is used in China for conservation of soil and water in dryland areas?

a Making ridges and furrows
b Digging ditches in fields
c Contour ploughing and planting
d All above

24 Which of the following practice (s) is used in China for conservation of soil and water in these areas?

a Digging pits for seeding
b Contour plough furrows
c Cultivation in pits or furrows
d All above

25 Which of the following practice (s) is adopted in China under conservation of soil and water in dryland areas?

a Narrow crop rows
b Intercropping and inter planting
c Dust mulching
d All above

26 Which of the following practice (s) is prominent in improving crop yields in India?

a Deep ploughing once in three years for better intake and storage of water
b Bunding to conserve soil and water
c Use of farmyard manure to supply plant nutrients.
d All above

27 Which of the following practice (s) is adopted in improving crop yields in India?

a Use of a low seeding rate | b Inter cultivation for weed and evaporation control

c Both a & b | d None above

28 Major dryland farming areas with a Mediterranean-type climate are located at

a Morocco | b Algeria,

c Iraq | d All above

29 Farming systems in Mediterranean region continue to evolve depending on

a Increasing population | b Increasing shortage of land

c Increasing mechanization, | d All above

30 Which of the following farming system was developed in the United States of America?

a Woodland | b Prairie

c Irrigated | d All above

31 Which of the following is a component (s) of NMSA have significant role in achieving SDGs?

a Rainfed Agriculture | b Soil Health Management

c Organic Farming | d All above

32 Which of the following is used to increase the organic residue in the soils?

a Livestock based farming system | b Integrated Farming System

c Horticulture based farming system | d All above

33 Pradhan Mantri Krishi Sinchayee Yojana (PMKSY) was launched in year

a 2015-16 | b 2016-17

c 2017-18 | d None above

34 The main motto of Pradhan Mantri Krishi Sinchayee Yojana (PMKSY) is

a Har Khet Ko Paani | b Jal Sichay

c Jal Sinchan | d All above

35 Accelerated Irrigation Benefit Programme(AIBP) was implemented by

a Ministry of Jal Shakti | b Department of Land Resources

c Department of Agriculture | d None above

36 PMKSY -Har Khet ko Pani was implemented by

a Ministry of Jal Shakti b Department of Land Resources

c Department of Agriculture d None above

37 PMKSY -Watershed was implemented by

a Ministry of Jal Shakti b Department of Land Resources

c Department of Agriculture d None above

38 PMKSY- Per Drop More Crop (PDMC) was implemented by

a Ministry of Jal Shakti b Department of Land Resources

c Department of Agriculture d None above

39 Micro Irrigation aims at

a Water saving b Reducing fertilizer usage

c Labour saving d All above

40 In which year, Trainer's Training Centre (TTC) was established as a part of AICRPDA?

a 1978 b 1977

c 1976 d None above

41 In which year, Operational Research Projects on Dryland farming were started?

a 1978 b 1977

c 1976 d None above

42 In which year, World Bank assisted watershed development program was started in India?

a 1984 b 1978

c 1976 d None above

43 In which year, 'Dry Land Development Board' was established in Karnataka?

a 1984 b 1978

c 1976 d None above

44 Central Research Institute for Dryland Agriculture (CRIDA) was established in Hyderabad in year

a 1984 b 1985

c 1976 d None above

45 National Watershed Development Project on Rainfed Agriculture (NWDPRA) was launched by Central Govt. in year

a 1990-91 b 1991-92

c 1992-93 d None above

46 National Rainfed Area Authority (NRAA) was established in year

a 3rd November 2006 b 4rd November 2006

c 6rd November 2006 d None above

47 Which of the following is a mandate of NRAA ?

a To prepare a perspective plan

b Outlining the national strategy and road map for holistic and sustainable development of rainfed farming areas

c To identify rainfed areas in different States

d All above

48 Which of the following is a mandate of NRAA ?

a To identify gaps in input supply, credit availability

b To guide the implementing agencies on priority setting and monitor the specific interventions required

c To develop plans/ programs for capacity building of Centre/ State Government functionaries in rainfed areas

d All above

49 Major dryland-farming areas with a Mediterranean-type climate are located at

a Tunisia b Tunisia Australia

c Turkey d All above

50 Summer fallow was an important practice in the development of dryland farming in

a North America b West Africa

c South America d All above

51 Yermosols, Cambisols and Luvisols soils are found in

a North America b West Africa

c South America d All above

52 Which of the following Soils are highly susceptible to wind and water erosion?

a Yermosols b Cambisols

c Luvisols d All above

53 Who concluded that ‘continuous cropping without any fallow period is technically achievable while increasing yield’?

a Pieri b Lal

c Gras d None above

54 The collaborative research work between India and Canada Govt. on dryland agriculture was initiated in year

a August 25, 1970 b August 26, 1970

c August 25, 1972 d None above

55 International Crop Research Institute for Semi arid Tropics (ICRISAT) was established in

a 1972 b 1973

c 1974 d None above

56 Indus Valley civilization is characterized with

a Sophisticated irrigation system b Water storage system

c Large granaries d All above

57 Mauryan dynasty is famous for

a Soil mapping b Meteorological observations

c Construction of dams and its maintenance d All above

58 Chola dynasty is popular for

a Tank irrigation b Canal irrigation

c Tube well irrigation d All above

59 Systematic dry farming was started first of all in

a USA b UK

c Africa d India

60 The term' dry farming' was first used in

a Utah, USA in 1973 b Pune, India in 1923

c Queensland in Australia d None above

61 Which of the following year of Bengal Famine is very famous?

a 1865 b 1869

c 1870 d 1871

62 'Royal Commission on Agriculture' was constituted in year

a 1928 b 1929

c 1930 d 1931

63 The systematic research work on dryland agriculture was started in year

a 1923 b 1924

c 1925 d 1926

64 Manjri farm, Pune is famous for

a Beginning of systematic work on dryland agriculture in India
b Watershed work
c Micro irrigation
d All above

65 Who was the first leader of Manjri farm, Pune?

a Dr. V. N. Tamhane
b Mr. N. V. Kanitkar
c Mr. C. V. Mehta
d Dr. H. H. Maan

66 Who was the responsible person (s) for initiation of dryland research in India?

a Dr. V. N. Tamhane
b Dr. H. H. Maan
c Mr. C. V. Mehta
d All above

67 In 1926, The responsibility of Manjri Farm was given to

a Dr. V. N. Tamhane
b Mr. N. V. Kanitkar
c Mr. C. V. Mehta
d Dr. H. H. Maan

68 The dryland centre and its corresponding year are given below. Which of the pair is correctly matched?

a Sholapur in year 1933
b Bijapir in 1933
c Raichur in 1934
d All above

69 Why did the farmers not adopt the recommendations made by dryland research centres?

a The crop yield was increased only 5-10% in 10 years research
b The yield increase was not significant
c The recommendations were not reached to the farmers
d all above

70 In which year ICAR established 8 soil conservation and training centres?

a 1951
b 1954
c 1956
d 1958

71 Dryland Farming Demonstration Centres were initiated in India during

a 1955-60
b 1960-65
c 1970-75
d 1975-80

72 Which of the following centre was opened under AICRPDA in the moisture index ranging 60-80%?

a Hisar
b Anantpur
c Jodhpur
d All above

73 Which of the following centre was opened under AICRPDA in the moisture index ranging 40-60%?

a Sholapur b Akola

c Hyderabad d All above

74 Which of the following centre was opened under AICRPDA in the moisture index ranging 20-40%?

a Bangalore b Ludhiana

c Indore d All above

75 Royal Famine Commission was launched in year

a 1880 b 1881

c 1882 d 1884

Answer

1. a	2. a	3. c	4. d	5. a	6. d	7. d	8. d	9. d	10. d
11. a	12. b	13. d	14. d	15. d	16. d	17. d	18. a	19. d	20. a
21. d	22. d	23. d	24. d	25. d	26. d	27. b	28. d	29. d	30. d
31. d	32. a	33. a	34. a	35. a	36. a	37. b	38. c	39. d	40. a
41. c	42. a	43. a	44. b	45. a	46. a	47. d	48. d	49. d	50. a
51. b	52. d	53. a	54. a	55. a	56. d	57. d	58. a	59. a	60. a
61. a	62. a	63. a	64. a	65. a	66. d	67. b	68. d	69. d	70. b
71. a	72. d	73. d	74. d	75. c					

3

Indian Climate

Outlines

Introduction

India comes in a tropical region and so it has a tropical monsoonal climate. The climate of India is governed by mountain ranges as well as direction of rain winds like south-west and north-east monsoon and alternate season is the main feature of Indian climate. Climate in South India is generally hotter and extremely humid than that of North India. South India is more humid due to nearby coasts. The southern half of the nation doesn't experience temperature below 10 °C in winter but it usually tends to exceed 40 °C during summer. 'Mawsynram' in Meghalaya receives the world's highest rainfall but on other hand, 'Jaisalmer' in Rajasthan may receive the lowest or no rain in a particular year. Similarly, the temperature in summer occasionally touches 50°C in some parts of the Rajasthan desert, whereas it may be around 20°C in Pahalgam in Jammu and Kashmir. On a winter night, temperature at Drass in Jammu and Kashmir may be as low as minus 45°C. Thiruvananthapuram, on the other hand, may have a temperature of 22°C. On the other hand, there is hardly any difference in day and night temperature in the Andaman and Nicobar islands or in Kerala.

Alternating season is the chief characteristic feature of Indian Climate. Most often two terms 'climate' and 'weather' are used. Climate refers to the sum total of weather conditions and variations over a large area for a long period of time (more than thirty years).The climate refers to average weather conditions, while weather to the state of the atmosphere over an area at any point of time. The elements of weather and climate are the same, i.e. , atmospheric pressure, wind, humidity and precipitation.

The word monsoon is derived from the Arabic word 'mausim' which literally means season. 'Monsoon' refers to the seasonal reversal in the wind direction during a year.

Seasons

Indian Meteorological Department (IMD) categorizes following major season of India;

i. ***Cold weather season (Winter from December to February***): The cool and dry winds blow from direction of north towards south and the temperatures remain low in Northern India. The average temperature is around 10-15^{0}C in the north-west and it rises towards equator, average around 20-25^{0}C in mainland India's south-east.

ii. ***Hot weather season (Summer from March to May):*** In the hot weather season, the sun rays more or less directly fall in this region. Therefore temperature becomes very high. Hot and dry winds called loo blow during the day. In western and southern regions, the month of April is hottest but May is the hottest in northern regions. Average temperature ranges from 32 to-40^{0}C in major parts of the north-west country.

iii. ***South-West monsoon season (Rainy season from month of June to September):*** This season is marked by the onset and advance of monsoon. The winds blow from Arabian Sea and Bay of Bengal towards the land. They carry moisture with them. When these winds strike the mountain barriers rainfall occurs. In this season humid south-west monsoon moves across the country causing rains in whole country but particularly in north-west India. Monsoon rains also begin to recede from north India at the beginning of October. South India typically receives more rainfall.

iv. ***Season of retreating monsoon (Autumn in months of October and November***): When winds move back from the mainland to the Bay of Bengal, this season is known by name of the retreating monsoons. The southern part of India, particularly Tamil Nadu and Andhra Pradesh receive rainfall in this season but in north-western India, months of October and November are generally cloudless.

Traditionally, India has six seasons with each being two months long in duration. They are based on the astronomical division of 12 months into 6 parts. These seasons are reflected in the ancient Hindu calender. These are listed with their traditional names below:

1. Spring (Vasanta)
2. Summer (Grishma)
3. Monsoon (Varsha)
4. Early Autumn (Sarada)
5. Late Autumn (Hemanta)
6. Winter (Sisira)

Classification of climate

The weather of an area represents the state of the atmospheric environment over a brief period. Integrated weather condition over 30 years is generally referred to as climate or more accurately, as macro-climate. Regions having similar climatic features are grouped under one climatic zone. In other words, weather is the momentary state of the atmosphere while climate refers to the average of the weather conditions over a longer period of time. Weather changes quickly, may be within a day or week but climate changes imperceptively and may be noted after 50 years or even more. The climate of India has many regional variations expressed in the pattern of winds, temperature and rainfall, rhythm of seasons and the degree of wetness or dryness. These regional diversities may be described as sub-types of monsoon climate. An analysis of the climate of a particular region can help in assessing the seasons or periods beneficial to crop production or human comforts. It further helps in identifying the climatic elements, as well as their severity, that cause discomfort.

The earliest known climatic classifications were those of Classical Greek times. Such schemes generally divided earth into latitudinal zones based on the latitudes, i.e. the Equator, the 'Tropics of Cancer' and 'Capricorn', and the Arctic and Antarctic circles, respectively and on the length of day. Modern climate classification has its origins in the mid-19th century, with the first published maps of temperature and precipitation over earth's surface, which permitted the development of methods of climate grouping using both variables simultaneously.

Types of climatic classification

Different schemes of climatic classification are broadly differentiated as either empiric or genetic methods. *Empirical methods* make use of observed environmental data, such as temperature, humidity, and precipitation, or simple quantities derived from them (such as evaporation). In contrast, *genetic methods* classify climate on the basis of its causal elements, the activity and characteristics of all factors (circulation systems, fronts, jet streams, solar radiation, topography etc.) that give rise to the spatial and temporal patterns of climatic data. Hence, while empirical classifications are largely descriptive of climate, genetic methods are explanatory. However, for all practical applications empirical classifications are widely adopted.

The climate is the pre requisite for developmental work, therefore the climatic classification of any region or area is quite important. Number of scientists and researchers tried to divide India into different climatic zones. First among these, Kendrew-Stamp classified Indian climate based on variation in temperature and humidity but it fails to demarcate the distinct boundaries of different climatic regions. Some important classifications are given below;

1. Koppen's classification

The most popular empirical classification is given by Wladimir Köppen, in 1900 and several revised versions thereafter. Koppen's scheme used certain critical values of temperatures of the warmest and the coldest months and of rainfall of the wettest and the driest months. His climatic divisions generally coincide with vegetational divisions. Koppen (1936) divided the world climate into the following 5 principal groups;

A. *Tropical Rainy Climate (Equatorial climate)*: Coolest month temperature does not exceed 18^{0}C.

i. Af-Tropical rainforest (Equatorial climate) – No dry season

1. Warm Temperature throughout year with. mean value > 27°C.
2. Abundant rainfall (annual average 250 cm).
3. Suitable for luxuriant vegetation.
4. Prevails over Amazon basin, Zaire basin and south-east Asia.

ii. Aw- Tropical Savanna –Dry season in winter

1. Mean annual temperature 23°C,
2. Wet summers (due to convectional rainfall) and dry winter,
3. Annual rainfall 160 cm.
4. Floods and droughts are common.
5. Vegetation is tropical grassland or savanna with scattered deciduous trees.
6. Prevails over Sudan, Veld plateau and the tropical grasslands of Australia.

iii. Am-Monsoon type

1. Seasonal reversal of winds, associated with alternate periods of rainfall and drought with a short dry season,
2. This climate is experienced over the Pacific coast of Colombia, Guinea coast of West Africa, south-east Africa, south and south-east Asia and northern Australia.

iv. As-Dry summer

1. A rare climatic type prevailing over some rain shadow areas along eastern coast of southern India in Tamil Nadu and Orissa that remain dry during summer monsoon, and
2. Receive winter rainfall from retreating monsoons.

B. *Dry Climates:* Potential evapotranspiration remains more than rainfall, so moisture deficiency prevails over the year.

i. ***Bwh-Desert (Low Latitude) climate***

1. Sub-tropical high pressure region with mean annual temperature is 38°C,
2. Scanty and erratic rainfall.
3. Vegetation varies with the soil type.
4. This climate prevails over southwest USA, north Africa (Sahara), west Asia, Thar desert and central Australia.

ii. ***Bwk-Mid Latitude Deserts climate***

1. It prevails over Takla Makan (China) and Gobi desert (Mongolia), and
2. Climate is similar to the low- latitude desert conditions.

iii. ***Bsh and Bsk-Semi arid and Steppe climate***

1. Mean annual temperature is around 21°C,
2. Rainfall is meager 30 cm,
3. Dry region due to an interior location and absence of mountain barriers across the path of prevailing winds.
4. These climatic conditions prevail over in the deep interiors of landmasses such as Eurasia and North America.

C. Humid mesothermal /or Warm Temperate Rainy Climate

Mild winters, mean temperature of coldest month is below 18°C but above -3°C and that of the warmest month is above 10°C.

i. ***Cfa - Humid subtropical or China type climate***

1. Hot and humid summer and mild winter with average annual temperature is 20°C.
2. Well distributed rainfall is around 100 cm.
3. Hurricanes and typhoons are very common.
4. It prevails within 25° to 45° latitude on east coast in both hemispheres e.g., south-east USA, southern Brazil, Uruguay, Argentina, and south-eastern Africa, eastern coastal belt of Australia, eastern China and Japan.

ii. ***Cfb - Marine west European climate***

1. It is characterized by on shore oceanic influences, short cool summers and mild winters with average annual temperature around 10°C,
2. Annual rainfall is 140 cm.
3. Weather is variable and unpredictable.
4. Prevails between 45° latitude and 65° latitude on west coast in both hemispheres e.g., Western Europe, narrow coastal belt in North and South America, south-eastern Australia and New Zealand.

iii. ***Cs* - *Mediterranean climate***

1. Warm and dry summer (mean temperature 20°C- 27°C) due to sub-tropical high pressure conditions,
2. Mild winter (temperature 4°C to 10°C),
3. Annual rainfall 40-60 cm,
4. It is found within 25° and 45° latitudes on west coasts in both hemispheres—over central California, central Chile, Mediterranean region, southern South Africa, southeastern and southwestern Australia. west coasts in both hemispheres over central California, central Chile, Mediterranean region, southern South Africa, southeastern and southwestern Australia.

D. Humid microthermal /or Cold Frost Climate

It is characterized by severe winters, temperature of the coldest month is < -3°C and warmest month >10°C.

i. ***Df* - *Coolest coast climate***

1. Hot and humid summer with mean temperature 25°C, influenced by tropical maritime air masses.
2. Cold winter (mean temperature -4°C to 0°C),
3. Variable precipitation - convectional rainfall during summer and snowfall in winter.
4. It prevails between 45° and 65° latitude on east coasts, over north-eastern USA, lower Danube plains, Korea, Japan, northern China.

ii. ***Ds* - *Taiga climate*** - 'Taiga' actually refers to the softwood coniferous forest Cover.

1. Short summer and temperature ranges between- 10°C and 15°C.
2. Long and cold winters.
3. Low precipitation as influenced by continental polar air masses.
4. It prevails over the belts from Alaska to Newfoundland and from Norway to Kamchetka peninsula sub-Arctic region.

iii. ***Dw* - *Continental type climate***

1. Short and cool summer and temperatures 10° to 21°C.
2. Long and cold winters (temperatures below 0°C).
3. Variable rainfall mostly during summers and snowfall during winter,
4. It prevails in deep interiors of the continents between Taiga and the mid-latitude deserts over Poland and the Baltic states, Russian plains, northern states of USA and the southern states of Canada.

E. Polar Climate

The polar climate experiences <10°C temperature in warmest month. There is no warm season in this climate. It is two types;

i. ***Et -Tundra climate***

1. Experienced over coastal fringes of the Arctic Ocean.
2. Short & cool summer, long & cold winter and meager precipitation that limit Taiga vegetation.

ii. ***Ef - Ice Cap***

1. Areas permanently covered with snow. Average temperature of the warmest month is below 0°C.
2. These conditions occur over the poles and the interiors of Greenland.

H. Highlands climate

1. This climate mainly prevails over the mountainous regions of Rockies, Andes, Alps and the Himalayas, and
2. High insulation, low temperature, low pressure, high precipitation and larger diurnal ranges at higher altitudes.

Weaknesses of Koppen classification

Koppen's classification is based on temperature, precipitation and their combined interaction with vegetation. He also described letter symbols to denote various characteristics, which is practical and convenient but ignores other factors like cloudiness, wind, rainfall intensity, currents and above all, the air masses which form the basis of modern climatology. It is also difficult to explain the existence of different vegetation types within the same climatic division and similar vegetation types in different climatic divisions.

2. Koppen-Gieger classification

Later Koppen-Gieger classification for Indian climate (1980-2016) was proposed with following 5 zones (Map.1);

1. **Tropical climate**
   i. Tropical rainforest (Af)
   ii. Tropical monsoon (Am)
   iii. Tropical Savannah (Aw)

2. **Arid climate**

i. Arid desert- hot (BWh)
   ii. Arid desert- cold (BWk)
   iii. Arid steppe- hot (BSh)
   iv. Arid steppe- cold (BSk)

3. **Temperate climate**
   i. Temperate, dry summer, hot summer (Csa)
   ii. Temperate, dry summer, warm summer (Csb)
   iii. Temperate, dry winter, hot summer (Cwa)

iv. Temperate ,dry winter, warm summer (Cwb)
v. Temperate, no dry season, hot summer (Cfa)
vi. Temperate, no dry season, warm summer (Cfb)

4. **Cold climate**
   i. Cold, dry summer, hot summer (Dsa)
   ii. Cold, dry summer, warm summer (Dsb)
   iii. Cold, dry summer, cold summer (Dsc)
   iv. Cold, dry winter, warm summer (Dwb)
   v. Cold, dry winter, cold summer (Dwc)
   vi. Cold, no dry season, hot summer (Dfa)
   vii. Cold, no dry season, warm summer (Dfb)
   viii. Cold, no dry season, cold summer (Dfc)
5. **Polar climate**
   i. Polar, tundra (ET)
   ii. Polar, frost (EF)

3. Thornthwait's Rational classification (1948): He introduced water balance concept in classification. He classified climate based on PET, seasonal variation of effective moisture and average annual thermal efficiency.

According to Thornthwaite's classification scheme,

Humidity Index = I_h = ***100 x s / n, and***

Aridity Index = Ia = 100 x d /n ,

Where

s = Monthly water surplus calculated as the sum of the monthly differences between precipitation and potential evapotranspiration when precipitation is greater than evapotranspiration

d = Monthly water deficit calculated as sum of monthly values of potential evapotranspiration for those months when precipitation is less than evapotranspiration

n = Water need

Thornthwaite assumed that a surplus of 6 inch of water will counteract a deficit of 10 inch and thus gave higher weightage to humidity index. Thus, Moisture index is given by

$I_{m} = I_h - 0.6 \times I_a = (100s - 60d) / n$

Table 1. Climate types and moisture index in Thornthwaite's classification

Symbol	Climate type	Moisture Index	Broad group
A	Per humid	100 and above	Moist climate
B4	Humid	80 to100	
B3	Humid	60 to 80	
B2	Humid	40 to 60	
B1	Humid	20 to 40	
C2	Moist sub humid	0 to 20	
C1	Dry Sub humid	-20 to 00	Dry climate
D	Semi arid	-40 to -20	
E	Arid	-60 to -40	

Table 2. Subdivisions of moist and dry climates in Thornthwaite's classification

Moist climate description		Aridity Index	Dry climate-description		Humidity Index
R	Little or no water deficiency	0 -10	D	Little or no water surplus	0 – 16.7
S_1	Moderate summer water deficiency	10 – 20	S_1	Moderate summer water deficiency	16.7 – 33.3
W_1	Moderate winter water deficiency	10 – 20	W_1	Moderte winter water deficiency	16.7 -33.3
S_2	Large summer water deficiency	≥ 20	S_2	Large summer water deficiency	≥33.3
W_2	Large winter water deficiency	≥ 20	W_2	Large winter water deficiency	≥33.3

Thornthwaite considered potential ET as thermal efficiency (TE index) and used as a thermal limit in his climatic classification as given below;

Table 3. Thermal limits in climatic classification of Thornthwaite

S.N	T-E Index	Thermal regime-Climate type*
1	0 – 14 cm	Frost (E)
2	14 – 28 cm	Tundra (D)
3	28 – 56 cm	Microthermal (D, C1, C2)
4	56 – 98 cm	Mesothermal (C2, B1, B2, B3, B4)
5	Above 98 cm	Megathermal (B4, A1)

Therefore, the climatic regions of India falls under both moist and dry climate as per the Thornthwaite classification as given below;

i. ***Perhumid (A):*** West coast of India, south of Goa and some parts of northeastern India

ii. ***Humid (B4 to B1):*** The humid climate (B) prevails all along the coast adjoining regions of west coast, northern part of West Bengal and the neighboring parts of northeastern India.

iii. ***Moist sub humid (C2):*** It covers along the Western Ghats and also over Orissa and West Bengal.

iv. ***Dry sub humid (C1):*** The Ganga valley and northeastern parts of Central India have a dry subhumid type of climate (C1).

v. ***Semi humid (D):*** The peninsular interior, western Madhya Pradesh, Haryana and Punjab have a semiarid type of climate (D).

vi. ***Arid (E):*** The regions of Saurashtra, Kuchchh and Rajasthan, on the other hand, experience an arid climate (E).

4. Mather classification

Mather classified Indian climate into five major zones after modifying Thornthwaite's classification. He took following three ratios to identify the micro variations in particular regions;

1. Moisture index.
2. Seasonal variation of moisture
3. Index of thermal efficiency
4. Summer concentration of thermal efficiency

i. **Arid zone:** North-Western India, comprising some parts of Rajasthan and Gujarat is classified as arid region. 50 cm isohyete line marks its boundary in eastern section. Arid or semi- arid condition prevails in this region throughout the year

ii. **Semi arid zone:** This climate zone perhaps covers the largest area in India. Entire peninsula falls under this category of climate. Besides the peninsula, some region in North India also may be classed as semi-arid. This includes Punjab and Haryana plain.

iii. **Semi-humid zone:** This zone covers the states of Punjab, Bihar, Madhya Pradesh and Orissa. Here, annual precipitation is comparatively higher than Western arid zone. One pronounced rainy season is visible during July-November. Coastal tract of Tamil Nadu also experiences this climate.

iv. **Humid:** Eastern and Western coastal area, West Bengal, Andaman and Nicobar islands and some other regions are considered as humid region.

v. **Per humid zone:** It is very difficult to delineate separation line between humid and per humid region. The Himalayan region, Assam, Meghalaya and some other heavy rainfall regions are included in this zone. It has been observed that some of the heaviest rainfall occurs in this region. Here lies the famous Mousinram, the heaviest rainfall area on earth.

5. Indian Meteorological Department (IMD)

IMD divided Indian climate into 4 zones based on Himalyan mountains, latitude, surface distance from sea, land and water distribution, relief and altitude as given below;

i. **Tropical wet**

1. Mainly found in regions with warm or high temperatures.
2. It covers Indian coastal states of Kerala, southern Assam and the Western Ghats, Andaman and Nicobar Islands and Lakshadweep islands.
3. This zone experience moderate to high-temperature conditions throughout the year.
4. It receives an annual rainfall of 2,000 mm ranging within the months of May to November.
5. The driest months are December to March.

ii. **Tropical dry**

1. The main feature of this zone is over evaporation and moisture loss.
2. The different states of India experience a tropical dry climate including Karnataka, Maharashtra, Tamil Nadu and Andhra Pradesh (mainly the states lying south of the tropic of Cancer).
3. The zone receives an annual rainfall of 400-750 mm.
4. March to May are massively hot and dry while October to December are comparatively cool with a substantial amount of rainfall.
5. Western Rajasthan, Punjab and Haryana also experience a tropical dry and arid climatic condition.
6. The annual rainfall in Punjab and Haryana amounts to around 35-65cm, comparatively less than the southern parts of India.

iii. **Subtropical humid**

1. Most of the parts of north and northeast India experience a subtropical climate.
2. This zone experiences extreme temperature conditions that are extremely hot in summers and extremely cold in winter.
3. The winter season advances some amount of precipitation in these areas.
4. High wind speeds prevail throughout these regions mainly due to their immediate proximity to Himalayan mountains.
5. Winters are mostly dry with occasional rainfall or snowfall or even thunderstorms in some parts.
6. Annual rainfall varies from 1000mm to 2500 mm.
7. Average temperatures may range between 24-27 degrees celsius.

iv. **Montane**

1. It is also known as the alpine climate and it is usually found in the northern most tip of India.

2. The climatic conditions in this region decrease with elevation (around 5 degrees celsius lowering down of the temperature after every single km).
3. The landscape of this zone has thick layers of snow.
4. Rainfall is variable. It means that no particular record or measurement can be accorded to annual rainfall in this belt. Most of the precipitation occurs during the winter and the spring seasons.
5. The areas of south Himalayas experience this type of montane climate.
6. December and January receive the maximum snowfall.
7. Elevations beyond 5000 m do not receive rainfall and only snowfall occurs beyond these heights.

6. Krishnan and Singh classification

Krishnan and Singh (1972) classified agro-climatic classification based on moisture and thermal indices. The moisture index (MI) and thermal index (TI) can be calculated as given below;

$$MI = (P - PET/PET) \times 100$$

$$TI = T_{max} - T_{min}/2$$

Where,

P = Precipitation

PET = Potential evapotranspiration.

Indian climate was classified into 8 and 5 zones on the basis of moisture index (Table. 4a) and thermal index (Table. 4b), respectively as given below;

Table 4a. Climatic zones based on Moisture index (MI)

S.N	Climatic zones	Moisture index value
1	Extremely dry	< - 80
2	Semi dry	- 80 to - 60
3	Dry	- 60 to – 40
4	Slightly dry	- 40 to – 20
5	Slightly moist	- 20 to 0
6	Moist	0 to 50
7	Wet	50 to 100
8	Extremely wet	> 100

Table 4b. Climatic zone on the basis of thermal index

S.N	Climatic zones	Thermal index value
1	Very hot	> 28°C
2	Hot	25 to 28°C
3	Mild	20 to 25°C
4	Cold	10 to 20°C
5	Very cold	< 10°C

The above climatic zonation based on MI and TI were superimposed on a soil map of India to demarcate the agro-climatic regions of India. Further, the major defect of this classification is non consideration of soil moisture, while temperature given in thermal index does have much significance in agriculture.

7. Common classification based on moisture index: 5 climatic zones (Table.5)

Table 5. Climatic zones based on moisture index

S.N	Climatic zones	Moisture index value	Area (%)
1	Arid	- 66.7	19.6
2	Semi arid	- 66.7 to 33.3	37.0
3	Dry sub humid	1.1.2 to 00	21.1
4	Moist sub humid	00 to +20	10.2
5	Humid	+20.1 to 99.9	7.8
6	Pre humid	100	8.3

FAO (1993) and UNEP (1992)

According to FAO (1993 and UNEP (1992) classified drylands in 4 zones based on P/PET ratio and rainfall as given below (Table.6);

Table 6. Climatic zones (FAO, 1993 and UNEP, 1992)

S.N	Zones	P/PET ratio	Rainfall (mm)	Area in world (mha)	Dryland area of total globe land (%) (FAO, 2016)
1	Hyper arid	<0.05	200	1.0	6.9
2	Arid	0.05 – 0.20	<200 winter <400 summer	1.62	12.4
3	Semi arid	0.20 – 0.50	200 – 500 winter 400-600 summer	2.37	17.5
4	Dry sub-humid	0.50 - 0.65	500 – 700 winter 600 – 800 summer	1.32	10.0

World Atlas of Desertification (WAD)

According to World Atlas of Desertification, the Indian monsoon was divided based on P/PET ratio and rainfall into 4 regions as given in table.7 below;

Table 7: Climatic zones of India (WAD)

S.N	Region	P/PET ratio	Rainfall (mm)	Inter annual variability rainfall (%)	Growing degree days
1	Arid region	<0.02	<200	50 -100	<75
2	Semi arid region	0.02 – 0.50	500 winter 800 summer	25 -50	75-120
3	Dry sub humid	0.50 – <0.65		<25	120-150
4	Humid	>0.65			>150

10. Ramkrishnan (1997) and Paul (1997) classification

Ramkrishnan (1997) and Paul (1997) classified the rainfed areas into 5 zones based on rainfall as given in table.8 below;

Table 8. Climatic zones of rainfed areas (Ramkrishnan, 1997 and Paul, 1997)

S.N	Rainfall (mm)	Zones	Net sown area (%)	Rainfall grade
1	<500	Arid	16	Very low
2	500-750	Semi arid	17	Low
3	750-1100	Dry sub humid	35	Medium
4	1100-1400	Moist sub humid	24	High
5	>1400	Humid mountains	8	Very high

Bansal & Minke (1988): 6 zones

i. Hot and dry.
ii. Warm and humid.
iii. Moderate.
iv. Cold and cloudy.
v. Cold and sunny, and
vi. Composite.

Table 9. Bansal & Minke climate zone criteria

Climatic zones	Mean monthly temperature (^{0}C)	Relative Humidity (%)	Precipitation (mm)	Number of Clear Days
Hot & Dry (Jaisalmer, Jodhpur, Sholapur etc.)	>30	<55	<500	>20
Warm & Humid (Mumbai, Kolkatta, Chennai, Goa etc.)	>30	>55	>500	<20
Moderate (Bengaluru, Pune etc.)	25-30	<75	<500	>20
Cold & Cloudy (Shimla, Srinagar,Shillong, Mahabeshwar etc.)	<25	>55	>500	<20
Cold & Sunny (Leh, Mount Abu etc.)	<25	<55	<5	>20
Composite (Delhi, Kanpur, Allahabad etc.)	When six months or more do not fall within any of the above categories			

National Building Codes (NBC, 2005): 5 zone

i. Hot and dry
ii. Warm and humid
iii. Moderate
iv. Cold, and
v. Composite

13. Singh and Venkateshvarulu (2009)

They classified Indian monsoon again into 4 regions based on rainfall and rainfed area as given below in table.10;

Table 10. Singh and Venkateshvarulu (2009) classification of Indian climate

S.N	Region	Rainfall (mm)	Rainfed area (mha)
1	Arid	<500	15.0
2	Semi Arid	500-750	15.0
3	Sub humid	750-1150	42.0
4	Humid	>1150	25.0

14. Bureau of India Standards (2005): 05 climatic zones (map.2).

i. Hot and Dry
ii. Warm and Humid
iii. Temperate
iv. Cold
v. Composite

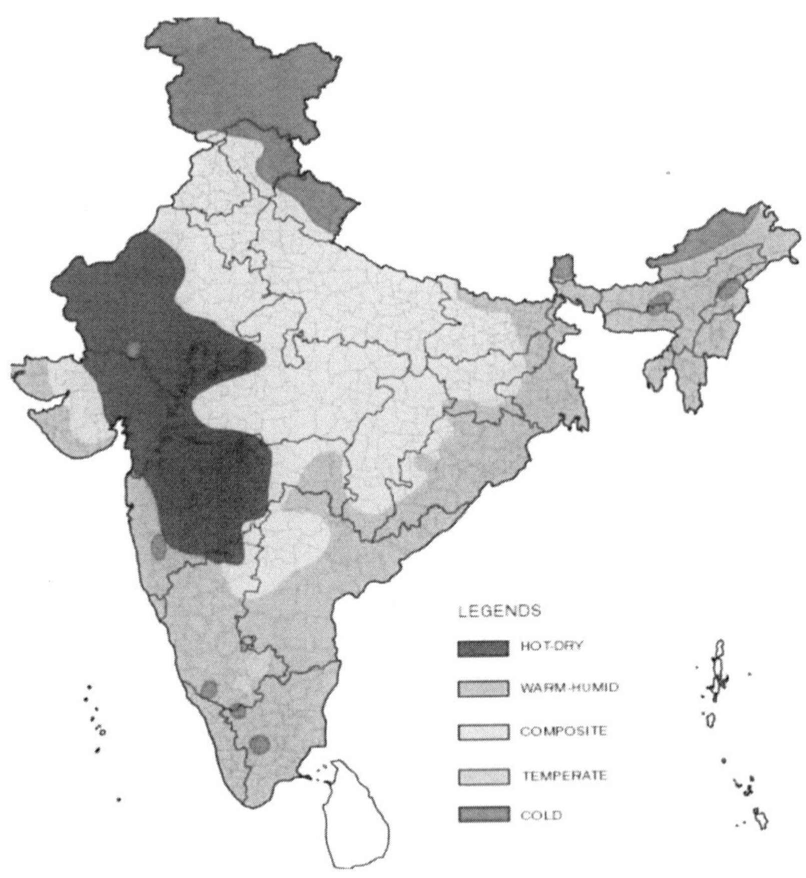

Map 2: Climatic zones (BIS, 2005)

15. CRIDA (2013) classification: 06 zones

CRIDA classified country monsoon at district level and grouped whole country into 6 groups (map.3);

1. Arid
2. Semi arid
3. Dry Sub humid
4. Moist sub humid
5. Humid
6. Per humid

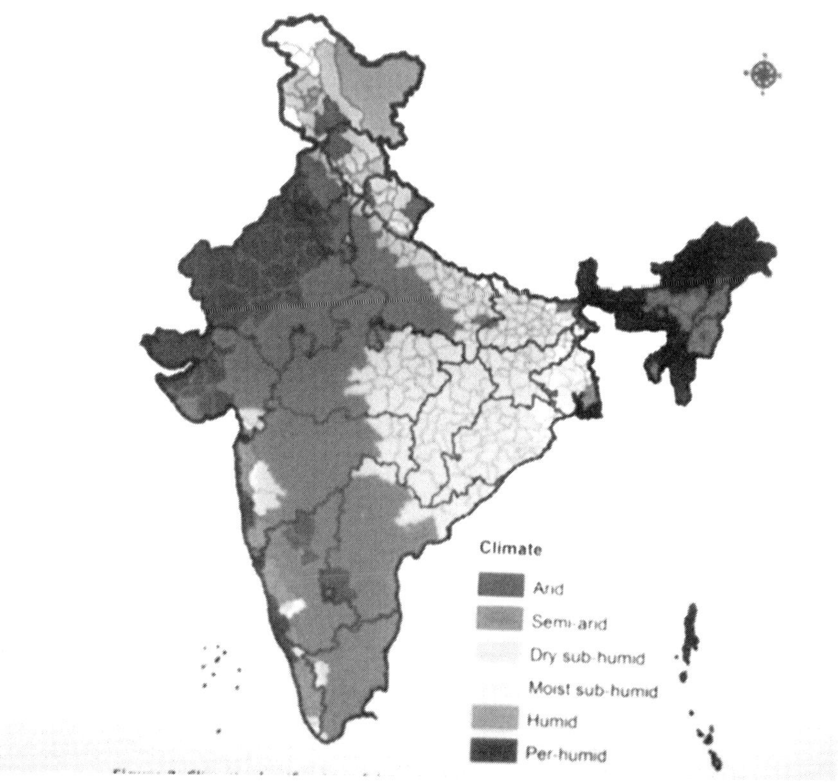

Map 3: Climatic zones of India (CRIDA, 2013)

16. Cocheme and Franquin classification

Cocheme and Franquin divided climate into 5 zones considering crop growing period and water balance based on precipitation and evapotranspiration of region as listed in table below in table.11;

Table 11. Climatic zones based on precipitation/Et ratio

S.N.	Precipitation/ET	Region
1	P > ET	Humid
2	P > ET/2	Moist
3	P = ET/2 to ET/4	Moderately dry
4	P = ET/4 to ET/10	Dry (Semi Arid)
5	P < ET/10	Very Dry (Arid)

Other classification

1. Highlands (Mountain)
2. Humid sub tropical
3. Tropical Wet
4. Tropical Wet & Dry

5. Arid
6. Semi arid

Map 4: Climatic zones of India

Aridity Index (AI)

An aridity index (AI) is a numerical indicator of the degree of dryness of the climate at a given location. A number of aridity indices have been proposed and these indicators serve to identify, locate or delimit regions that suffer from a deficit of available water, a condition that can severely affect the effective use of the land for such activities as agriculture or stock-farming.

Historical background and indices

At the turn of the 20th century, Wladimir Köppen and Rudolf Geiger developed the concept of a climate classification where arid regions were defined as those places where the annual rainfall accumulation (cm) is less than R/2 where:

- R = 2 x T - if rainfall occurs mainly in the cold season,
- R = 2 x T + 14 - if rainfall is evenly distributed throughout the year, and
- R = 2 x T +28 - if rainfall occurs mainly in the hot season.

Where, R and T are the annual rainfall and mean annual temperature in Celsius, respectively.

This was one of the first attempts at defining an aridity index, one that reflects the effects of the thermal regime and the amount and distribution of precipitation in determining the native vegetation possible in an area. It recognizes the significance of temperature in allowing colder places such as northern Canada to be seen as humid with the same level of precipitation as some tropical deserts because of lower levels of potential evapotranspiration in colder places. In the subtropics, the allowance for the distribution of rainfall between warm and cold seasons recognizes that winter rainfall is more effective for plant growth that can flourish in the winter and go dormant in the summer than the same amount of summer rainfall during a warm-to-hot season. Thus a place like Athens, Greece that gets most of its rainfall in winter can be considered to have a humid climate (as attested in lush foliage) with roughly the same amount of rainfall that imposes semi-desert conditions in Midland, Texas, where rainfall largely occurs in the summer.

1. C.W. Thornthwaite (1948)

Where,

Thornthwaite proposed aridity index as below;

AIT = 100 x d/n = (A – ET/ET) x 100

AI_T = Thornthwaite aridity index

A = Precipitation

ET = Evapotranspiration

where the water deficiency d is calculated as the sum of the monthly differences between precipitation and potential evapotranspiration for those months when the normal precipitation is less than the normal evapotranspiration; and where n stands for the sum of monthly values of potential evapotranspiration for the deficient months (after Huschke, 1959). This AI was later used by Meigs (1961) to delineate the arid zones of the world in the context of the UNESCO Arid Zone Research program.

2. United Nations Environment Program (UNEP)

In the preparations leading to the UN Conference on Desertification (UNCOD), the United Nations Environment Programme (**UNEP**) issued a dryness map based on a different aridity index, proposed originally by Mikhail Ivanovich Budyko (1958) and defined as follows:

AI_B=100 x R/LP

Where

R : Mean annual net radiation (also known as the net radiation balance),

P : Mean annual precipitation,

L : Latent heat of vaporization for water. Note that this index is dimensionless and that the variables R, LR and P P can be expressed in any system of units that is self-consistent.

3. UNEP (1992)

UNEP proposed aridity index as a ratio of annual precipitation to potential evaportanspiration, as given below

$AI_U = P/PET$

Where, PET is the potential evapotranspiration (mm) and P is the average annual precipitation (mm). The global monsoon climate was divided into 4 groups based on aridity index as given in table 12.

Table 12. Classification of climate based on aridity index

Classification	Aridity Index	Global land area
Hyper arid	AI < 0.05	7.5%
Arid	0.05 < AI < 0.20	12.1%
Semi-arid	0.20 < AI < 0.50	17.7%
Dry sub humid	0.50 < AI < 0.65	9.9%

4. Indian Meteorological Department (IMD)

Aridity index = (PE-AE) / PET = (1-MAI)

Where,

PE – Potential Evaporation

AE – Actual Evaporation

PET – Potential evapotranspiration

MAI- Moisture Availability Index

It is an index to assess the fortnightly drought over India. The water deficiency of drought is estimated on the basis of difference between actual and normal values of aridity index. IMD issues fortnightly drought bulletins based on this approach as given below;

Table 13. Drought severity based on IMD classification

S.N	Drought severity	values
1	Non arid (NA)	0 or less
2	Mild (M)	1 – 25
3	Moderate (MO)	26 -50
4	Severe (S)	>50

5. National Remote Sensing Agency (NRSA)

NRSA has developed a National Agricultural Drought Assessment and Monitoring System (NADAMS) based on Advanced Very High Resolution Radiometer(AVHRR) data and developed Normalized Difference vegetation Index (NDVI) from red wavelength of visible radiation (VIS) infra red radiation (NIR) as given below;

NDVI = NIR – VIS/NIR + VIS

NRSA also issues periodic bi weekly drought bulletins and detailed monthly reports at district level for many states. The detail is given below;

Table 14. NRSA classification of drought severity level based on NDV anomaly

S.N	Drought severity	NDVI anomaly values compared to normal year (%)
1	Normal	Up to 10
2	Mild	10 - 25
3	Moderate	25 -50
4	Severe	>50

6. Palmer Drought Severity Index (PDSI)

Wayne C. Palmer developed drought severity index known as Palmer drought severity index (PDSI in 1965 using data on precipitation, temperature and local soil available water content and calculates the difference between Climatically Appropriate For Existing Conditions (CAFEC) rainfall and actual rainfall as a drought indicator. It is widely used in USA and other countries. PDSI quantifies abnormality of weather in a region including in historical past. It can be used for spatio-temporal variability of drought but sometimes it did not match in some cases like extremely rainfall and mountains areas and also does not match with hydrological impacts. However, the index is most effective measuring impacts sensitive to soil moisture conditions, such as agriculture; is useful as a drought monitoring tool and has been used to trigger actions associated with drought contingency plans. The PDSI varies from 4 to – 4 as given below;

Table 15. Palmer index classification

S.N	Condition	PDSI values
1	Extremely wet	≥4.0
2	Very wet	3.0 – 3.99
3	Moderately wet	2.0 – 2.49
4	Slightly wet	1.0 – 1.99
5	Incipient wet spell	0.5 to 0.99
6	Near normal	0.49 to -0.49
7	Incipient dry spell	-0.5 to -0.99
8	Mild drought	- 1.0 to -1.99
9	Moderate drought	-2.0 to -2.99
10	Severe drought	- 3.0 to -3.99
11	Extreme drought	≤ - 4.0

7. Moisture adequacy index (MAI)

MAI is also used to quantify the agricultural drought. It is obtained from weekly water balance using soil characteristics, crop growth period and water requirement of major crops. Drought impact is related to moisture availability at certain crop growth stages. It is calculated as;

MAI = (AE/PE) x 100

AE = Actual evaporation

PE = Potential evapotranspiration

Normally it is calculated on short period basis like weeks or 5 days of crop growing season. The detail is given below;

Table 16. Drought classification based on MAI

S.N	Drought severity	MAI values
1	Severe drought	< 25%
2	Moderate drought	25 – 49.9%
3	Crop growing season	>50%
4	Humid period (good moisture status	> 75%

Agro-climatic Regions of India

The agricultural production depends mainly on local climate in addition to soil type and management therefore, it requires scientific management of regional resources to meet the food, fiber, fodder, fuel and wood without adversely affecting the status of natural resources and environment (FAO, 1983). It was realized to divide India into smaller zones having similar type of climate and soil so that specific recommendation may be sought for better utilization of natural resources and sustainable agricultural production and these zones were termed as 'Agro Climatic Zones'. So, the 'Agro-climatic zone' refers to a land unit in terms of major climates, suitable for a certain range of crops and cultivars. Later Agro-climatic zonation of India has been accomplished by different institutions e.g., Indian Council of Agricultural Research (ICAR), Planning Commission, National Agricultural Research Project (NARP) under ICAR, National Bureau of Soil Survey and Land Use Planning (Agro -ecological regions by the NBSS & LUP).

i. ***Agro-Ecological Zoning (AEZ):*** It refers to the division of an area of land into smaller units, which have similar characteristics related to land suitability potential production and environmental impact.

ii. ***Agro-Ecological Zone:*** It is a land resource mapping unit, having a unique combination of land form, soil and climatic characteristics and/or land cover having a specific range of potentials and constraints for land use (FAO, 1996).

iii. ***Agro-Ecological Cell (AEC):*** It is defined by a unique combination of land form, soil and climatic characteristics.

1. DPAP and DDP districts

Two important programs i.e. Drought Prone Area Program (DPAP) and Desert Development Program (DDP) were launched by Ministry of Rural Development, Govt. of India in 1973-74 and 1977-78, to tackle the severity of rainfed areas chronically affected by drought as well as hot and cold desert areas, respectively.

Later a High Level Technical Committee was constituted in April 1993 under the chairmanship of Prof C.H. Hanumantha Rao to review the programs and suggest suitable measures for implementation of above programs effectively. The committee developed the criteria to identify districts covered under both programs mainly based on climatic zones and percent net irrigated area. The moisture index (P – PE / P) x 100) was used as criteria, where P is precipitation and PE is potential evapotranspiration as given below. Accordingly 183 districts in 16 states were covered under DPAP and 40 districts in seven states under DDP.

S.N	Zone	MI	Net irrigated area (%)
1	Arid Ecosystem (DDP)	> - 66.7	Not more than 50%
2	Semi Arid Ecosystem (DPAP)	-66.7 to – 33.3	Not more than 40%
3	Dry Sub Humid Ecosystem (DPAP)	< -33.3	Nor more than 30%

Source: Ministry of Rural Development, 1994)

2. Planning Commission

The Planning Commission (Khanna, 1989) after examining the earlier studies on the regionalization of the agricultural economy has recommended that agricultural planning be done on the basis of agro-climatic regions. The country has been broadly divided into 15 agricultural regions based on agro-climatic features, particularly soil type, climate including temperature and rainfall and its variation and water resources availability. Out of 15 agro-climatic zones, 14 are in the main land and one covering the islands of Bay of Bengal and the Arabian sea (map.5). The detail is given below;

- Western Himalayan division
- Eastern Himalayan division
- Lower Gangetic plain region
- Middle Gangetic plain region
- Upper Gangetic plain region
- Trans-Gangetic plain region
- Eastern plateau and hill region
- Central plateau and hill region
- Western plateau and hill region
- Southern plateau and hill region
- East coast plain and hill region
- West coast plain and hill region
- Gujarat plain and hill region
- Western plain and hill region
- Island region.

I. Western Himalayan Region

i. The Western Himalayan Region covers Jammu and Kashmir, Himachal Pradesh and the hill region of Uttarakhand,

ii. Great variation in topography and temperatures. The average temperature in July ranges between 5°C and 30°C, while in January it ranges between 5°C and -5°C.

iii. Mean annual rainfall varies between 75 cm to 150 cm; in Ladakh, however, it is less than 30 cm. There is alluvial soil in the valleys of Kashmir, Kullu and Dehradun and brown soil in the hills.

iv. Rice is major crop in valleys, while maize in hilly tracts in *Kharif* season. Winter crops are barley, oats and wheat. The region supports horticulture, especially apple orchards and other temperate fruits such as peaches, apricot, pears, cherry, almond, litchis, walnut, etc. Saffron is also grown in this region.

v. The high altitude alpine pastures, locally known as 'dhoks' or 'margs' or 'Bugyals' (Uttarakhand), are used by the Gujjars, Bakarwals and Gaddis to rear their sheep, goats, cattle and horses.

vi. Poor accessibility, soil erosion, landslides, inclement weather, inadequacy of marketing and storage facilities are main problems of this region.

vii. The population is generally rural-based and poor and economy is largely agrarian.

II. Eastern Himalayan Region

i. It includes Arunachal Pradesh, the hills of Assam, Sikkim, Meghalaya, Nagaland, Manipur, Mizoram, Tripura, and the Darjeeling district of West Bengal.

ii. The topography is rugged.

iii. Temperature varies 25-30°C in July and 10-20°C in January.

iv. Average rainfall is 200-400 cm.

v. The red-brown soil is not highly productive. Jhuming (shifting cultivation) prevails in the hilly areas.

vi. The main crops are rice, maize, potato, tea. There are orchards of pineapple, litchi, oranges and lime.

vii. Infrastructural facilities in the region need to be improved and shifting cultivation controlled by developing terrace farming.

III. Lower Gangetic Plain Region

i. West Bengal (except the hilly areas), eastern Bihar and the Brahmaputra valley lie in this region.

ii. Average annual rainfall lies between 100 cm-200 cm.

iii. Temperature in July varies from 26°C to 41°C and for January from 9°C to 24°C.

iv. The region has adequate storage of ground water with high water table.

v. Rice is the main crop which at times yields three successive crops (Aman, Aus and Boro) in a year. Jute, maize, potato, and pulses are other important crops.

IV. Middle Gangetic Plain Region

i. The Middle Gangetic Plain region includes large parts of Uttar Pradesh and Bihar.

ii. The average temperature in July varies from 26°C to 41°C and that of January 9°C to 24 °C

iii. Average annual rainfall is between 100 cm and 200 cm.

iv. It is a fertile alluvial plain drained by the Ganga and its tributaries.

v. Rice, maize, millets in kharif, wheat, gram, barley, peas, mustard and potato in rabi are important crops.

vi. Alternative farming systems and utilising chaur lands for pisciculture are some measures to boost agricultural production.

vii. Reclamation of user lands, wastelands, and fallow lands for agriculture and allied activities (agro-forestry, silviculture, floriculture etc.) should be done.

V. Upper Gangetic Plains Region

i. The central and western parts of Uttar Pradesh and the Haridwar and Udham Singh Nagar districts of Uttarakhand cover in this region.

ii. The climate is sub-humid continental with temperature in July between 26°C to 41°C and temperature in January between 7°C to 23°C.

iii. Average annual rainfall is 75-150 cm.

iv. The soil is sandy loam.

v. Canal, tube-well and wells are the main source of irrigation.

vi. Intensive agriculture is followed with wheat, rice, sugarcane, millets, maize, gram, barley, oilseeds, pulses and cotton are the main crops.

VI. Trans-Ganga Plains Region

i. This region is also called the Satluj-Yamuna Plains extends over Punjab, Haryana, Chandigarh, Delhi and the Ganganagar district of Rajasthan.

ii. Climate is Semi- arid with July's mean monthly temperature 25-40°C and that of January between 10-20°C.

iii. The average annual rainfall varies between 65 cm and 125 cm.

iv. The soil is alluvial which is highly productive.

v. Canals and tube-wells and pumping sets have been installed by the cultivators and the governments.

vi. The intensity of agriculture is the highest in the country.

vii. Important crops grown are wheat, sugarcane, cotton, rice, gram, maize, millets, pulses and oilseeds etc.

viii. The region has the credit of introducing Green Revolution in the country and has adopted modern methods of farming with greater degree of mechanisation.

ix. The region is also facing the menace of waterlogging, salinity, alkalinity, soil erosion and falling water table.

VII. Eastern Plateau and Hills

i. Chhotanagpur Plateau, Jharkhand, Orissa, Chhattisgarh and Dandakaranya come in this region.

ii. The region enjoys 26-34°C of temperature in July, 10-27°C in January

iii. Annual rainfall is 80-150 cm.

iv. Soils are red and yellow with occasional patches of laterites and alluviums.

v. The region is deficient in water resources due to plateau structure and non-perennial streams.

vi. Rainfed agriculture is dominant.

vii. Important crops are rice, millets, maize, oilseeds, ragi, gram and potato.

VIII. Central Plateau and Hills

i. It is spread over Bundelkhand, Baghelkhand, Bhander Plateau, Malwa Plateau, and Vindhyachal Hills with high density of ravine lands.

ii. Climate is semi-arid with temperature in July 26-40°C, in January 7-24°C.

iii. Average annual rainfall is 50 -100 cm.

iv. Soils are mixed red, yellow and black.

v. There is scarcity of water.

vi. Important crops are millets, wheat, gram, oilseeds, cotton and sunflower.

IX. Western Plateau and Hills

i. It comprises southern part of Malwa plateau and Deccan plateau (Maharashtra),

ii. It is a region of the regur (black) soil with July temperature between 24-41°C and January 6-23°C.

iii. Average annual rainfall of 25-75 cm.

iv. Wheat, gram, millets, cotton, pulses, groundnut, and oilseeds are the main crops of rainfed areas while sugarcane, rice and wheat of irrigated areas. Fruits like oranges, grapes and bananas are also grown at large scale.

X. Southern Plateau and Hills

i. This region falls in interior Deccan and includes parts of southern Maharashtra, the greater parts of Karnataka, Andhra Pradesh, and Tamil Nadu uplands from Adilabad District in the north to Madurai District in the south.

ii. The mean monthly temperature of July varies between 25-40°C and the mean January temperature is between 10-20°C.

iii. Annual rainfall is between 50 cm and 100 cm.

iv. It is an area of dry-zone agriculture where millets, oilseeds and pulses are grown. Coffee, tea, cardamom and spices are grown along the hilly slopes of Karnataka plateau.

XI. Eastern Coastal Plains and Hills

i. It falls under Coromandal and northern Circar coasts of Andhra Pradesh and Orissa.

ii. The mean July temperature is 25-35°C and the mean January temperature 20-30°C.

iii. The mean annual rainfall ~ varies between 75-150 cm.

iv. The soils are alluvial, loam and clay and are troubled by the problem of alkalinity.

v. Main crops include rice, jute, tobacco, sugarcane, maize, millets, groundnut and oilseeds.

XII. Western Coastal Plains and Ghats

i. It is extended over the Malabar and Konkan coastal plains and the Sahyadris.

ii. Climate is humid with the mean July temperature varies 25-30°C and mean January temperatures 18-30°C.

iii. The mean annual rainfall is more than 200 cm.

iv. The soils are laterite and coastal alluvial.

v. Rice, coconut, oilseeds, sugarcane, millets, pulses and cotton are the main crops.

vi. The region is also famous for plantation crops and spices which are raised along the hill slopes of the Western Ghats.

XIII. Gujarat Plains and Hills

i. It includes the hills and plains of Kathiawar and the fertile valleys of Mahi and Sabarmati rivers.

ii. It is an arid and semi-arid region with the mean July temperature reading 30°C and that of January about 25°C.

iii. The mean annual rainfall varies between 50-100 cm.

iv. Soils are regur in the plateau region, alluvium in the coastal plains and red and yellow soils in Jamnagar area.

v. Important crops are groundnut, cotton, rice, millets, oilseeds, wheat and tobacco are the main crops. It is an important oilseed producing region.

XIV. Western Dry Region

i. Rajasthan and West of the Aravallis fall under this region,

ii. Erratic rainfall is very common and an annual average is less than 25 cm.

iii. Climate is dry/desert and causes high evaporation.

iv. Temperature may vary from 28 to 45°C in June and 5 to 22°C in January.

v. Bajra, jowar and moth are main crops of *Kharif* and wheat and gram in *Rabi.*

vi. Livestock contributes greatly in desert ecology.

XV. Island Region

i. It covers Andaman-Nicobar and Lakshadweep.

ii. Climate is typically equatorial climate.

iii. Annual rainfall is less than 300 cm.

iv. Mean July and January temperature of Port Blair is 30°C and 25°C, respectively.

v. The soils vary from sandy along the coast to clayey loam in valleys and lower slopes.

vi. The main crops are rice, maize, millets, pulses, arecanut, turmeric and cassava. Nearly half of the cropped area is under coconut. The area is covered with thick forests and agriculture is in backward stage.

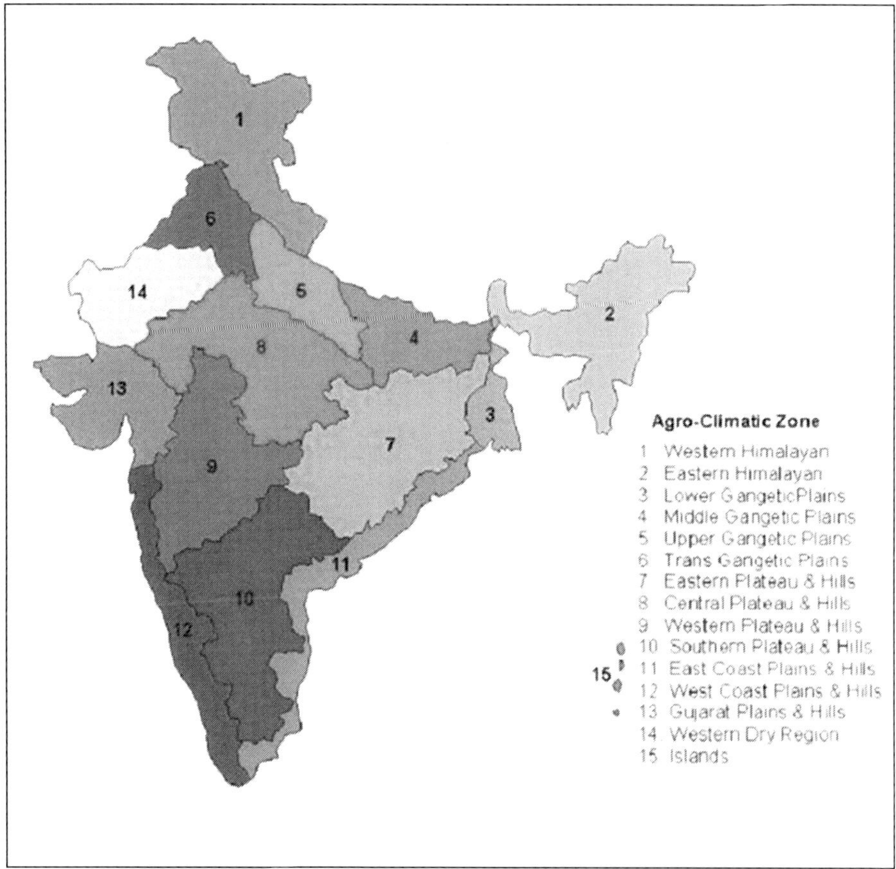

Map 5: Agro-climate zones of India (Planning Commission)

3. Indian Council of Agricultural Research (ICAR)

The concept of homogenous agro-ecological zones was initiated by ICAR in 1979 under the National Agricultural Research Project (NARP) with support from International Development Agency (IDA) of World Bank. The concept of zoning under NARP was based on soil, climate, topography, crops, vegetation etc as major influencing factors. The zones were selected as contiguous areas within the state boundary and to the possible extent zones have homogenous physical characteristics such as topography, rainfall, soils, cropping pattern and irrigation availability. Normally each NARP zone cover 2-4 districts and is spread over an area as high as 40-50 thousand sq. km.

NARP has divided country into 127 agro-climatic zones falling under 17 major states and 6 states/Union Territories of North Eastern Hill Region. Out of 127 NARP zones, 73 are predominantly rainfed (map.6).

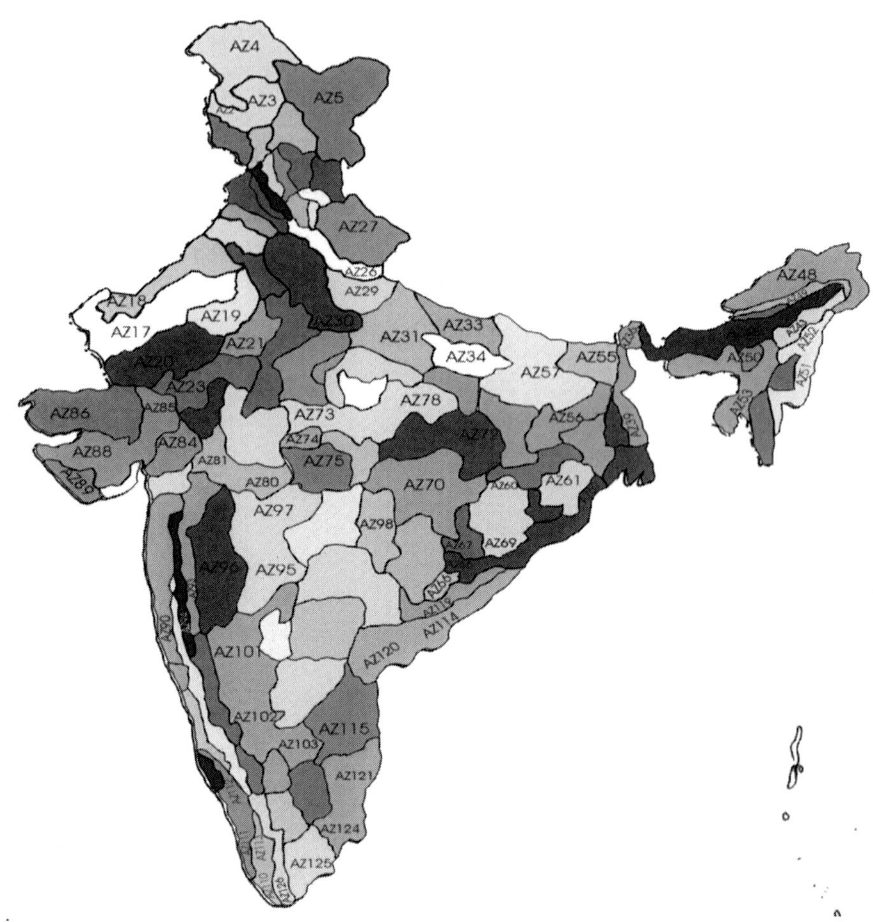

Map 6: Agro-climatic zones of India (NARP-ICAR)

4. National Bureau of Soil Survey & Land Use Planning (NBSS & LUP)

The National Bureau of Soil Survey & Land Use Planning (NBSS&LUP, 1999) came up with twenty agro-ecological zones based on the growing period as integrated criteria of effective rainfall, soil groups, delineated boundaries adjusted to district boundaries with a minimal number of regions. Subsequently, these 20 agro-ecological regions (Sehgal *et al*, 1992 and 1995) were sub- divided into 60 agro-ecological sub-zones. Later in 2015, it was revised with again 20 agro-climatic regions (AESR) to 127 AESR. The major criteria for delineation of regions were length of growing period based on effective rainfall, soil groups, delineated boundaries adjusted to district boundaries and number of regions as minimal as possible. The list of 20 zones is given in table.17 and map.7 below;

Table 17: Agro ecological zones of India (NBSS&LUP)

S.N	Agro-Ecological Regions	Climatic region	Soil types	Growing Period (Days)
1	Western Himalayas,	Cold arid region	Shallow skeletal soils	<90
2	Western Plain, Kachchh, and part of Kathiwara Peninsula	Hot arid eco region	Desert and saline soils	<90
3	Deccan Plateau	Hot arid eco region	Red and black soils	<90
4	Northern Plain and Central Highlands including Aravallis	Hot semi arid eco region	Alluvium derived soils	90-150
5	Central Malwa Highlands, Gujarat Plains, and Kathiawar Peninsula	Hot semi arid eco region	Medium and deep black soils	90-150
6	Deccan Plateau,	Hot semi-arid eco region	Shallow, medium and deep black soils	90-150
7	Deccan (Telengana) Plateau and Eastern Ghats	Hot semi-arid eco region	Red and black soils	90-150
8	Eastern Ghats, Tamil Nadu Plateau and Deccan (Karnataka)	Hot semi-arid eco region	Red loamy soils	90-150
9	Northern Plain	Hot sub-humid (dry) eco region	Alluvium derived soils	150-180
10	Central Highlands (Malwas, Budelkhand, and Eastern Satpura)	Hot sub humid eco region	Black and red soils	150-180 (upto 210)
11	Eastern Plateau (Chattisgarh),	Hot sub-humid eco region	Red and yellow soils	150-180
12	Eastern (Chotanagpur) Plateau and Eastern Ghats	Hot sub-humid eco region	Red and lateritic soils	150-180 (upto 210)
13	Eastern Gangetic Plain	Hot sub humid (moist) eco region	Alluvium derived soils	180-210
14	Western Himalayas	Warm sub humid (Humid to per humid)	Alluvium derived soils	180-210
15	Bengal and Assam Gangetic and Brahmaputra plains	Hot sub humid (moist) to humid (per humid) eco region	Alluvium derived soils	>210

S.N	Agro-Ecological Regions	Climatic region	Soil types	Growing Period (Days)
16	Eastern Himalayas	Warm per humid eco region	Brown and red hill soils	>210
17	North Eastern Hills (Purvanchal)	Warm per humid eco region	Red and lateritic soils	>210
18	Eastern Coastal Plain	Hot sub humid to semi arid eco region	Coastal alluvium derived soils	190-210+
19	Western Ghats and Coastal Plain	Hot humid to per humid eco region	Red, lateritic and alluvium derived soils	>210
20	Island of Andaman Nicobar and Lakshadweep	Hot humid to per humid eco region	Red loamy and sandy soils	>210

Source: NBSS&LUP (1992)

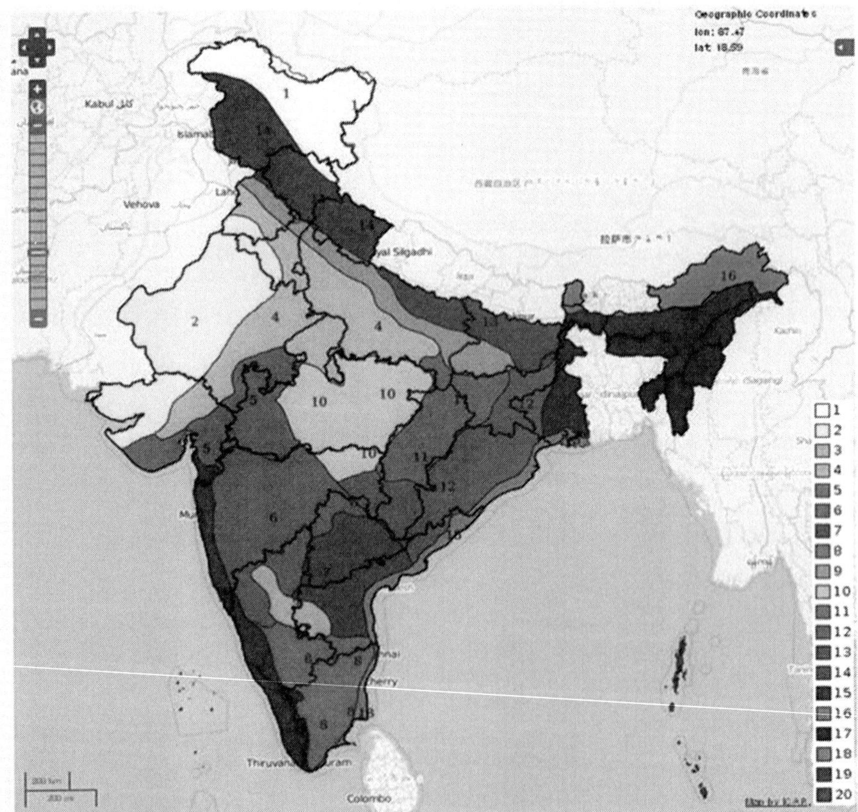

Map 7: Agro ecological zones of India (NBSS & LUP)

5. NATP Classification

National Agricultural Technology Project (NATP), ICAR proposed the production system approach to delineate the rainfed districts of India based on crop, livestock, socio economic parameters and irrigated area <30%. Districts with highest area under a given crop/cropping system but stagnant, declining or low yield were takes as high priority districts. Finally the rainfed area was grouped into 5 major agro-ecological zones as given in table.18.

Table 18: Agro-ecological zones of India (NATP-ICAR)

S.N	Agro ecological zones	States	Old zones	Soil types	Growing period (Days)
1	Northern Gangetic Plains	Punjab, Haryana, UP, Bihar	9, 4 (Gangetic plains areas)	Alluvial soils	90-180
2	Eastern Gangetic Plains	UP, Bihar, WB	13, 15	Alluvial soils	150-210+
3	Central & Eastern Highlands	MP, Orissa, Bihar & WB	10,11,12,18	Black, red & Lateritic soils, coastal areas of Orissa with alluvial soils	150-210
4	Central Highlands, Gujarat plains & Deccan plateau	Rajsthan, Gujarat, MP, Maharashtra, Karnataka & AP	5,6,4 (upland areas of Rajasthan and Gujarat)	Black soils	90-150
5	Deccan Plateau, Eastern Ghats and TN uplands	AP, Karnataka, TN	7,8,18	Red soils, coastal areas of AP & TN	90-150

Note: Desert areas, the Himalayas, the north east, the Western Ghat and the west coast all drop out of the sample because they do not fall in rainfall bounds. Zones omitted completely are 1,14,16,17,19, zones 2 and 15 are omitted partially and zones 3 and 18 are combined with neighboring zones

6. Koppen's classification

Koppen classified India into 8 climatic regions (Map.8) as given below;

1. Monsoon type with short dry season (Amw),
2. Monsoon type with dry season in summers (As),
3. Tropical savannah type (Aw),
4. Semiarid steppe climate (Bshw),
5. Hot desert type (Bwhw)
6. Monsoon type with dry winters (Cwg),
7. Cold humid winter type with short summers (Dfc)
8. Polar type (E).

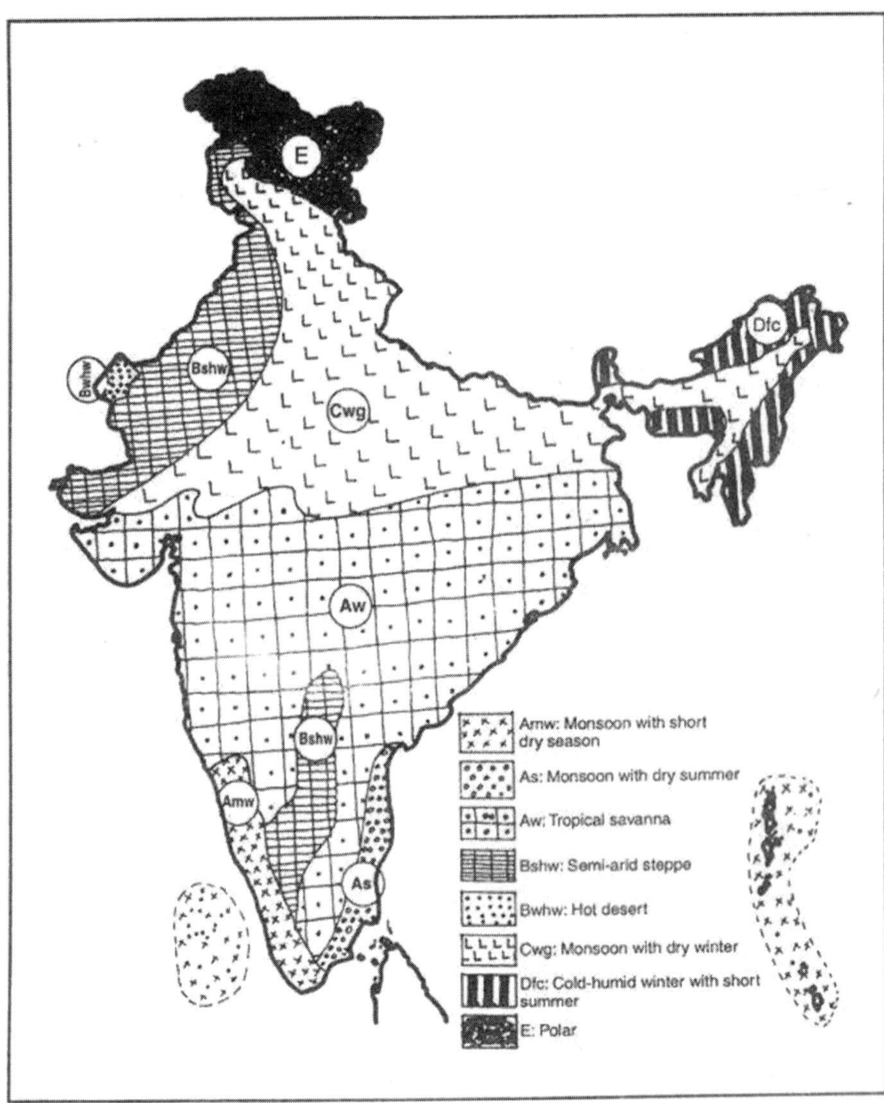

Map 8: Climatic regions based on Koppen classification

Main features

1. This is clear from above map that the Amw type of climate prevails over the western coast of India and south of Goa.
2. The dry climate is found at two parts in India i.e. the interior peninsula and Rajasthan
3. Haryana has Bshw type of climate, while the extreme western Rajasthan experiences Bwhw type of climate.

4. Most of the peninsular plateau has tropical savannah type of climate (Aw).
5. The plain of India falls under the warm temperate type of climate with dry winters (Cwg).
6. The northeastern India falls under Dfc type of climate. Here, the winters are cold and humid, while the summers are short.
7. Kashmir and the adjoining mountain ranges have a polar type of climate (E).

7. Climatic regions based on Thornthwaite

As per the Thornthwaitc classification, India has both moist and dry climates. He divided Indian climate into 5 climatic regions (Map.9) as listed below;

1. Perhumid (A) - West coast of India, south of Goa and some parts of northeastern India
2. Humid (B4 to B1)- West coast, northern part of West Bengal and the neighbouring parts of north-eastern India.
3. Moist sub-humid (C2) - Western Ghats and also over Orissa and West Bengal.
4. Dry sub-humid (C1)- Ganga valley and northeastern parts of Central India.
5. Semiarid (D)- Peninsular interior, western Madhya Pradesh, Haryana and Punjab
6. Arid (E)- Saurashtra region, Kuchchh and Rajasthan.

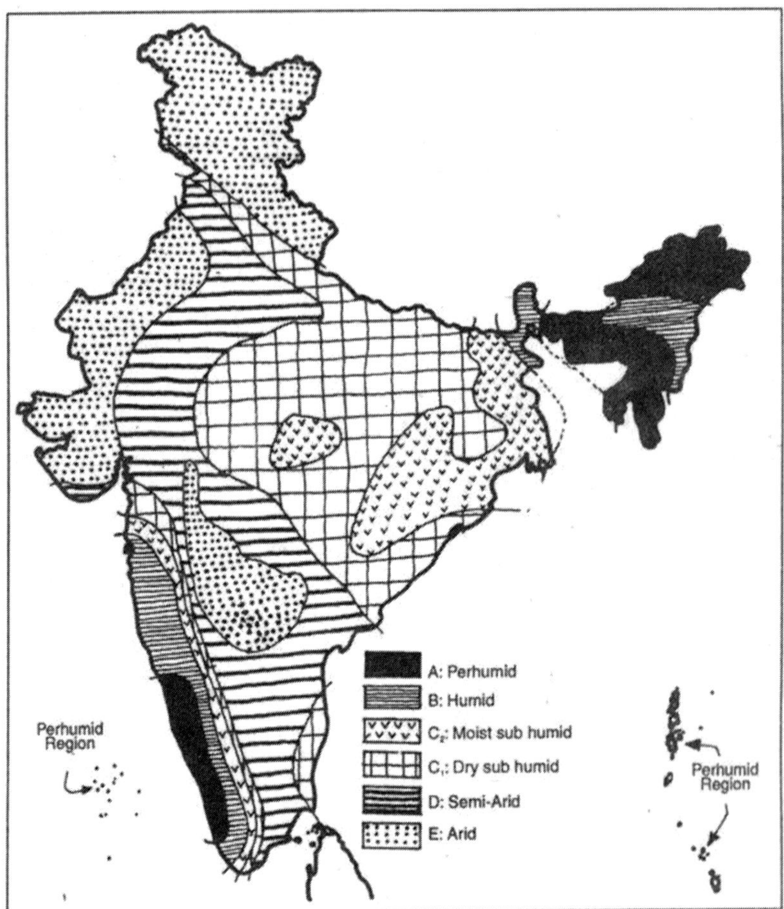

Map 9: Climatec region based on Thronthwaite scheme

Rainfall Pattern in India

Precipitation is the process in which the local air becomes saturated with vapor and starts to pour in form of liquid, hail, dew and snow. Rainfall can be defined as the precipitation in the liquid form.

In India, the rainfall is mainly contributed by two monsoon i.e. South-West and North–East monsoon but S-W monsoon provides rainfall in whole country, while the N-E monsoon is concentrated mainly in AP, TN and Odisha, therefore Indian climate is classified as a Monsoon climate. The South-West monsoon is known as summer monsoon, contributes nearly 74%, while North-East monsoon only 10% of total rainfall in the country. The contribution of winter season and pre-monsoon rains was only 3 and 13%, respectively (Table.19). The rainfall is highly variable year to year, so the departure from the normal's or average rainfall may also be assessed as it is important for Indian agriculture point of view (Table.20).

Table. 19. Rainfall pattern in India

S.N	Season/Period	Total Rainfall (mham)	Percentage
1	Winter (Jan-Feb)	12	3
2	Pre monsoon (Mar-May)	52	13
3	South-West Monsoon (June-Sept)	296	74
4	North-East Monsoon (Oct-Dec)	40	10
5	Total	400	100

Table 20. Category of rainfall based on departure from normal rainfall

S.N	Category	Departure from Normal's
1	Large Excess (LE)	60% or more
2	Excess (E)	20 to 59%
3	Normal	- 19 to +19%
4	Deficient	- 20 to - 59%
5	Large Deficient (LD)	- 60 to - 99%
6	No Rain	- 100%
7	No Data	Data not available

The easterly jet streams and tropical cyclones are responsible for monsoon or S-W monsoon or summer rains with higher intensity covering whole country, while westerly jet streams and tropical cyclones are responsible for N-E monsoon or winter rains. Most parts of the India do not receive rainfall during the winter season. The tropical cyclones originated from Way of Bengal and Indian Ocean is most destructive because of heavy rains and high winds particularly in states like TN, AP and Orissa. Some areas which receive rainfall during the winters are –

1. Central and northern parts of India get occasional rainfall during winter.
2. Weak temperature cyclones lead to rainfall in Delhi, Haryana, Punjab and western Uttar Pradesh. This rainfall benefits the *Rabi* crops.
3. Northeastern areas of the country also receive winter rainfall.
4. In October and November, northeast monsoon receives heavy moisture while blowing over the Bay of Bengal and cause enormous rainfall over the coast of Tamil Nadu and the southern tip of Andhra Pradesh.

Facts about Indian Monsoon

1. S-W monsoon rains mainly occur in the months of July, August and September.
2. The Indian rains are chiefly orographic in nature. As a result, the regions situated on the windward side receive greater rainfall than the regions located on the leeward side.

3. Only a scant amount of rainfall is received from cyclones and convectional rainfall.
4. Rainfall is erratic and unpredictable.
5. Rainfall is unevenly distributed and so flood and drought are very common.
6. Rainfall is deficient.
7. Recently significant shift in S-W monsoon has been observed in both arrival and retreat of monsoon.

Types of Rainfall

a. Based on origin

i. Convectional rainfall
ii. Orographic or relief rainfall
iii. Cyclonic or frontal rainfall

i. Convectional rainfall

1. The convectional rains are heavy with lightning and thunder and do not last long,
2. The heated air becomes light and rises in convection currents and later it expands and temperature drops and subsequently condensation takes place and cumulus clouds are formed,
3. These rain occurs usually in the summer or the hotter part of the day,
4. It generally takes place in the equatorial regions and internal parts of the continents, predominantly in the northern hemisphere, and
5. It is usually associated with hail and snow.

ii. Orographic rainfall/Relief rainfall

1. It is most common rainfall,
2. The saturated air mass is forced to rise when comes across a mountain leading to temperature falls and condensation of moisture.
3. The principal characteristic of this type of rain is that the windward slopes get more rainfall and
4. The leeward slopes remain dry and rainless, so this side is known as rain-shadow area.

iii. Cyclonic Rainfall

1. Cyclonic activity causes cyclonic rain and it occurs along the fronts of the cyclone.
2. It is formed when two masses of air of unlike density, temperature and humidity meet.

3. The layer that separates them is known as the front.
4. A warm front and the cold front are the two parts of the front.
5. At the warm front, the warm lighter wind increases slightly over the heavier cold air.
6. As the warm air rises, it cools, and the moisture present in it condenses to form clouds.
7. This rain falls gradually for a few hours to a few days.

a. Based on rainfall intensity

1. ***Light rain*** – Rate of rain varies between 0 to 2.5 millimeters/hr
2. ***Moderate rain*** – Rate of rain varies between 2.6 millimeters to 7.6 millimeters/hr
3. ***Heavy rain*** – Rate of rain is beyond 7.6 millimeters/hr

Factors affecting distribution of Rainfall

1. Direction of moisture laden winds
2. Pressure conditions
3. Cyclonic depressions determined by pressure gradients
4. Direction of relief features

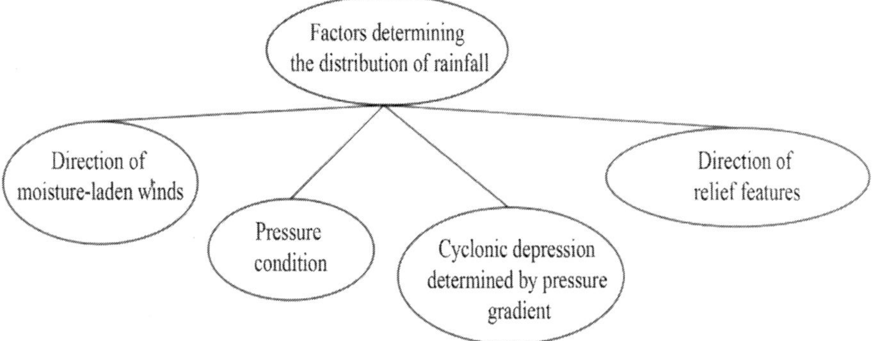

Regions receiving hefty rainfall (more than 200 cm)

1. Slopes of the Western Ghats and the Western Coastal Plains.
2. Meghalaya Hills (Garo, Khasi and Jaintia) and
3. The southern slopes of the Eastern Himalayas, Assam, Arunachal Pradesh and West Bengal.

Regions receiving moderate rainfall (100-200 cm)

1. The southern part of Tamil Nadu and the northern parts of Andhra Pradesh.
2. Middle Ganges Valley and

3. Some parts of the Western Ghats, Eastern Maharashtra, Madhya Pradesh and Odisha.

Regions receiving low rainfall (50-100 cm)

1. Parts of the Deccan plateau consisting of the regions of Karnataka, Andhra Pradesh and Tamil Nadu.
2. Eastern Rajasthan, Punjab, Haryana and Kashmir.

Regions receiving scanty rainfall (50 cm and less)

1. Northern parts of Kashmir, southern Punjab and western Rajasthan.
2. The rain shadow regions of the Western Ghats lying in the Deccan Plateau.

The average annual rainfall of the country is 1200 mm amounting to 400 million ha meter of rain water over the country's geographical area (329 m ha). However, the distribution across the country varies from less than 100 mm in extreme arid areas of western Rajasthan to greater than 3600 mm in North East states and 1100 mm from east coast 2500-3000 mm in the west coast. The broad area of the summer monsoon activity extends between 30^0 N to 30^0 S and from 30^0 W to 16.5^0 E. The detail information on rain fall and monsoonal pattern in India has been summarized in the following table.21.

Table 21. Average annual rainfall of India from 2012 to 2019

S.N	Year	Annual average rainfall (mm)
1	2011	1083.2
2	2012	1073.5
3	2013	1216.2
4	2014	1033.7
5	2015	1093.2
6	2016	1083.1
7	2017	1127.1
8	2018	1020.8
9	2019	1284.1
10	2020	954.0
	Average	1096.89

Rainfall pattern in fall

The long period average (LPA) of annual rainfall of the country also varies with the period in consideration. For example, the LPA of the season rainfall over the country as a whole for period 1901-1970 is 117cm, while for period 1961-2010 is 88 cm. The rainfall to an extent of 94-104% of the LPA is considered'normal'. If it is considered for period of last 10 years w e.f. 2011 to 2020, an average rainfall is 1096.89mm (110cm) (Table.22). Rainfall is highly variable in time and space. Over 75% of the annual rainfall is received in the four rainy months of June to September only thereby leading to large variations on temporal scale. The areas on the Western Ghats and the Sub-Himalayan areas in North East and Meghalaya

Hills receive heavy rainfall of over 250 cm annually, whereas the Areas of Northern parts of Kashmir and Western Rajasthan receive rainfall less than 40 cm. In India, 'Mawsynram' in Meghalaya receives the world's highest rainfall but on other hand, 'Jaisalmer' in Rajasthan may receive the lowest or no rain in a particular year. Rainfall Normals have been computed using rainfall records of 50 years (1951-2000) of a network of 2412 stations all over the India. The two significant features of India's rainfall is that, in the north India, rainfall decreases westwards and in the Peninsular India, it decreases eastwards and then increases in the coastal region (IMD, 2016). The broad area of the summer monsoon activity extends between 30^0 N to 30^0 S and from 30^0 W to 16.5^0 E.

Table 22. Average annual rainfall of India from 2012 to 2019

S.N	Year	Annual average rainfall (mm)
1	2011	1083.2
2	2012	1073.5
3	2013	1216.2
4	2014	1033.7
5	2015	1093.2
6	2016	1083.1
7	2017	1127.1
8	2018	1020.8
9	2019	1284.1
10	2020	954.0
	Average	1096.89

IMD (2001) published the seasonal rainfall report of different meteorological sub divisions of India (Table.23). It indicates following major points;

1. S-W monsoon or summer monsoon is the main rainy season of India and accounts nearly 74 -75% of total annual rainfall.
2. Nearly 60-95% rainfall is contributed by S-W monsoon over different sub divisions except J&K, Andaman & Nicobar Islands, Coastal AP, Rayalseema and TN.
3. In TN, N-E monsoon accounts for 48% of annual rainfall compared to 33% by S-W monsoon.
4. The contribution of N-E monsoon is nearly 20% in coastal AP, Rayalaseema, south Karnataka, Andman & Nicobar islands and Lakshdweep and 16% in Kerala.
5. Winter rainfall occurs in J&K, HP, Uttarakhand, UP, Punjab and Haryana and its contribution differs from 7% in Haryana and 17% in J&K.

Table 23. Meteorological sub division wise percentage of the normal contribution of S-W monsoon rainfall to annual rainfall (IMD, 2001)

S.N	S-W monsoon contribution (%)	State/Region
1	>90	East Rajasthan, West MP, Gujrat region, Saurashtra, Kutch & Konkan
2	80-90	Jharkhand, Bihar, East UP, West UP, Haryana, West Rajasthan, East MP, Vidharbaha, Madhya Maharashtra, Marathwada, Telangana and Coastal Karnataka.
3	70-80	Sub Himalayan and Gangetic West Bengal, Orissa, Uttarakhand, Punjab.
4	70-60	Arunachal Pradesh, Assam, NMMT, HP, North Interior Karnataka, South Interior Karnataka, Kerala and Lakshadweep.
5	50-60	Andaman & Nicobar Islands, Coastal AP, Rayalseema
6	<50	TN, J&K

Arrival of monsoon

i. South-West (S-W) monsoon enters in Indian sub-continent through Andaman and Nicobar Islands around 20th May and covers central Bay of Bengal by 25th May (Map.10a).

ii. The normal date of monsoon onset over Kerala is 1st June and then it continues to cover entire country from south to north and from east to west.

iii. Once arrived in Kerala, it also covers parts of its adjacent states like Tamil Nadu and Rayalaseema and also some parts from north-eastern states like Tripura, Nagaland, Manipur and Mizoram.

iv. By 5th June, monsoon covers parts of Karnataka, Andhra Pradesh and reaches to south Maharashtra and entire north east India.

v. By 10th June, it advances up to Mumbai along with regions from West Bengal, Sikkim, Assam, Meghalaya and most parts from the Arunachal Pradesh.

vi. By 15th June, it covers further north and north-west parts of the country (parts of Gujarat and Kutch, Madhya Pradesh, Bihar and east Uttar Pradesh).

vii. On 1st July, it reached to most parts of east Rajasthan, Punjab, Haryana, Delhi, Uttarakhand, west Uttar Pradesh and Jammu-Kashmir.

viii. Climatologically, S-W monsoon covers entire country by 15th July.

ix. S-W monsoon normally stays in the country for four months (June to September).

Withdrawal of monsoon

i. The monsoon starts withdrawing from extreme western parts of north-west Indian region by 1st September (Map-10b).

ii. By 15th September, it withdraws from most parts of west Rajasthan and Punjab and some of the surrounding regions.

iii. Thereafter, monsoon withdraws from most parts of the Saurashtra and Kutch, adjoining parts of Maharashtra like Konkan and Goa, parts of Madhya Maharashtra, west Madhya Pradesh, remaining parts of Haryana, Chandigarh and Delhi, whole of Himachal Pradesh, Jammu and Kashmir and Uttarakhand and west Uttar Pradesh at relatively faster pace.

iv. Later it withdraws from most parts of central, east and north east India (subdivisions like Marathwada, Vidarbha, east Madhya Pradesh, Chhattisgarh, Odisha, Jharkhand, West Bengal, east Uttar Pradesh, Bihar, Sikkim, Assam, Meghalaya and most parts from the Arunachal Pradesh as well as Nagaland, Manipur, Mizoram and Tripura (NMMT)) around 15th October.

v. Subsequently the south-west monsoon retreats from the country and north-east monsoon get established over south Peninsula.

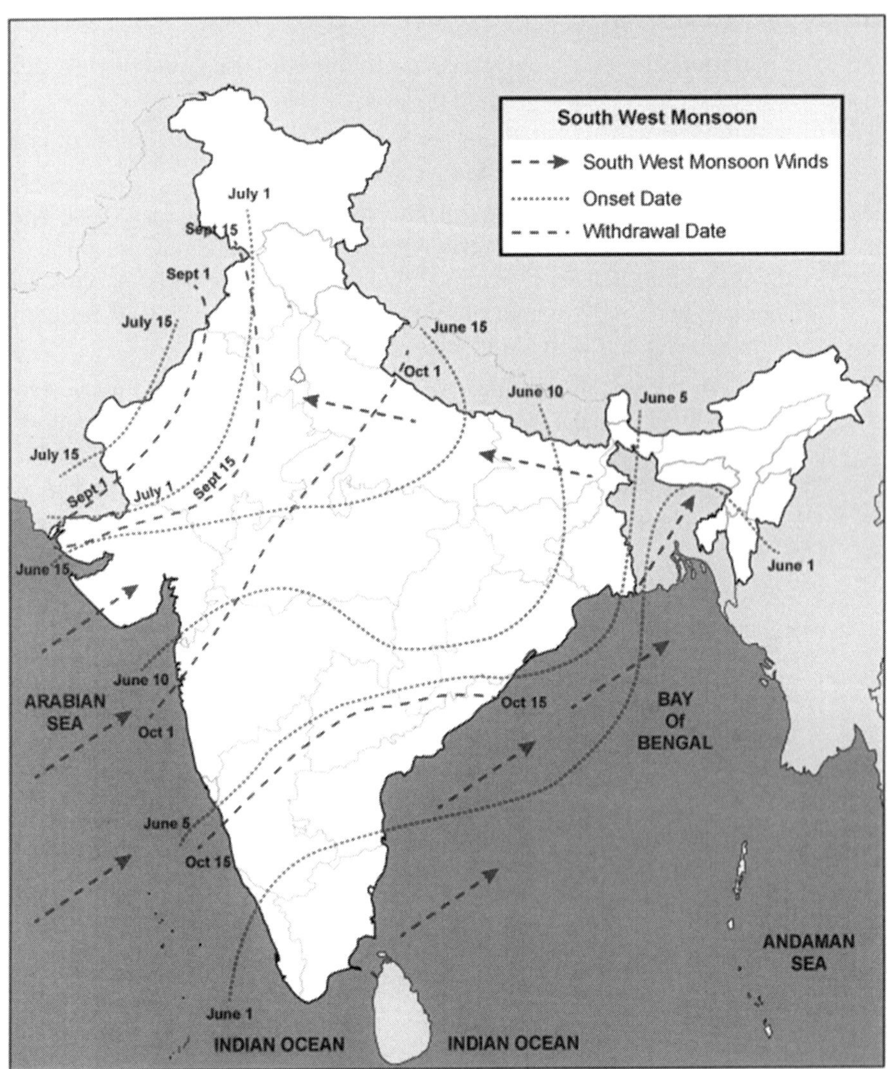

Map 10a: Rainfall distribution in India.

Shift in onset/retreat of monsoon

1. ***Onset of monsoon***

   i. Based on IMD data on monsoon onset (1961-2019), the onset of monsoon indicates that the S-W monsoon arrives over Andaman and Nicobar Islands around 22nd May about 2 days delay but covers central Bay of Bengal by 26th May with only one day delay compared to existing normal date.

   ii. Monsoon advanced over Kerala and south Tamil Nadu close to the existing normal date, i.e. around 1st June.

iii. By 5th June, it covers coastal Karnataka, most areas of south interior Karnataka and some southern areas of north interior Karnataka, parts of Tamil Nadu, south Rayalaseema, south CAP and North-east India.

iv. By 10th June, monsoon enters into Goa, south Maharashtra state, most parts of north interior Karnataka and Rayalaseema, south Telangana and some more areas of CAP.

v. By 15th June and it advances over more parts of Maharashtra including Mumbai, remaining areas of Telangana, south Chhattisgarh, remaining areas of CAP, most areas of Odisha, most parts of West Bengal, Sikkim and some eastern areas of Bihar and Jharkhand. Southwest monsoon advances over southern parts of Gujarat and Kutch, remaining parts of Maharashtra, south Madhya Pradesh state, remaining parts of Odisha.

vi. By 20th June, it covers most parts of north Chhattisgarh, most areas of Jharkhand and Bihar.

vii. Thereafter monsoon progresses further north and westwards and covers most parts of Gujarat and Kutch, Madhya Pradesh and Uttar Pradesh, Uttarakhand, Ladakh, most parts by Jammu & Kashmir, Punjab and Haryana, Delhi and east Rajasthan by 30th June.

viii. Monsoon advances over the remaining areas of the north-west India by 8th July and thus covers entire country nearly one week before the existing normal date.

ix. Presently, there is a delay 0-2 days over Andaman & Nicobar Islands, 0-4 days over north-east India, southern Peninsula and western parts of the Peninsula up to south Maharashtra. The delay is 4-7days in the remaining parts of the Peninsula.

x. Over central India, the duration of delay showed an increase as the monsoon progress from eastern to western parts.

xi. The delay in the new normal monsoon onset based on the objective criteria is 2-5 days over east central India, which increases to 7-15 days as the monsoon progress towards Gujarat and neighboring areas. However, monsoon showed rapid progress from west central India to extreme north-west India in the new normal resulting early monsoon onset/advance over extreme north-west India by about 1-9 days compared to existing normal.

xii. The standard deviation of the dates of monsoon onset/progress based on new rainfall criteria is about 5-8 days over most parts of the country except in some areas of north-west and north India where the standard deviation increases slightly to 9-10 days indicating relatively higher year to year variation of monsoon onset dates over these areas.

2. ***Withdrawal of monsoon***

i. On the basis of IMD average of monsoon withdrawal (1971-2019) indicates that monsoon starts withdrawing from north-west parts of the country around 17th September with a significant delay of more than 2 weeks compared to the existing normal date of 1st September. However, thereafter monsoon withdrawal takes place much faster.

ii. By 20th September, the monsoon withdraws from more parts of Rajasthan and some parts of north Gujarat state and some western areas of Punjab and Haryana with a delay of only around 5 days compared to existing normal date of 15th September.

iii. By 1st October, monsoon withdraws from most areas of Gujarat, Rajasthan, Jammu and Kashmir, Ladakh, Punjab, Haryana, Delhi, Himachal Pradesh, Uttarakhand, west Uttar Pradesh (UP) and some parts of east UP and west Madhya Pradesh, with a delay of only about 2-3 days compared to existing normal dates.

iv. By 10th October, monsoon withdrawal covers most parts of Maharashtra, east MP, north Chhattisgarh, east UP, most parts of Bihar and Jharkhand, and some parts of Odisha and Gangetic West Bengal.

v. By 15th October, monsoon withdrawal line nearly coincides with that of existing date and withdrawal covers most parts of the country including Maharashtra, Goa, north Karnataka, some parts of coastal Karnataka, most parts of Telangana, Chhattisgarh, Odisha and entire north east India.

vi. Subsequently southwest monsoon retreats from the remaining areas and north-east monsoon gets established over south Peninsula.

vii. Thus, it is clear that though normal date of start of withdrawal of S-W monsoon from north-western most parts of the country as per the operational data during 1971-2019 is delayed by more than 2 weeks compared to existing normal date.

viii. By 15th October, the complete monsoon withdrawal was found from the whole country, consistent with the existing normal date. The withdrawal was observed faster compared to normal monsoon.

ix. Significant delay was observed in the new normal monsoon withdrawal over extreme north-west India (11-16 days), west central India (4-10 days), north-west Peninsula (7-12 days), north India (3-7 days) and north-east (5-8 days).

x. Over other areas, the delay in the new normal is 1-3 days indicating close match with the existing normal. The standard deviation of the dates of monsoon withdrawal based on IMD operational criteria is about 7-9 days over most parts of the country except in some parts of Gujarat where it is 10 days and over north-east India where it is 6 days.

xi. The existing and new normal dates of monsoon arrival and its withdrawal of some important cities and deviation from normal dates is given in table.1 just to compare the deviation and changes in new normal monsoon dates.

The monsoon arrival and withdrawal of last 5 years is given below. It indicates a shift in both arrival and withdrawal of monsoon. The arrival of monsoon is delayed only 3-4 days but withdrawal is delayed nearly >15 days and this delayed monsoon has positive impact on agricultural production (Table.24).

Table 24. Onset and withdrawal of monsoon from 2015 to 2019

S.N	**Normal Monsoon Onset**	**Normal Monsoon Withdrawal**
	June 1	September 1
	Actual Monsoon Onset	Actual Monsoon Withdrawal
1	June 8, 2019	October 9, 2019
2	May 29, 2018	September 19, 2018
3	May 30, 2017	September 27, 2017
4	June 8, 2016	September 15, 2016
5	June 5, 2015	September 4, 2015
Average	3 days delayed	16 days delayed

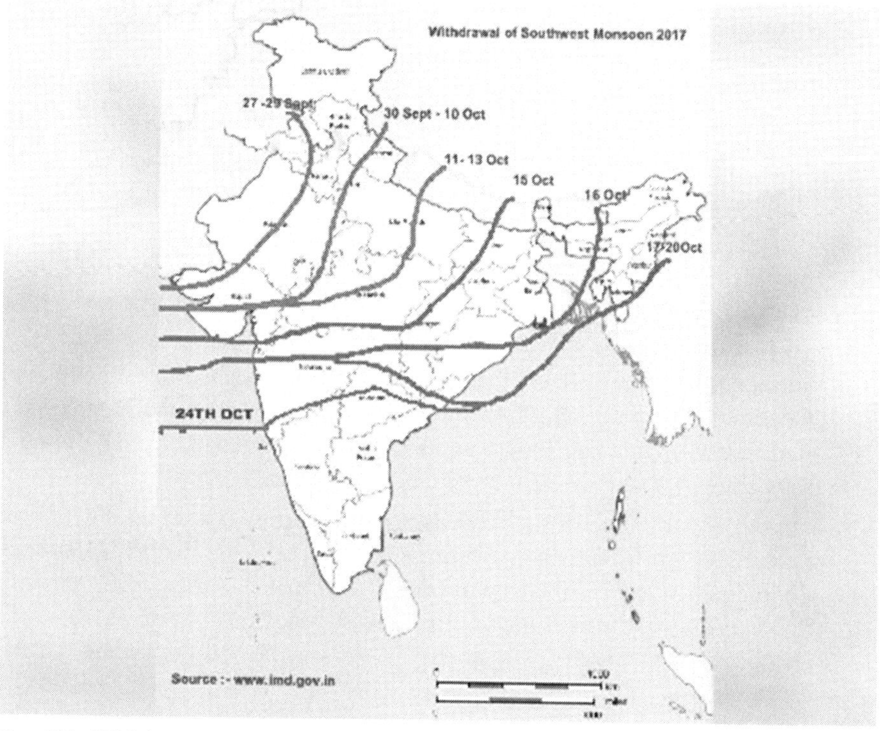

Map 10b: Withdrawal of monsoon in 2017

Salient features of new normal onset and withdrawal of monsoon

IMD (2020) reported following salient features of these new normal onset/progress and withdrawal dates of the southwest monsoon over the country;

1. The Southwest monsoon advances over Andaman and Nicobar Islands around 22nd May with a delay of about 2 days compared to existing normal date. But it arrives over central Bay of Bengal by 26th May close to existing normal date (with a one-day delay).
2. In general, there is a delay in the new monsoon onset/progress normal dates compared to the existing normal dates over most parts of the Indian monsoon region except Lakshadweep islands, a few grids from western part of north-east India and western parts of south Peninsula and some areas of north and extreme northwest India where monsoon advance is relatively faster in the new onset dates than the existing normal dates.
3. The Monsoon sets for Kerala around 1st June, nearly same as the existing normal date.
4. Monsoon advances over most parts of North-East India by around 5th June with a delay of about 1-4 days compared to the existing normal.
5. Monsoon covers most parts of the country except extreme north-west India by end of June close to the existing normal date. But covers the entire country by 8th July exactly one week before the existing normal date of 15th July.
6. Monsoon onset/ progress over most parts of central India between 2nd week of June and early July in the new normal dates is delayed by 1-8 days compared to existing normal dates.
7. As per the new normal, monsoon start its withdrawal from north-west India around 17th September, which is a delay of more than 2 weeks compared to the existing normal date (i.e. 1st September).
8. Further monsoon withdrawal is much faster. By 20th September, it withdraws from more parts of Rajasthan and some parts of north Gujarat state, and some western areas of Punjab and Haryana with a delay of only around 5 days compared to existing normal date of 15th September.
9. Monsoon withdraws from most areas of Gujarat, Rajasthan, Jammu & Kashmir,Gilgit, Baltistan, Muzzafarabad, Ladakh, Punjab, Haryana, Chandigarh, Delhi, Himachal Pradesh, Uttarakhand, west Uttar Pradesh (UP) and some parts of east UP and west Madhya Pradesh (MP) by 1st October with a delay of only about 2-3 days compared to existing normal dates.
10. Monsoon withdraws from most parts of the country except south Peninsula and some parts of neighboring central India by 15th October coinciding with the existing normal dates of withdrawal.

11. After 15th October, southwest monsoon retreats from the country and northeast monsoon gets established over south Peninsula.
12. The new normal dates of monsoon onset and retreat indicates that monsoon takes 38 days to cover whole country starting from 1st June from Kerala to reach last part of Rajasthan on 8th July and also retreats within only 28 days from the country beginning from 17th July to 15th October in compared to existing normal monsoon dates that took 45 days each for covering (1st June to 15th July) and retreating (1st September to 15th October) from whole country (Table.25).

Table 25. Normal and new normal dates of monsoon onset and retreat in India

S.N	Place / Region	Normal date (1901-40))		New normal dates		Deviation	
		Onset	Retreat	Onset (1961-2019)	Retreat (1971-2019)	Onset	Retreat
1	Andaman and Nicobar Islands	20 May	-	22 May	-	-02 days	-
2	Central Bay of Bengal	25 May	-	26 May	-	-01 days	-
3	Onset of monsoon in Kerala	29May	-	01 June	-	-02 days	-
4	Bengaluru	02 June	-	03 June	-	-01 day	-
5	Coimbatore	02 June	-	05 June	-	-03 days	-
6	Vishakhapatnam	08 June	10 Oct		10 Oct	+01 day	-
7	Gulbarga (Kar)	08 June	03 Oct	07 June	13 Oct	+01 days	+06 days
9	Tezpur (Assam)	03 June	08 Oct	05 June	14 Oct	-02 days	+06 days
8	Patna (Bihar)	11June	03 Oct	16 June	08 Oct	-05 days	+05 days
9	Ranch (Jharkhand)	10 June	03 Oct	14 June	09 Oct	-04 days	+05 days
10	Raipur (Chh.)	10 June	03 Oct	16 June	09 Oct	-04 days	+05 days
11	Ahmedabad (Guj)	14 June	18 Sept	26 June	28 Sept	-12 days	+10 days
12	Bhopal (MP)	13 June	26 Sept	20 June	03 Oct	- 07 days	+07 days
13	Jabalpur (MP)	13 June	03 Oct	19June	05 Oct	-06 days	+02 days
14	Mumbai	10 June	29 Sept	11 June	08 Oct	-01 day	+09 days
15	Pune	08 June	30 Sept	10 June	09 Oct	-02 days	+ 09 days
16	Cuttack (Odisha)	09 June	11 Oct	13 June	13 Oct	-04 days	+02 days
17	Chandigarh	22 June	21 Sept	28 June	23 Sept	-06 days	+02 days
18	Delhi	29 June	21 Sept	27 June	25 Sept	+02 days	+04 days
19	Ludhiana	26 June	13 Sept	01 July	21 Sept	-05 days	+07 days
20	Kargil, J&K	24 June	18 Sept	24June	22 Sept	No change	+04 days
21	Pokharan (Raj)	08 July	05 Sept	06 July	17 Sept	+02 days	+12 days

22	Agra	20 June	23 Sept	27 June	28 Sept	-07 days	+05 days
23	Jhansi	16 June	23 Sept	24 June	30Sept	-08 days	+07 days
24	Varanasi	15 June	06 Oct	23 June	04 Oct	-07 days	-02 days
25	Dehradun (UK)	20 June	21 Sept	20 June	27 Sept	No change	+06 days
	Country coverage	01 June-15 July	01 Sept-15 Oct	01June-08 July	17 Sept-15 Oct	+07 days advance	+16 days

Important terminology

Rain - It refers to water vapor that condenses around condensation nuclei (such as dust) and falls when the droplet is heavy enough.

Sleet – When rain droplets freeze somewhere along its path from the clouds to the ground. Sleet is mixture of snow and rain.

Snow - Water droplets and then freezing occur slowly, allowing for the development of uniquely designed ice crystals known as snow. Snow is formed when water vapour is deposited in the higher reaches of the atmosphere at a temperature less than zero degrees centigrade and falls to the ground. Snowflakes are typically symmetrical, hexagonally shaped groups of ice crystal that from while failing in below clouds. Simply snow forms if the air in cloud is below freezing. The water vapour then turns to ice instead of rain and the tiny ice crystal stick together until they form snowflakes. When they get enough to fall, they drop out of the clouds. At this points though, we still don't know whether they travel through on the way down to the ground. It gets warmer, they turn into rain, but if the air stays close to freezing all the way down, then the snowflakes will make it without melting and so fall as snow. If they occur in mountain area, it is possible for snow to be falling on the mountain top while lower in the valley the air is warmer and so it is raining instead.

Hail – Water droplets are carried high into the atmosphere by thunderstorm updraft, which cause them to freeze. Multiple drops tend freeze together, which is why the diameter of hail can be large.

Dew - water vapor condenses on the object such as blades of grass when the surface temperature is equal to the dew point.

Frost - Dew forms and then it freezes. This commonly occur when night time. Radiational cooling drops the ground temperature down enough. There are three kinds of frost; delicate crystallized ice on windowpanes is called **Hoar frost.** **Glazed frost** consist of thick coating of ice on cold surfaces. **Rime frost** is formed when super- cooled water droplets freeze on contact with cold surface. It often occurs when freezing fogs or drizzle blanket the ground.

Freezing rain – Forms and falls as rain. At the surface or near the surface the temperature is at or below freezing, which causes the rain to on contact.

Fog - A thin cloud of varying size formed at the surface of the freeze earth by condensation of atmosphere vapour.

Mist - It refers to a very thin fog.

Drizzle – It refers to a light steady rain in fine drops (0.5mm) and intensity< 0.5 mm/hr.

Moderate rain- Rain intensity >0.5mm/hr but less than 4mm/hr. Heavy rain - Rain intensity >4mm/hr but less than 8 mm/hr.

El-Nino and the Indian Monsoon

El Niño is a complex weather pattern resulting from variations in ocean temperatures in the Equatorial Pacific region along South America. The term El Niño refers to the **large-scale ocean-atmosphere climate interaction linked to a periodic warming in sea surface temperatures across the central and east-central Equatorial Pacific.** It appears once every three to seven years, bringing drought, floods and other weather extremes to different parts of the world.

The system involves oceanic and atmospheric phenomena with the appearance of warm currents off the coast of Peru in the eastern pacific and affects weather in many places including India. It is merely an extension of the warm equatorial current which gets replaced temporarily by cold Peruvian current or **Humbolt Current**. This results in mainly two things:

1. The distortion of equatorial atmosphere circulation, and
2. Irregularities in the evaporation of sea water.

La Nina effect

i. La Nina refers to a little girl in the Spanish language.

ii. As opposed to El Nino, La Nina refers to a decrease in the oceanic temperatures across the equatorial Pacific and so the water of the Pacific ocean turns colder than usual resulting into strong and sturdy wind flow.

iii. La Nina creates a low-pressure condition under its presence around the southern Pacific and causes below-average precipitation in the southern Pacific region.

iv. The Coriolis force, in this case, is fairly high.

v. La Nina has a massive tendency to trigger tropical cyclones.

vi. Both El Nino and La Nina impact global temperatures periodically.

vii. Both of these comprise the El Nino- southern oscillation (ENSO) cycle with a neutral phase.

viii. El Nino's are more frequent as compared to La Nina's.

Factors affecting the Climate of India

The Climate in India is affected by following factors

- Latitude
- Himalaya Mountains
- Altitude

- Distance from the sea
- Geographical limits like western disturbances, conditions in the regions surrounding India, Conditions over the ocean.
- Jet Streams.

1. ***Latitude*** : The Tropic of Cancer divides India into two parts i.e. North and South India. Thus, northern part of the India lies in sub-tropical and temperate zone and the south of the Tropic of Cancer falls in the tropical zone. The tropical zone being nearer to the equator, experiences high temperatures throughout the year with small daily and annual variation range. Area north of the Tropic of Cancer being away from the equator, experiences extreme climate with high daily and annual range of temperature.
2. ***The Himalayan Mountains*** : The extension of great Himalaya from J & K to Aurnachal Pradesh acts like a shield and protect subcontinent from the cold northern winds. These cold and chilly winds originate near the Arctic circle and blow across central and eastern Asia. In addition, the Himalayas also trap the monsoon winds, forcing them to shed their moisture within the sub continent.
3. ***Distance from the Sea*** : The coastal area has an equable climate but in contrast, the areas in the interior of India experience extremes of climate mainly because of far away from the moderating influence of the sea. For example, the people of Chennai, Kanyakumari, Goa, Mumbai and the Konkan coast have hardly any idea of extremes of temperature and the seasonal rhythm of weather but interior places like Delhi, Amritsar, Jaipur, Kanpur, Prayagraj have extremes of climate with hot summer and chilly winter seasons.
4. ***Distribution of Land and Water*** : India is flanked by the Indian Ocean on three sides in the south and girdled by a high and continuous mountain-wall in the north. As compared to the landmass, water heats up or cools down slowly. This differential heating of land and sea creates different air pressure zones in different seasons in and around the Indian subcontinent. Difference in air pressure causes reversal in the direction of monsoon winds.
5. ***Altitude*** : The temperature decreases with height, so the mountains are cooler than plains. For example, Agra and Darjiling are located on the same latitude, but temperature of January in Agra is 16°C whereas it is only 4°C in Darjiling.
6. **Relief** : The temperature, air pressure, direction and speed of wind and the amount and distribution of rainfall are affected by the physiography or relief. The windward sides of Western Ghats and Assam receive high rainfall during June-September whereas the southern plateau remains dry due to its leeward situation along the Western Ghats.

7. **Air Pressure and Wind** : Air pressure and wind are correlated to each other and play greater role to influence the Indian climate, so the mechanism of following are important to understand for both winter and summer seasons;
   i. Distribution of air pressure and winds on the surface of the earth.
   ii. Upper air circulation caused by factors controlling global weather and the inflow of different air masses and jet streams.
   iii. Inflow of western cyclones generally known as western cyclonic disturbances during the winter season and tropical depressions during the south-west monsoon period into India creating weather conditions favorable to rainfall.

Mechanism of Weather in Winter season

i. ***Surface Pressure and Winds*** : The winter weather of India is generally influenced by the distribution of pressure in Central and Western Asia. A high pressure centre in north of the Himalayas develops during winter leading to flow of air at the low level from the north towards the Indian subcontinent, south of the mountain range. The surface winds blowing out of the high pressure centre over Central Asia reach India in the form of a dry continental air mass. These continental winds come in contact with trade winds over north western India. The position of this contact zone is not, however, stable. Occasionally, it may shift its position as far east as the middle Ganga valley with the result that the whole of the north western and northern India up to the middle Ganga valley comes under the influence of dry northwestern winds.

ii. ***Jet streams and Upper Air Circulation:*** Jet streams are the bands of strong wind that generally blow from west to east all across the globe, so they are ate also known as westerly winds. There are four types of primary jet streams: two polar jet streams, near the north and south poles, and two subtropical jet streams closer to the equator. The strongest jet streams are the polar jets, at 9–12 km (30,000–39,000 ft) above sea level, and the higher altitude and somewhat weaker subtropical jets at 10–16 km (33,000–52,000 ft). The Northern Hemisphere and the Southern Hemisphere each have a polar jet and a subtropical jet. The Jet streams move north and south too, following the boundary between warmer and colder air. The wind direction in the jet stream can change from the normal west to east to almost north to south. The climate of whole western and central Asia is affected by these westerly winds along the altitude of 9-13 km from west to east. These winds blow across the Asian continent at latitudes north of the Himalayas roughly parallel to the Tibetan highlands. These are known as jet streams. Tibetan highlands act as a barrier in the path of these jet streams. As a result, jet streams get bifurcated. One of its branches blows to the north of the Tibetan highlands, while the southern branch blows in an eastward direction, south of the Himalayas. It has its mean position at 25°N in February at 200-300 mb level. It is believed that this southern branch of the jet stream exercises an important influence on the winter weather in India (Map.11).

iii. ***Western Cyclonic Disturbance and Tropical Cyclones***: The western cyclonic disturbances which enter the Indian subcontinent from the west and the north-west during the winter months, originate over the Mediterranean Sea and brought into India by the westerly jet stream. An increase in the prevailing night temperature generally indicates an advance in the arrival of these cyclones disturbances.

Tropical cyclones originate over the Bay of Bengal and the Indian ocean. These tropical cyclones have very high wind velocity and heavy rainfall and hit the Tamil Nadu, Andhra Pradesh and Orissa coast. Most of these cyclones are very destructive due to high wind velocity and torrential rain that accompanies it.

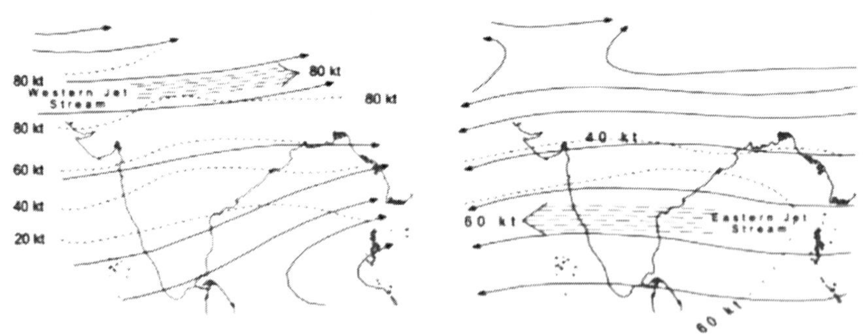

Westerly jet streams Easterly jet streams

Map 11: Jet streams and Upper Air circulation

Mechanism of Weather in Summer season

i. ***Surface Pressure and Winds***: The sun shifts northwards in summer season and the wind circulation over the subcontinent undergoes a complete reversal at both, the lower as well as the upper levels. By the middle of July, the low pressure belt nearer the surface termed as Inter Tropical Convergence Zone (ITCZ)] shifts northwards, roughly parallel to the Himalayas between 20° N and 25° N. By this time, the westerly jet stream withdraws from the Indian region. The ITCZ being a zone of low pressure, attracts inflow of winds from different directions. The maritime tropical air mass (mT) from the southern hemisphere, after crossing the equator, rushes to the low pressure area in the general south westerly direction. It is this moist air current which is popularly known as the S-W monsoon.

ii. ***Jet Streams and Upper Air Circulation***: An easterly jet stream flows over the southern part of the Peninsula in June at maximum speed of 90 km per hour. In August, it is confined to 15° N latitude, and in September up to 22° N latitudes. The easterlies normally do not extend to the north of 30° N latitude in the upper atmosphere.

iii. ***Easterly Jet Stream and Tropical Cyclones***: The easterly jet stream steers the tropical depressions into India. These depressions play a significant role in the distribution of monsoon rainfall over the Indian subcontinent. The tracks of these depressions are the areas of highest rainfall in India. The frequency at which these depressions visit India, their direction and intensity are deciding factors of rainfall pattern during the S-W monsoon period.

Measurement of Weather parameters

Meteorological observations are required for following purposes;

i. Preparation of real- time preparation of weather charts and maps,

ii. Weather forecast and severe weather warnings,

iii. Study of climate,

iv. Local weather dependent operations like agricultural operation, drought and flood management, crop contingent planning, soil and water conservation measures etc.

v. Research in meteorology and climatology.

The meteorological applications have their own preferred time scales and space scales for averaging, station density and resolution of phenomena. It may be small for agricultural meteorology and large for global long-range forecasting. The dimension of small-scale or local applications is 10 km or less. The specific purpose of different scale observations is given below;

i. Microscale (less than 100 m): for agricultural meteorology, for example, evaporation;

ii. Toposcale or local scale (100m–3 km): for example, air pollution, tornadoes.

iii. Mesoscale (3–100 km): for example, thunderstorms, sea and mountain breezes.

iv. Large scale (100–3000 km): for example, fronts, various cyclones, cloud clusters.

v. Planetary scale (larger than 3000 km): for example, long upper tropospheric waves.

In general, following meteorological observations are recorded from meteorological observatory or automatic weathers stations;

1. Rainfall / precipitation
2. Humidity / Relative humidity

3. Air temperature (minimum and maximum)
4. Soil temperature
5. Sunshine hours / Solar radiation
6. Wind speed
7. Wind direction
8. Atmospheric pressure
9. Evaporation
10. Snow cover
11. Visibility
12. Cloud: cloud amount, cloud type and cloud base height.

Number of instruments is available to measure different meteorological observations except the cloud type. Some meteorological stations take upper-air measurements, measurements of soil moisture, ozone and atmospheric composition.

Automatic weather stations Most of the elements required for synoptic, climatological or aeronautical purposes can be measured by automatic instrumentation. Meteorological observing stations are designed so that the representative measurements (or observations) can be taken according to the type of the station involved. The ground should be covered with short grass or a surface representative of the locality, and surrounded by open fencing to exclude entry of unauthorized persons. Within the enclosure, a bare patch of ground of about 2 m x 2 m is reserved for observations of the state of the ground and of soil temperature at depths of equal to or less than 20 cm.

1. Rainfall: *Rain Gauge / Standard raingauge/ Automatic raingauge*

a. An instrument that measure the amount of rain that falls during a given time interval usually in millimetres or inches. The total amount of rainfall which reaches the ground in a stated period is expressed as the depth to which it would cover the earth if it were all flat and level, and if there were no loss by evaporation, or if there was no runoff or infiltration.

b. The standard rain gauge takes the form of a collector above a funnel leading to a receiver with some sort of measuring device. Sometimes the amount of rain is poured carefully into a measuring cylinder or dip rods or a ruler dipped into the receiver to determine the amount.

c. Three types of automatic precipitation recorders are in general use: the weighing recording, the tilting or tipping-bucket type and the float type. Only the weighing type is satisfactory for measuring all kinds of precipitation (Fig.1). The use of the other two types for the most part is limited to the measurement of rainfall.

A. Raingauge **B. Standard Raingauge** **C. Automatic Raingauge**

Fig 1: Raingauge for measurement of rainfall

2. Humidity: Hygrometer / Electrical hygrometer / Infrared hygrometer

An instrument used to measure the humidity or amount of water vapour in the atmosphere. The sensing mechanism of the instrument can be hair (hair hygrometer), a plate coated with carbon (electrical hygrometer) or an infrared sensor (infrared hygrometer) (Fig.2).

Humidity measurement instruments usually rely on measurements of some other quantity such as temperature, pressure, mass or a mechanical or electrical change in a substance as moisture is absorbed.

a. ***Gravimetric hygrometry*** : This method uses the absorption of water vapour by a desiccant from a known volume of air. The gravimetric hygrometer is used for this purpose.

b. ***The psychrometric method*** : The measurement of atmospheric humidity is an important requirement in most of the areas of meteorological studies. Psychrometry is defined as the measurement of the moisture content of air. Nowadays many humidity measuring devices are available. A psychrometer composed of two similar thermometers. The bulb of one thermometer is kept wet (by means of a thin, wet cloth wick) so that the cooling that results from evaporation makes it register a lower temperature than the dry-bulb thermometer. When readings are taken simultaneously, it is possible (with the use of psychrometric tables) to determine the relative humidity and dew-point temperature of the air.

c. ***Sorption methods:*** Certain materials interact with water vapour and undergo a change in a chemical or physical property that is sufficiently reversible for use as a sensor of ambient humidity. Absorption of electromagnetic radiation by water vapour (ultraviolet and infrared absorption hygrometers) is used for this measurement purpose.

A

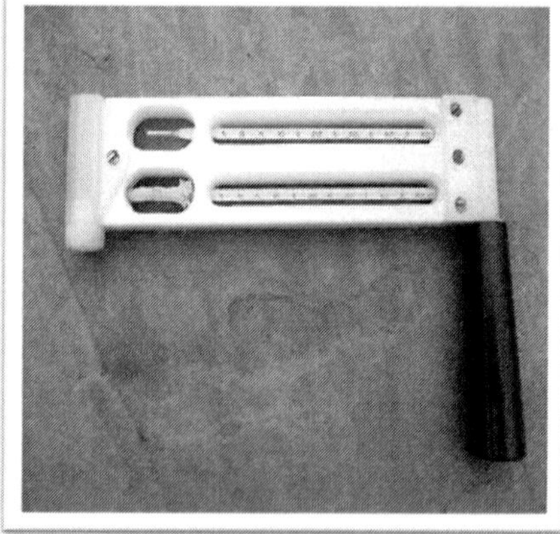

B

Fig. 2: A Hygrometer and *B*. Psychrometer

3. *Evaporation: Evaporimeter*

A evaporation gauge or evaporimeter is a pan used in the measurement of the evaporation of water into the atmosphere over specified periods of time (Fig.3)

Fig 3: Pan Evaporimeter

4. Wind Speed: Wind vane / cup or propeller wind *Anemometer / Wind sensors*

Normally cup or propeller type anemometers are available in observatory *to* measure wind speed (Fig.4). A lot of wind sensors are used for this purpose. Some of them include: 1) Pitot tube anemometers, 2) Sonic anemometers Hot-disc anemometers are recently developed solid-state instruments, 3) Hot-wire anemometers and 4) Remote wind-sensing techniques with sound (sodar), light (lidar) or electromagnetic waves (radar). The wind speed is measured over a leveled open terrain 10 m above the ground.

Fig. 4: Wind Anemometer

5. Wind Direction: *Wind Vane* and Wind Sock

a. Wind Vane: A device that measures the direction of the wind (Fig.5). It spins on a rod and points in the direction from which the wind blows. The part of the vane that turns into the wind is usually shaped like an arrow. The other end is wide so it will catch the smallest breeze.

b. Wind Sock: A light, flexible cylinder or cone mounted on a mast to show the direction and strength of the wind. Wind sock is commonly used at an airport or chemical plant. The cone-shaped bag opened at both ends so that it extends horizontally as the wind blows through it and indicates the wind direction by pointing away from the wind.

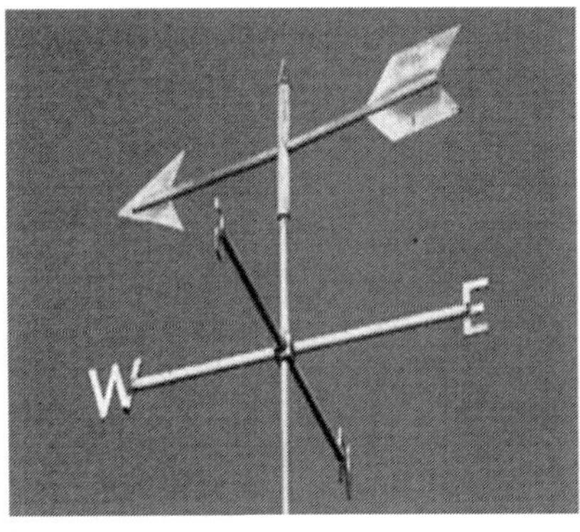

A

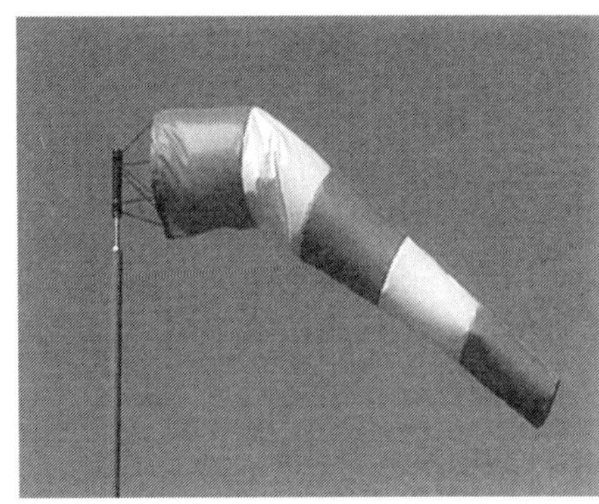

B

Fig. 5: A. Wind vane and B. Wind sock

6 Temperature and humidity: Stevenson Screen is used to measure both temperature and humidity (Fig.6). A standard shelter for housing four types of thermometers dry, wet, maximum and minimum. The screen shields the instruments from direct sunlight.

Fig. 6: Stevenson Screen

Temperature: Thermometers / Liquid-in-glass thermometers / Ordinary thermometers/mercury-in-glass- type thermometer / Meteorological thermometers / Radiometric thermometers.

The atmospheric temperature may be measured in (Celsius (C) or Fahrenheit (F), or Kelvin (K)). A traditional thermometer consists of mercury, red spirit or green spirit in a glass tube and operates on the principle that the liquid expands more than the glass does when heated.

Maximum temperature: Maximum temperature has historically been measured with a mercury-in-glass thermometer which has a constriction in the neck of the thermometer tube. As the air temperature rises mercury is forced past the constriction. However, as the temperature falls the constriction prevents the mercury from returning to the bulb of the thermometer. The height of mercury in the tube remains at that reached at the hottest time of day. The thermometer is reset by gentle shaking.

Minimum temperature: Alcohol-in-glass thermometers containing a moveable index are used to manually record minimum temperatures. When the temperature falls, the liquid and index move down the column, but when the temperature rises the index remains in the lowest position while the liquid expands up the tube. The position of the index indicates the lowest temperature reached since the last reset - which is achieved by tilting the thermometer, bulb end upwards (Fig.7).

Maximum temperature

Minimum temperature

Fig. 7: Air temperature measurement

7. Humidity: ***Humidity Sensor or Wet and Dry Thermometer***

8. Atmospheric Pressure: Barograph / Aneroid *Barometer / Digital Barometer / Pressure Sensor (Digital piezoresistive barometers)*

i. ***Mercury barometers /Electronic barometers***: A barometer is the instrument used to measure atmospheric pressure. Pressure tendency can forecast short term changes in the weather.

ii. ***Aneroid displacement transducers***: The aneroid displacement transducer contains a sensor with electrical properties (resistance or capacitance) that changes as the atmospheric pressure changes.

iii. ***Digital piezoresistive barometers***: Today's mostly preferred pressure sensor is the piezoresistive sensor. It is cheap, and still delivers a good result. But it has drawbacks—significant power requirements, low output signal, large offset and temperature dependence. Many sensors are therefore internally compensated for temperature effects.

iv. ***Cylindrical resonator barometers***: A cylindrical resonator barometer (or vibrating cylinder air-pressure transducer) is designed to measure absolute air pressure using the vibrating element principle, providing a frequency output from which pressure is computed.

v. ***Aneroid barometers:*** Aneroid barometers have lower accuracy than mercury barometers. These are compact and portable equipments. Aneroid barometers are easier to handle and use and are suitable for self-recording.

vi. ***Bourdon-tube barometers***: Bourdon Tubes are known for its very high range of differential pressure measurement in the range of almost 100,000 psi (700 MPa). It is an elastic type pressure transducer. Bourdon-tube barometers usually consist of a sensor element.

vii. ***Automatic digital barometers***: Automatic digital barometers make use of a combination of sensor and microprocessor techniques. In fact these instruments are composed of sensors, transducers, a micro-processor and digital electronics. Electrical signals generated by the pressure sensor(s) are interpreted by the micro-processor using a more or less sophisticated algorithm (Fig.8).

1. Barograph

B. Aneroid Barometer

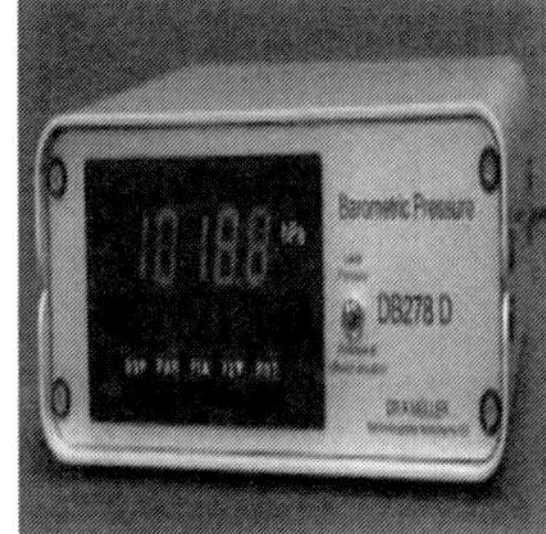

C. Digital Barometer

Fig. 8: Type of Barometers for measurement of atmospheric pressure

8. Solar radiation: Absolute Pyrheliometers, Pyranometer, Sun photometer Spectral direct solar irradiance and measurement of optical depth: Broadband pyrheliometry Sun radiometry (photometry) and aerosol optical depth.

9. Sunshine duration: Campbell-Stoke Sunshine Recorder (fig.9)

Campbell-Stokes Pattern Sunshine Recorder employs a glass sphere to focus the sun's rays to an intense spot, which will burn a mark on a curved card mounted concentrically with the sphere. As the earth rotate the position of the spot moves across the card. When the sun is obscured, the trace is interrupted. At the end of the day the total length of the burn or trace, less gaps, equals the duration of sunshine hours.

Fig. 9: Campbell-Stoke Sunshine Recorder

10. Soil temperature: *Soil thermometer/Digital Soil Thermometer/Analog Soil Thermometer/Long Stem Thermometer*

The soil temperature plays significant role for germination of crop seed. The soil temperature is measured at different soil depth i.e. 5, 10 and 15cm, so different soil thermometers are required. The thermometers are placed at optimum soil depth at right angles (Fig.10).

1. **Digital Soil Thermometer:** It measures soil temperature from 0^0 to 160^0F (-17 to 71^0C). It is a pocket sized soil meters built of tough, corrosive resistant stainless steel. It takes only 1 to 15 seconds to record the soil temperature. It is accuracy is $\pm 2^0$.
2. **Analog Soil Thermometer:** It measures soil temperature in Fahrenheit from 25 to 125^0F with $\pm 1^0$.
3. **Long Stem Digital Thermometer:** It has 20cm long stem and measures soil temperature from -58 to 302^0F (-50 to 150^0C) with accuracy of $\pm 1.8^0$.

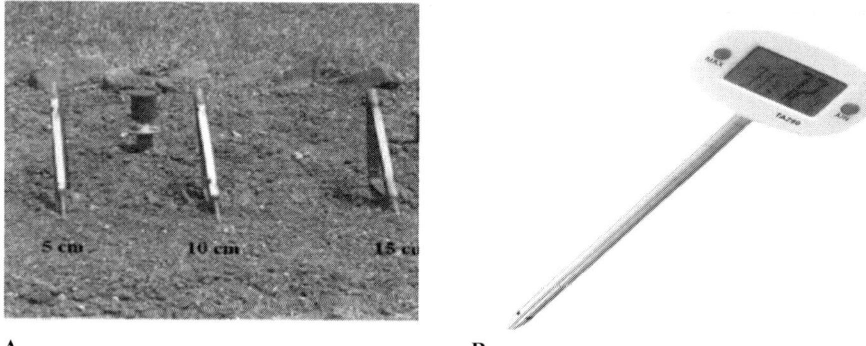

A B

Fig. 10: A. Normal Soil Thermometer and B. Digital Soil Thermometer

7. Cloud height: Ceilometer

A ceilometer is a device that uses a laser or other light source to determine the height of a cloud ceiling or a cloud base from the ground. Ceilometers can also be used to measure the aerosol concentration within the atmosphere (Fig.11).

Fig. 11: Ceilometer

Other measurable parameters are cloud cover, sunshine hours and evapo-transpiration. Some of this information is gathered using satellite data for weather forecasting. The weather parameters can be measured at one place through the installation of a weather station and can be manual or automatic. In a manual station, all the parameters are measured and noted manually while in an automatic station the data is measured and stored automatically in the data logger.

Impact of Monsoon on the Indian Economy

i. The monsoon has a direct impact on the country's agricultural GDP. The planting of key *Kharif*, or summer, crops like rice, sugar cane, pulses and oilseeds begins with the arrival of monsoon rains in June.

ii. Summer crops account for almost half of India's food output and a delayed or poor monsoon means supply issues and acceleration in food inflation, a key metric which influences Reserve Bank of India's decision on interest rates.

iii. A deficit monsoon could also lead to a drought-like situation, thereby affecting the rural household incomes, consumption and economic growth.

iv. A poor monsoon not only leads to weak demand for fast-moving consumer goods, two-heelers, tractors and rural housing sectors but also increases the imports of essential food staples and forces the government to take measures like farm loan waivers, thereby putting pressure on finances.

v. Most of the Indian power project installed on the Perennial Rivers. If monsoon fails, it would lower the water levels that have detrimental effects on the power generation as well as irrigation facilities.

vi. Weak Monsoon is strongly correlated with receding water table which has worsened the farmer distress.

To ensure food and farmer's income security in case of a weak monsson requires comprehensive and broad approach as with climate change monsoon variability is bound to increase which need holistic transformation which include-

1. Development Initiatives- Some recent initiatives like fasal bima yojana, krishi sichani yojana, promotion to micro/drip irrigation, Organic farming etc need to be expedited especially to small and marginal farmers.
2. Interlinking of river is a strong initiative with high potential to raise farm income, but require time bound implementation. Eg Godavari-Krishna River.
3. Making agriculture Climate resilient which includes enhanced carbon sink in soils, precise weather forecasts, access to market intelligence for timely decision making by farming community and efficient water, energy and input use. Eg-NICRA
4. There is a need for capacity development in frontier areas to bridge the knowledge gap (linking with Skill India).
5. Promotion of weather-resistant crop varieties like flood and salinity resistant rice and drought-resistant pulses should be promoted and further improved. Likewise short-cycle crops of particular region which can withstand heat and ripen before heavy rains should be given due recognition. Eg Narendra 97 is a good example of a short cycle paddy crop grown in Uttar Pradesh that can deal with intense heat and ripens in a shorter period than the commonly cultivated paddy.
6. Promotion to Agro-forestry and Mixed cropping like growing paddy with pulses, millets and vegetables should be promoted to ensure food security in case one of the crops fail due to drought or floods.

7. Promotion to agriculture allied sectors like Animal husbandry, Fisheries, Apiculture etc.

Conclusion

The need of hour is to use best technologies and services to boost production, income and reduce climate risks, it will also contribute to Intended Nationally Determined Contributions (INDCs). Therefore, a realistic evaluation of constraints and opportunities is critical for improvising ecological, technological, social and economic factors affecting farming community.

References

Bansal, N. K., & Minke, G. (1988). Climatic zones and rural housing in India, (199004). https://doi.org/citeulike-article-id:10601847.

Bureau of India Standards. (2005). National Building Code of India(2005th ed.). Bureau of India Standards.

CRS Research Report. 2020. New Normal Dates of Onset/Progress and Withdrawal of Southwest Monsoon over India. RR No. 3/2020, Office of the Head, Climate Research and Services, Pune (MS), India.

Krishnan, A. and Singh, M. 1972. Soil climate zones in relation to cropping patterns. Proceedings of the Syposium on Cropping Patterns in India. ICAR, New Delhi. pp. 172-185.

NARP.https://www.nbsslup.in/assets/uploads/clinks/Delineating%20Agro-Ecological%20Regions.pdf

Singh, A. and Venkatesharlu, B. 2009. Climate change and rainfed agriculture. Indian Journal of Dryland Agricultural Research and Dvevelopment 24 (2): 01-09.

Thornthwaite, C. W. 1978. An approach towards a rational classification of climate. Geographical Review 38: 55-94.

UNEP. 1972. World Atlas of Desertification. Nairobi.

Questions

a. Short and Descriptive Questions

1. Write short notes on following;
    a. Monsoon
    b. Aridity index
    c. Moisture index
    d. Moisture availability index (MAI)
    e. NDVI
    f. El Nino effect
    g. La Nino effect
2. Name the instrument to measure the following climate parameters;
    a. Rainfall
    b. Wind velocity
    c. Evaporation

   d. Solar radiation
   e. Relative humidity
   f. Sunshine hours
   g. Cloud height
   h. Air temperature
   i. Soil temperature
   j. Jet stream

3. What is climate? Discuss the climate classification proposed by following;
   a. Koppen
   b. Koppen and Geiger
   c. Thornthwaite
   d. Ramakrishnan (1997) and Paul (1997)
   e. Singh and Venkateshvarulu (2009)
   f. FAO (1993)
   g. Indian Meteorological Department
   h. Bansal and Minke (1988)
   i. Palmer
   j. IMD
4. What do you mean by Agro climatic zonation? Describe the agro climatic regions of India proposed by following;
   a. Planning commission
   b. NARP
   c. NBSS &LUP 91999)
   d. NATP
5. Describe the main features of rainfall and its pattern. Also throw light on present scenario of monsoon arrival and withdrawal in India.
6. What are the factors influencing Indian monsoon ? Also briefly describe the significance of monsoon on Indian economy.

b. Objective questions

1 Climate refers to
   a Sum total of weather conditions and variations over a large area for a long period of time
   b State of the atmosphere over an area at any point of time
   c Sum total of weather conditions and variations over a large area for a short period of time
   d None above

2 Weather refers to

a Sum total of weather conditions and variations over a large area for a long period of time

b State of the atmosphere over an area at any point of time

c Sum total of weather conditions and variations over a large area for a short period of time

d None above

3 Monsoon hs been derived from 'mausim' that is a

a Arabic b Greek

c Latin d Sanskrit

4 IMD classified Indian seasons into

a 02 b 03

c 04 d 06

5 How many seasons India has as per Ancient Hindu Calender?

a 02 b 03

c 04 d 06

6 Empirical method of climate classification is based on

a Use of environmental data, such as temperature, humidity, and precipitation, or simple quantities derived from them (such as evaporation).

b Causal elements, the activity and characteristics of all factors (circulation systems, fronts, jet streams, solar radiation, topography etc.)

c Only Temperature, Jet streams and solar radiation

d All above

7 Generic method of climate classification is based on

a Use of environmental data, such as temperature, humidity, and precipitation, or simple quantities derived from them (such as evaporation).

b Causal elements, the activity and characteristics of all factors (circulation systems, fronts, jet streams, solar radiation, topography etc.)

c Only Temperature, Jet streams and solar radiation

d All above

8 Empirical method of climate classification is of

a Descriptive in nature
b Explanatory in nature
c Both 'a' & 'b'
d None above

9 Generic method of climate classification is of

a Descriptive in nature
b Explanatory in nature
c Both 'a' & 'b'
d None above

10 Kendrew-Stamp classified Indian climate based on

a Temperature
b Humidity
c Both 'a' & 'b'
d None above

11 Koppen has divided world climate into

a 4 regions
b 5 regions
c 6 regions
d 8 regions

12 If the region has higher evapotranspiration than rainfall , it is named by Koppen as

a Tropical Rainy Climate
b Dry climate
c Warm Temperate Rainy Climate
d Cold Frost Climate

13 In which of the following region the temperature in coolest months does not exceed 18^{0}C as per Koppen classification of Monsoon?

a Tropical Rainy Climate
b Dry climate
c Warm Temperate Rainy Climate
d Cold Frost Climate

14 As per Koppen classification of monsoon, Polar climate refers to

a <10°C temperature in warmest month
b Most of the area is covered by snow
c Short cool summer and long cold winter
d All above

15 What is the weakness of Koppen classification of monsoon?

a Ignores factors like cloudiness, wind, rainfall intensity, currents & air mass
b It did not explain type of vegetation
c Both 'a' & 'b'
d None above

16 Thornthwaite classification of monsoon is based on

a Potential evapotranspiration
b Moisture index
c Thermal efficiency
d All above

17 Thornthwaite classified monsoon into

a 4 types
b 5 types
c 6 types
d 7 types

18 As per the Thornthwaite classification of monsoon, the arid climate has moisture index equal to

a 0 to 20 b -20 to 00

c -40 to -20 d -60 to -40

19 As per the Thornthwaite classification of monsoon, the semi arid climate has moisture index equal to

a 0 to 20 b -20 to 00

c -40 to -20 d -60 to -40

20 As per the Thornthwaite classification of monsoon, moist sub humid climate is found in

a Peninsular interior, western Madhya Pradesh, Haryana and Punjab

b Saurashtra, Kuchchh and Rajasthan

c Ganga valley and northeastern parts of Central India

d Western Ghats and also over Orissa and West Bengal.

21 As per the Thornthwaite classification of monsoon, dry sub humid climate is found in

a Peninsular interior, western Madhya Pradesh, Haryana and Punjab

b Saurashtra, Kuchchh and Rajasthan

c Ganga valley and northeastern parts of Central India

d Western Ghats and also over Orissa and West Bengal.

22 As per the Thornthwaite classification of monsoon, semi arid climate is found in

a Peninsular interior, western Madhya Pradesh, Haryana and Punjab

b Saurashtra, Kuchchh and Rajasthan

c Ganga valley and northeastern parts of Central India

d Western Ghats and also over Orissa and West Bengal.

23 As per the Thornthwaite classification of monsoon, arid climate is found in

a Peninsular interior, western Madhya Pradesh, Haryana and Punjab

b Saurashtra, Kuchchh and Rajasthan

c Ganga valley and northeastern parts of Central India

d Western Ghats and also over Orissa and West Bengal.

24 Mather has classified climate into

a 5 regions b 6 regions

c 7 regions d 8 regions

25 IMD classified Indian climate into

a 4 regions
b 5 regions
c 6 regions
d 8 43gions

26 IMD classification on climate is based on

a Himalyan Mountains
b Latitude
c Surface distance from sea
d All above

27 Montane climate is also known as

a Alpine climate
b Tropical climate
c Arid climate
d Semi arid climate

28 Krishnan and Singh (1972) classified climate based on

a Moisture index
b Thermal index
c Both 'a' & 'b'
d None above

29 Thermal index is equal to

a $(T_{max} - T_{min})/2$
b (P –PET/PET) x 100
c 100 x (Monthly water surplus / Water need)
d 100 x(Monthly water deficit / water need)

30 Moisture index refers to

a $(T_{max} - T_{min})/2$
b (P –PET/PET) x 100
c 100 x (Monthly water surplus / Water need)
d 100 x(Monthly water deficit / water need)

31 Humidity index refers to

a $(T_{max} - T_{min})/2$
b (P –PET/PET) x 100
c 100 x (Monthly water surplus / Water need)
d 100 x(Monthly water deficit / water need)

32 Aridity index refers to

a $(T_{max} - T_{min})/2$
b (P –PET/PET) x 100
c 100 x (Monthly water surplus / Water need)
d 100 x(Monthly water deficit / water need)

33 Krishnan and Singh (1972) classified climate based on moisture index into

a 5 zones
b 6 zones
c 8 zones
d 10 zones

34 Krishnan and Singh (1972) classified climate based on thermal index into

a 5 zones
b 6 zones
c 8 zones
d 10 zones

35 As per the climate classification of Krishnan and Singh (1972), the climate zones paired with thermal index values are given below. Which of the following pair is wrongly matched ?

a Very hot - > 28^0C
b Hot – 25-28^0C
c Cold – 10-20^0C
d Very cold - < 0^0C

36 Major weakness of Krishnan and Singh (1972) classification of monsoon is

a Non consideration of soil moisture
b Over estimation of temperature
c Non consideration of solar energy
d All above

37 The basis of dryland climate classification of FAO (1993) was based on

a Moisture index
b Humidity Index
c Thermal Index
d All above

38 Which of the following pair is correct as per classification of dryland climate by FAO?

a Hyper arid - <0.05MI
b Arid - 0.05-0.2 MI
c Semi arid – 0.20-0.50
d All above

39 What is the basis of Indian monsoon classification proposed by World Atlas of Desertification (WAD)?

a Moisture Index
b Rainfall
c Growing degree days
d All above

40 In Indian monsoon climate classification proposed by WAD , the growing degree days were identified in different regions. Which of the following pair is correct?

a Arid - < 75 days
b Semi arid – 75-120 days
c Humid - >150 days
d All above

41 In Indian monsoon climate classification proposed by WAD , the moisture indices were paired with different regions. Which of the following pair is correct?

a Arid - < 0.02 MI
b Semi arid – 0.02-0.50 NI
c Humid - >0.65 MI
d All above

42 Ramkrishnan (1997) and Paul (1997) classified the rainfed areas into 5 zones. The zones and rainfall pairs are given below. Find out the correct pair?

a Arid - <500mm b Semi arid – 500 -750mm
c Dry sub humid – 750 -1100mm d All above

43 Ramkrishnan (1997) and Paul (1997) classified the rainfed areas into 5 zones. The zones and rainfall pairs are given below. Find out the correct pair?

a Arid - <500mm b Semi arid – 500 -750mm
c Moist sub humid – 1100 - 1400 d All above

44 Bansal and Minke (1988) classified Indian monsoon into

a 5 zones b 6 zones
c 7 zones d 8 zones

45 Bansal and Minke (1988) classified Indian monsoon based on

a Mean monthly temperature b Relative humidity
c Precipitation d All above

46 Which of the following places come under 'Composite' climate as per classification of Bansal and Minke (1988)?

a Kanpur b Delhi
c Allahabad d All above

47 Which of the following places come under 'Hot and Dry' climate as per classification of Bansal and Minke (1988)?

a Jaisalmer b Sholapur
c Jodhpur d All above

48 Which of the following places come under 'Warm & Humid' climate as per classification of Bansal and Minke (1988)?

a Mumbai b Chennai
c Kolkatta d All above

49 Which of the following places come under 'Cold & Cloudy' climate as per classification of Bansal and Minke (1988)?

a Shimla b Srinagar
c Shillong d All above

50 Which of the following places come under 'Cold & Sunny' climate as per classification of Bansal and Minke (1988)?

a Leh b Mount Abu
c Both 'a; & 'b' d None above

51 Which of the following places come under 'Moderate' climate as per classification of Bansal and Minke (1988)?

a Bengaluru b Dehradun
c Pune d All above

52 Singh and Venkateshvarulu (2009) classified Indian monsoon into 4 regions based on rainfall and rainfed area. Which of the following region receives the highest rainfall?

a Arid b Semi arid
c Sub humid d Humid

53 Singh and Venkateshvarulu (2009) classified Indian monsoon into 4 regions based on rainfall and rainfed area. Which of the following region has the highest rainfall?

a Arid b Semi arid
c Sub humid d Humid

54 Singh and Venkateshvarulu (2009) classified Indian monsoon into 4 regions based on rainfall and rainfed area. How much does arid region receive rainfall?

a <500 mm b 500-750 mm
c 750 – 1150 mm d >1150 mm

55 Singh and Venkateshvarulu (2009) classified Indian monsoon into 4 regions based on rainfall and rainfed area. How much does semi arid region receive rainfall?

a <500 mm b 500-750 mm
c 750 – 1150 mm d >1150 mm

56 Singh and Venkateshvarulu (2009) classified Indian monsoon into 4 regions based on rainfall and rainfed area. How much does sub humid region receive rainfall?

a <500 mm b 500-750 mm
c 750 – 1150 mm d >1150 mm

57 Singh and Venkateshvarulu (2009) classified Indian monsoon into 4 regions based on rainfall and rainfed area. How much does humid region receive rainfall?

a <500 mm b 500-750 mm
c 750 – 1150 mm d >1150 mm

58 CRIDA (2013) classified country monsoon on district levels and grouped into 6 zones. Which of the following is correct combination?

a Arid, Semi arid, Dry Sub Humid, Moist Sub Humid, Humid & Per Humid.

b Arid, Semi arid, Humid, Moist Sub Humid , Per Humid & Highlands

c Arid, Semi arid, Dry Sub Humid, Moist Sub Humid, Humid & Highlands.

c Arid, Semi arid, Dry Sub Humid, Moist Sub Humid, Humid & Composite.

59 Cocheme & Franquin classified climate into 5 zones based on precipitation and ET. What were the other considerations they made in their classification ?

a Crop growing period b Water balance

c Both 'a' & 'b' d None above

60 Cocheme & Franquin classified climate into 5 zones based on precipitation and ET. Which of the following region has P >ET?

a Humid b Moist

c Moderately dry d Dry

61 Cocheme & Franquin classified climate into 5 zones based on precipitation and ET. Which of the following region has P >ET/2 ?

a Humid b Moist

c Moderately dry d Dry

62 Cocheme & Franquin classified climate into 5 zones based on precipitation and ET. Which of the following region has P = ET/2 to ET/4 ?

a Humid b Moist

c Moderately dry d Dry

63 Cocheme & Franquin classified climate into 5 zones based on precipitation and ET. Which of the following region has P = ET/4 to ET/10 ?

a Humid b Moist

c Moderately dry d Dry

64 Cocheme & Franquin classified climate into 5 zones based on precipitation and ET. Which of the following region has P < ET/10 ?

a Humid b Moist

c Dry d Very Dry

65 UNEP (1992) classified global monsoon mainly based on

a Moisture index | b Aridity index
c Bothe 'a' & 'b' | d None above

66 Drought Prone Area Program was launched in year

a 1973-74 | b 1977-73
c 1982-83 | d 1985-86

67 Desert Development Program was launched in year

a 1973-74 | b 1977-78
c 1982-83 | d 1985-86

68 Planning commission divided India into how many agro-climatic zones;

a 10 | b 12
c 15 | d 20

69 NBSS&LUP-ICAR divided India into how many agro climatic zones?

a 10 | b 12
c 15 | d 20

70 NARP-ICAR divided India into how many agro climatic zones?

a 15 | b 20
c 127 | d 141

71 NATP-ICAR grouped rainfed districts if India into how many agro ecological regions?

a 04 | b 05
c 06 | d 07

72 Which of the following states are covered under Western Himalayan Region ?

a J&K | b HP
c Hill region of Uttarakhand | d All above

73 Which of the following states are covered under Eastern Himalayan Region?

a Arunachal Pradesh & Hills of Assam, Darjeeling district of WB | b Sikkim, Meghalaya, Nagaland
c Manipur, Mizoram and Tripura | d All above

74 Which of the following states are covered under Western Dry Region?

a Rajasthan & Western Aravallis
b Andaman-Nicobar & Lakshadweep

c Hills and Plains of Kathiawar & fertile valleys of Mahi and Sabarmati rivers

d None above

75 Which of the following states are covered under Island Region?

a Rajasthan & Western Aravallis

b Andaman-Nicobar & Lakshadweep

c Hills and Plains of Kathiawar & fertile valleys of Mahi and Sabarmati rivers

d None above

76 Which of the following states are covered under Gujarat Plains and Hills?

a Rajasthan & Western Aravallis

b Andaman-Nicobar & Lakshadweep

c Hills and Plains of Kathiawar & fertile valleys of Mahi and Sabarmati rivers

d None above

77 Which of the following considerations wer taken in agro climatic zonatio of India by NBSS&LUP?

a Growing period b Effective rainfall

c Soil types d All abobe

78 What is growing period of Western Himalayas region

a < 90 days b 150-180 days

c 180 - 210 days d >210 days

79 What is growing period of Northern Plains region?

a < 90 days b 150-180 days

c 180 - 210 days d >210 days

80 What is growing period of Eastern Gangetic plains region?

a < 90 days b 150-180 days

c 180 - 210 days d >210 days

81 What is growing period of Eastern Himalayas region?

a < 90 days b 150-180 days

c 180 - 210 days d >210 days

82 Which of the following is a normal arrival date of South-West monsoon in India ?

a 1st June b 1st September

c 5th June d 15th October

83 Which of the following is a normal date of final withdrawal of South-West monsoon from India ?

a 1st June
b 1st September
c 5th June
d 15th October

84 What is the normal date of starting retreat South-West monsoon from India?

a 1st June
b 1st September
c 5th June
d 15th October

85 Indian monsoon is

a Unimodal
b Bimodal
c Both 'a' & 'b'
d All above

86 Which of the following is correct for Indian monsoon?

a Erratic
b Uneven
c Deficient
d All above

87 What is starting date of North-East monsoon in India?

a 1st June
b 1st September
c 5th June
d 15th October

88 The highest rainfall is received by

a S-W monsoon
b N-E monsoon
c Pre monsoon
d Winter rains

89 Which of the following rainfall is common in India?

a Convectional rainfall
b Orographic rainfall
c Cyclonic rainfall
d All above

90 What is the intensity of light rainfall?

a 0-2.5mm/hr
b 2.6-7.6 mm/hr
c >7.6 mm/hr
d Not known

91 What is the intensity of moderate rainfall?

a 0-2.5mm/hr
b 2.6-7.6 mm/hr
c >7.6 mm/hr
d Not known

92 What is the intensity of heavy rainfall?

a 0-2.5mm/hr
b 2.6-7.6 mm/hr
c >7.6 mm/hr
d Not known

93 If rainfall of a particular year is less than 20 to 59%, the rainfall may be categorized as

a Deficient
b Normal
c Excess
d None above

94 If rainfall of a particular year is 20 to 59%, the rainfall may be categorized as

a Deficient
b Normal
c Excess
d None above

95 If rainfall of a particular year is -19 to +19%, the rainfall may be categorized as

a Deficient
b Normal
c Excess
d None above

96 Average annual rainfall of India is

a 85mm
b 105 cm
c 119 cm
d 205 cm

97 Normal feature of India's rainfall is

a It decreases westward in North India
b It decreases eastward in Peninsular India
c It increased in coastal regions
d All above

98 Winter rainfall occurs normally in

a North India
b East India
c South India
d All Above

99 By which date the S-W monsoon covers entire country?

a 1st July
b 15th July
c 15th September
d 15th October

100 Which of the following is correct in present scenario of Indian Monsoon?

a 5-8 days delay in onset
b 15-16 days delay from retreating
c Monsoon advances fast in different parts of the country after delayed onset
d All above

101 Pune receives much lesser rainfall mainly due to

a Located on the leeward side of Western Ghats
b Located on the windward side of Western Ghats
c Continentality
d Distance from the sea

102 In winter the western cyclonic disturbances originate from which sea?

a Mediterranean sea
b Indian sea
c Caspian sea
d Arabian sea

103 In which one of the following places could winter temperature drop to -45^{0}C?

a Shillong
b Pahalgam
c Drass
d Thiruvanthapuram

104 What generally happens to air temperature as we move from equator to the poles?

a Increase
b Decrease
c Remains constant
d None above

105 Which of the following latitudes passes through the middle of our country giving it the characteristic of tropical as well as sub tropical climate?

a Tropic of Capricorn
b Equator
c Tropic of Cancer
d 82^{0}3'N

106 The Indian sub continent experiences comparatively milder winter as compared to central Asia mainly due to

a The Himalayas
b Ocean currents
c Tropic of Cancer
d Surrounding sea

107 Which of the following atmospheric conditions govern the climate and associated weather conditions of India?

a Pressure and surface winds
b Western cyclonic disturbances and tropical cyclones
c Upper air circulation
d All above

108 Which of the following place receives the highest rainfall?

a Silchar
b Mombasa
c Mawsynram
d Guwahati

109 The wind flowing in northern plains is summer is known as

a Kaal Baisakhi
b Loo
c Trade winds
d None above

110 Which one of the following causes winter rainfall in northern-western parts of India?

a Cyclonic depression
b Retreating monsoon
c Western disturbances
d S-W monsoon

111 Monsoon arrives in India approximately in

a Early June
b Early July
c Early May
d Early August

112 Which of the following is not an element of weather and climate

a Atmospheric pressure
b Humidity
c Temperature
d Altitude

113 Which of the following place experiences highest diurnal range of temperature?

a Pahalgam
b Thiruvanthapuam
c Leh
d Jaisalmer

114 Which of the following factors affect the climate?

a Latitude
b Jet streams
c Altitude
d All above

115 Which of the following factors affect the climate?

a Himalayan mountains
b Distance from sea
c Western disturbances
d All above

116 El Nino effect is responsible for

a Floods
b Drought
c Humbolt current
d All above

117 La Nina effect is responsible for

a Strong wind flow
b Tropical cyclones
c Turn Pacific Ocean water colder
d All above

118 Westerly jet stream are responsible for

a S-W monsoon
b N-E monsoon
c Winter rains in India
d None above

119 Which of the following is correct for tropical cyclones?

a Originate from Bay of Bengal and Indian Ocean
b Bring high wind velocity
c Cause heavy rains in TN, AP and Orissa
d All above

120 Easterly jet streams are responsible for

a Winter rains in India
b Monsoon rains in India
c Both 'a' & 'b'
d None above

121 As per the new normal of monsoon data, monsoon covers the entire country by

a 1 July
b 8 July i.e. one week earlier
c 15 July
d 30 July

122 As per the new normal of monsoon data, monsoon starts withdrawing from country from around

a 1 September
b 17th September i.e. 2 weeks delay
c 20 September
d None above

123 As per the new normal dates of onset and retreat of monsoon, it takes total days for both onset and retreat around

a 28 days
b 45 days
c 51 dates
d 55 days

124 Rainfall is measured by

a Rainguage
b Evaporimeter
c Hygrometer
d anemometer

125 Humidity is measured by

a Rainguage
b Evaporimeter
c Hygrometer
d Anemometer

126 Evaporation is measured by

a Rainguage
b Evaporimeter
c Hygrometer
d Anemometer

127 Wind velocity is measured by

a Rainguage
b Evaporimeter
c Hygrometer
d Anemometer

128 Wind direction is measured by

a Wind Vane
b Wind sock
c Both 'a' & 'b'
d None above

129 Temperature and humidity are measured by

a Stevenson Screen
b Sunshine recorder
c Ceilometer
d Barograph

130 Sunshine hours are measured by

a Stevenson Screen
b Sunshine recorder
c Ceilometer
d Barograph

131 Cloud height is measured by

a Stevenson Screen
b Sunshine recorder
c Ceilometer
d Barograph

132 Atmospheric pressure is measured by

a Stevenson Screen
b Sunshine recorder
c Ceilometer
d Barograph

133 Atmospheric pressure is temperature is measured by

a Barograph
b Barometer
c Pressure sensor
d All above

134 Solar radiation is measured by

a Pyrheiometer
b Pyranometer
c Sun photometer
d All above

135 Soil temperature is measured by

a Digital Soil Thermometer
b Analog Soil Thermometer
c Long Stem Thermometer
d All above

136 Air temperature is measured by

a Thermometer
b Radiometric thermometer
c Meteorological thermometer
d All above

137 Soil temperature is important mainly for

a Germination of crop seed
b Growth of crop plants
c Crop maturity
d All above

138 Which of the following soil depths are set to measure the soil temperature?

a 5, 10 & 15 cm soil depth
b 10, 15 & 20 cm soil depth
c 15, 20 & 25 cm soil depth
d All above

139 Monsoon is important for

a Agricultural production
b Ground water recharge
c Economic growth
d All above

140 Which of the following agency is responsible for study of Indian monsoon?

a Indian Meteorological Department
b Indian Institute of Remote Sensing
c Ministry of Human Resource Development
d Ministry of Water Resource Development

141 Palmer drought severity index values vary in between

a 1 to -1 b 2 to -2

c 3 to -3 d 4 to -4

142 As per the PDSI, the value of extreme drought varies in between

a - 1.0 to -1.99 b -2.0 to -2.99

c - 3.0 to -3.99 d ≤ - 4.0

143 As per the PDSI, the value of severe moderate drought varies in between

a - 1.0 to -1.99 b -2.0 to -2.99

c - 3.0 to -3.99 d ≤ - 4.0

144 As per the PDSI, the value of moderate drought varies in between

a - 1.0 to -1.99 b -2.0 to -2.99

c - 3.0 to -3.99 d ≤ - 4.0

145 As per the PDSI, the value of mild drought varies in between

a - 1.0 to -1.99 b -2.0 to -2.99

c - 3.0 to -3.99 d ≤ - 4.0

146 Moisture Availability Index value of severe drought is

a < 25% b 20-50%

c >50% d >75%

147 Moisture Availability Index value of moderate drought is

a < 25% b 20-50%

c >50% d >75%

148 Moisture Availability Index value of crop growing season is

a < 25% b 20-50%

c >50% d >75%

149 Moisture Availability Index value of humid region is

a < 25% b 20-50%

c >50% d >75%

150 NDVI value under severe drought is

a < 25% b 20-50%

c >50% d >75%

Answers

1. **a**	2. **b**	3. **a**	4. **c**	5. **d**	6. **a**	7. **b**	8. **a**	9. **b**	10. **c**
11. **b**	12. **b**	13. **a**	14. **d**	15. **c**	16. **d**	17. **b**	18. **d**	19. **c**	20. **d**
21. **c**	22. **a**	23. **b**	24. **a**	25. **a**	26. **d**	27. **a**	28. **c**	29. **a**	30. **b**
31. **c**	32. **d**	33. **c**	34. **a**	35. **d**	36. **a**	37. **a**	38. **d**	39. **d**	40. **d**
41. **d**	42. **d**	43. **d**	44. **b**	45. **d**	46. **d**	47. **d**	48. **d**	49. **d**	50. **c**
51. **d**	52. **d**	53. **c**	54. **a**	55. **b**	56. **c**	57. **d**	58. **a**	59. **c**	60. **a**
61. **b**	62. **c**	63. **d**	64. **d**	65. **a**	66. **a**	67. **b**	68. **c**	69. **d**	70. **c**
71. **b**	72. **d**	73. **d**	74. **a**	75. **b**	76. **c**	77. **d**	78. **a**	79. **b**	80. **c**
81.**d**	82. **a**	83. **d**	84. **b**	85. **b**	86. **d**	87. **d**	88. **a**	89. **b**	90. **a**
91. **b**	92. **c**	93. **a**	94. **c**	95. **b**	96. **c**	97. **d**	98. **a**	99. **c**	100. **d**
101.**a**	102. **a**	103. **c**	104. **b**	105. **c**	106. **a**	107. **d**	108. **c**	109. **c**	110. **b**
111. **a**	112. **d**	113. **d**	114. **d**	115. **d**	116. **d**	117. **d**	118. **c**	119. **d**	120. **b**
121. **b**	122. **b**	123. **a**	124. **a**	125. **c**	126. **b**	127. **d**	128. **d**	129. **a**	130. **b**
131. **c**	132. **d**	133. **d**	134. **d**	135. **d**	136. **d**	137. **a**	138. **a**	139. **d**	140. **a**
141.**d**	142. **d**	143. **c**	144. **b**	145. **a**	146. **a**	147. **b**	148. **c**	149. **d**	150. **c**

4

Soils of Rainfed/Dryland Areas

Outlines

Introduction

Soil is generally defined as the upper layer of earth in which plants grow. It is regarded as a source of nourishment to all animals, moving or static. From agriculture or crop production point of view, soil is loose earth surface that supports vegetation, its quality, type and density. It is generally treated as a dynamic living entity and its physic-biochemical properties are responsible for its nutritional index. Precision production technology with suitable management is the key factors for its longer sustainability.

In dryland regions, nearly 30% soils are covered by Alfisols and associated soils, 35% by Verisols and associated soils and 10% by Entisols of the alluvial areas. Most of the Alfisols are sandy loams with a sub soil alluvial layer of clay, less water retentive and prone to drought. The Vertisols are deep (150-240 cm) with a clay content of 30-70% and rich in bases. The clay types are Smectute, Montmorillonite and Beideite. These soils have high water holding capacity, frequent fertile and less drought prone.

General characteristics of rainfed soils

i. Low soil fertility with multiple nutrient deficiencies,
ii. High erodibility,
iii. Fragile,
iv. Shallow depth,
v. Poor buffering capacity,
vi. Coarse textured soils,

vii. Highly degraded soil,
viii. Low WHC,
ix. Not suitable for intensive cultivation,
x. Excess salts (Saline and Alkaline soils) in arid and semi arid regions.

General characteristics of Dryland soils

i. Low soil fertility with multiple nutrient deficiencies,
ii. Low in OM,
iii. Alkaline to slightly acidic in nature,
iv. $CaCO_3$ accumulation in upper 150 cm soil layer,
v. Weak to moderate soil profile development,
vi. Coarse to medium texture with low biological activities,
vii. Poor WHC of soil,
viii. Crops prone to drought to nutrient deficiencies,
ix. Prone to wind erosion.

Major soil types

The nature of parent material, climate, vegetation, soil organisms and topography are the principle factors that affect soil formation in boith dry and humid regions. However the physical weathering, erosion and redeposition are most important factors in soil formation in arid region. Rainfall again plays important role and also affects the soil type and other properties of soil. The soils in semi arid areas of Deccan plateau are mainly Verisols (black soil) and Alfisols (red soils), while in the arid zone of Rajasthan, they are Arid soils and Entisols. Inceptisols and Mollisols are found in humid and sub humid regions (Srinivas *et al.*, 1999). The detail of taxonomical soil orders of rainfed area are given in table.1a & 1b and general characteristics in table. 2 below:

Table 1a. Major soil orders in India and total area

S.N	Soil order	Region/State	Area	
			Actual area (mha)	% value
1	Entisols (Alluvial soil)	Rajasthan, Gujrat, Haryana	80.1	24.4
2	Inceptisols (Alluvial soil)	UP	95.8	29.1
3	Vertisols (Black soil)	Kovilpatti (TN), Rajasthan, Gujrat, Maharashtra, MP, Karnataka, AP	26.3	8.02
4	Aridisols (Desert soil)	Haryana, West Rajasthan, Gujrat, Anantapur	14.6	4.45
5	Alfisols (Red soil)	AP, Jhansi (UP), TN, Karnataka	79.7	24.3

6	Oxisols (Laterite soil)	Ranchi (Jharkhand), Bhubaneswar (Orisa)	0.3	0.1
7	Mollisols	Tarai area of Uttarakhand, UP, Bihar	8.00	2.43
8	Ultisols (Lateritic soil)		0.80	0.24
9	Non Classified soils		23.1	7.01
	Total		328.7	100.00

Source: Prasad and Biswas (1979) & Srinivas et al. (1999)

Table 1b. Soil resources of India and soil related constraints (ICAR Classification)

S.N	Soil type	Area (mha)	% of total area	Soil related constraints
1	Entisols (Alluvial soil))	92.13	28.08	Erosion, nutrient depletion, low soil organic matter
2	Inceptisols (Alluvial soil))	130.37	39.74	Erosion, nutrient imbalance, low soil organic matter
3	Vertisols (Black soil)	27.96	8.52	Massive structure, poor tilth, drought stress, water erosion
4	Alfisols (Red soil)	44.29	13.55	Weak soil structure, crusting, compaction, water erosion
5	Aridisols (Desert soil	14.07	4.28	Drought stress, nutrient depletion, wind erosion, desertification, secondary salinization
6	Ultisols	8.25	2.51	Water erosion, nutrient imbalance, acidification, P fixation
7	Mollisols	1.32	0.40	Soil and water erosion, acidification, land slides
8	Histosols	0.002	-	High organic matter
9	Others	9.67	2.92	
	Total	328.06	100	

Source: Soils of India, National Bueau of Soils Survey and Land Use Planning, Publication No. 94 and Velayutham and Bhattacharya (2000)

Chief characteristics soil orders

- **Entisols** – Immature soils that lack the vertical development of horizons. These soils are often associated with recently deposited sediments from wind, water, or ice erosion. Given more time, these soils will develop into another soil type.
- **Inceptisols** – young soils that are more developed than entisols.
- **Vertisols** – heavy clay soils that show significant expansion and contraction due to the presence or absence of moisture. These are common in areas that have shale parent material and heavy precipitation.
- **Aridisols** – soils that develop in very dry environments.

- **Ultisols** – associated with humid temperate to tropical climates. Warm temperatures and the abundant variability of moisture enhance the weathering process and increase the rate of leaching in these soils.
- **Mollisols** – soils common to grassland environments.

Table 2. Major soil orders in India and their characteristics

A	**Entisols and Inceptisols (Alluvial soils)**		
	Entisols		Soil order comprising mineral soils that have not yet differentiated into distinct horizons (USDA Soil taxonomy)
	Inceptisols		• Soil order comprising draining soils in which the formation of distinct horizons is not far advanced ssuch as brown soils. • Found in all over world but prominent in mountain area • Good for crop production
	Type	1	**Light loam/Recent alluvium (Khadar)**
		i	Light in color, course texture, less or no horizon, less gravel (kankari)
		ii	Silt content 70-80%, moderate clay at top layer
		2	**Heavy loam (Old alluvium or Bhangar)**
		i	More clay content about 40%
		ii	No lime on top surface but accumulated in sub surface layers known as 'kankar' that varies with locations as Rohtak centre has 41.6% kankar in sub soil.
	Distribution		Indo-Gangetic plaians of India i.e. Punjab, Rajasthan, UP, plains of Uttarakhand, Bihar, WB, North West MP etc.
	Dryland Centres		Agra and Varanasi
	Crops		Most of the crops can be grown
	General Characteristics		
		i	Most important group of Indian soil
		ii	Formed from parent material transported by major rivers i.e. Indus, Ganga, Brahmputra and their tributaries.
		iii	High depth and more storage of water
		iv	Moderate WHC with high porosity
		V	High residual soil moisture
		vi	Low to medium OM and neutral to moderate alkalinity
		vii	Deficient N and P and Zn
		viii	Moderate lime but high at deeper soil alyers
		ix	Ratio of Fe-oxides and Al-oxides equal but some cases Al-oxides>Fe oxides
		x	Mn low at top soils and increases with depth
		Xi	Soils are deep and variable in texture
		xii	High agronomic potential
		xiii	Developed from sand and qartz.

	Crop production related problems		
		i	In Agra - high lime accumulation and sometimes presence of Kankar on upper surface that interferes with root development and other practices.
		ii	In Varanasi region – Salinity in low areas.
		iii	Crusting in upland light texure soil is a general problem.
B	**Alfisols (Red soil)**		
	Type	**i**	**Red loam:** cloddy structure, low percentage of sesquioxides
		ii	**Red earth with loose top soil:** Friable, high % of sesquioxides (red color)
	Distribution		AP, Bihar, Assam, TN, WB, Goa, Daman & Diu, parts of Kerala, Maharashtra, Karnataka and Bundelkhand region.
	Dryland Research centres		Jhansi, Rewa, Bijapur, Anantpur, Hyderabad, Bhubaneswar and Ranchi
	Major Crops grown		Groundnut, wheat, millets, pigeon pea etc.
	Characteristics	i	Red- reddish brown or yellowish brown color due to coating of ferric oxides on soil particles. Red color is due to diffusion of iron rather than high proporton of Fe.
		ii	Moderately weathered and derived from granites, gneiss and other metamorphic rocks or decomposed rock materials.The dominant clay material is *kaolinite* with an admixture of *illite*.
		iii	Light in texture, friable structure, moderate clay (10-20%), absence of lime and $CaCO_3$, low soluble salts, low WHC, shallow depth.
		iv	Slightly acidic to slightly alkaline, pH 6-7,medium CEC, near base saturated, low OM, N, P, S, Ca, Zn, sufficient K, dominant clay mineral is Kaolinite with mixture of Illite (Chemical properties)
		v	Excessive gravelliness, well drained with moderate permeability
		vi	Surface crust formation, compact sub surface
		vii	Susceptible to soil erosion at high slopes, produce large volume of runoff.
		viii	Rainfall: 750-2000mm.
	Crop production related problems		
		i	Restricted root growth due to presence of gravel and weathered rock fragments.
		ii	Crops are prone to suffer frequent drought spells.
		iii	Soil crusting on drying is a common feature of these soils, so crop growth suffers heavily during dry spells.
		iv	Tillage is difficult on drying of soil.
		V	Rainfall is poorly distributed and uncertain, so good crop establishment is always risky.
		vi	High surface runoff and erosion under naked soil conditions.

		vii	Leaching of nutrients is a common phenomenon due to its well draining capacity and requires frequent application of nutrients for good yield.
C	**Vertisols/Entisols (Black soil): 22% of total geographical area, Inceptisols - 95.8 mha (29.1%) and Vertisols - 26.3 mha (8.02%)**		
	Type	**1**	**Shallow black soils** – comes under order of Entisols and Inceptisols, normally found on slopes.
		i	<30cm rainfall
	Distribution	ii	Found mainly in Maharashtra, Bijapur and Gulbarga (Kanataka), Wardha, Nagpur, Ahmednagar and Madhy Pradesh-Nimach, Sagar and Jabalpur districts and parts of Andhra Pradesh, Gujarat and Tamil Nadu.
		iii	Silty loam to clay
		iv	Dark brown to yellow brown color
		v	Granular structure
		vi	Lime present in form of fine grains or nodules.
		2	**Medium Black soil (Entisols/Inceptisols)**
		i	30-100cm rainfall
		ii	Found in Bellary and Dharwar (Karnataka), parts of Mararashtra and MP, Coimbatore area.
		iii	Silty clay to clay texture, moderately rich OM and fairly well drained.
		iv	Lime present in form of thick layers of Calcarious nodules (20-30cm thick) in C horizon.
		3	**Vertisols (Deep Black soil)**
		i	>100 cm rainfall
		ii	Found at large part of Deccan plateau and Vidharb region and other parts of Maharashtra, Malva region and Bhopal (Nabada valley-MP), Bellary (Karanaka) and Bundelkhand region of UP
		iii	Generally known as Regur. Vertosols are more productive than Alfisols.
	Distribution		Central and Southern parts of the country i.e. Maharashtra, MP, parts of Gujrat, AP and TN
	Dryland Research centres		Bellary and Bijapur (Kanataka), Indore and Rewa (MP), Akola and Sholapur (MS) and Udaipur (Rajasthan)
	Major Crops grown		Cotton, maize, groundnut, millets, pulses, sunflower and safflower.
	Characteristics	i	Black soils are formed from Deccan basalt trap rocks either in-situ ot transported parent material.
		ii	**Physical properties:** Dark grey to dark black color, pastic and sticky when wet and very hard at drying, heavy texture with clay content 40-60%, high WHC.
		iii	**Chemical properties**: soil neutral to slightly alkaline, pH 7.5-8.5, low OM, low N, P, S and Zn and high K, Ca, Mg,Fe, Al high CEC.

		iv	Soil depth varies from a low of 50 cm to several meters and many cases it overlies decompose rock (parent material known as murram).
		v	High runoff 10-40%, prone to water erosion-sheet erosion.
		vi	Clay content from *Montmorillonite* group.
		vii	Swelling at wetting and shrinking and cracking at drying and blocks developed known as 'Self mulching/self ploughing. It is true black cotton soils.
		viii	Low infiltration rate leading to waterlogging, so drainage is required.
		ix	Lime present in form of nodules and accumulation below 150 m depth and develop 'hard pan'.
		x	Light rain doesn't penetrate to greater depth leading to waterlogging, so drainage is required.
		xi	Problem of salinity and alkalinity.
		xii	Normal 50-150 cm rainfall.
	Crop production related problems		
		i	More problem than red soils
		ii	Swelling and shrinking are common at wetting and drying, respectively, so at drying, large cracks or blocks are formed leading to loss of soil moisture and also creates problem in soil management.
		iii	High plasticity and stickiness are problematic in ploughing and intercultural operations.
		iv	Fallow soils are more prone to sheet erosion in Kharif season and run off loss may vary from 10-40% depending on rainfall intensity and slope.
		v	Waterlogging is a common problem in deep black soils as their infiltration rate is very low.
		vi	Salinity and alkalinity is also a problem in these soils.
		vii	Shallow black soils- crop production is possible in post rainy season.
		viii	Vertisols: double cropping is possible in rainy season as well as post rainy season.
		ix	Coarse grains, cotton, oilseeds and pulses may be grown.
		x	Soils are deficient in N, P, K and S and also other micronutrients.
D	**Oxisols (Laterite) and Ultisols (Lateritic soil): Oxisols-0.3 m ha (0.1%) and Ultisols -23.1 mha (7.01%) = 23.4 m ha (7.11%)**		
	Oxisols		Soil of an order comprising stable highly weathered tropical mineral soils with highly oxidized subsurface horizons. Important crops grown are coca, rubber and rice.
	Ultisols		Soil order comprising a highly weathered leached red or redish –yellow acid soil with a clay rich B horizon occurs in warm humid climate, CEC <35%, Ca, Mg, nitrate and K leached down.

	Type	1	**High level laterites -** found in high hills, shallow and gravely and unfit for agriculture.
		2	**Low level of laterites** –found in low hills and coastal plains, high clay and loam, better soil depth and good for agriclutre.
	Distribution		Summits of Western ghats, Kerala, Karnataka, Ghat of Orissa, Maharashtra , West Bangal, Tamil Nadu, Western Assam, Bihar, HP, Uttarakhand and parts of MP.
	Crops grown		**At lower elevation**-Rice, minor millets, trees, shrubs **High elevation**- plantation crops like tea, coffee, cinchona, rubber, cashew etc.
	Research centres		Dehradun, Bhubaneswar, Banagaluru.
	General Characteristics		
		i	Formed through process of laterization under high rainfall, silca is leached down and upper horizon becomes rich in Fe and Al oxides.
		ii	Reddish, yellow and grayish color, poor in plant nutrients in the top layer but lower layers have fine texture, darker and rich in organic matter and plant nutrients.
		iii	Mostly weathered soils
		iv	Low clay but no sticky – Kaolinite and Illite in traces.
		v	Low CEC, poor Ca and Mg., acidic in nature
		vi	Well drained with good hydraulic conductivity, acidic in nature
		vii	5th most important group.
		viii	**Upper soil** – poorly fertile, low OM and light color. **Lower soil** – Highly fertile, high OM and dark color.
		ix	Soil and water conservation measures required in low and medium rainfall regions
		x	Generally found in humid region under tropical and sub tropical region.
	Crop production related problems		
		i	In high level laterites (high elevation) like in Maharashtra and Kerala, plantation crops like tea, coffee, cinchona, rubber, cashew etc, are grown.
		ii	In low level laterites (low elevation) in Kerala, konkan region of Maharashtra and some parts of WB, paddy , minor millets, trees and shrubs are grown.
		iii	High level of fertilizer application is required for good crop.
		iv	Soil and water conservation measures are to be adopted.
E	**Aridisols (Desert soils) – 14.6 mha (4.45%)**		
	Type	1	Saline soil (Solonchak) or Black alkali soils.
		2	Alkaline (Solonetz) or white alkali soils.
		3	Saline & Alkaline soils.
	Distribution		Arid and semi arid regions : Rajasthan, Gujrat, Haryana and Punjab.

	Crops grown		Maize, Bajra, Guar, Sesame, Groundnut, Castor, Nigerseed, Rape & Mustard.
	Research centre		Hisar, Jodhpur, Dantiwala.
	General Characteristics		
		i	Desert, reddish desert Sierozem, Solonchalk, Solonetz groups **(Old classification)**
		ii	Rainfall-50-400mm (S-W monsoon)
		iii	PET and E are very high, high aridity index, major parts are covered by sand dunes and undulated soils.
		iv	Lime accumulation up to 60-120 mm deep.
		v	Very low clay content i.e. 2-8%, clay type is illite with small amount of kaolinite.
		vi	High Na% cause soil dispersion – poor permeability.
		vii	Soil pH 8-9, Low WHC, OM, N and P leading to poor soil fertility.
		viii	Found in low rainfall region.
		ix	Soils are fit for alternate land use systems.
		x	Least scope to increase crop productivity.
	Crop production related problems		
		i	Light textured, Low WHC and poor soil fertility, so low water and nutrient requiring crops are grown.
		ii	Salinity and alkalinity, so reclamation of these soils is essential for crop production, so tolerant crops are grown.
		iii	$CaCO_3$ in subsoils causing hard pan.
		iv	Deep soils-Kharif crops are possible.
		v	Fit for alternate land use systems.
F	**Sub Montane soils**		
	Distribution		Undulating land of sub Himalayan region under coniferous forest.
	Crops grown		Rice, wheat, ragi, madira, maize, amaranthus, rice bean, toria, soybean, buckwheat, tea, peach, plum, apple, apricot, citrus etc.
	Research centres		Dehradun, Ranichauri, Almora (Uttarakhand), Ooty (TN), Barapani (Assam).
	General Characteristics		
		i	Spread in hills and foothills of India.
		ii	High rainfall – 750-2000mm, so prone to erosion and runoff.
		iii	Adoption of soil and water conservation measures are essential.
		iv	Accumulation of OM at top soils.
		v	Acidic in nature.
		vi	Absence of free lime.
		vii	Landslides and soil erosion are common problems.
		ix	Leaching of bases-acidic reaction.
		x	Deficient in N and P

	Problems in crop production		
		i	Medium and high rainfall leading waterlogging problem.
		ii	Drainage is required.
		iii	Single and double cropping only
		iv	Soil moisture storage varies from 20-30% per metre of soil profile.
		v	Soil and water conservation measures are followed.
G	**Mollisols (Tarai soils)**		
	Distribution		Tarai belt of Uttarakhand, UP, Bihar.
	Crops grown		Rice, wheat, sugarcane, maize, sorghum, soybean, berseem.
	Research Centre		Dehradun and Almora.
	General characteristcis	i	Mollisols derived from latin word '*Mollis*', means soil of grassland ecosystems.
		ii	Characterizd a thick and dark surface horizon. The fertile surface horizon, known as Mollic Epipedon' is the result of long term deposition of orgnic materials derived from plant roots.
		iii	Derived from the materials washed down by the erosion of mountain (alluvial origin).
		iv	Heavy clay, coarse, sand and garvel, high water holding capacity.
		v	Sandy loam to silt loam in texture.
		vi	Rich in organc matter, plant nutrients but recently deficient in N and Zn.
		vii	Most of the crops can by grown with proper drainage.
		viii	Prone to water erosion due to heavy rainfall.
H	**Saline-Alkaline soils**		
	Type	i	***Saline /Solonchalk / White alkali soil***): ESP(%) - <15, EC (ds/m) - > and pH - <8.5, more Cl^-, SO_4^- and HCO_3^- ions, free carbonates absent., SAR <13
		ii	***Alkaline l/Solonetz / Black alkali soil***): ESP(%) - >15, EC (ds/m) - <4 and pH - >8.5, More Na^+ ions, more $CaCO_3$ at soil pH >9.5, more free carbonates, SAR >13
		iii	***Saline alkaline soils***: ESP(%) - >15, EC (ds/m) - >4 and pH - >8.5, SAR >13
	Distribution		8.5 mha area spreading in arid and semi arid region of the country where the temperature remains high and ET is higher than rainfall.
	Crops grown		Maize, pearlmillet, clusterbean, lentil, khesari, niger, raper & mustard, sesame, safflower, linseed.
	Research centres		Jhansi, Ayodhya, Jodhpur, Udaipur and Hisar .
	General Characteristics	i.	Accumulation of salts on soil surface like chlorides and sulphates of Na, Mg and Ca.
		ii	Poor organic matter.
		iii	Deficient in N and Zn.
		iv	Waterlogging due to low permeability.

		v	High aridity index.
		vi	Saline ground water.
		vii	Surface crusting, hard at drying and soft at wet conditions.
	Problems in crop production	i	Poor plant growth due to salt deposition.
		ii	Deficiency of plant nutrients.
		iii	Salt tolerant crops and varieties are not available.
		iv	Low yield.
I	**Acid soils**		
		i	49 m ha (pH >5.6 = 26 m ha and pH 6.5 = 23 mha).
		ii	Acid soil occupies nearly 8% of total geographical area in India.
	Definition	i	Acid soils are those that have a pH value of less than 5.5 for most of the year. They are associated with a number of toxicities (Aluminum) as well as deficiencies (Molybdenum) and other plant restricting conditions. Many of the acid soils belong to Acrisols, Alisols, Podzols and Dystric subgroups of other soils.
		ii	Acid soils are developed from acidic rocks like granite, gneiss, quartz and silica.
		iii	Laterization and podzolization in areas with sub temperate to temperature climate are responsible for acidic soils.
	Types	i	Ultra acidic soil: soil pH = 3.3
		ii	Extremely acidic soil: soil pH = 3.5 – 4.5
		iii	Very strong acidic soil: soil pH = 4.5 – 5.0
		iv	Strongly acidic soil: soils pH = < 5.1 – 5.5
		v	Moderately acidic soils: Soil pH = 5.6 – 6.0
		vi	Slightly acidic soil: soil pH = 6.1 – 6.5
	Distribution	i	Heavy rainfall regions like North-East states, J&K, Uttarakhand hills
		ii	Laterite, lateriric, peat, podzols and brown forest soils are acidic in nature.
	Crops grown		Oat, timothy, triticale, rye and wheat, raddish, sweet potato, beans, cabbage, carrot
	Research centres		Simla, Barapani, Imphal, Almora, Dehradun.
	General Characteristics	i	Light soil texture from sandy to loam
		ii	High permeability
		iii	Deficient in P, Ca, Mg
		iv	Low organic matter
		v	Poor water holding capacity
		vi	Kaolinitic and illitic are major clay minerals.
		vii	Low CEC and high base saturation
		viii	Toxic effects of Al, Fe, Mn, Zn, Cu and Co at soil pH.

	Problems in crop production		
		i	Soil acidity affects plant growth and soil fauna like Rhizobium adversely
		ii	Acidity caused deficiency of P, Ca, Mg and Mo (Fig.1).
		iii	Mineralization and nitrification processes reduced and decreased availability of N and other essential nutrients.
			Development of fungi more in acid soils and may cause root diseases.
	Management	i	Reclamation is required with addition of liming materials like dolomite, calcitic lime stones, brunt lime, hydrated lime, lime slug and stags but ground limestone (Ca CO_3) is common.
		ii	Avoid application of acidic fertilizers like Ammonium sulfate, Ammonium nitrate, sulfur coated urea etc.

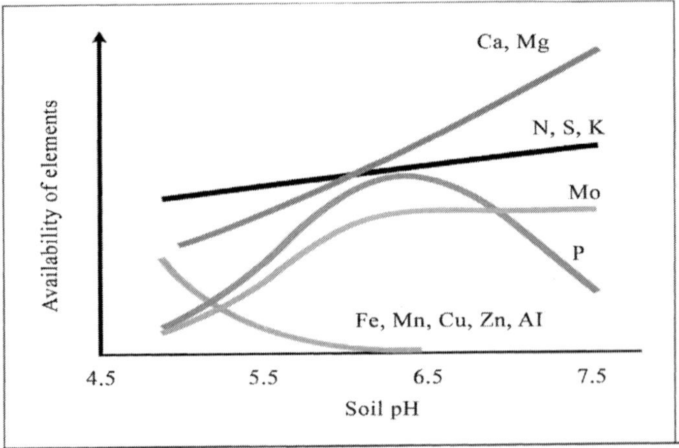

Fig. 1: Availability of nutrients vs. soil pH

(*Source*: https://www.slideshare.net/AbhilashPanju/acid-soil-and-their-management)

Table 3. Tolerant crops and tree species to saline-alkaline and acidic soils

S.N	Type of crop/plant	Saline soil	Alkaline soil	Acid soil
1	Highly sensitive	Legumes like green gram, black gram, lentil, gram and peas	Oat, Triticale, rye,	Alfalfa, sweet clover, barley, sugar beet
2	Moderately sensitive	Major and minor millets		Canola, wheat, corn

3	Highly tolerant	Sugar beet, rapeseed, cotton, kale, bermuda and rhode grass, date palm	Asparagus, sugar beet, barley, Wheat, cotton, , rice, grasses like rhodes grass, trees like amla, ber, eucalyptus hybrid, *Dalbergia sissoo*. *Terminalia arjuna*	Rice, oat, millets, field beans, soybean, pea, lentil, groundnut, cotton, timothy, triticale, rye, raddish, sweet potato, beans, cabbage, carrot, pepper, plum, strawberry, goose beery
4	Moderately tolerant	Alfalfa, flax, sunflower, sorghum, barley, oat, wheat, rice, oat, clover, barley, sudan grass, fig, grape, pomegranate, olive	Pea, cauliflower, kale, beans	Sugarcane, wheat, onion, tomato, jack fruit, orange, pineapple, litchi
5	Slightly tolerant	Pea, celery, cabbage, sweet potato, beans, egg plant, white and red clover, grape, plum, pear, apple, almond, orange, peach, lemon etc.	Tomato, pumpkin, garlic	Maize, sorghum, barley, mango, guava, banana, cashew

Soil classification and management in Ancient India
(Ref: Ancient History of Indian Agriculture)

1. **In Vedas**, soil has been described as the 'Mother' and human beings as her sons. 'Mother Earth' referred to as soil which is rich in energy, was thought to be activated by '*Parjanya*'(cloud) and '*Dyus*' or the sun.
2. **In Atharvaveda**, soil has been described as;
   i. Something in which the vegetation and agricultural crops grow and ripen,
   ii. Something upon which all the living beings are nourished, and
   iii. Something which provides man plenty of enjoument.
3. **Kashyapa (c. 800 AD) in his treatise ' Kashyapiyakrishisukti'stated**;
   i. Soil should have the substantial mass, devoid of bones and stones,
   ii. Soil should consist of clay with reddish and black hue,
   iii. Soil should not be deep or shallow,
   iv. Soil should be conductive to speedy seeding and emergence,
   v. Soil should be easy for ploughing,

vi. Soil should be capable of absorbing moisture easily, and

vii. Soil should be inhabited by beneficial living organisms.

Thus the significance of soil and its capacity to produce food/fuel/fodder etc. have been realized and understood from ancient times. Therefore soil is not only the medium of agriculture and forestry but it is also helpful in filtering, buffering and transformation between atmosphere and ground water, protecting the food chain and drinking water from pollution and also broadening biodiversity.

Questions

a. Short and Descriptive

1. Write short notes on following;
   a. Regur
   b. CRIDA
   c. CAZRI
   d. Self mulching in black soils
2. Differentiate between;
   a. Khadar vs. Bhangar
   b. Solonetz vs. Solonchalk
   c. Red loam vs. Red Earth
   d. Oxisols vs. Ultisols
   e. Entisols vs. Inceptisols
   f. Shallow black soils vs. Medium black soils
3. What are the major types of rainfed soils? Describe in brief the major problems of crop production in rainfed area.
4. What are major constraints of black soils in crop production? Also discuss how to improve crop production in black soils.
5. What are major constraints of alfisols in crop production? Also discuss how to improve crop production in alfisols.
6. Write down major constraints of aridisols in crop production. Also mention research institutes working on dryland areas.
7. Describe the process of laterization. Also mention the important characteristics of laterite and lateritic soils.

b. Objective(MCQs)

1 Which of the following is an important characteristic of rainfed soil?

a	Low soil fertility with multiple nutrient deficiencies	b	Low WHC
c	High erodibility	d	All above

2 Which of the following is an important characteristic of dryland soil?

a Low soil fertility with multiple nutrient deficiencies
b Low in OM
c Alkaline to slightly acidic in nature
d All above

3 Entisols in rainfed region occupies an area of

a 4.45%
b 8.02%
c 24.4%
d 29.1%

4 Inceptisols in rainfed region occupies an area of

a 4.45%
b 8.02%
c 24.4%
d 29.1%

5 Aridisols in rainfed region occupies an area of

a 4.45%
b 8.02%
c 24.4%
d 29.1%

6 Vertisols in rainfed region occupies an area of

a 4.45%
b 8.02%
c 24.4%
d 29.1%

7 Mollisols in rainfed region occupies an area of

a 0.1%
b 0.24%
c 24.4%
d 2.43%

8 Rainfed Ultisols occupies area of

a 0.1%
b 0.24%
c 4.45%
d 24.4%

9 Entisols are found mainly in

a Rajasthan, Gujarat, Haryana
b Kovilpatti (TN), Rajasthan, Gujarat, Maharashtra, MP, Karnataka, AP
c Haryana, West Rajasthan, Gujarat, Anantapur
d UP

10 Inceptisols are found mainly in

a Rajasthan, Gujarat, Haryana
b Kovilpatti (TN), Rajasthan, Gujarat, Maharashtra, MP, Karnataka, AP
c Haryana, West Rajasthan, Gujarat, Anantapur
d UP

11 Vertisols are found mainly in
- a Rajasthan, Gujarat, Haryana
- b Kovilpatti (TN), Rajasthan, Gujarat, Maharashtra, MP, Karnataka, AP
- c Haryana, West Rajasthan, Gujarat, Anantapur
- d UP

12 Aridisols are found mainly in
- a Rajasthan, Gujarat, Haryana
- b Kovilpatti (TN), Rajasthan, Gujarat, Maharashtra, MP, Karnataka, AP
- c Haryana, West Rajasthan, Gujarat, Anantapur
- d UP

13 Alfisols are found mainly in
- a AP, Jhansi (UP), TN, Karnataka
- b Tarai area of Uttarakhand, UP, Bihar
- c Ranchi (Jharkhand), Bhubneswar (Orisa)
- d Rajasthan, Gujarat, Haryana

14 Oxisols are found mainly in
- a Tarai area of Uttarakhand, UP, Bihar
- b AP, Jhansi (UP), TN, Karnataka
- c Rajasthan, Gujarat, Haryana
- d Ranchi (Jharkhand), Bhubaneswar (Orisa)

15 Mollisols are found mainly in
- a Tarai area of Uttarakhand, UP, Bihar
- b Kovilpatti (TN), Rajasthan, Gujarat, Maharashtra, MP, Karnataka, AP
- c Rajasthan, Gujarat, Haryana
- d UP

16 Khadar is also known as
- a Heavy loam
- b Light loam/Recent alluvium
- c Both a &b
- d None above

17 Bhangar is also known as
- a Heavy loam
- b Heavy loam /Old alluvium
- c Light loam
- d Red loam

18 Alfisols are known as
- a Red loam
- b Light loam
- c Heavy loam
- d All above

19 Vertisols are also known as
- a Shallow black soils
- b Black soil
- c Deep Black soil
- d None above

20 Oxisols are also known as

a Laterite soil b Lateritic soil

c Desert soil d All above

21 Ultisols are also known as

a Laterite soil b Lateritic soil

c Desert soil d None above

22 Aridisols are also known as

a Desert soils b Lateritic soil

c Laterite soil d All above

23 In Atharvaveda, soil has been described as;

a Something in which the vegetation and agricultural crops grow and ripen,

b Something upon which all the living beings are nourished

c Something which provides man plenty of enjoument.

d All of above

24 Kashyapa in his treatise ‘ Kashyapiyakrishisukti’stated soil as;

a Soil should have the substantial mass, devoid of bones and stones

b Soil should consist of clay with reddish and black hue

c Soil should not be deep or shallow

d All of above

25 High & Low level Laterites are type of

a Ultisols b Oxisols

c Both a & b d None above

26 Which of the following characteristic is found in Montane soils?

a Spread in hills and foothills of India. b Acidic in nature

c Deficient in N and P d All of above

27 Light loam and Heavy loam are types of

a Inceptisols b Alfisols

c Both a & b d None above

28 Which of the following characteristic (s) suits to Alfisols?

a Rainfall: 750-2000mm.

b Surface crust formation, compact sub surface

c Light in texture, friable structure, moderate clay (10-20%), absence of lime and $CaCO_3$, low soluble salts, low WHC, shallow depth.

d All of above

29 Which of the following is the type of Vertisols/Entisols?

a Shallow black soils
b Medium Black soil
c Deep Black soil
d All of above

30 Which of the following type belongs to Aridisols?

a Black alkali soils
b White alkali soils
c Both a & b
d None above

31 Which of the following research centre is working on Entisols and Inceptisols?

a Agra
b Varanasi
c Both a& b
d None above

32 Entisols and Inceptisols are deficient in

a N
b P
c Zn
d All above

33 Which of the following is a general problem of crop production in Entisols and Inceptisols?

a Crusting in upland light texure soil
b Salinity in low areas of Varanasi region
c High lime accumulation and sometimes presence of Kankar on upper surface in Agra region
d All above

34 Old alluvium or Bhangar has clay content around

a 10%
b 20%
c 30%
d >40%

35 The silt content in recent alluvium (Khadar) is around

a 20-30%
b 40-50%
c 50-60%
d 70-80%

36 Which of the following characteristics suit to Entisols and Inceptisols?

a Most important group of Indian soil
b Moderate WHC with high porosity
c High residual soil moisture
d All above

37 Which of the following research centre is working on Alfisols?

a Jhansi
b Bijapur
c Rewa
d All above

38 Alfisols receives rainfall approximately

a 650-2000mm b 750-2000mm

c 850-900mm d None above

39 Major Crop grown in Alfisols is

a Groundnut, b Wheat,

c Pigeonpea d All above

40 pH of Alfisols is

a 6-7 h 7-8

c 5-4 d None above

41 Which of the following is a general problem of crop production in Alfisols?

a Restricted root growth due to presence of gravel and weathered rock fragments

b Crops are prone to suffer frequent drought spells

c Tillage is difficult on drying of soil.

d All above

42 Shallow black soils receives rainfall approximately

a <30cm rainfall b <42cm rainfall

c <52cm rainfall d None above

43 Granular structure is found mainly in

a Shallow black soils b Medium Black soil

c Both a & b d None above

44 Shallow black soils are found in

a Bijapur and b Gulbarga (Karnataka),

c Nagpur d All above

45 Medium black soils receives rainfall approximately

a 40-100cm rainfall b 30-100cm rainfall

c 50-100cm rainfall d 60-100cm rainfall

46 Medium Black soils are found in

a Karnataka b Maharashtra

c Coimbatore d All above

47 Moderately rich OM soil is

a Shallow black soil b Medium Black soil

c Deep Black soil d None above

48 Deep black soils are generally known as

a Regur b Khadar

c Bhangar d All above

49 Deep black soil receives rainfall approximately

a >100 cm rainfall b >200 cm rainfall

c >300 cm rainfall d >400 cm rainfall

50 Which of the following dryland research centre is working on Vertisols?

a Bellary in Karnataka b Indore in MP

c Udaipur in Rajasthan d All of above

51 Major crops grown on Vertisols is

a Cotton b Groundnut

c Maize d All of above

52 Which of the following physical property is found in Vertisols?

a Dark brown to dark black color b Plastic and sticky

c Both a & b d None above

53 The clay content of Vertisols is approximately

a 30-60% b 40-60%

c 50-70% d 70-90%

54 Which of the following chemical property is found in Vertisols

a High OM b Low N, P, S and Zn

c High K, Ca, Mg d All above

55 pH of Vertisols ranges in between

a 7.5-8.5 b 4.6-5.5

c 6.5-7.9 d None above

56 Which of the following is a general problem of crop production in Vertisols?

a Salinity and alkalinity b Swelling and shrinking

c Waterlogging d All above

57 Which of the following research centre is working on Oxisols & Ultisols?

a Dehradun b Bangaluru

c Bhubaneswar d All above

58 Major crop grown in Oxisols & Ultisol is

a Cinchona, b Rubber

c Both a & b d None above

59 Which of the following characteristic suits to Oxisol & Ultisol?

a Humid region under tropical and sub tropical region.
b Reddish, yellow and grayish color.
c Acidic in nature
d All above

60 Which of the following is a general problem of crop production in lateritic soils?

a Poor soil fertility
b Soil and water erosion
c Acidic soils
d All above

61 Which of the following research centre is working on Aridisols?

a Hisar
b Jodhpur
c Dantiwala
d All above

62 Major Crops grown in Aridisols is

a Bajra,
b Sesame
c Groundnut
d All above

63 The clay content in Aridisols is around

a 2-8%.
b 3-6%
c 2-6%
d None above

64 Aridisols receives rainfall approximately

a 100-400mm
b 50-400mm
c 50-300mm
d All above

65 Which of the following characteristic belongs to Aridisols?

a Lime accumulation up to 60-120 mm deep.
b High Na% cause soil dispersion – poor permeability
c Soil are fit for alternate land use systems
d All above

66 Aridisols are found in

a High rainfall region
b Low rainfall region
c Both a & b
d None above

67 Aridisols are narmally

a Saline soil & Alkaline
b Acidic soils
c Wet soils
d All above

68 Which of the following is a general problem related to crop production in Aridisols ?

a Light textured, Low WHC and poor soil fertility
b Soil moisture stress
c $CaCO_3$ in subsoils causing hard pan
d All above

69 Which of the following research centre is working on Sub Montane soils?

a Dehradun b Ranichauri

c Almora d All above

70 Which of the following is a major crop of Sub Montane soils?

a Rice b Apple

c Both a & b d None above

71 Sub Montane soils are deficient in

a N b P

c Both a & b d None above

72 Sub Montane soils are

a Basic in nature b Acidic in nature

c Both a & b d None above

73 Which of the following is correct for Sub Montane soils?

a Spread in hills and foothills of India. b Accumulation of OM at top soils.

c Absence of free lime d All above

74 Sub Montane soils receive rainfall approximately

a 750-2000mm b 550-2000mm

c 650-2000mm d 450-2000mm

75 Which of the following is a general problem of crop production in Sub Montane soils?

a Waterlogging b Erosion

c Drainage d All above

76 Which of the following is an important characteristic of rainfed soil?

a Shallow depth b Poor buffering capacity

c Coarse textured soils d All above

77 Which of the following is an important characteristic of rainfed soil?

a Highly degraded soil b Low WHC

c Not suitable for intensive cultivation d All above

78 Which of the following is an important characteristic of dryland soil?

a $CaCO_3$ accumulation in upper 150 cm soil layer b Weak to moderate soil profile development

c Coarse to medium texture with low biological activities d All above

79 Which of the following is an important characteristic of dryland soil?

a Poor WHC of soil
b Crops prone to drought to nutrient deficiencies
c Prone to wind erosion
d All above

80 Which of the following is an important characteristic of dryland soil?

a Low soil fertility with multiple nutrient deficiencies
b Low in OM
c Alkaline to slightly acidic in nature
d All above

81 In Vedas, soil has been described as

a Mother
b Son
c Both a & b
d None above

82 Which of the following is not correct statement about alluvial soils?

a Alluvial soil is the most abundant in India
b Alluvial soil is generally fertile
c Alluvial soil lacks N and tend to be phosphoric
d Alluvial soil generally comprises high percentage of clay and retain moisture for long time

83 Which of the following soil is predominantly found in Rarh Region of West Bengal

a Entisols
b Vertisols
c Alfisols
d Oxisols

84 Under which climatic conditions do the laterite soils develop?

a Wet tropical climate
b Cold temperature climate
c Hot and dry climate
d Mediterranean type of climate

85 Which one of the following soil types of India is infertile by the presence of excess iron?

a Alluvial soils
b Lateritic
c Podzolic soils
d Desert soils

86 The transfer of minerals from top soil to subsoil through soil water is called

a Leaching
b Seepage
c Percolation
d Conduction

87 Black soils are ideal for cultivation of cotton because of

a Higher moisture retention capacity
b Found on plateau regions
c Made of lava
d Black color

88 Which of the following is not a tropical type of soil?

a Red soils | b Brown soils
c Black soils | d Alluvial soils

89 The another name of acid or alkaline soil is

a Regur | b Kallar
c Bangar | d Khadar

90 The soil of northern plain of India is formed by

a Aggradation | b Weathering in situ
c Degradation | d Erosion

91 Black cotton soil of the Deccan region of India is associated with

a Volcanic rocks | b Sedimentary rocks
c Plutonic rocks | d Hypabysal rocks

92 Which of the following statements regarding red soils of India is/are correct?,1. The colour of the soil is-red due to ferric oxide content.,2. Red soils are rich in lime, humous and potash.,3. They are porous and have friable structure.

a 1 only | b 1 and 3 only
c 2 and 3 only | d 1, 2 and 3 only

93 The composition of laterite soil in higher areas is

a Neutral | b Acidic
c Alkaline | d Saline

94 Which of the statements given are correct about the soils of India :,1. High temperature during most of the year retards formation of humus, 2. Soils which have humus require regular application of chemical fertilizers to remove nitrogen deficiency, 3. Titanium salts in basalt impart black, colour to the soils of the Indian plateau,4. Red soils are confined to the peripheral areas of the Indian plateau.

a 1 and 2 | b 2 and 3
c 2 and 4 | d 1, 3 and 4

95 Which of the following is the most important factor in soil formation?

a Climate | b Relief
c Natural vegetation | d Clay minerals

96 Red soils develop a reddish colour due to

a Presence of oxidized ferric oxides | b The presence of potash and magnesia
c Deforestation and over-grazing | d All above

97 Ca rich soils are known as

a Podzol | b Pedalfer

c Pedocal | d Laterite

98 Which of the following types of soils is found throughout India in all climatic zones?

a Aridisols | b Oxisols

c Alfisols | d Entisols

99 Which of the following statement related to soil types is incorrect?

a Vertisols are highly retentive of moisture and very productive

b Oxisols are found in highly rainfall areas

c Alfisols are rich in N, P and lime content

d Alluvial soils are easy to plough

100 Petrology is the study of

a Rock | b Soil

c Mineral | d Land

101 Which of the following clay mineral is dominated in alfisols?

a Kaolinite | b Montmorillonite

c Illite | d All above

102 Alfisols are derived from

a Granites | b Gneiss

c Metamorphic rocks | d All above

103 The color of alfisols is red due to presence of

a Coating of ferric oxides on soil particles | b Coating of Al-oxides on soil particles

c Coating of lead oxides on soil particles | d None above

104 Which of the following clay mineral is dominated in laterite soils?

a Kaolinite | b Montmorillonite

c Illite | d All above

105 Which of the following clay mineral is dominated in vertisols?

a Kaolinite | b Montmorillonite

c Illite | d All above

106 Which of the following clay mineral is dominated in aridisols?

a Kaolinite | b Montmorillonite

c Illite | d All above

107 Which of the following soil type is well drained with high hydraulic conductivity?

a Alfisols | b Vertisols
c Oxisols | d Aridisols

108 Soil crusting at drying is a major problem in

a Alfisols | b Vertisols
c Ultisols | d Aridisols

109 Which of the following is true for 'Laterization' process?

a Silica is leached downwards due to heavy rainfall | b Upper horizons are rich in oxides of Fe and Al
c Both 'a and 'b' | d None above

110 Which of the following crop is generally grown in Laterite soils at high altitudes?

a Plantation crops | b Cereals
c Oilseeds | d Pulses

111 Mollisols are characterized by

a High water holding capacity | b High organic matter
c Sandy loam to silty loam in texture | d All above

112 Sub –Montane soils are characterized by

a Undulated lands of sub Himalayan region | b Receives high rainfall
c Landslides and erosion | d All above

113 Which of the following is true for saline soils?

a pH < 8.5 | b EC > 4 ds/m
c ESP < 15 | d All above

114 Which of the following is true for alkaline (sodic soils?

a pH > 8.5 | b EC < 4 ds/m
c ESP > 15 | d All above

115 Which of the following is true for saline-alkaline soils?

a pH > 8.5 | b EC > 4 ds/m
c ESP > 15 | d All above

116 Which of the following is correctly matched?

a Solonchalk-Saline soils | b Solonchalk- White alkali soils
c Saline soils-White alkali soils | d All above

117 Which of the following is correctly matched?

a Alkali soils-Solonetz
b Solonetz-Alkali soils
c Solonetz-Black alkali soils
d All above

118 Which of the following is true for saline soils?

a Formed by accumulation of soluble salts
b SAR < 13
c Absence of free carbonates
d All above

119 Which of the following is true for alkaline soils?

a Formed by adsorption of Na^+ on exchange complex in place of Ca^{++} and Mg^{++}
b SAR > 13
c Presence of free carbonates
d All above

120 Which of the following crop is highly tolerant to saline soils?

a Barley
b Oat
c Pearlmillet
d All above

121 Which of the following crop is moderately tolerant to saline soils?

a Sunflower
b Tamarind
c Cotton
d All above

122 Which of the following crop is tolerant to alkaline soils?

a Barley
b Wheat
c Rice
d All above

123 Which of the following crop is highly sensitive to saline soils?

a Green gram
b Gram
c Lentil
d All above

124 Which of the following crop is moderately sensitive to saline soils?

a Sorghum
b Finger millet
c Maize
d All above

125 Which of the following tree plant is tolerant to alkaline soils?

a Amla
b Ber
c Eucalyptus
d All above

126 Which of the following characteristic(s) acid soils possess?

a Adsorption of H^+ and Al^{+3} on soil particles
b Low pH
c Low CEC
d All above

127 Which of the following crop is tolerant to acidic soils?

a Oat b Cowpea

c Triticale d All above

128 Acidic soils are in general deficient in

a Nitrogen b Phosphorus

c Molybdenum d All above

129 Which of the following fruit tree is tolerant to acidic soils?

a Pineapple b Mango

c Citrus d All above

130 Which of the following is an acidic fertilizer?

a Urea b Anhydrous ammonia

c Ammonium nitrate d All above

Answer

1. d	2. d	3. c	4. d	5. a	6. b	7. d	8. b	9. a	10. d
11. b	12. c	13. a	14. d	15. a	16. b	17. b	18. a	19. c	20. a
21. b	22. a	23. d	24. d	25. a	26. d	27. a	28. d	29. d	30. c
31. c	32. d	33. d	34. d	35. d	36. d	37. d	38. b	39. d	40. a
41. d	42. a	43. a	44. d	45. b	46. d	47. b	48. a	49. a	50. d
51. d	52. c	53. b	54. d	55. a	56. d	57. d	58. c	59. d	60. d
61. d	62. d	63. a	64. b	65. d	66. b	67. a	68. d	69. d	70. a
71. c	72. b	73. d	74. a	75. d	76. d	77. d	78. d	79. d	80. d
81. a	82. d	83. c	84. a	85. b	86. a	87. a	88. b	89. b	90. a
91. a	92. b	93. b	94. d	95. a	96. a	97. c	98. a	99. c	100. a
101. a	102. d	103. a	104. a	105. b	106. c	107. c	108. a	109. c	110. a
111. d	112. d	113. d	114. d	115. d	116. d	117. d	118. d	119. d	120. d
121. d	122. d	123. d	124. d	125. d	126. d	127. d	128. c	129. d	130. d

5

Aberrant Weather Situations and Contingency Crop Planning

Bread, cloth and house (Roti, Kapra and Makan) are the basic need of human life and linked directly with agriculture because agriculture provides livelihood for nearly two-thirds of the population in India and it currently accounts for 14.2% of the GDP and also employs 55% of country's total workforce. The sharp fluctuations in agricultural growth are mainly attributed to the vagaries of monsoon. Rainfed agriculture is spread over 55% of net cultivated area, contributing over 40% to the national agricultural production. Thus Monsoon rainfall is crucial for agriculture production and food security of the country. South West Monsoon (June to September) rainfall contributes to about 73% of total rainfall in the country. Timely onset and spatial distribution of rainfall is crucial for cultivation of *Kharif* crops that accounts for about 55% of upland rice, 87% of coarse cereals and pulses, 70% oilseed production and 65% of cotton of the country. Rainfall especially during the months of June and July are crucial for sowing of *Kharif* crops. Coastal areas of peninsular India, in particular Tamil Nadu receive bulk of their annual rainfall from the North-East Monsoons, between October and December. Resource poor farmers, the backbone of rainfed agriculture, were largely bypassed by green revolution whose benefits were mostly reaped by the already resourceful farmers holding irrigated lands. It is estimated that even after achieving the full irrigation potential, barring for successful completion of river linking project, nearly 45 to 50% of the total cultivated area will remain dependent on rain. Undoubtedly, the rainfed agriculture would continue to occupy a prominent place in Indian agriculture for a long time to come (Venkateswarlu *et al.*, 2011).

Drought is a common phenomenon of Indian agriculture as it is based on arrival and departure of the S-W monsoon including its total rainfall, intensity, distribution and sowing of *Kharif* crops. India experiences severe or partial drought every year in different parts of the country. Therefore the intensity of drought is monitored and based on its severity, the government declared relief packages to the affected regions. The Crop Weather Watch Group in the Department of Agriculture, Cooperation and Farmers' Welfare (DAC & FW&FW) and the Drought Monitoring Centres (DMCs) under the State Disaster Management Authorities should meet frequently, and at least once a week during the South-West Monsoon season to monitor portents of drought. Normally the central and State Governments monitor following parameters that may indicate an onset of drought;

1. Rainfall,
2. Progression of sowing,
3. Remote sensing based vegetative indices,
4. Soil moisture based indices,
5. Hydrological indices.

Advance preparation for drought by Central and State Governments

The state government is solely responsible for managing the drought. Both state and central governments to take following necessary action after an assessment of vulnerability / risk of drought at district / sub-district levels:

i. Updated District-wise Crop contingency plans (CAP) kept in readiness and available with extension functionaries at Block /Tehsil/Mandal/Taluka levels.

ii. Sufficient provisioning for drought resilient varieties of seeds at appropriate hubs.

iii. Cultivation of drought resistant fodder crops & crop varieties.

iv. Sufficient shelf of projects and funds under MGNREGA.

v. Contingency Plans for cattle care, water, health, energy and food kept in readiness at the operational/functional levels.

vi. Make available the fodder, feed and water for each Livestock.

vii. Repair of Hand Pumps and drinking water schemes through flexi funds & state's own resources.

viii. Adequate no. of water tankers and water kiosks are to be kept in readiness.

ix. Repair of Transformers in rural areas.

x. Desiltation /repair of canals.

xi. Identification of sites / organizations for cattle camps.

xii. Increasing and strengthening the number of Fodder Banks.

xiii. Collect the storm water in nallahs and store in natural or artificial ponds, tanks, recharge pits etc. to recharge ground water.

xv. Ensure issue of advisories to farmers from time to time.

Climate Variability and Agriculture Impacts

The climate change and globalization are two important factors affecting world economy but its impact on agriculture is witnessed all over the world, but countries like India are more vulnerable in view of the high population depending on agriculture and excessive pressure on natural resources. The warming trend in India over 0^0C the past 100 years (1901 to 2007) was observed to be 0.51^0C with accelerated warming of 0.21^{0}C per every 10 years since 1970 (Krishna Kumar, 2009). The projected impacts are likely to further aggravate yield fluctuations of many crops with impact on food security and prices. Climate change impacts are likely to vary in different parts of the country. Parts of western Rajasthan, Southern Gujarat, Madhya Pradesh, Maharashtra, Northern Karnataka, Northern Andhra Pradesh and Southern Bihar are likely to be more vulnerable in terms of extreme events. The rainfall analysis of India showed that significant negative trends of rainfall were observed in the eastern parts of Madhya Pradesh, Chhattisgarh and parts of Jharkhand, U.P. and northeast India. Lal *et al.* (2001) reported that annual mean area-averaged surface warming over the Indian sub-continent is likely to range between 3.5 and 5.5^{0}C by 2080s. These projections showed more warming in winter season over summer. The spatial distribution of surface warming suggests a mean annual rise in surface temperatures in North India by 3^{0} C or more by 2050. The study also indicated that during winter, the surface mean air temperature could rise by 3^{0}C in Northern and Central parts, while it would rise by 2^{0}C in Southern parts by the same period. In case of rainfall, a marginal increase of 7 to10% in annual rainfall is projected over the sub-continent by 2080.

Rainfall is the key variable influencing crop productivity in rainfed farming. Intermittent and prolonged droughts are a major cause of yield reduction in most crops. Long term data for India indicates that rainfed areas witness 3-4 drought years in every 10 year period. However, no definite trend is seen on the frequency of droughts as a result of climate change so far. For any R&D and policy initiatives, it is important to know the spatial distribution of drought events in the country. Along term analysis by CRIDA of rainfall trends in India (1901 to 2004) using 'Mann Kendall Test of Significance' indicate significant increase in rainfall trends in West Bengal, central India, coastal regions, south western Andhra Pradesh and central Tamil Nadu. Significant decreasing trend was observed in central part of Jammu Kashmir, northern MP, central and western part of UP, northern and central part of Chattisgarh. Analysis of number of rainy days based on the IMD grid data from 1957 to 2007 showed declining trends in Chattisgarh, Madhya Pradesh, and Jammu Kashmir. In Chattisgarh and eastern Madhya Pradesh, both rainfall and number of rainy days are declining which is a cause for concern as this is a rainfed rice production system supporting large tribal population who has poor coping capabilities (Rao, 2013).

Since the beginning of 21st Century, India has experienced droughts in quick succession, of which the 2009 was the most recent one that significantly affected Kharif crop production. It was the second largest all India monsoon rainfall deficit since 1972 (23% below normal). Incidentally, 2009 also achieved the distinction of being the warmest year in past several centuries across the world. However, after 2009, 2010 proved to the warmest year on record since 1850. 2011 is now the 11 warmest year on record since 1850. Apart from that, 1998 was one of the warmest years; 2003 experienced unprecedented heat and cold waves across the globe; occurrence of high temperature in March 2004 adversely affected crops like wheat, apple, potato etc. across northern India; 2005 witnessed destructive hurricanes/cyclones across the globe and again, 2007 was as warm as 1998 in the entire northern hemisphere and unusual summer rains and floods were experienced in many parts of India.

Similarly the number of studies has projected the decline trend of most of the crops in future but on the other hand it was also stated that the yield of winter crops particularly in western countries and high hills may increase due to climate change as the temperature will rise. The trend of projected yield of few crops is given below;

Table 1. Projected crop yield in 2035, 2065 and 2100 AD

S.N	Crop	2035	2065	2100
1.	Rice	-7.1	-11.5	-15.4
2.	Maize	-1.2	-3.7	-4.2
3.	Sorghum	-3.3	-5.3	-7.1
4.	Pigeon pea	-10.1	-17.7	-23.3
5.	Groundnut	-5.6	-8.6	-11.4
6.	Wheat	-8.3	-15.4	-22.0
7.	Barley	-2.5	-4.7	-6.4
8.	Chickpea	-10.0	-18.6	-26.2
9.	Rapeseed-mustard	0.3	0.7	0.5

Note: Maximum changes in temperature and rainfall are .30 and 7% by 2035, 2.50C and 26% by 2065 and 3.50C and 27% by 2100, respectively (*Source*: Birthal *et al.*, 2014).

Recently it has been observed that the amount and distribution of rainfall is becoming more and more erratic which is causing greater incidences of droughts and floods globally. From last few years, it is a clear-cut indication that monsoon has been delayed by 15-20 days and the number of rainy days has also decreased in most parts of India. Again the distribution of rainfall is very erratic and most of the rain is coming in few days with high intensity causing heavy loss of soil and water. It is a serious concern for agricultural production.

Impacts of weather variability on Agriculture

1. Rainfall drives water availability and determines sowing time (rainfed crops),
2. Temperature drives crop growth and duration,
3. Temperature and Relative Humidity influence pest and diseases incidence on crops,
4. Radiation influences the photosynthetic productivity,
5. Wet and dry spells cause significant impact on standing crops, physiology, loss of economic products (eg. fruit drop) and
6. Extreme events (eg. high rainfall/floods/heat wave/cold wave/cyclone /hail/ frost) cause enormous losses of standing crops.

Aberrant weather

Unfavorable deviated weather/monsoon from the normally expected one is called aberrant weather. Drought, cyclones, hailstorm, flood, heat waves and sea water inundation are the main causes of aberrant weather condition which is a common feature of rainfed agriculture. Crop production in dry areas is mainly dependent upon rainfall, which is received during the south west monsoon. The rainfall is seasonal, erratic and highly variable with space & time. Vagaries of monsoon rains and breaks of varying duration are most common in India. Anomalies in the distribution of south west monsoon rainfall are well known.

Main features of rainfall distribution in India

1. There is large variation in the dates of commencement of the southwest monsoon rainfall from year to year in different parts of the country.
2. The monsoon rainfall is of sequential nature with dry spells interspersed with wet spells, dry spells extending from few days to several weeks.
3. There is a large year to year variation in the dates of cessation of south west monsoon rainfall in different parts of country.
4. There is variation in the quantum of rainfall received from year to year and the coefficient of variation of rainfall increases with decrease in monsoon rainfall.
5. High intensity rainfall occurs in association with the movement of cyclones/ depressions resulting in sizeable loss of rain water through run-off and deep drainage.

Drought, floods, cyclones, heat waves, hailstorms and sea water inundation are the major factors or aberrant weather that will affect field crops, horticulture, livestock, poultry and fisheries as given in fig.1 below;

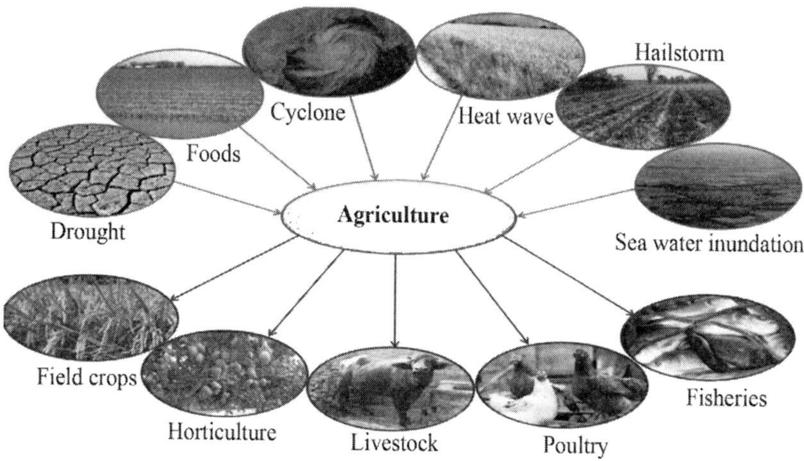

Fig. 1: Weather Aberration and affected agriculture and allied sector

Three important aberrations in the rainfall behavior have been commonly observed as given below;

1. The commencement of rains may be quite early or considerably delayed,
2. There may be prolonged breaks during the southwest monsoon season during which most of the dryland crops are grown, and
3. Rains may terminate earlier than normal cessation date or may continue beyond the normal rainy season.

Weather is the most variable factor in rainfed crop production. Therefore the contingent plans have been evolved for different types of aberrant weather conditions for efficient weather management. The infrastructure requirements have to be developed and established by the government and other agencies as people themselves will not be in a position to adopt contingency measures more effectively because of their poor resource base. Some of the management options that can be adopted to mellow down the adverse effects of aberrant weather are discussed here.

Contingency Crop Planning

Contingency planning refers to mitigate any unexpected, unusual, unfavorable and hence unwanted accidental or aberration weather situations occurring without prior knowledge at any time during crop period i.e. before the crop sown or after crop sown. The contingency crop planning therefore is proposed to mitigate such situation through the choice of appropriate crop and varieties, cropping systems or other necessary relevant crop management interventions. To develop a contingency plan for an area, a detailed study of the rainfall data should be done first. Based on this rainfall data, the planning should be done.

The District Agricultural Contingency Plans (DACP) are technical documents aimed to be ready reckoner for line departments and farming community on prevailing farming systems and technological interventions to manage various weather aberrations such as droughts, floods, cyclones, hailstorms, heat and cold waves addressing different sectors of agriculture including horticulture, livestock, poultry, fisheries. The contingency plans are useful for preparedness and real time implementation towards sustainability agriculture production system in the events of weather aberrations and extreme climatic events. The district contingency plans are prepared by ICAR-CRIDA (nodal institute) along with other institutes of Natural Resource management Division of Indian Council of Agricultural Research (ICAR) and State Agricultural Universities and KVKs under the overall guidance and supervision of ICAR and Department of Agriculture and Cooperation (DAC). Total 614 district agriculture contingency plans are placed in the 'farmer portal' of the Ministry of Agriculture and Farmers Welfare, Government of India (http://www.farmer.gov.in) and also in the ICAR / CRIDA website (http://www.crida.in) for downloading the full plan by stakeholders for operational use. The district crop contingent plans have been proposed under following weather aberrations;

1. Delay in monsoon onset
2. Breaks in monsoon leading to early, mid and late-season droughts
3. Delayed or limited release of water for irrigation
4. Floods, Unseasonal rains
5. Extreme weather events: Heat wave, Cold wave, Frost, Hailstorm, Cyclone

To make it more simple and adoptable, the crop contingent plan for following weather aberrations has been proposed in this chapter;

1. Early onset of monsoon.
2. Late/Delayed onset of monsoon.
3. Normal onset of monsoon followed by immediate prolonged dry spell.
4. Early cessation of rainfall.
5. Extended monsoon.
6. Long dry spells during crop period.
7. Long gap in rainfall,
8. Inadequate and uneven rainfall.
9. Submergence/Flood and waterlogging due to heavy rain and water accumulation in low lands.
10. Extreme weather like heat wave/cold wave/hailstorm/frost.

1. Crop Contingent Plan-1: Early onset of monsoon

1. Grow long duration crops like sorghum, bajra, sesame, maize etc.,
2. Follow soil and water conservation measures,
3. Recycle water for life saving irrigation,
4. Adopt mid season correction if required,

2. Crop Contingent Plan-2: Delayed/Late onset of monsoon

During some of the years the onset of south west monsoon gets delayed so that the crops/varieties which are regularly grown in the region cannot be sown in time. Delayed sowing of the crops can lead to reduced and even uneconomical crop yields. Under these situations, following measures should be adopted;

1. **Transplantation-** Under delayed monsoon, the crops may be grown in advance in small pocket with assured irrigation so that the seedling may be ready by the delayed onset of monsoon. The ready seedlings are transplanted just after onset of monsoon. It will save the time and also crop will get full growing period. This is the surer way of compensating the delay in the commencement of monsoon. Rice, bajra, fingermillet etc. can be transplanted without much effort. The maize may be grown in small polythene bags and 20 days old seedlings may be transplanted under optimum soil moisture conditions followed by light irrigation.

2. **Sowing of alternate Crops/Varieties** - Certain crops/varieties can perform better even if sown late in the season. Depending on the receipt of rains, such crops and varieties can be selected. Maize, Bajra, Black gram, green gram, Cluster bean, cowpea, Groundnut, Castor (Aruna), sunflower, nutricereals etc. may be sown varying with receipt of how much delayed monsoon. Some of the short duration varieties are given below;
3. **Dry sowing**: Some hardy crops like rice are sown after ploughing the field. The crops germinate after onset of monsoon. In hills, Chetki Dhan is sown in April just after harvesting of Rabi crops and it is germinated after pre or monsoon rainfall. It is a very common practice in Uttarakhand hills.
4. Increase seed rate to compensate low field emergence,
5. Complete weed control to avoid water loss,
6. Life saving irrigation in standing crops including drip and sprinkler irrigation.
7. Mixed/Intercropping is again a good option.
8. Addition of organic manures (FYM/compost/vermicompost).

Table 2. Important short duration and drought tolerant varieties of some crops

S.N	Crop	Short duration varieties
1	Blackgram	PU-35, Shekhar, Pusa Vishal, CO-6, Samrat, Azad Urd-1
2	Green gram	PDM-139, VBN-3, BM-2003-2, Pratap
3	Groundnut	TG 37A, TG 38, ICGV-91114, JGN-23
4	Pigeon pea	TJT-501, ICPL-88039, BRG-2, BRG-4, Asha
	Pearlmillet	GHB-558, ICTP-8203
5	Castor	NPH-1, GAUCH-1, YRCH-1, TMV-5, TMV-6, CO-1, TMVCH., GCH-7
6	Sesame	Shekhar
7	Clusterbean	GG2

3. Contingent plan-3: Dry spell immediately after normal sowing

Dry spells are common in dryland agriculture. Due to dry spell after sowing of crop it might result in poor germination due to soil crusting, withering of seedlings and poor establishment of crop stand. The adverse effect of moisture stress should be mitigated to avoid total crop failure. Under such conditions, following measures should be followed;

1. If germination is very poor and plant population is not optimum, the field should be ploughed and go for resowing of the crop after receipt of rainfall,
2. If the dry spell is long, try to choose short duration crops like legumes and oilseeds,

3. Foliar spray of urea @1-2%, Thiourea@0.5-1.0%, KNO3@2% and KCL@1% in pearlmillet, soybean and maize, respectively may be recommended under water stress condition.
4. Life saving irrigation may be given in mid season dry spells and terminal drought.
5. Zero tillage is also effective to avoid terminal drought at zero tillage has 15-20 days early sowing , low weed population and also saves 10-15% irrigation water.

4. Contingent plan 4: Early cessation of monsoon

Early withdrawal of monsoon is responsible for mid season and terminal drought that may cause severe crop loss because the crops may suffer water stress at flowering and grain filling stages. Further the sowing of *Rabi* season crops may also be affected adversely. Therefore it is required to adopt all possible measures to mitigate the adverse effect of prolonged dry spells during crop period so that standing crop may be saved from severity of water stress. Following measures may be adopted;

1. Sowing of short duration crops,
2. Reduce plant population by thinning,
3. Mulching,
4. Use of antitranspirants, and
5. Life saving irrigation.
6. Harvest standing crops as early as possible and plant Rabi season crops quickly so that maximum residual soil moisture can be tapped.

5. Crop Contingent Plan 5: Extended monsoon

Extended monsoon is seldom a problem in dryland agriculture. It benefits long duration *Kharif* crops taken as a component crops in intercropping system and is also advantageous for *Rabi* crops. The prediction of extended monsoon is little difficult, therefore following measures can be adopted;

1. Harvest *Kharif* crops as early as possible,
2. Early sowing of *R*abi crops to utilize residual soil moisture,
3. Stored water can be recycled for life saving irrigation.

6. Crop Contingent Plan 6: Long dry spells during crop period

The crop production in rainfed areas is always very risky and uncertain mainly because of we are not assure about the rainfall, its distribution and also dry spells. The rainfall may be very high, very low at any time of crop period after seedling emergence to crop maturity. Similarly the dry spells may come or may not come. The dry spells may come after sowing, at tillering or peak growth stage, flowering stage or even at grain filling stage. The duration of dry spells may be of 10 to 30 days. Under such type of weather aberrations, specific strategy varying with agro-climatic conditions is to be worked for mid season corrections to minimize the severity of drought as given below;

a. ***If dry spells comes within 10 days of sowing*** - Under this condition, the seedling emergence may be poor or the plant population will not be optimum, hence the expected yield will not be economic. Therefore resowing of same or other alternate crop is recommended.

b. ***If mild moisture stress at 30-35 days after sowing***- Under prevailing conditions, the soil moisture is supposed to be not sufficient to support the standing crop population, therefore thinning of alternate rows of sorghum and pearlmillet is recommended to reduce the plant stand. Intercultural practices, mulching, spray of water or 2% urea spray may be recommended to maintain the soil moisture regime for supporting crop growth.

c. ***If sever moisture stress at 30-35 days after sowing***- If the crop experiences severe drought at 30-35 days after sowing, it means the crop is at its peak vegetative stage or pre flowering stage. If the crop is drying due to drought and there is no hope of rain or life saving irrigation, following crop plan is recommended;

   i. Cutting of sorghum and pearlmillet is recommended for green fodder and allow for ratooning. It has been observed that sugarcane crop was harvested for green fodder due to severe drought in Pune region of Maharashtra.

   ii. Early Rabi crops like toria, lentil, sunflower and some vegetables if possible may be opted for sowing so that residual soil moisture can be utilized.

d. ***If moisture stress at blooming stage***- Same as above.

In general, following contingent plan or mid season corrections is recommended for prolonged dry spells;

a. **Planting of drought tolerance crops and varieties** (Table.3).

b. **Ratooning** - If the drought occurs at 40- 50 DAS, reduction in leaf area either by ratooning or thinning plant population can mitigate the adverse effects of drought. Under above case, following contingent crop plan may be followed;

   i. The crops like sorghum, maize, pearlmillet may be harvested and use as a green fodder.

   ii. Sorghum and pearl millet may be allowed for ratooning.

   iii. Spray water or 2 % urea at 10-15 days intervals may save the crop like castor, pigeon pea and groundnut.

iv. Remove the lower old leaves up to half of the plant so that transpirational losses may be minimized without affecting the yield.

c. **Mulching** – Mulching is a effective tool to reduce evaporation losses. Repeated intercultural practices have been found very effective in black soils where cracks develop under drying the soil surface. The crop residues or straw or polythene may be used for mulching.

d. **Thinning** – Thinning or uprooting plant population up to 30-50% is recommended under severe drought at advance crop stage like pre-flowering stage or flowering stage. Therefore every third plant or alternate row may be removed to reduce the soil moisture loss. It has been observed that the soil moisture can be saved upto 35-50 days by proper thinning of the plant population.

e. **Weed Control** - Weed control is very essential to save the crop from onslaught of drought as weeds rob soil moisture and nutrients as well.

f. **Rain water harvesting and recycling** – Rainwater harvesting is always recommended as it will not only supply safe water for crop and human beings but also reduce the runoff losses. Therefore *in situ* i.e. inter plot and intra plot water harvesting should be followed for more retention of rainwater, recharging of ground water and also life saving irrigation.

g. **Removal of Stripping of lower leaves** - Generally, photosynthesis occurs in the upper 3-4 leaves. Hence the defunct lower leaves from the plant up to 50-60% of the plant height may be removed to reduce the transpirational losses.

h. **Intercropping and risk distribution** – Mixed or intercropping is insurance in dryland agriculture, so the crop combinations are chosen based on pattern of rainfall, growing period and socio-economic status of the region. The rainfall pattern of the region may fall in following categories;

1. Rainfall may be more uncertain in the early part of season.
2. Rainfall may be more uncertain in the later part of season.
3. Rainfall may be more or less uniformly distributed (in other words, the risk is uniformly distributed.

Table 3. Drought resistant varieties of some important crops

S.N	Crop	Important drought resistant varieties
1	Rice	Sahabhagi Dhan, Vandana, Anjali, Satyabhama, DRR Dhan-2 (IR64 Drt 1), DRR Dhan-43, Birsa Vikas Dhan-203, Birsa Vikas Dhan-111, Rajendra Bhagwati, Jaldi Dhan-6
2	Wheat	PBW-527, HI-1531, HI-8627, HD-2888, HPW-349, PBW-644, WH-1080, HD-3043, PBW-396, K-9465, K-8962, MP-3288, HD-4672, NIAW-1415, HD-2987
3	Maize	Pusa Hybrid Makka-1, HM-4, Pusa Hybrid Makka-5, DHM-121, Buland,

4	Sorghum	CSH-19 R, CSV-18, CSH-15R
5	Pearl Millet	HHB-67 improved, GHB-757, GHB-719, Dhanshakti, HHB-234, Mandor Bajra Composite-2, HHB-226, RHB-177, Pusa Composite-443,
6	Barley	RD-2660, K-603
7	Chickpea	Vijay, Vikas, RSG-14, RSG-888, ICCV-10, Pusa-362, Vijay
8	Groundnut	Ajaya, Girnar-1, TAG-24, Kadiri 6, ICGV 91114
9	Soybean	NRC-7, JS 95-60
10	Sugarcane	Co-98014 (Karan-1), Co-0239, Co-0118, Co-0238, Co-06927, Co 0403, Co-86032
11	Cotton	HD-324, CICR-1, Raj DH-7, Jawahar Tapti, Pratap Kapi, Suraj, Surabhi, Veena, AK- 235,
12	Jute	JBO-1 (Sudhangsu), JRO-204, JRO-524, JRC-80

Source: NARS-ICAR (2015)

Therefore, following intercropping systems have been found more remunerative (Table.4);

Table 4. Remunerative intercropping systems for different dryland areas

S.N	Region	Intercropping systems
1	Hisar	Pearlmillet + Cowpea
2	Agra	Pearlmillet+ Greengram
3	Banglore	Pigeonpea + Cowpea
4	Akola	Sorghum + Greengram
5	Jodhpur	Pearlmillet+Green gram
6	A P and T N	Pearlmillet+Pigeon pea
7	Maharsthra and Gurat	Sorghum + Pigeon pea (2:1)
8	Ranchi	Ragi + Soybean
9	Aurangabad (MS)	Soybena +Pigeonpea (4:2), Corron+Greengram (1:1), Cotton+Blackgram (1:1)

7. Contingent crop plan 7: Long gap in rainfall

If long gap in rainfall is a general trend of monsoon, the following contingent plan may be adopted;

1. Increase in seed rate to obtain more population,
2. Spraying of urea and water at 10-15 days interval,
3. Provide life saving irrigation if possible, and
4. Follow weed control and intercultural operations.

8. Contingent crop plan 8: Inadequate and uneven rainfall distribution of rainfall

Following contingent plan is recommended;

1. Cultivation of low water required crops like nutri cereals, millets, pulses and oilseeds,

2. Growing of short duration crops,
3. Provide life saving irrigation,
4. Rainwater harvesting and recycling,
5. Use of mulches, and
6. Application of antitranspirants.

9. Contingent crop plan 9: Submergence/Flooding and unseasonal rains

The exchange of gases is reduced under submergence/flooding resulting into leaf dehydration, reduced root permeability, reduced photosynthesis, stomatal closure, leaf senescence and accumulation of harmful gases like CH_4, NO, C_2H_2,H_2S, CO etc., reducing water and nutrient uptake and finally wilting of crops. These changes in leaves, shoots or roots are due to anaerobic respiration and production of toxic substances like acetaldehyde and ethyl alcohol and harmful gases. The endogenous levels of plant growth hormones (PGR) like auxin, gibberellic acid (GA) and cytokinins (CK) are reduced in the roots and enhanced levels of ABA and ethylene in the shoots causing stomatal closure and early onset of senescence respectively. The precursor for the ethylene biosynthesis i.e. 'Aminocyclopropane -1-Carboxylic Acid' (ACC), is increased under flooding stress.The specialized action of different PGRs under waterlogged/flood conditions is described below in table.5.

Table 5. Effect of flooding stress on the production of PGRs and their action on plants

S.N	Type of PGR	Effects on plants under water logging
1	Auxins	Reduced auxins causes "Hypertrophy" (Swelling of stem base by collapse or enlargement of cells in cortex)
2	GA	Decreased GA reduced cell enlargement and stem elongation
3	Cytokinin	Decreased CK causes early leaf senescence and reduced rate of assimilate partitioning to the sinks
4	ABA	Increased level of ABA enhances stomatal closure and reduced gas exchanges during photosynthesis, respiration and transpiration; resulted into in efflux of K+ from the guard cells; decreases ion transport due to lower rate of transpiration; decrease the starch formation in the guard cells resulting in stomatal closure.
5.	Ethylene	High concentration of causes "Epinasty" of leaves (uneven growth of leaves due to more cell elongation on upper side than the lower side of the leaf); induces senescence and Hypertrophy in plants.

Following contingent plan under waterlogging/flood conditions may be adopted;

1. Select crops and varieties tolerant to water logging and flooding (table.6),
2. Early planting with higher seed rates,
3. Facilitate bio-drainage by inclusion of legumes and trees (agroforestry systems)in the cropping systems,
4. Provide safe drainage of surface runoff,

5. Diversion of water towards water bodies or dryland areas,
6. Adopt soil and water conservation measures,
7. Try to utilize residual soil moisture for crop production.

In addition, following other options may also be adopted as and when require as they were reported useful under flood stress conditions;

1. Spray of growth retardant cycocel @ 500 ppm for arresting apical dominance and thereby promoting growth of laterals,
2. Foliar spray of 2% DAP + 1% KCl (MOP) to strengthen the stem and avoid lodging,
3. Nipping terminal buds to arrest apical dominance and thus promoting growth sympodial branches (as in cotton) for increasing productivity,
4. Spray of 40ppm NAA for controlling excessive pre-mature fall of flowering/ buds/young developing fruits and pods,
5. Spray of Brassinolide @ 0.5 ppm for increasing photosynthetic activity,
6. Foliar spray of Salicylic acid@100 ppm for increasing stem reserve utilization under high moisture stress,
7. Foliar spray of 0.3 % Boric acid + 0.5 % $ZnSO_4$ + 0.5 % $FeSO_4$ + 1.0 % urea during critical stages of the stress.

Table 6. Rice varieties tolerant to waterlogging and flooding

S.N	Flooding & submergence	Varieties
1	Water level <1 m and submergence <10 days	Swarna sub-1, Dishang, Joymoti, MTU-1010, MTU-1100, MTU-1064, MTU-1140, MTU-7029, BPT-5204, Dehangi, Gitesh, Shasharang
2	Water level >1 m and submergence >10 days	Jalashree, Jalkuwari, Rajashree, Karjat-2, Karjat-6, GAR-13, Lalat, Luit
3	Sunhemp	K-12 (Yellow), K-12 (Black), SUIN 053, SH-4, JRJ 610 (Prankur)
4	Jute	RO 7835, JRO 878, JRC 321, JRC 7447, JRC 532, JRC-517, Bidhan Pat-1
5	Sugarcane	Co-98014 (Karan-1), Co-0239, Co-0118, Co-0238, Co-0233, Co-05009
6	Maize	HM-5, Seed Tech-2324, HM-10, PMH-2
7	Fodder crops	Sorghum, Maize, Teosinte, Berseem, Oat, Barley, Para grass

10 Crop contingent plan.10: Extreme weather like heat wave/cold wave/ hailstorm/frost

A. Heat waves

i. Provide light irrigation.
ii. Provision of wind breaks,
iii. Spray of thiourea @ 1000 ppm in wheat and mustard at grain filling stage and 2% KNO_3 in cotton at boll formation stage, and
iv. Harvest crop at physiological stage.

B. Cold waves and frost

i. Apply light irrigation,
ii. Smoking during night ,
iii. Grow seedlings of fruits and vegetables in polyhouses,
iv. Harvest crop as early as possible,
v. Horticultural produce and vegetables may be preserved in cold storage.

C. Hailstorm

i. Apply light irrigation,
ii. Resowing of crop in case of severe damage at early stage,
iii. Harvest crop green for fodder if severely damaged at later stage of crops,
iv. Adopt required measures in advance if early warning/forecast issued,
v. Harvest horticultural products timely and shift at safer places.
vi. Use anti hail nets and
vii. Grow vegetables in polyhouses.

Impacts of crop contingent planning

The benefits of crop contingent planning and its timely execution are enormous by reducing the adverse effects of aberration weather. In addition, the adoption of crop management as well as soil and water conservation measures help to either maintain to improve the crop productivity and soil moisture regime during crop period. The rainfall distribution and its pattern was assessed and crop contingent plan was adopted in different parts of the country during 2015-16 and its impacts were evaluated as given below;

1. The productivity of soybean, total oilseeds and total pulses was increased from 0.733 to 0.937, 0.52 ti 0.60 and 0.75 t/ha 0.94 t/ha, respectively after adoption of crop contingent plan during 2015-16 despite the deficit rainfall (-20 to -59% of normal) in 30 districts out of 51 districts in Madhya Pradesh.
2. The rainfall receipt was 77 to 88% of normal in 2014-15 in East Gujrat, Central Gujrat, Kutch and South Gujrat but the impact of deficit rainfall

on area and production was negligible except groundnut, sorghum and pearlmillet.

3. In Karnataka, the deficit monsoon was managed by;
   I. Alternate crops like maize, pearlmillet and cotton were sown in unsown area of sorghum, pigeonpea, groundnut and sunflower,
   II. Adoption of land configuration and soil and water management practices,
   III. Utilizing common property resources for fodder production and drought safety net programs eg. crop insurance.
4. In Jharkhand, growing of community rice nursery was promoted to cope up delayed monsoon and encouraging effective water storage by lining of ponds and field bunding were adopted during deficit monsoon in 2014-15.
5. In Telangana, in situ water conservation, additional power supply for irrigation, promotion of short duration rice varieties and soybean, greengram, groundnut in cotton growing areas under delayed monsoon were the major intervention taken up during 2014-15.
6. In Maharashtra, *in situ* water conservation, micro irrigation, high density cotton planting, soybean with broad bed furrow technology, mulching in horticultural plants, sprinkler irrigation in soybean, hybrid pigeonpea promotion and green gram as substitute for soybean were the major interventions promoted during 2014-15 to overcome delayed monsoon.
7. In Andhra Pradesh, following contingent measures were proposed under delayed monsoon or delay in release of canal water;
   i. Sowing of crops after receipt of good rain,
   ii. Wide publicity of crop contingent plan,
   iii. Promotion of direct seeded rice/drum seeded rice,
   iv. Sowing of horsegram in place of groundnut in Anantpur district,
   v. Growing of fodder crops as intercrop in horticultural plantations and
   vi. Flood water diversion to Rayalseema region to irrigate horticultural crops.

Conclusion

Contingent crop planning is paramount need to avoid severity of aberrant weather conditions that are the common features of South-West monsoon in India. Our country experiences flood and drought or drought like conditions every year resulting heavy loss of crops and animals. Hence timely execution of crop contingent plan with monitory support by government is the pre requisite for benefit of not only farming communities but as a whole to nation. Number of schemes/programs have been launched by state and central government like **National Mission for Sustainable Agriculture (NMSA),** Pradhan Mantri Fasal

Bima Yojana (PMFBY), Pradhan Mantri Krishi Sinchai Yojana (PMKSY), PM-KISAN (Pradhan Mantri Kisan Samman Nidhi) Scheme, Pradhan Mantri Kisan Pension Yojana (PMKPY) etc. to support farmers and their agriculture so that our farming communities may contribute shoulder to shoulder in the development of the nation.

Questions

a. Short and Descriptive

1. Write short notes on following:
   a. Aberrant weather
   b. Contingent crop plan
   c. High yielding varieties for dryland areas
   d. High yielding varieties for waterlogged areas
   e. Important intercropping systems for dryland areas
2. What do you mean by plant hormones? Describe its role in crop production dryland areas.
3. Describe crop contingent plan in early arrival and late cessation of rainfall.
4. Describe the crop contingent plan during long dry spells after crop sowing.
5. Describe the crop contingent plan under extreme weather conditions like heat wave, cold wave and hailstorm/frost.
6. What is crop contingent plan? Explain its impacts on crop production in rainfed areas.
7. What do you mean by aberrant weather conditions? What measures you may suggest for successful crop production under submergence/waterlogged areas and unseasonal heavy rainfalls?
8. How can you utilize the extended monsoon for crop production in rainfed areas? Suggest important strategies with examples.
9. Suggest crop contingent plan for late arrival of monsoon.
10. Explain the importance of crop contingent planning in dryland areas.

b. Objective (MCQs)

1 Aberrant weather is

a Unfavorable weather b Unusual monsoon

c Both a & b d None above

2 Which of the following is the main cause of aberrant weather condition?

a Drought b Hailstorm

c Cyclones d All above

3 Which of the following is the main cause of aberrant weather condition?

a Flood
b Sea water
c Heat waves
d All above

4 Which of the following is the main feature of rainfall distribution in India?

a Heavy rain during S-W and N-E monsoon
b Bimodal nature
c Uneven and erratic rainfall
d All above

5 Aberrant weather will affect most?

a Field crops
b Livestock
c Horticulture
d All above

6 Which of the following will be affected most under aberrant weather conditions?

a Poultry
b Fisheries
c Both a & b
d None above

7 Which of the following is an important weather aberration?

a The commencement of rains may be quite early or considerably delayed
b Rains may terminate earlier than normal cessation
c Both a & b
d None above

8 Which of the following is the most variable factor in rainfed crop production?

a Weather
b Soil
c Management
d All above

9 District crop contingent plan may be proposed under

a Delay in monsoon onset
b Delayed or limited release of water for irrigation
c Floods and unseasonal rains
d All above

10 Which of the following measure can be adopted under early onset of monsoon?

a Grow long duration crops like sorghum, bajra, sesame, maize etc
b Recycle water for life saving irrigation
c Follow soil and water conservation measures
d All above

11 Which of the following measure can be adopted under late onset of monsoon?

a Transplanting of crops | b Sowing of alternate Crops
c Dry sowing | d All above

12 Which of the following measure can be adopted under delayed/late onset of monsoon?

a Increase seed rate to compensate low field emergence, | b Complete weed control to avoid water loss
c Mixed/Intercropping is again a good option. | d All above

13 Which of the following crop can be transplanted?

a Rice | b Bajra
c Finger millet | d All above

14 Which of the following conditions should be followed under dry spell?

a If the dry spell is long, try to choose short duration crops like legumes and oilseeds
b Life saving irrigation may be given in mid season dry spells and terminal drought
c Both a & b
d None above

15 Which of the following is responsible for mid season and terminal drought?

a Early cessation of monsoon | b Extended monsoon
c Heavy rainfall | d None above

16 Which of the following measure can be adopted under early cessation of monsoon?

a Sowing of short duration crops | b Reduce plant population by thinning
c Mulching | d All above

17 Which of the following measure can be adopted under early cessation of monsoon?

a Use of antitranspirants | b Life saving irrigation
c Harvesting of standing crop | d All above

18 Which of the following measure can be adopted under dry spells occurred just after normal sowing?

a Resowing | b Life saving irrigation
c Foliar spray of Urea@1-2% | d All above

19 Which of the following can be used as a mulch material?

a Dust b Crop residues
c Polythene d All above

20 Which of the following is a draught resistant variety of black gram?

a PU-35 b Shekhar
c Azad Urd-1 d All above

21 Which of the following is a draught resistant variety of black gram?

a Pusa Vishal b Samrat
c CO-6 d All above

22 Which of the following is a draught resistant variety of green gram?

a PDM-139 b BM-2003-2
c Pratap d All above

23 Which of the following is a draught resistant variety of green gram?

a VBN-3 b Samrat
c CO-6 d None above

24 Which of the following is a draught resistant variety of gram?

a Pant Kabuli gram-3 b Pant gram-3
c Pant gram-4 d All above

25 Which of the following is a draught resistant variety of gram?

a JG-36 b JGK-1
c JGK-3 d All above

26 Which of the following is a draught resistant variety of groundnut?

a TG 37A, b ICGV-91114
c TG 38 d All above

27 Which of the following is a draught resistant variety of groundnut?

a JGK-1 b JGN-23
c JG-36 d None above

28 Which of the following is a draught resistant variety of pigeon pea?

a TJT-501 b ICPL-88039
c BRG-2 d All above

29 Which of the following is a draught resistant variety of pigeon pea?

a BRG-4 b Asha
c Both a & b d None above

30 Which of the following is a draught resistant variety of pearlmillet?

a GHB-558, b ICTP-8203

c Both a & b d None above

31 Which of the following is a draught resistant variety of castor?

a NPH-1 b GAUCH-1

c YRCH-1 d All above

32 Which of the following is a draught resistant variety of castor?

a TMV-5 b TMV-6

c CO-1 d All above

33 Which of the following is a draught resistant variety of castor?

a GCH-7 b TMVCH

c Both a & b d None above

34 Which of the following is a draught resistant variety of sesame?

a Shekhar b TC-25

c Patan-64 d All above

35 Which of the following is a draught resistant variety of sesame?

a Gujrat Til-2 b Gujrat Til-1

c Krishana d All above

36 Which of the following is a draught resistant variety of cluster bean?

a GG2 b GCH-7

c TMV-5 d None above

37 Which of the following is a draught resistant variety of mustard?

a Varuna b RH-819

c Vaibhav d All above

38 Which of the following is a draught resistant variety of mustard?

a Prakash, b Durgamani

c RH-30, d All above

39 Which of the following is a draught resistant variety of toria?

a M-27 b TS-29

c V-54 d All above

40 Which of the following is a draught resistant variety of toria?

a VR- 23 b Dk-1

c T-9 d All above

41 Which of the following is a draught resistant variety of brown Sarson?

a BSH-1 | b KVS-3
c Pusa Kalyani | d All above

42 Which of the following is a draught resistant variety of yellow sarson?

a Binoy (B-9) | b Subinoy
c Both a & b | d None above

43 Which of the following is a draught resistant variety of taramira (Eruca sativa)?

a TMH-52 | b ITSA
c T-27 | d All above

44 Which of the following is a draught resistant variety of Karan Rai (Barssica carinata)?

a DLSC-1, | b PVC-9221
c Both a & b | d None above

45 Which of the following is a draught resistant variety of linseed?

a Jawahar-17 | b Jawahar-23
c Jawahar-552 | d All above

46 Which of the following is a draught resistant variety of linseed?

a Sweta | b Shubhra
c T-397, | d All above

47 Which of the following measure can be adopted under extended monsoon?

a Harvest Kharif crops | b Recycling of stored water
c Utilization of residual soil moisture | d All above

48 Which of the following contingent crop plan may be followed under long dry spells?

a Sorghum, maize, pearlmillet may be harvested | b Remove the lower old leaves
c Spray water | d All above

49 Thinning is recommended under severe drought at

a Pre-flowering stage | b Flowering stage
c Both a & b | d None above

50 Thinning may save crop for approximately

a 35-50 days | b 35-60 days
c 45-50 days | d 35-70 days

51 Inter plot and intra plot water harvesting help in

a Retention of rainwater
b Recharging of ground water
c Life saving irrigation
d All above

52 Which of the following is remunerative intercropping system in Hisar region?

a Pearlmillet + Cowpea
b Pigeonpea + Cowpea
c Pearlmillet+ Greengram
d None above

53 Which of the following is remunerative intercropping system in Agra region?

a Pigeonpea + Cowpea
b Pearlmillet+ Greengram
c Pearlmillet + Cowpea
d Sorghum + Greengram

54 Which of the following is remunerative intercropping system in Banglore region?

a Pearlmillet + Cowpea
b Sorghum + Greengram
c Pigeonpea + Cowpea
d All above

55 Which of the following is remunerative intercropping system in Akola region?

a Sorghum + Greengram
b Ragi + Soybean
c Pigeonpea + Cowpea
d None above

56 Which of the following is remunerative intercropping system in Ranchi region?

a Sorghum + Greengram
b Ragi + Soybean
c Both a & b
d None above

57 Which of the following is a remunerative intercropping system in Aurangabad (MS) region?

a Soybean +Pigeonpea (4:2),
b Cotton+Greengram (1:1)
c Cotton+Blackgram (1:1)
d All above

58 Which of the following is a remunerative intercropping system in Jodhpur region?

a Pearlmillet+Green gram
b Sorghum + Greengram
c Ragi + Soybean
d Pigeonpea + Cowpea

59 Which of the following is a remunerative intercropping system in A.P. and T.N. region?

a Pearlmillet+Green gram
b Pearlmillet+Pigeon pea
c Sorghum + Pigeon pea
d All above

60 Which of the following is a remunerative intercropping system in Maharashtra and Gujarat region?

a Sorghum + Pigeon pea
b Sorghum + Greengram
c Ragi + Soybean
d Pearlmillet+Green gram

61 Which of the following measure can be adopted under long dry gaps?

a Increase seed rate
b Crop saving irrigation
c Spraying of urea and water
d All above

62 Which of the following contingent plan is recommended under inadequate and uneven rainfall?

a Growing short duration crops
b Provide life saving irrigation
c Cultivation of low water requiring crops
d All above

63 Which of the following contingent plan is recommended under inadequate and uneven rainfall?

a Rainwater harvesting and recycling,
b Use of mulches
c Application of antitranspirants
d All above

64 Which of the following hormone is responsible for hypertrophy under waterlogging?

a Auxins
b GA
c Cytokinin
d None above

65 Gibberellic acid (GA) is responsible for

a Cell enlargement
b Stem elongation
c Both a & b
d None above

66 Decrease cytokinin (CK) is main cause of plant under water logging?

a Early leaf senescence
b Hypertrophy
c Stem elongation
d None above

67 Which of the following effects of Abscisic acid (ABA) on plants under water logging?

a Reduced gas
b Stomatal closure
c Decreases ion transport
d All above

68 Which of the following hormone is responsible for Epinasty in plant under water logging?

a Ethylene
b GA
c Auxins
d All above

69 Which of the following measure can be adopted contingent plan under waterlogging/flood conditions?

a Early planting with higher seed rates b Diversion of water towards water bodies

c Bio-drainage d All above

70 Which of the following variety of sunhemp is tolerant to waterlogged condition?

a K-12 (Yellow) b K-12 (Black)

c SUIN 053 d All above

71 Which of the following variety of sunhemp is tolerant to waterlogged condition?

a SH-4, b JRJ 610 (Prankur)

c Both a & b d None above

72 Which of the following variety of jute is tolerant to waterlogged condition?

a RO 7835 b JRO 878

c JRC 321 d All above

73 Which of the following variety of jute is tolerant to waterlogged condition?

a Bidhan Pat-1 b JRC 532

c JRC-517 d All above

74 Which of the following variety of sugarcane tolerant to waterlogged condition?

a Co-98014 (Karan-1) b Co-0239

c Co-0118 d All above

75 Which of the following variety of Maize tolerant to waterlogged condition?

a HM-5 b HM-10

c Seed Tech-2324 d All above

76 Which of the following fodder crop is tolerant to rainfed condition?

a Sorghum, b Teosinte

c Barley d All above

77 Which of the following crop contingent plan may be adopted under heat wave condition?

a Light irrigation b Wind breaks

c Harvest crop at physiological stage d All above

78 Which of the following crop contingent plan may be adopted under cold waves and frost condition?

a Apply light irrigation
b Smoking during night
c Grow seedlings of fruits and vegetables in Polyhouses
d All above

79 Which of the following crop contingent plan may be adopted under cold waves and frost condition?

a Crop harvest after frost
b Vegetables may be preserved in cold storage.
c Both a & b
d None above

80 Which of the following crop contingent plan may be adopted under hailstorm condition?

a Apply light irrigation
b Harvest crop green for fodder
c Resowing of crop
d All above

81 Which of the following crop contingent plan may be adopted under hailstorm condition?

a Harvest horticultural products timely
b Use anti hail nets
c Grow vegetables in polyhouses
d All above

82 Which of the following fodder crop is tolerant to waterlogged condition?

a Berseem
b Para grass
c Oat
d All above

83 Which of the following is an important drought resistant variety of rice?

a Sahabhagi Dhan
b Rajendra Bhagwati
c Vandana
d All above

84 Which of the following is an important drought resistant variety of rice?

a Anjali
b Birsa Vikas Dhan-203
c DRR Dhan-2 (IR64 Drt 1)
d All above

85 Which of the following is an important drought resistant variety of wheat?

a PBW-527
b HI-1531
c HI-8627
d All above

86 Which of the following is an important drought resistant variety of wheat?

a HD-2888
b HPW-349
c WH- 1080
d All above

87 Which of the following is an important drought resistant variety of maize?

a Pusa Hybrid Makka-1 b HM-4
c DHM-121 d All above

88 Which of the following is an important drought resistant variety of maize?

a Buland b Pusa Hybrid Makka-5
c Both a & b d None above

89 Which of the following is an important drought resistant variety of sorghum?

a CSH-19 R b CSH-15R
c CSV-18 d All above

90 Which of the following is an important drought resistant variety of pearl millet?

a HHB-67 b GHB-757
c Pusa Composite-443 d All above

91 Which of the following is an important drought resistant variety of pearl millet?

a GHB-719 b Dhanshakti
c Mandor Bajra Composite-2 d All above

92 Which of the following is an important drought resistant variety of barley?

a RD-2660 b K-603
c Both a & b d None above

93 Which of the following is an important drought resistant variety of soybean?

a NRC-7 b JS 95-60
c Both a & b d None above

94 Which of the following is an important drought variety of groundnut?

a Ajaya, b TAG-24
c Girnar-1 d All above

95 Which of the following is an important drought variety of sugarcane?

a Co-98014 (Karan-1) b Co-0239
c Co-0238 d All above

96 Which of the following is an important drought variety of sugarcane?

a Co-0118
b Co-06927
c Both a & b
d None above

97 Which of the following is an important drought variety of cotton?

a HD-324
b CICR-1
c Raj DH-7
d All above

98 Which of the following is an important drought variety of cotton?

a Jawahar Tapti
b Pratap Kapi
c AK- 235
d All above

99 Which of the following is an important draught variety of jute?

a JBO-1 (Sudhangsu)
b JRO-204
c JRO-524
d All above

100 Which of the following is an important draught variety of jute?

a JRC-80
b Jawahar Tapti
c Raj DH-7
d CO-06927

Answer

1. c	2. **d**	3. **d**	4. **d**	5. **d**	6. **c**	7. **c**	8a	9. **d**	10. **d**
11. **d**	12. **d**	13. **d**	14. **c**	15. **a**	16. **d**	17. **d**	18**d**	19. **d**	20. **d**
21. **d**	22. **d**	23. **a**	24. **d**	25. **d**	26. **d**	27. **b**	28**d**	29. **c**	30. **c**
31. **d**	32. **d**	33. **c**	34. **d**	35. **d**	36. **a**	37. **d**	38**d**	39. **d**	40. **d**
41. **d**	42. **c**	43. **d**	44. **c**	45. **d**	46. **d**	47. **d**	48. **d**	49. **c**	50. **a**
51. **d**	52. **a**	53. **b**	54. **c**	55. **a**	56. **b**	57. **d**	58. **a**	59. **b**	60. **a**
61. **d**	62. **d**	63. **d**	64. **a**	65. **c**	66. **a**	67. **d**	68. **a**	69. **d**	70. **d**
71. **c**	72. **d**	73. **d**	74. **d**	75. **d**	76. **d**	77. **d**	78. **d**	79. **c**	80. **d**
81. **d**	82. **d**	83. **d**	84. **d**	85. **d**	86. **d**	87. **d**	88. **c**	89. **d**	90. **d**
91. **d**	92. **c**	93. **c**	94. **d**	95. **d**	96. **c**	97. **d**	98. **d**	99. **d**	100. **a**

6

Soil and Water Conservation

Outline

Introduction

Out of 329 m ha geographical area in the country, 187.8 m ha or 53% is affected by various forms of degradation process which include 148.9 m ha affected by water erosion, 13.5 m ha by wind erosion, 13.8 m ha by chemical deterioration and 11.6 m ha by physical deterioration. The annual loss of topsoil is nearly 16.0 t/ha/year resulted in to 8.4 m t/year total nutrient loss. The fragmentation in land holdings caused by high population pressure and degradation of soil loss, therefore, the per capita availability of arable land has steadily declined from 0.48 ha in 1950-51 to 0.15 m ha in 2000 and is expected to further reduce to 0.12 ha by 2015 A.D. Similarly, per capita forest area availability is only 0.28 ha against the world average of 10 ha. Hence, land is one of the most important natural resources and it needs to be conserved properly for future generations. The sloppy hills and rain fed areas are more prone to wind and water erosion.

The loss of fertile top soil by water erosion is estimated to be about 5334 mt every year, out of which 29% soil is lost primarily to sea, 10% deposited in irrigation reservoirs and 59% deposited as alluvium in different places (Narayan and Babu, 1983).The annual water erosion rate varies from < 5t/ha in arid region of Rajasthan to >80 t/ha in Shivalik hills (Singh *et al.*, 1992). Wind erosion is a serious problem in arid and semi arid region including the states of Rajasthan, Haryana, Gujrat and Punjab. It is to be noted that 1cm top soil layer is formed in 130 to 400 years. Recently, Velayutham and Bhattacharya (2000) reported the loss of crop productivity up to 50% due to water erosion in different soils (Table.1).

Table 1. Expected loss of productivity due to water erosion in different soils

Soil erosion class	Soil loss (t/ha)	Loss in productivity (%)		
		Inceptisols (Alluvial soil)	Vertisols (Black soil)	Alfisols (Red soil)
Nil to very slight	<5	Nil	<5	<10
Slight	5-10	<5	5-10	10-25
Moderate	10-20	5-10	10-25	25-50
Strong	20-40	10-25	25-50	>50
Severe	>40	25-30	>50	N.A.

Soil erosion

Defnition

Soil erosion is the geological process in which earthen materials or soil cover are worn away and transported by natural forces such as wind or water, so it is a gradual process that occurs when the impact of water or wind detaches and removes soil particles, causing the soil to deteriorate.

or

Soil erosion is defined as the wearing away of topsoil. Top soil is the top layer of soil and is the most fertile because it contains the most organic, nutrient-rich materials. It involves detachment, transporation and deposition of soil particles by action of these natural forces.

Types

There are two types of soil erosions;

1. *Natural or normal or geological soil erosion*

Natural or geological soil erosion is a very common and continuous process of soil erosion. The physical, chemical and biological weathering of rocks and parent material are pre-requisites for soil formation. Therefore the geologic erosion maintains equilibrium between the removal and formation of soil. If the balance between soil removal and formation is disturbed, the extent of natural erosion in form of soil depth and its characteristics will be changed.

2. *Accelerated soil erosion*

Accelerated soil erosion is removal of the surface soil from any area denuded of its natural protective cover as a result of human and animal interference. It involves higher rate of soil removal than soil formation. Therefore it is quite serious erosion type that makes soil unproductive and even barren if not controlled timely. At normal geological pace, nature requires 1000 years to build up 2.5 cm of top soil but unhealthy farming practices may erode top soil within few years. It is reported that a fraction of millimeter (about 1 ton soil/ha/annum) is added under normal conditions.

Adverse effects of soil erosion

i. ***Severe land degradation:*** The soil erosion is responsible for degradation of our fertile soil. About 32.7% of total geographical area of India is degraded and both water and wind erosion accounts for nearly 65% of land degradation.

ii. ***Loss of fertile top soil:*** The top soil cover is weared away by both water and wind erosion. It is estimated that 5,334 million tones of soil (16.35 tones/ha/ annum) is eroded annually, which is four fold higher than the permissible soil loss (4-5 tones/ha/annum). About 25% of the eroded material is permanently lost to the sea, 61% is dislocated from one to another place and 10% is deposited in the reservoirs resulting in the loss of storage capacity by 1-2% per annum. The data on the river valley projected on 17 medium and small reservoirs in India showed that the rate inflow of annual sediment rate three fold (9.17 ha-m/100km^2) when compared with their designed rate of 2.93 ha-m/100km^2/ annum, thus reducing their life expectancy and hydro-electric power generation to only one-third.

iii. ***Flood:*** Flood is a common feature during monsoon around the world. Siltation of water bodies, reduced carrying capacity of river belts due to again siltation and also choking and blocking of natural drains are the main cause of flood that not only damage human and animal life and property but also productive agricultural land at large scale.

iv. ***Siltation of water bodies:***The run off water along with eroded soils is dumped either in water bodies or sea. The soil particles later settled in the bottom of water bodies and slowly the water storage capacity of water bodies is reduced and need huge investment to clean the water bodies. The water bodies are clogged and also affecet adversely drinking water supply mainly in drought prone areas.

v. ***Reducing land productivity:*** Soil erosion reduces the productivity as well as fertility of soil as the most fertile upper soil cover is washed away from even sheet erosion, however the extent of erosion varies from sheet erosion to stream bank erosion that causes maximum soil degradation. It is reported that 160 m ha land accounting >11% of the world crop land is under severe soil erosion and need immediate attention. It was reported that the productivity of lightly to moderately degraded lands had 10% lower crop productivity, while severly degraded lands reduced crop productivity to the tune of 18-20% than non degraded lands

vi. ***Environmental pollution :*** The runoff water along with toxic elements of sewage and sludge flow over the wide range of stream beds due to siltation of river beds and polluting the whole flowing beds. Moreover the part of the polluted water remains all along the flowing ways due to undulated surface. After receding the flood, this store water smells badly due to rotting of wastes resuting into soil, ground water as well as the crop land polluted.

vii. ***Change in forest cover:*** India has 2% land, 10% forests and 0.5% grazing area of the world but has the burden of supporting 18% human and 15% cattle population. The recorded forest area of the country is 77 million ha constituting nearly 19% of geographical area of India. Out of this 11.73% is under dense forest with more than 40% canopy, 7.61% under open forest, 0.13% under mangrove forest (State of forest report, 1997). About 39 million ha area is under degraded forests. The forest cover is declining due to degradation of permanent pasture and open forests in the desert and arid regions. Besides, the composition of forest cover is also changed due to soil erosion.

viii. ***Loss of biodiversity:***The natural habitat is disturbed due to erosion and flood causing loss of biodiversity. About 45,000 species of wild plant and over 77,000 of wild animal were recorded, which comprise 6.5% of the world's known wildlife. An assessment of the loss of wildlife habitat in tropical Asia in 1986 showed that the country had only 615,095 km^2 out of the original wildlife habitat of 3,017,009 km^2, i.e. loss of 20%. The plant seeds and propagules are lost due to erosion.

Agents of soil erosion

Water, wind and seas waves are three major agents of soil erosion. The detail is given below;

1. Water erosion

The soil erosion caused by water or flood water is known as water erosion. Its different types are discussed below;

i. **Sheet erosion:** Sheet erosion refers to uniform washing away of top thin layer of soil by run off water but remains unnoticed. It happens every year during monsoon season and larger crop field are affected. It deteriorates the soil fertility and productivity to the greater extent. It is a first stage of water erosion. The breaking action of rain drops combined with surface flow is the major cause of sheet erosion. The flowing of muddy run off water in the field during rains is the characteristics of sheet erosion. It is most extensive as it occurs at light to medium rainfall.

ii. **Rill erosion:** It is a second stage of water erosion in which small and uniform finger shaped channels or grooves are formed over the entire field at rainfall. These channels are known as rills and this type of erosion is called rill erosion. These rills are dismantled after ploughing but are very harmful as the large fertile land is washed away from the field with run off water. The flow rate of run off water > 0.3-0.7 mm/second causes rill erosion.

iii. ***Gully erosion:*** It is an advanced stage of water erosion in which rills or narrow channels become wider and deeper, known as gullies and type of erosion is called gully erosion. These gullies cut up large fields into small fragments and in course of time, make them unfit for cultivation provided these gullies are left unchecked. The gully erosion is very common on sloppy land under heavy rainfall conditions.

v. ***Ravines:*** A ravine refers to a small narrow steep-sided valley that is larger than a gully and smaller than a canyon and that is usually worn by runoff water, so ravines are typically classified as larger in scale than gullies, although smaller than valleys. When gully become wider and deeper, the land become almost unfit for cultivation. Ravines are typically found in deep alluvial soils. Ravines are found in both sides of Yamuna and Chambal rivers particularly in Bundelkhand region of UP and MP.

vi. ***Landslides:*** Landslides are very common in mountainous region where slope exceeds 20%. The heavy rains and cloud bursts are main cause of landslides that cause road blockage and loss of property.

vii. **Stream –bank erosion:** The small streams, rivers and torrents (hill streams) are the main cause of stream bank erosion in which the sides of river banks are cut off and washed away in very unregular shapes with violant flowing water. The land, trees, boulders etc are also rolling with flowing water moving from one place to other and later they are scattered in the larger flowing beds making land area unpleasant and polluted. Besides, the land area becomes flooded and growing crops are damaged.

Mechanism of water erosion

Water erosion caused by rainfall. The falling rain drops have a energy and detach the soil particle from the land and these detached soil particles make a suspension with rain water and move from one place to other along with run off water. The land topography changes the run off flow rate and also severity of water erosion.

Factors affecting swater erosion

Water erosion is affected by climate, topography, vegetation and soil. It can be summarized as below;

E = f (C, T, V, S)

Where

E = Water erosion

f = factor

C= Climate

T= Topography

V= Vegetation

S = Soil

2. Wind erosion

Wind erosion is very common in North West Rajasthan, Haryana, Punjab and Gujarat. The part of UP, TN, Karnataka, AP and Maharashtra also suffer wind erosion mainly before monsoon when dry airs blow during May-June. The Rajasthan desert covers an area of nearly 2.1 km^2 and the major affected divisions are Bikaner and Jodhpur. The entire region is experiencing strong wind and the mean wind velocity varies from 3.2 to 32km/ha that causes the wind erosion. The arid and semi arid region devoid of vegetation is more prone to wind erosion. The soil particles on the land surface are lifted and blown off dust storms. When the velocity of the dust bearing wind is retarded, coarse soil particles are deposited in the form of dunes sometimes rendering fertile lands into unfit cultivation. The finer particles are carried away to farther distance. The fertile soil is blown away by winds and the subsoil is exposed, as a result the productive capacity of the soil is considerably reduced. Areas susceptible to wind erosion were delineated on global as well as country level. Thar desert of the western Rajasthan shows high losses due to wind erosion. Deep ploughing of sand plains lost more than 3000 tones soil/ha during sand storm of 1987 whereas similar areas with 10-12% plants cover or with cloudy surface suffered negligible soil loss.

Wind erosion mechanism

There is following three mechanisms occur during wind erosion as given below;

i. ***Saltation***: The major portion of the soil carried by the wind moves in a series of short bounces or jumps called 'saltations'. Normally the size of soil particles varying from 0.05 to 0.50 mm in diameter is prone to saltation. It causes by the direct pressure of the wind on the soil particles and their collision with other particles. After being pushed along the ground surface by the wind, the particles leap almost vertically in the first stage of saltation. Some grains rise only as short distance, other leap 30 cm or higher. The jumping height of soil particles varies with size and density of soil particles, roughness of soil surface and velocity of the wind. These soil particles later either roll or slide or rise upward and become the part of the saltation movement.

ii. ***Suspension***: Suspension refers to floating of small size soil particles in the air. The fine dust particles < 0.05 mm and particularly < 0.01 mm is extremely resistant to move by direct action of wind velocity but due to adhesion and cohesion forces make soil particles little bigger (< 0.1 mm) under moisture action and these paricles are picked up by the saltation particles and become the part of turbulent air and then they may be lifted high into the air and move in kilometers before being redeposited. The wind velocity upto 3-5 mm/ha is considered sufficient to lift very fine sand, silt and clay and also much of the soil organic matter. The dust particles remain in lower part of air layers and settled after wind subsides or the rain washed them down.

iii. ***Surface creep***: Surface creeping refers to rolling or sliding of large soil particles along the ground surface. These soil paricles are too heavy or bigger to be lifted by wind moves by saltation. It causes maximum erosion as the size of soil particles are of 0.5 to 2 mm in diameters. The soil moved in saltation and surface creeps usually remain within the eroded area, especially when winds are from different direction but by surface creep the soil paricles move from main field to other fields so the erosion is higher than saltation and suspension.

About 90% soil movement in wind erosion is below to the height of 30cm and about 50% of it is within 5cm in ground level. The proportion of soil movement under different types of wind erosion varies with soil types, however the soil loss varies from 50-75% by weight under saltation, 3-40% under suspension and 5-25% under surface creep. The saltation and suspension are responsible for wind erosion of pulverized and surface creep for coarse soils or cloddy surface. The control of wind erosion is mainly based on the reduction or elimination of movement in saltataion. The surface roughness, cloddy field, crop stubbles and grasses are very effective in reducing wind erosion.

3. Wave erosion

The coastal area is eroded or lost due to displacement of land or the long term removal of sediment and rock along the coastal line mainly due to sea waves, currents, tides, wind driven water, water borne ice or impacts of storms. Therefore the waves erode sediments from cliffs and shorelines. The sediment in ocean water acts like sandpaper. Bigger waves carry more sediment and cause more erosion. When waves reach the shore, they can form deposits such as beaches, spits, and barrier islands. Groins, jetties, breakwaters, and seawalls are structures that protect the shore from breaking waves. Wave erosion can be transformed into following types;

i. ***Landforms*** - Different unique landforms are created by wave erosion.
ii. ***Wave - cut cliffs form*** – This form is found when wave erode a rocky shoreline. They create a vertical wall of exposed rock layers.
iii. ***Wave cut plat forms*** – Wave cut platforms are level areas formed by wave erosion. Since these platforms are above the sea level, it means that either sea level was higher or the rock was lower.
iv. ***Sea arches form*** – The sea arches are formed when waves erode both sides of a cliff. They create a hole in the cliff.
v. ***Sea stacks form*** - This form is created when waves erode the top of a sea arch. This leaves behind pillars of rock.

Sand dunes

The sand dunes are the common examples of wind erosion. A dune is a mound of sand formed by the wind, usually along the beach or in a desert. Dunes form when wind blows sand into a sheltered area behind an obstacle. Dunes grow as grains of sand accumulate. Every dune has a windward side and a slipface. There are five different shapes of sand dunes i.e. crescentic, linear, dome, parabolic and star. The typical sand dune has a gentle slopes of 5-11% to windward and 33% to leeward side. Sand dunes are very famous in Bikaner and Jodhpur region of Rajastahan. The threshold wind velocity for wind erosion though varies with soil particles, is 16 km//hr at a height of 30 cm for 0.1 mm diameter soil particles and it increases with increasing size of soil particles. The threshold wind velocity refers to the minimum wind velocity to initite the movement of soil particles.

Types of sand dunes

There are five types of sand dunesa as listed below;

i. ***Transverse sand dunes*** - This type of sand dune is a dune elongated perpendicular to the direction of the wind and formed when the wind is constant and there is a large supply of sand (Fig.1).

ii. ***Linear or Longitudinal sand dunes*** – It is a dune elongated in the general direction of the wind and formed from slightly different wind directions with a limited supply of sand.

iii. ***Star or complex sand dunes*** – Star dune is a dune formed when the direction of the wind is variable.

iv. ***Crescentic or Barchan sand dunes*** -It is a crescent shaped dune with the convex portion facing the wind direction and formed when the wind direction is constant and the sand supply is limited.

v. ***Parabolic or Blow out sand dunes*** – It is also known as U-shaped or hairpin dunes. It tends to form where vegetation covers the sand. Winds may erode a section, pushing the sediment leeward. The vegetation will hold back the arms of the dune, so that the dune points in the leeward direction.

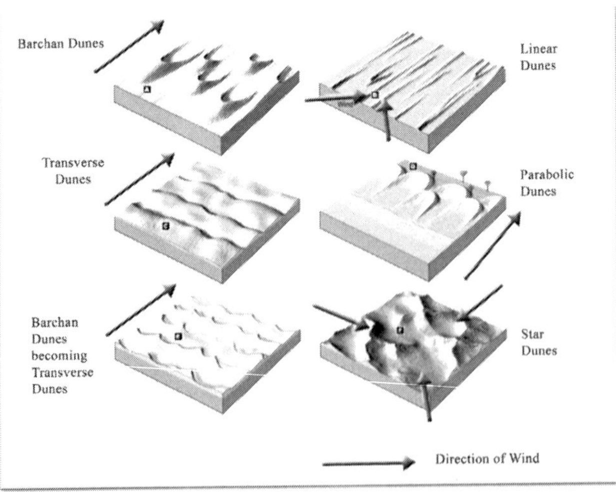

Fig 1. Types of sand dunes
Source: https://socratic.org/questions/what-are-the-different-types-of-sand-dunes (August 2021)

Soil and water conservation

Definition

Soil and water are the integral part of production system and both are prone to different types of losses. Soil is lost in form of either water and wind erosion, chemical deterioration (acidification, salinization and alkalinization), physical deterioration like water logging or fertility loss due to nutrient leaching, conversion and uptake by crop plants. Soil without water or vice versa is no use. The physical properties of soil like texture and structure decide the water retention capacity of soil. The major part of the top soil is lost by water erosion that dissociates the most fertile top layer of the soil and takes away from its original place and makes soil less fertile. Therefore, if the water is conserved, the top soil layer may be conserved. But the case will differ where wind erosion removes top soil layer. Here, the soil conservation measures will save water as it is held by soil particles. Hence, if one (soil) is conserved, other (water) will be conserved. Under present scenario of global production system coupled with increasing population, decreasing trend of natural resources including soil and water, high inflation rate, and uncertainty of global warming, soil and water conservation is a paramount need for food and nutritional security of the nation.

In mathematical terms, erosion is a function of the *erosivity* and the *erodibility* of the soil as given below;

Erosion = f (Erosivity x Erodibility)

The erosivity may be defined as a potential ability of the rain to cause erosion, while erodibility as the vulnerability of the soil to erosion. Based on the factors responsible for erosion, the universal soil-loss equation (USLE) was evolved and now it is widely accepted in India too to predict the soil erosion from agricultural lands and watersheds as given below;

A = R x K x L x S x C x P

Where,

A = Predicted soil loss (t/acre/year)

R = Rainfall and runoff factor

K = Soil erodibility

L = Slope length

S = Slope gradient or steepness

C = Soil cover and management

P = Erosion control practices

Similarly, wind erosion (E) is the result of ;

E = I x R x K x F x C x W x D x B

Where,

E = Soil loss by wind erosion

I = Soil cloudiness factor

R = Surface cover factor

K = Surface roughness factor

F = Soil textural class factor

C = Local wind factor

W = Field width factor

D = Wind direction factor

B = Wind Barrier factor

The soil and water conservation may be defined by many ways as given below;

- *'Soil conservation measures imply using lands as it should be used i.e. treating land according to its need and using it according to its capacity'. He further indicates that the soil conservation is not merely terracing, contouring, strip cropping, filling gullies or planning excessively steep or erodible land with grass and trees, it is that and whatever or make it more productive. It means drainage if the land is too wet and irrigation if it is too dry, and it means addition of fertilizer and organic matter if the soil is deficient in these constituents'.*

  (Bennet, 1945)

- *'Soil conservation connotes all methods of soil management and other measures required to preserve soil and soil fertility from total or partial losses caused in one or other way'. The basic aim of soil conservation is to maintain the soils at a high level to ensure sustained level of production'.*

  (National Planning Committee, 1948)

- *Soil conservation means, 'the status quo ante i.e. the prevention of deterioration or protection from destruction rather than a positive change for the better'.*

  Thirumalai (1951)

Principles of soil and water conservation

Following principles are involved in proper and efficient soil and water conservation measures;

1. Land use based on its capability,
2. Conservation of top soil,
3. Avoid soil erosion,
4. Control flooding,
5. Increase soil moisture regimes,
6. Regeneration of ground water,

7. Increase crop productivity,
8. Improve soil health ,
9. Reclamation of saline, alkaline and arid soils, and
10. Improve the social-economic status of the society.

Soil erosion is a serious problem limiting crop production on eroded soils. The national commission on Agriculture (1975) estimated 150 m ha total area was prone to serious water and wind erosion (Table.2a) but Sehgal and Abrol (1994) reported 148.9 m ha area under water erosion (Table. 2b). ICAR (2010) reported that total land under degradation accounts 120.72 m ha (Table.2c). The recent estimate of land degradation of India as published in *The Times of India* on September 2019 indicated that 96.4 mha is under degradation that constitutes 29.3% of the total geographical area of India while 13.2 billion ha land is under soil degradation in world (Fig.2).

Table 2a. Area under different types erosion in India

S.N	Parameters	Area (m ha)
1.	Serious water and soil erosion	150
2.	Critical stage of deterioration due to erosion	69
3.	Wind erosion	32
4.	Gullies & ravines	4
5.	Shifting cultivation	3
6.	Rain fed farming	10

Table 2b. Severity and extent of soil degradation in India (m ha)

Degradation process	Severity of degradation				Total area
	Low	Medium	High	Very high	
Water erosion					148.9
a. Loss of top soil	27.3	99.8	5.4		132.5
b. Terrain deformation		11.8		4.6	16.4
Wind erosion					13.5
a. Loss of top soil	0.3	5.5	0.4		6.2
b.Terrain deformation		4.6			
c.Terrain deformation/ over blowing			2.7		2.7
Chemical deterioration					13.8
a. Loss of nutrients		3.7			3.7
b. Salinization	2.8	7.3			10.1
Physical deterioration					11.6
a. Waterlogging	6.4	5.2			11.6
Total area	36.8	137.9	8.5	4.6	187.8

Source: Sehgal and Abrol (1994).

Table 2c. Extent of soil erosion in India

S.N	Degradation type	Arable land (mha)	Open forest (mha)
A	**Soil Erosion**		
1	Water erosion	73.27	9.30
2	Wind erosion	12.40	-
	Sub total	85.67	9.30
B	**Chemical degradation**		
1	Exclusively salt affected soil	5.44	
2	Salt affected and water eroded soils	1.20	0.10
3	Exclusively acidic soils (<5.5 soil pH)	5.09	
4	Acidic and water eroded soils	5.72	7.13
	Sub total	17.45	7.23
5	Physical degradation	1.07	16.53
	Total	**104.19**	-
	Grand total	**120.72 (104.19+16.53)**	

Source: ICAR (2010)

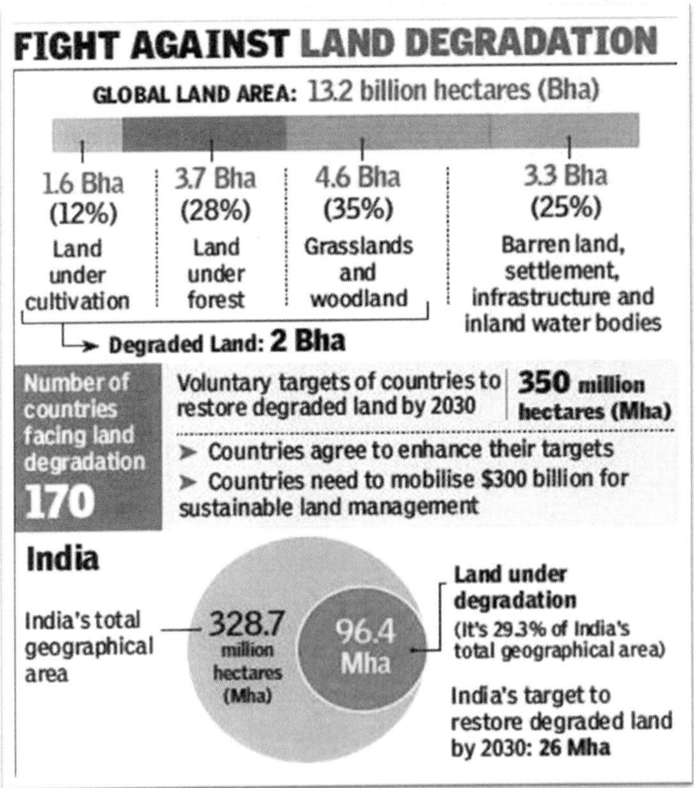

Fig. 2. Land degradation in India and world

Source: The Times of India dated 19th September 2019

History of conservation work

The soil and water conservation is as ancient as civilization because deep tillage, small dams, ponds, khadin, Naula (hills) etc. are age old practices of soil and water conservation. In modern era, a German scientist Woolny used vegetative and soil mulch to control soil and water erosion during 1877-1895. In USA, soil and water conservation work was started in 1850 but the conservation work was seriously taken up after Govt. policy launched on soil conservation in 1907. Dr. H.H. Bennet established soil conservation service in 1935. He did commendable work on soil conservation so he is known as father of soil conservation.

In India, soil conservation work was initiated with implementation of Soil Conservation Act in Punjab in 1904 through which the works on bunding contour trenching, gully control, terracing, afforestation etc. were emphasized. The systematic work was initiated with the establishment of 5 regional research centres, located at Solapur and Bijapur (Maharashtra), Hagari and Raichur (Karnatika) and Rohtak (Haryana) on dry farming in 1933. Later in 1953, Central Soil Conservation Board was established and 10 soil conservation research demonstration and training centres were opened at Dehradun and Agra (UP), Kota and Jodhpur (Rajasthan), Chandigarh (Pb), Oatacamund (TN), Vasad, Bellary and Ibrahimpatnam (AP) etc and major emphasis was given on contour bunding. Later, other centres in different states were opened to train people and government officials. The Dehradun centre was upgraded to '*Central Soil and Water Conservation and Training Institute*' and other centres except Jodhpur became the sub centres of Dehradun. The Jodhpur centre was also upgraded to '*Central Arid Zone Research Institute* (CAZRI)'. Besides, other centres and institutes like '*Central Research Institute for Dryland Agriculture (CRIDA)*', Hyderabad and other SAU's, *ICRISAT*, Hyderabad are also working on soil and water conservation work.

Conservation measures

A. Temporary Structures

Soil conservation is essentially required to protect the top soil against water and wind erosions. The selection of conservation measures may vary according the land, practicability colonies, extent of problem etc. Soil and water conservation measures (Table.4) may be grouped into two categories:

1. Mechanical or Engineering measures
2. Biological measures
    a. Agronomic practices
    b. Vegetative measures

Again, the soil and water conservation measures may also be grouped in two categories;

1. Conservation measures for arable land,
2. Conservation measures for non arable land.

1. Arable Lands

A. Mechanical or Engineering measures

The main objectives of the engineering measures are to provide the conditions for higher infiltration rate and to reduce the runoff velocity. Following engineering approaches are practiced:

a. ***Basin listing (Pitting)*** – Small basins or pits are dug along the contour on mild sloppy lands with the help of basin lister. These basins collect rain water and reduce the runoff velocity.

b. ***Sub soiling*** – The subsoil hard pan is broken with the help of subsoiler mainly to increase infiltration rate and conserve the soil moisture. The subsoiling also helps in reducing runoff velocity and conserve soil and water.

c. ***Bunding***: The main objectives of bunding are to conserve runoff water *in situ* and to conserve soil and water by compartmentalization of the fields. The bunding on contour line is called *contour bunding* that are made across the slope, so that runoff flow may be checked at different contours. The *contour bunding* is useful in low rainfall (<600 mm) and low slope (≤6%) areas, having high infiltration and permeability rates. *Graded bunding* may be useful in high rain fall (> 600 mm) and high slope (8%) areas with fine textured soils. The normal slope of bunds is maintained from 0.1 to 0.5%. Different field studies revealed that soil loss decreased and crop yields increased by many folds by graded bunding.

d. ***Terracing***: A terrace is an earth embankment, constructed across the slope to control runoff and minimize soil erosion. Terracing is good practice for conserving hilly and sloppy lands under crop production. The terraces, made on parallel to contour lines, are known as *contour terracing*. The main objectives of terracing are to convert sloppy lands into arable lands, check runoff flow and encourage infiltration rate. Terracing consists of construction of step like fields by half cutting and half filling. The terraces are mainly two types;

1. Broad base ridge terraces: They consist of a channel with a ridge on its lower side to drain or to absorb surface runoff. They may be of two types i.e. graded terrace (drainage type) and level terrace (absorption type). They are suitable for land up to 10 % slope with low permeability. They are made either to retain or absorb runoff water on slopes. The horizontal and vertical distance between two terraces may be determined by under given formula;

$$H.I. = \frac{200}{S} + 10$$

$$V.I. = 2 + \frac{S}{10}$$

Where

H. I. = Horizontal interval between two terracing,

V. I. = Vertical interval (20 % variation is accepted).

S = Land slope

2. Bench terraces: These are the most popular conservation measures in hills. The bench terracing is suitable up to 33% land slope but farmers are adopting up to 60-70% slopes. The steps of bench type structures are made along the contours by cut and fill method to break the length and degree of slope. They are either '*level bench terrace*', '*outward sloping bench*' or '*inward sloping bench terraces*'. The level bench terraces are useful in medium rainfall areas with permeable soils so that all rainfall may be absorbed. The outward sloping type terraces are useful in low rainfall areas with highly permeable soils. The bench has slope towards the general slope of the field. The inward sloping terraces have inward slope and the drainage channel is made so that the excess rain water may be drained off. They are suitable in heavy rainfall areas with low soil infiltration rates.

3. Contour trenching: Trenches are made across the stream flow or runoff along a uniform level or at a particular contour. These trenches reduce the runoff velocity, conserve moisture and retard its scouring action and carrying capacity. The size of trenches varies with slope, rainfall and depth of soil available. The trenches are usually 60 cm×48 cm. in size and spacing varies from 10 to 30 m. The contour trenching has been widely used in the Aravalli region for soil and water conservation and the trenching sides were used for afforestation.

e. ***Diversion channels***: The diversion channels are made across the slope or flow to divert either full or a part of it mainly for reducing the flow and diverted water may be used for irrigation. In arid areas, this practice is used either to store the runoff in the reservoirs or to recharge aquifers by providing anicut/percolation tank etc.

f. ***Gully or nalah control structures***: Some temporary and permanent structures are made to reclaim gullies. Generally gully plugging is carried out either with live hedges, boulders, gabions earth or concrete made check dams.

g. ***Percolation tanks***: The percolation tanks are made on small streams or rivulets for impounding surface runoff, therefore, these tanks will not only divert the runoff but also to recharge groundwater table.

B. Agronomic measures

Agronomic practices are quite important not only to intercept rain drops and reduce the splash effect but also to reduce run off flow. Following agronomic practices may be adopted scientifically for improving soil and water management:

1. ***Contour farming***: It means tilling and planting at right angles to the natural slope of the land. The tillage operations should be parallel to the terraces.

Generally, contour farming may be used to control water erosion during high intensity rainfall at moderately slope of 2-8 percent. The reports reveal that contour tillage combined with strip cropping on terraces across the slope are more effective than contour tillage or strip cropping alone.

2. ***Strip cropping***: It refers to plant in alternate strips of intensively cultivated crops with strips of sod farming crops. The erosion sediments and runoff from the clean tilled strips are filtered out and retained on sod like crop strips. There are four types of strip cropping;
   a. *Field strip cropping*: The practice in which crops are planted in different strips uniformly across the slope, is called field strip cropping.
   b. *Contour strip cropping*: In this system the two types of crops are grown on contour line across the slope. It decreases the soil erosion by reducing the runoff velocity and promotes water intake in the soil.
   c. *Wind strip cropping*: Two types of crops like tall and small are grown in alternate strips against the wind flow to check the wind velocity.
   d. *Permanent buffer strip cropping*: Generally the perennial planting is encouraged in alternate strips with pastures, legumes and grasses in the highly eroded or in steep slope areas particularly to protect the eroded areas. The buffer strip cropping may be permanent or temporary in nature but it is not a part of rotational cropping.
3. ***Crop rotation***: The suitable crop rotations are adopted according to soil type, slope, nature and severity of erosion for sustaining the soil health and its productivity. Mixed cropping including different types of crops have been found very useful in controlling the water and wind erosion of eroded soils.
4. ***Mulching***: The grasses, weeds, green leaves straw, crop residues etc. are used to cover soil surface to prevent soil from blowing and being washed away, reduce evaporation, increasing infiltration, reducing weed population, improve organic matter and soil structure and maintain soil health. Sometimes, intercultural practices are carried out to check the weed infestation and the disturbed soil works as a good mulch. The soil mulch saves soil moisture induces water intake rate and reduces runoff flow. Similarly, the stubble mulching (crop residue) is also considered good technique to control wind and water erosion particularly in rainfed areas.
5. ***Minimum/Zero tillage***: The tillage operations make top soil loose and fine that can be eroded easily by wind and water, so minimum or zero tillage may be beneficial for soil and water conservation in high rainfall and windy areas.
6. ***Cover cropping***: In both rainfed or sloppy mountain areas, growing of cover crops may protect top soil from heavy wind and high rainfall. Hence, the

crops particularly grasses and pulses suitable to the regions should be grown in eroded areas for conserving soil and water.

C. Vegetative or agrostological measures

Growing of grasses, trees, vegetables, fruit trees etc. with field crops on highly eroded areas should be encouraged as an alternate land use systems, not only to utilize the land resources efficiently but also to stabilize the eroded areas. Therefore following vegetative approaches or agrostological systems may be adopted;

i. Agri-silvi culture (crops + trees)
ii. Agri-horticulture (crops + fruit trees)
iii. Silvi-horticulture (trees + fruit trees)
iv. Horti-silvi-pastoral system (fruit trees + trees + grasses)
v. Silvi-pasture system (Trees + grasses)
vi. Agri-silvi-pasture system (crops + trees + grasses)
vii. Horti-pasture system (fruits + grasses)
viii. Agri-horti-silviculture (crops + fruits +trees)
ix. Ley farming (crops + grasses)
x. Afforestation (free farming) (trees-multi purpose trees)

The vegetative approaches differ with soil, slope, climate and socio-economic conditions. Therefore, following approaches may be adopted for different conditions as given below;

i. Degraded Lands

In the shiwalik foot hills, agri-horticulture and horti-pastoral systems such as intercropping of guar, cowpea or pearlmillet with peach, turmeric with papaya, *Chrysopogon fulvus* or *napier grass* with Aonla or ber have been identified as an economically viable and eco friendly systems for rehabilitation of these lands. In semi arid regions, fruit plants like guava sapota, Kinnow and ber, fuelwood plantation of *Leucaena leucoephala, Acacia nilotica* and *Eucalyptus* and grass species of *Cymbopogon martini*, *Vetivaria zizanoides*, *Dicanthium annulatum* and *Cenchrus ciliaris* were found economically suitable.

In Doon valley, intercropping of turmeric with kinnow or peach, cowpea/urdbean and toria with mango and litchi were found viable systems.

ii. Ravine lands

The ravinous areas of Yamuna, Chambal and Mahi rivers can be rehabilitated with planting of *Acacia nilotica* in the gully bottom and *Acacia* tortilis in top location along with *Cenchrus ciliaris*. Ber, aonla and lemon are common fruit trees to be planted in the eroded areas of Chambal river. The grass species like *Bracharia mutica* and tree species *Prosopis juliflora, Eucalyptus tereticornis* and

Tamarix dioca may be planted in saline river banks. Similarly, intercropping of *Dandrocalamus strictus* + *Cenchrus ciliaris* was found suitable for gully plugging in the ravines of Mahi river. The *Leucaena* based alley cropping is a good example for most of the eroded arid and semiarid areas.

iii. Desert lands (Desertification control)

Prosopis juliflora, P-cineraria, Acacia tortilis, A. radiana, Zizyphus mauritiana, Colligonum polygonides Crotalaria burhia, Z. nummularia, Lasiurus indicus, Panicum turgidum and *P. antidotale* may be used to stabilize the sand dunes in Rajasthan. Shelter belts on field boundaries and wind breaks across the wind velocity may protect the crops, soil and water losses. For effective wind breaks, the combination of trees and shrubs having centre row of *Albizia lebbek* with single row of *Acacia tortilis, Cassia siamea* or *Prosopis juliflora* on both sides of centre row were found very effective. The shelter belt should be maintained continuously without any gap. In one of the studies, it has been observed that pearlmillet gave 70% higher grain yield grown in leeward side of the wind breaks and soil retained 14% higher moisture.

iv. Sodic land

Ber(*Zizyphus mauritiana),* guava *(Psidium guajava)*, karunda (*Carisa carandus*), aonla (*Emblica officinalis*), pomegranate (*Punica granatum*) and jamun fruits (Syzygium cumini), popular, eucalyptus and Acacia spp. (trees) may be planted in sodic lands. Studies indicated that popular in association with rice-berseem system improved soil organic matter and available N beside soil and water conservation.

v. Saline lands

Prosopis juliflora, Tamarix troupii, Tamarix articulata, Acacia farnesiana, A. nilotica, A. tortitlis, Casurina glauca, Eucalyptus camaldulensis and Leucaena leucocephala etc., could be grown successfully on saline soils for higher benefit per unit land area.

vi. High water table areas

Date palm, kokum, eucalyptus, popular, ber, banana, jamun, water chestnut (T*rapa natans*), gorgan nut (*Euryale ferox*), Palmyra palm (*Borassus flabellifer*), mahua, serewpine (*Pandanus tectorius)*, nipa palm etc. are tolerant to high water table areas. Besides, surface and subsurface drainage facilities should be created before planting of trees in flooded and water logged areas.

vii. Arid and semi arid areas

Some of the tree speeds like *Albizia spp., Acacia* spp. *Aegle marmelos, Ailanthus excelsa. Dalbergia siszo, Bulia monospanma, Hardwickia binato* and *Anogeissus pendula* are suitable for different agroforestry systems and help in drought moderation under highly degraded lands. *Albizia amara, Aceacia tortilis* based silvi pastoral systems and *Hardwikia binata* based agri-silvipasture system were found promising agroforestry systems semi arid regions. In another study, *Acacia amara* + *Leucaena leueocephale* + *D. cinerea* + *C. fulvus* + *Stylos* with natural grass lands based very remunerative silvipastoral system.

viii. Bouldery wastelands

About 1/3rd of the Himalayan region is bouldery wastelands formed due to excessive erosion and depreciation in the hilly slopes. Under these conditions, silvi-pasture systems like *Dalbergia sissoo* + *Chrysopogon fulus*, *Accacia catechu* + *Eulaliopsis binata*, *Eucatyptus* hybrid + *Chrysopogon* fulvs and Leucaena *leucocephala* + Napier grass may be adopted. Some multi purpose trees (MPT's) ie. Albizia Lebbek, *Grewia optiva*, *Bauhinia purpurea* and *Leucaena Leucocephala* may be grown with two grasses i.e. *Chrysoprgon fulvus* and *Eulaliopus binata*. In Shiwalik hills, *Eulaliopsis binata*, *Saccharum munja* or *Vertivaria zizanoids* with *Accacia nilotica*, *Eulaliopsis biuata* with *Eucatyptus tereticorsis* were found promising silvi-pasture systems (Table.3).

Table 3. List of multipurpose tress (MPTs) for different regions/conditions

S.N	Region/condition	MPTs
1.	Himalayan	*Grewia optiva, Morus alba, Celtis australis, Eucalyptus ss., Alnus nepalensis, Acacia spp., Pine spp., Paraserianthes falcataria, Albizia spp.and terminalia spp.*
2.	Indo-Gangetic plains	*Pupulus spp., Acacia spp., Dalbergia sisso, Anthocephalus cadamba, Casuarina equisetifolia, Prosopis chilensis, Eucalyptus spp.*
3.	Arid & Semi arid	*Prosopis chilensis, Azadirachita indica, Dalbergia sosso, Leucaena leucocephala, Acacia spp., Bamboos and Harwickia binata.*
4.	Humid & Sub humid	*Gmelina arborea, Acacia spp., Dalbergia sisso and Bamboos.*
5.	Tropical & Islands	*Tamarindus indica, Azadrachita indica, Ceiba pentandra, Tectona grandis and Gliricidia spp.*
6.	Saline lands	*Prosopis juliflora, Tamarix troupii, Tamarix articulata, Acacia farnesiana, A. nilotica, A. tortitlis, Casurina glauca, Eucalyptus camaldulensis and Leucaena leucocephala.*
7.	Sodic lands	Popular, Eucalyptus, Karunda, Guava, Ber, *Acacia* spp. Jamun and Aonla.

Source: Dhyani (2002)

2. Non-arable lands

Mine spoils, landslides, torrents, gully and roadside erosion are the major problems of soil and water conservation in non arable soils. The soil erosion may be as high as 550 t/ha/annum at such degraded sites compared to 3 t/ha/annum from a well-forested watershed (Jynal *et al*., 1995). They cause considerable damage to the ecosystem and pose a constant threat to the life and property of the inhabitants.

1. Landslides

The landslide is downward movement of a rock or its part or soil due to slope and gravity force. The landslide is a major problem of hilly areas. It causes severe damage to national property as roads, dams, power generation units, reservoirs, human settlements, agriculture, orchard, trade, tourism, forest, pastures etc. damaged and wastage of time and money in rejuvenation.

Causes of landslides

The causes can be grouped into two categories:

a. Man-made or preventable causes

i. Overgrazing
ii. Deforestation
iii. Unscientific agriculture on steep stapes
iv. Unscientific mining and road construction
v. Forest fires etc.

b. Natural or non preventable causes

i. Seismic disturbances
ii. Weak and unstable geological formation
iii. Hydrological – pore water pressure in soil or seepage water flow.

2. Mine spoils

Mining is essential for development but unscientific mining causes havoc to nature and human settlement as it is variably entails removal of all plant cover on land surface and productive topsoil, destroying the habitat for all flora and fauna. Mining is spread over an area of 9, 43, 380 ha under 7365 mining leases in India; more than 90% being surface mines. Mining operations leave behind huge overburden or debris dumps that are subjected to erosion and causes revert pollution by dissolved toxic substances heavy sedimentation and floods and also degrade the water courses, both surface and underground.

On the basis of rainfall intensity, slope and erosion intensity, the soil and conservation measures can be grouped into three categories:

i. Slope stabilization measures
ii. Gully channel stabilization measures
iii. Torrent control measures

i. *Slope stabilization*: Following measures can be adopted to stabilize the slope for further use;

a. ***Diversion drains:*** The diversion drains are made across the slope or flow to divert run off water away from main stream to desired places particularly for irrigation purposes. The gradient of diversion drain should be preferably be 0.5% and may be increased to 2% in exceptional cases.

b. ***Trenches:*** Generally, trenches are made across the slope mainly to reduce the velocity of run off water and recharge the groundwater. The trenches are made with a cross section of 0.3 × 0.3 m, while staggered trenches are made up to 15 m. in length with inter space between them. The trenches are used to collect the whole run-off water or a part of it. If the run off water is excess than the capacity of trenches, excess water is drained with opening at downside the slope. The trenches may be designed to collect run off expected from storms of 5-10 years recurrence interval.

c. ***Geojute:*** The jute mats are used to stabilize the land slides; mine spoil areas and constructed slopes. These geojutes check the runoff flow and also air and water erosion. Afforestation on denuded soils and sand dunes can be quickly rehabilitated with the help of geojute. Juyal *et al.* (1994) reported that vegetation could be established on mine spoils within 3 years with the help of geojutes.

d. ***Retaining walls(Gabions):*** The retaining walls are constructed across the slopes of precipitous hills to check the landslides. The width of wall at base should be two third of the wall height. Such type of wall is called *gabion.* The width of the retaining wall is reduced in steps to 1 m at the top if there is no surcharge. If there is a surcharge, 2 m top width is required.

e. ***Kutta-crate structures:*** The vacant cement gunny bags, filled with mine spoil, known as kutta-crate structures, are used to construct the check dams to stabilize the limestone mine spoil slopes. The filled gunny bags are laid in a row over one another in three layers to make a height of about 0.6 m. These structures have been used, successfully in Doon valley by U.P. forest department for stabilizing the limestone mine spoil slopes.

ii. Gully/Channel stabilization structures

The gullies or drainage channels are the carriers of runoff and sediment and cause soil and water erosion. Hence, the velocity of gully stream needs to be reduced up to permissible limit so that gully erosion may be checked. The check dams are mainly constructed to check the gully erosion. The check dams may be temporary dams or gabion check dams.

a. ***Temporary check dams*** - The temporary check dams are made across the small streams and made of loose stone masonry brushwood, log wood etc. It's life is usually 3-5 years. Later on, the vegetation could establish and provide protection to the surface.

b. ***Gabion check dams***- The gabion check dams may be constructed in second and third order (main) gullies/channels for retention of debris and soil accumulation without ponding. Normally, the gabions are made across the larger streams with the help of stones/boulders packed closely in G-1 wire mesh cages and thickness of wire should be 10 gauges. The boxes may be fabricated with a dimension of 3m×1m×1m or any other suitable dimensions with a mesh opening size of 10-20 cm depending on the stone size. These gabions check the flow and carrying debris settled in the bottom of stream and also in sides of gully that facilitate stabilization of gullies and promote need vegetation. The cross barriers are suitable in the head water reaches in higher order streams/main drainage channels receiving relatively large quantities of run off and debris flow. The gabions are preferred in soil conservation works mainly due to

i. Flexible (bend without breaking),
ii. Porous(water can seep through them),
iii. Stable, and
iv. Economical compared to cement structures.

Gully control structures may also be classified as;

1. Temporary gully control structures

They are temporary in nature and lost for a very short period of 2-4 years till the vegetation is established. Various types of temporary gully control structures are as follows:

(i) One row or single row wooden post check dam,
(ii) Two row or double row wooden post check dam filled with brush wood,
(iii) Two row or double row wooden post check dam filled with boulders,
(iv) Woven wire-wooden post check dam,
(v) Loose boulder check dam,
(vi) Sand filled plastic bag check dam, and
(vii) Wooden plank-post check dam.

2. Semi-permanent gully control measures

These measures usually have a longer life, say of 10-15 years, and normally do not require regular maintenance. They do not require the assistance of vegetative growth for controlling the gully but helps the vegetation to come up on gully sides. They are constructed in places where materials are available in sufficient quantity. These are very effective in steep gullies traversing hilly and mountainous regions. They are also useful in ravines where passage for livestock has to be provided. These are very much suited in second and third order streams with a runoff contributing area of 40 to 100 ha. The depth of gully is usually less than 4 m. Various types of semi-permanent gully control structures are:

a. Loose boulder check dam,
b. Log check dam, and
c. Gabion check dam.

3. Permanent Gully Control Structures

Concrete and masonry structures are efficient supplemental control measures. Good vegetative practices together with proper land use are indispensable in a sound soil and water conservation program. But there are many instances where vegetative measures and simple practices alone are inadequate to handle a concentration of water. In such cases, permanent structure of masonry and concrete play an important part in reinforcing or supplementing the other practices.

There are also instances where a high degree of safety and permanence are desirable. The conservation measures may be required which will be good insurance against loss of life or destruction of property. Vegetative control measures are subjected to the influences of such uncertain factors as climate, insects, etc., and therefore, are not too dependable. On the other hand, properly designed and installed structures are of longer life and dependability.

The structure's main function is to safely dispose the peak rate of runoff for a given frequency from a higher elevation to a lower elevation. It should have arrangements to dissipate the kinetic energy of discharge within the structure in a manner and degree that will protect both the structure and down stream channel from damage.

The permanent structures are used in medium to large gullies with medium to large drainage areas. Mainly three types of permanent structures are constructed;

a. Drop spillways

b. Chute spillways, and

c. Drop inlet spillways.

Table 4. Soil and water conservation measures for arable and non arable land

S.N	Measures	Most suitable in	
		Plain land and mild slope	**Hilly land**
A.	**Suitable for arable land**		
A.1.	**Mechanical or Engineering**		
(1)	Bunding		
(a)	Narrow based terracing		
(a1)	Contour bunding	*	
(a2)	Graded bundig (uniformly and variable graded)	*	
(b)	Broad based terracing		
(b1)	Level or absorptive type	*	
(b2)	Graded or disposal or channel type	*	
(2)	Trenching		
(a)	Contour trenching		
(a1)	Continuous	*	*
(a2)	Staggered	*	*

(b)	Graded trenching	*	*
(3)	**Trenching (Bench)**		
(a)	Level		*
(b)	Inward sloping		*
(c)	Outward sloping		*
(d)	Puertorican type		*
(4)	Stonewall		
(a)	Contour	*	*
(b)	Graded	*	*
(5)	Zing terracing	*	*
(6)	Conservation bench terracing	*	*
(7)	Land leveling/grading	*	
(8)	Silt detention tank	*	*
(9)	Water ways or disposal drain	*	*
(10)	Revetment		*
A.2	Biological measures		
A.2.1	**Agronomical Practices**		
(1)	Contour cultivation	*	*
(2)	Strip cropping (contour, field, wind, permanent or buffer)	*	*
(3)	Inter or mixed cropping	*	*
(4)	Green manuring	*	*
(5)	Crop rotation	*	*
(6)	Use of crop residue	*	*
(7)	Mulching	*	*
(8)	Relay cropping	*	*
(9)	Seasonal crop cover	*	*
(10)	Improved crop management practices	*	*
(11)	Tillage	*	*
A.2.2	**Vegetative measures**		
(1)	Grassed contour barrier	*	
(2)	Grassed waterway or grassed disposal drain	*	*
(3)	Agroforestry	*	*
(4)	Agri-pasture	*	*
(5)	Agri-horticulture	*	*
B	Suitable for non-arable land		
B.1	**Mechanical or Engineering or structural**		
(1)	**Trenching**		
(a)	Contour trenching		
(a.1)	Continuous	*	*
(a.2)	Staggered	*	*
(b)	Graded trenching	*	*

(2)	Stone wall		
(a)	Contour	*	*
(b)	Graded	*	*
(3)	Silt detention tank	*	*
(4)	Diversion drain	*	*
(5)	Toe wall/toe drain	*	*
(6)	Interceptor drain	*	*
(7)	Channelisation of flow or stream training (spurs, retard, check dam, desilting)	*	*
(8)	Waterways	*	*
(9)	Retaining wall		*
(10)	Revetment		*
(11)	Wattling		*
(12)	Check dam	*	*
(13)	Drop structure		*
(14)	Drop inlet structure		*
(15)	Chute structure		*
B.2	Biological measures		
(1)	Affforestation/reforestation	*	*
(2)	Social forestry	*	*
(3)	Silvi-pasture	*	*
(4)	Pasture development	*	*
(5)	Leguminous pasture plant	*	*
(6)	Farm and range plant	*	*
(7)	Meadow grasses	*	*
(8)	Check dam	*	*
(9)	Check dam	*	*
(10)	Wattling	*	*
(11)	Mulching	*	*

*Refers to adoption of the measures.

The selection of a suitable structural measure for a given condition is a specialized job and a specific measure is opted on the basis of following factors;

1. Land slope,
2. Soil depth and texture,
3. Crops to be raised, and
4. Rainfall.

Based on above parameters, the suitability of various structural control measures is given below in table.5.

Table 5. Suitability of erosion control measures and related land use pattern.

Erosion control measures	**Suitability**			
	Land slope (%)	**Soil depth**	**Rainfall (mm)**	**Crops or land use pattern**
1. Bunding				
A. Narrow based terracing				
(a) Contour bunding	<8	Shallow to deep	<800	Small millets, pulses, oilseeds, coarse grain, root crops.
(b) Graded bunding	<8	-do-	800-1500	-do- and wheat, vegetables.
B. Broad based terracing				
(a) Level or absorption type	<8	-do-	<800	Small millets, pulses, oil seeds, coarse grain.
(b) Graded or disposal or channel type.	<8	-do-	800-1500	Small millets, pulses, oilseeds, coarse grain, wheat, vegetables.
2. Trenching				
(a) Contour trenching				
(a1) Continuous	<8	Medium to deep	<1500	Tapioca, ginger, turmeric and similar annual crops.
(a2) Staggered	<8	Shallow	<800	Papaya, banana with irrigation.
	<33	Medium to deep but well drained	<2000	Tea, coffee, arecanut, coconut, black paper, nutmeg,Cinnamon papaya, banana, etc.
(b) Graded bunding	<33	Medium to deep soil but well drained	2000-3000	-do-
3. Terracing				
(a) Level	<33	Medium to deep	<2500-3000	i) Paddy (high rainfall). ii) Small millets, pulses, oil seed, coarse grain, vegetables in low rainfall.
(b) Inward sloping	<33	-do-	-do-	Potato, other vegetables, maize, oats, etc.
(c) Outward sloping	<33	Shallow	<1200	Small millets, oats, barley etc.
(d) Puertorican type	<33	Medium to deep	<1500	Root crops, vegetables, oats, small millets, etc.
4. Stonewall				
(a) Contour	<33	Shallow to medium	<1500	Annual crops like tea, coffee, spices, etc.

(b) Graded	<33	-do-	1500-2500	-do-
5. Zing terracing	<33	Shallow to deep	<1200	Paddy, maize, pulses, oil crops, sorghum, bajara root crops, vegetables etc.
6.Contour bench terracing	<33	-do-	<1200	-do-
7. Silt detention tank	<33	Shallow to deep	<2500	-
8. Waterways or disposal drain	<33	-do-	<2500	-
9. Revetment	<33	-do-	<2500	-

A simple comparison between terracing i.e. bench terracing, broad based terracing, narrow based terracing and trenching is given in table.6. It is clear that broad based and narrow based terracing are better than bench and trenching because of minimum area occupied and less soil disturbance.

Table 6. Comparison between terracing and trenching.

S.N	Item	Bench terracing	Broad based terracing	Narrow based terracing	Trenching
1.	Initial cost	High	Low	Low	Low
2.	Soil movement/disturbance	High	Less	Less	Less
3.	Skill involved	High	Low	Low	Low
4.	Reduction in area for cultivation	High	No	Less	Marginal
5.	Interference with farming	Yes	No	Yes	Yes
6.	Maintenance	More	More	Less	Less

iii. Torrent control measures

Following structures are used to control torrents;

1. Spurs,
2. Retards, and
3. Gabions.

1. Spurs

The spurs are also known as spur dikes or groyne. Spurs are structures constructed at an angle to the stream flow extending from the bank into the stream (Fig.3). The main objectives are to guide the stream flow along a desired alignment and to prevent the bank erosion. The spurs may be of different types;

i. Repelling type (pointing upstream),
ii. Attracting typing (pointing down stream), and
iii. Deflecting type.

The repelling as well as deflecting type of spurs is of good use as they induce sedimentation at both upstream and down stream sides but the nose is subjected to greater scouring. The attracting type of spurs does not induce sedimentation on upstream side but they have lesser scour at the nose and hence provide greater stability than repelling and deflecting type of spurs. The projected length of the spurs should not exceed 30% of the stream width. A spur should protect a certain length of the bank, hence, a series of spurs provide continuous effect. The spurs may be spaced 3-4 times its length in a straight reach, where as on a curved reach the spacing may be reduced to about 2 times the spur length. Later, the plantation may follow on siltation on banks.

2. Retards

Retards are used to reduce the stream velocity sufficiently near the bank to prevent bank erosion or scour its toe by inducing deposition. Generally, following types of retards or checks are used.

i. ***Live hedges-*** The live hedges of narkul (*Arundo donax*), *Ipomoea carnea*, hybrid Napier *Jatropha carcas* etc. are planted in 60cm wide trenches to reduce the torrent flow resulting deposition of debris and gravels on banks. The narkul is very strong and suits well to resist the flood flows and also protect the banks. The live hedges are mainly useful in small streams and torrents.

ii. ***Jacks-*** The jacks are wooden posts of 8 cm diameter and 2-3 m length, which are placed in series along the eroded bank and are tied together.

iii. ***Jetted posts-*** These are green posts of 2-2.5 m. in length and 6-12cm diameter. They are erected along the actively eroded banks in two rows having 1 m distance.

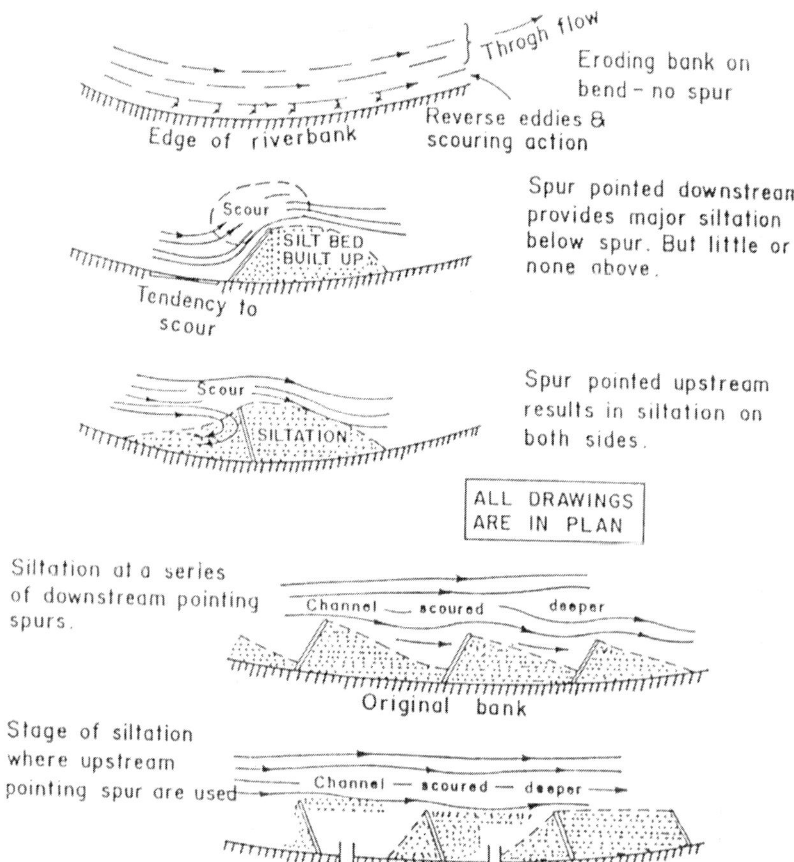

Fig. 3. Principles of spurs (Storm, 1962)

3. Gabions

The small stones and boulders are placed in wire nets and are placed along the banks particularly to prevent the banks of river or torrents. China is a good example, where the river banks near the big cities have been successfully protected.

After treatment of degraded lands, fuel, fodder, fibre, horticultural plantation, energy plantation etc. can be made on stabilized banks. Besides, water regime improves, lean flow period extends, debris flow reduces, vegetation cover improves and channel slope reduces after treatment (Table.7). The conservation measures also help to improve the water quality of the sites (Table.8).

Table 7. Eco-restoration of mined watershed at Sahastradhara (Dehradun)

S.N	Particulars	Before treatment	After treatment
1.	Vegetation cover (%)	10	80
2.	Debris flow (t/ha/hr)	550	08
3.	Monsoon run-off (%)	57	37
4.	Lean period flow (days)	60	240
5.	Channel slope (%)	38	20

Source: Juyal *et al.* (1995).

Table 8. Water quality improvement of treated mine spoil area

S.N	Sites	Nutrients (ppm)		
		Ca	**Mg**	**SO_4**
1	Treated mines	74	34	138
2	Untreated mines	389	120	756
3	Water quality standards	75	50	250

Source: Juyal *et al.* (1995).

B. Permanent structures

Following are the permanent structures used to control run off and conserve the soil land water resources;

a. Drop spillways
b. Chute spillways, and
c. Drop inlet spillways.

a. ***Drop spillways:*** It is a weir constructed across the flow to control the flow and check the erosion. The flow passes through the weir opening, drops to an approximately level apron or stilling basin and then passes in to the down stream channel. The drop spillways are placed at intervals along the channel. They stabilize the banks of up streams slowly. They are also used as outlets for tile and surface water at the upper ends of drainage ditches. They are limited to drops of 3-4m. They are unsuitable where greater drops are required. They are made of masonry materials like stone, bricks, concrete and concrete blocks with or without reinforcing.

b. ***Chute spillways:*** The chute spillways or flumes are used where a channel is to be brought down a steep slope. The flow water enters through inlet and down a lined steep channel to the flow of the outlet. They are frequently in combination with earth dams to drop water farther than is feasible with drop structures. A chute spillway usually requires less masonry or concrete than a drop inlet structure of the same capacity and drop. The chute spill ways are preferred over drop spillways when the drop exceeds 4m. It is also superior to drop inlet spillways when large discharges and no temporary storage are required.

c. ***Drop inlet spillways:*** The drop inlet spillways are preferred when there is a need to control the downstream channel flow by providing a temporary upstream storage. It consists of earth dam and a pipe spillway. The earth dam provides the upstream temporary storage of run off, while the pipe spillway controls the discharge downstream. They are commonly used to control the flood, constructing canals for irrigation, draining water in to reservoirs and lakes, silt detention reservoirs and settling basins.

Questions

a. Short and Descriptive

1. a. Define soil and water conservation?

   b. Discuss the extent of problem related to soil and water conservation in India?

1. Discuss the history of soil and water conservation in India?

2. a. What are the principles of soil and water conservation?

   b. What is erosivity and erodibility and how both are important for soil erosion?

   c. What is the mathematical formula of soil loss?

4. Discuss the following;

   a. Terracing

   b. Basin listing

   c. Contour bunding

   d. Retards

   e. Spurs

   f. Gabions

   g. Strip cropping

   h. Check dams

   i. Silvi-pasture system

   j. Social forestry

   k. Geojute

l. Diversion drains.

5. Describe the mechanical measures of soil and water conservation in arable land?

6. What do you mean by biological measures of soil and water conservation?

7. Discuss the agronomic manipulations for better soil and water conservation?

8. Describe the agrostological measures for following conditions;

a. Saline lands

b. Sodic lands

c. Waterlogging conditions

d. Hill regions.

e. Degraded lands

f. Bouldery wastelands.

9. Discuss the permanent measures for gully stabilization.
10. Describe the soil and water conservation measures under following heads;

a. Slope stabilization

b. Torrent control

b. Objective questions (MCQs)

1 Soil conservation refers to

a Protection of top layer of soil

b Save soil from water and wind erosion

c Save soil from chemical deterioration

d All above

2 Chemical deterioration of soil refers to

a Salinization

b Alkalinization

c Acidification

d All above

3 Physical deterioration of soil means

a Waterlogging

b Hardening of soil

c Mechanical impedance

d All above

4 Under moderate to strong soil erosion condition, the loss of crop productivity in inceptisols is expected equal to

a 5-25%

b 10-50%

c >50%

d None above

5 Under moderate to strong soil erosion condition, the loss of crop productivity in vertisols is expected equal to

a 5-25%

b 10-50%

c >50%

d None above

6 Under strong soil erosion condition, the loss of crop productivity in alfisols is expected equal to

a 5-25%

b 10-50%

c >50%

d None above

7 If the soil pH is in between 4.5 to 5.5, the loss of productivity is expected equal to

a Nil | b <10%
c 10-25% | d 25-50%

8 If the soil pH is in between 5.5 to 6.5, the loss of productivity is expected equal to

a Nil | b <10%
c 10-25% | d 25-50%

9 If the soil pH is in between 3.5 to 4.5, the loss of crop productivity is expected equal to

a Nil to very slight | b <10%
c 10-25% | d 25-50%

10 If the soil pH is <3.5, the loss of crop productivity is expected equal to

a Nil to very slight | b <10%
c 10-25% | d 25-50%

11 Annual loss of top soil is expected

a 5 t/ha | b 10 t/ha
c 16 t/ha | d 24 t/ha

12 Annual loss of nutrients due to loss of top soil is expected to

a 2.6 m t | b 3.8 m t
c 6.5 m t | d 8.4 m t

13 Water erosion is mainly influenced by

a Rainfall | b Soil erodibility
c Slope length | d All above

14 Who is known as father of soil conservation?

a Dr. H. H. Bennet | b Dr. M. S. Swaminathan
c Dr. R.S. Paroda | d None above

15 Which of the following is not a principle of soil and water conservation?

a Conservation of soil and water | b Land use as per capability classes
c Improvement of problematic soils | d None above

16 Wind erosion is influenced by

a Soil cloudiness | b Soil cover
c Local wind velocity | d All above

17 Which of the following is a mechanical measure for soil and water conservation?

a Terracing
b Bunding
c Sub soiling
d All above

18 Which of the following is a mechanical measure for soil and water conservation?

a Contour trenching
b Percolation tanks
c Basin listing
d All above

19 Which of the following is a agronomic measure for soil and water conservation?

a Contour farming
b Strip cropping
c Crop rotation
d All above

20 Which of the following is a agronomic measure for soil and water conservation?

a Mulching
b Cover cropping
c Conservation tillage
d All above

21 Which of the following is a vegetative measure for soil and water conservation?

a Agri-silvi culture
b Agri-silvi-pasture system
c Ley farming
d All above

22 Which of the following intercropping system is suitable in ravine soils?

a *Dandrocalamus strictus* + *Cenchrus ciliaris*
b *Leucaena leucocephala* based intercropping systems
c Pearlmillet + Pigeon pea
d All above

23 Which of the following fruit plant may be grown in ravine lands?

a Ber
b Aonla
c Lemon
d All above

24 Which of the following multipurpose tree can be grown on sodic soils?

a Ber
b Aonla
c Pomegranate
d All above

25 Which of the following multipurpose tree can be grown on high water table areas?

a Kokum
b Date palm
c Jamun
d All above

26 Which of the following multipurpose tree can be grown in Himalayan region?

a *Grewia optiva* b *Morus alba*

c *Eucalyptus* spp. d All above

27 Ley farming refers to

a Growing of crops and grasses b Growing of crops and trees

c Growing of aromatic and medicinal plants d All above

28 Silvi-pasture system refers to

a Growing of crops and grasses b Growing of crops and trees

c Growing of trees and grasses d All above

29 Which of the following silvi-pasture system can be grown in bouldery wastelands of Himalayan region?

a *Dalbergia sissoo* + *Chrysopogon fulus*

b Leucaena l*eucocephala* + Napier grass

c *Eucatyptu*s hybrid +*Chrysopogo*n fulvs

d All above

30 Which of the following is a major problem of soil and water conservation in non arable lands?

a Mine spoils b Land slides

c Torrents d All above

31 Which of the following is a major cause of landslides?

a Deforestation b Overgrazing

c Unscientific mining and construction d All above

32 Which of the following is a non preventable cause of landslides?

a Seismic disturbances b Weak and unstable geological formation

c Hydrological problems d All above

33 Which of the following is a slope stabilization measure?

a Diversion drains b Trenching

c Retaining walls d All above

34 Which of the following is a gully stabilization measure?

a Temporary check dams b Gabion check dams

c Diversion ditches d All above

35 Which of the following is not a type of retard?

a Live hedges
b Jacks
c Jetted posts
d Spurs

36 Spur refers to

a A structure constructed at an angle from river bank to the stream,
b A tree species
c Type of terrace
d None above

37 The main function of spur is

a Guide the stream flow
b Check the bank erosion
c Induce the siltation
d All above

38 Spurs are commonly used to control

a Torrents
b Wind erosion
c Gully erosion
d All above

39 Most popular soil and water conservation measure in hills is

a Terracing
b Contour trenching
c Bunding
d All above

40 Which of the following is a permanent gully stabilizing measure?

a Drop spillways
b Chute spillways
c Drop inlet spillways
d All above

41 Contour bunding is useful for controlling erosion in areas of

a Low rainfall (<800 mm)
b Moderate rainfall (800-1500 mm)
c Heavy rainfall (>1500 mm)
d All above

42 Graded bunding is useful for controlling erosion in areas of

a Low rainfall (<800 mm)
b Moderate rainfall (800-1500 mm)
c Heavy rainfall (>1500 mm)
d All above

43 Leveling terracing is useful for controlling erosion in areas of

a Low rainfall (<800 mm)
b Moderate rainfall (800-1500 mm)
c Heavy rainfall (>1500 mm)
d All above

44 Inward slopping terracing is useful for controlling erosion in areas of

a Low rainfall (<800 mm)
b Moderate rainfall (800-1500 mm)
c Heavy rainfall (>1500 mm)
d All above

45 Graded stone wall is useful for controlling erosion in areas of

a Low rainfall (<800 mm) | b Moderate rainfall (800-1500 mm)

c Heavy rainfall (>1500 mm) | d All above

46 Bench terracing is suitable up to the land slope of

a <10% | b <20%

c ≤ 33% | d >33%

47 Bunding is suitable up to the land slope of

a <8% | b <20%

c ≤ 33% | d >33%

48 Which of the following is a cause of soil loss/erosion?

a Deforestation | b Overgrazing of pastures

c Shifting cultivation | d All above

49 Which of the following disadvantage is not related to soil erosion?

a Heavy flood | b Silting of water bodies

c Destruction of vegetation | d All above

50 Which of the following is not a type of water erosion?

a Sheet erosion | b Rill erosion

c Gully erosion | d Saltation

51 Which of the measure is to be adopted for gully stabilization when temporary storage is required upstream?

a Drop spillways | b Chute spillways

c Drop inlet spillways | d All above

52 Chute spillways are suited under

a Steep flow | b Drop is > 4m

c Temporary storage upstream is no required | d All above

53 Drop inlet spillways are useful for

a Flood control | b Construction of canals

c Settling basins | d All above

54 Drop spillways have drop approximately

a 1-2m | b 2-3m

c 3-4m | d 4-5m

55 Which of the type of spur induces the siltation at both upstream and downstream sides?

a Repelling type
b Attracting type
c Both 'a' and 'b'
d None above

56 Soil erosion can be calculated by

a f (Erosivity) x (Erodibility)
b R x K x L x S x C x P
c I x R x K x F x C x W x D x B
d None above

57 Predicted soil loss (t/acre/year) as proposed by Wischmeier (1959) can be calculated by

a f (Erosivity) x (Erodibility)
b R x K x L x S x C x P
c I x R x K x F x C x W x D x B
d None above

58 Wind erosion (E) as proposed by Wischmeier (1959) can be estimated by

a f (Erosivity) x (Erodibility)
b R x K x L x S x C x P
c I x R x K x F x C x W x D x B
d None above

59 How much area is subjected to soil erosion in India?

a 175 mha
b 180 mha
c 190 mha
d 210 mha

60 How much percentage of total geographical area of India is affected by soil erosion?

a 44.5%
b 53.3%
c 58.2%
d 60.5%

61 The process of detachment of soil particles from the top soil and transportation of detached soil particles by wind or water is called

a Water erosion
b Wind erosion
c Soil erosion
d All above

62 The soil erosion caused by disturbance in natural equilibrium by activities of man and animal is called

a Geological erosion
b Accelerated erosion
c Wave erosion
d None above

63 Soil erosion occurs due to soil forming processes is called

a Geological erosion
b Accelerated erosion
c Wave erosion
d None above

64 Rill erosion occurs when the run off exceeds

a 0.2-0.6 mm/s
b 0.3-0.7 mm/s
c 0.4-0.8 mm/s
d 0.5-1.0 mm/s

65 The removal of uniform this soil upper layer by water is known as

a Sheet erosion | b Rill erosion
c Gully erosion | d Ravines

66 The water erosion in form of small and irregular channels is known as

a Sheet erosion | b Rill erosion
c Gully erosion | d Ravines

67 The water erosion in form of bigger and wider channels particularly in sloppy lands is called as

a Sheet erosion | b Rill erosion
c Gully erosion | d Ravines

68 The water erosion in form of deep and wide gullies is known as

a Sheet erosion | b Rill erosion
c Gully erosion | d Ravines

69 Landslides occur in mountainous region if the slope exceeds

a 10% | b 15%
c 20% | d 30%

70 When the flood water erodes the both banks of streams/rivulets/torrents, this type of water erosion is known as

a Rill erosion | b Landslides
c Ravines | d Stream bank erosion

71 Which of the following factor is responsible for water erosion?

a Climate | b Topography
c Vegetation | d All above

72 The process of bouncing, lifting and making suspension of fine soil particles with air from soil surface due to the effect of wind is called as

a Saltation | b Surface creep
c Suspension | d All above

73 Rolling and sliding of soil particles along the ground surface due to impact of particles descending and hitting during saltation is called as

a Saltation | b Surface creep
c Suspension | d All above

74 Movement of find soil particles from one place to other due to wind is called as

a Saltation | b Surface creep
c Suspension | d All above

75 Which of the following size of soil particles is moved by saltation effect of wind erosion?

a 0.10-0.50 mm   b 0.25-0.50mm

c 0.40-0.60 mm   d 0.50-0.070 mm

76 How much percentage of saltation may cause to total wind erosion?

a 25-30%   b 40-45%

c 50-70%   d 75-80%

77 What is the size of coarse particles that moves in the process of surface creep?

a 0.1-0.5 mm   b 0.5-1.0 mm

c 0.5-2.0 mm   d 1.0-2.0 mm

78 How much percentage of surface creep may cause to total wind erosion?

a 5-10%   b 5-15%

c 5-20%   d 5-25%

79 What is the size of soil particles that suspended in air?

a <0.1 mm   b 0.1-0.2 mm

c 0.2-0.3 mm   d 0.3-0.4 mm

80 How much percentage of suspension may cause to total wind erosion?

a 10%   b 15%

c 20%   d 20%

81 Which of the following measure is adopted to control soil erosion?

a Agronomic measures   b Mechanical measures

c Agrostological measures   d All above

82 Agronomic measures can be adopted for soil erosion if the slope of land is

a <1%   b <1.5%

c <2%   d <2.5%

83 Which of the following is a erosion resistant crop?

a Cowpea   b Groundnut

c Green gram   d All above

84 Which of the following is a erosion permiiting crop?

a Sorhgum   b Pearlmillet

c Maize   d All above

85 What is the ratio of erosion resistant and erosion permitting crops under strip cropping for controlling soil erosion?

a 1:1 b 1:2

c 1:3 d 1:4

86 What is the percentage of crop residue cover is necessary to keep soil loss within acceptable limit?

a 15% b 20%

c 30% d 40%

87 Which of the following increases the water stable aggregates and infiltration capacity of soil?

a Bitumen b NAA

c IAA d Polyvinyl alcohol

88 Growing of erosion permitting and erosion resisting crops in alternate strips along the contours is known as

a Buffer strip cropping b Contour strip cropping

c Wind strip cropping d Field strip cropping

89 Growing of erosion permitting and erosion resisting crops in alternate strips across the slope is known as

a Buffer strip cropping b Contour strip cropping

c Wind strip cropping d Field strip cropping

90 Growing of erosion permitting and erosion resisting crops in alternate strips across the direction of wind irrespective of contours is known as

a Buffer strip cropping b Contour strip cropping

c Wind strip cropping d Field strip cropping

91 Growing of perennial legumes/grasses or shrubs and field crops in alternate strips in sloppy land is known as

a Buffer strip cropping b Contour strip cropping

c Wind strip cropping d Field strip cropping

92 Which of the following type of spur is considered good for torrent control?

a Repelling type b Attracting type

c Deflecting type d Both 'a' and 'c'

93 The size of spur should be

a Not more than 10% of the stream width b Not more than 20% of the stream width

c Not more than 30% of the stream width d Not more than 40% of the stream width

94 Spur refers to

a Dyke or groyne

b Spur a are constructed at angle to the stream flow extending from bank into the stream

c Spur controls bank erosion

d All above

95 Which of the following type of retard is used to reduce the stream velocity near the river banks?

a Live hedges
b Jacks
c Jetted posts
d All above

96 Jacks refer to

a Wooden posts of 8cm diameter and 2-3 in length
b They are placed in series along the eroded stream banks
c They check the bank erosion
d All above

97 Jetted posts refer to

a Green posts of 2-2.5 m in length and 6-12 cm diameter

b They are erected along the actively eroded stream banks in two rows having 1 m distance

c They check the bank erosion

d All above

98 Gabion refers to

a Small stones and boulders are placed in wire nets

b They are placed along the stream banks

c They prevent bank erosion or streams or torrents

d All above

99 Kutta-crate refers to

a Vacant cement gunny bags filled with mine spoils

b They are used to construct the check dams to stabilize the limestone mine spoil slopes

c Filled gunny bags are used to erect a wall like structure along the mine spoil slopes to check the erosion

d All above

100 Geojute refers to

a It is jute mats used to stabilize the land slides

b They check the run off

c They are also useful to stabilize the denuded soils and sand dunes

d All above

101 Which of the following is a form of wave erosion?

a Landforms
b Wave cut cliffs
c Wave stacks
d All above

102 Sand dunes refer to

a Mound of sand formed by the wind
b Mound of lava
c Mound of rocks
d All above

103 Transverse sand dunes are

a Elongated perpendicular to the direction of the wind
b Elongated in the general direction of the wind
c Elongated in opposite direction of wind
d All above

104 Longitudinal sand dunes are

a Elongated perpendicular to the direction of the wind
b Elongated in the general direction of the wind
c Elongated in opposite direction of wind
d All above

105 Star type sand dunes are formed

a Constant wind direction and large sand supply
b Constant wind direction and limited sand supply
c Variable wind direction
d None above

106 Crescentic sand dunes are formed under

a Constant wind direction and large sand supply
b Constant wind direction and limited sand supply
c Variable wind direction
d None above

107 Parabolic sand dunes are also known as

a U shaped sand dunes
b Hair pin sand dunes
c Both 'a' and 'b'
d None above

108 Which of the region is more sensitive to sand dunes?

a West Rajasthan
b North Haryana
c West Gujarat
d All above

109 Which of the following is a man made soil erosion?

a Normal soil erosion
b Natural soil erosion
c Geological soil erosion
d Accelerated soil erosion

110 Which of the following is responsible for water erosion?

a Climate b Topography

c Vegetation d All above

Answer

1. **d**	2. **d**	3. **d**	4. **a**	5. **b**	6. **c**	7. **b**	8. **c**	9. **d**	10. **d**
11. **c**	12. **d**	13. **d**	14. **a**	15. **d**	16. **d**	17. **d**	18. **d**	19 .**d**	20. **d**
21. **d**	22. **d**	23. **d**	24. **d**	25. **d**	26. **d**	27. **a**	28. **c**	29. **d**	30. **d**
31. **d**	32. **d**	33. **d**	34. **d**	35. **d**	36. **a**	37. **d**	38. **d**	39. **a**	40. **d**
41. **a**	42. **b**	43. **c**	44. **c**	45. **c**	46. **c**	47. **a**	48. **d**	49. **d**	50. **d**
51. **c**	52. **d**	53. **d**	54. **c**	55. **a**	56. **a**	57. **b**	58. **c**	59. **a**	60. **b**
61. **c**	62. **b**	63. **a**	64. **c**	65. **a**	66. **b**	67. **c**	68. **d**	69. **c**	70. **d**
71. **d**	72. **a**	73. **b**	74. **c**	75. **a**	76. **c**	77. **c**	78. **c**	79. **a**	80. **b**
81. **d**	82. **c**	83. **d**	84. **d**	85. **c**	86. **c**	87. **a**	88. **b**	89. **d**	90. **c**
91. **a**	92. **d**	93. **c**	94. **d**	95. **d**	96. **d**	97. **d**	98. **d**	99. **d**	100. **d**
101. **d**	102. **a**	103. **a**	104. **b**	105. **c**	106. **b**	107. **c**	108. **a**	109. **d**	110. **d**

7

Water Harvesting Structures

Land, water and weather are natural resources that are critical for survival. Modern agriculture requires precise information on various agro-climatic parameters like soil, crop, climate and water resources for formulating efficient agricultural planning. Out of these, water is considered most scarce input in agriculture because its availability is reducing drastically for agriculture as well as drinking purposes. Besides, declining water tables and deteriorating quality of water resources caused pollution and sewage and effluents have further multiplied the problems. Presently, India has generated an irrigation potential of 106.61 m ha but the utilizable potential is 94.39 m ha at the end of 2002 (Target). Thus 33.6 % area of total cultivable land is still under rainfed cultivation but in true sense, nearly 50 % of total cultivable land is still under rainfed agriculture. Out of total precipitation 400 m ha m received over the country, 20 m ha m is lost through evaporation, 215 m ha m soaked by soil and 115 m ha m constitutes surface run-off. National Bureau of Soil survey and Land Use Planning (NBSSLUP) has estimated that 52% of the land area in India is potentially exposed to various degradation forces of which 45% is affected by water and 41% by wind erosion and 4.2 and 3.6 % by chemical and physical deterioration, respectively. Water erosion poses most serious problem resulting the soil loss of about 16.35 t/ha/yr. (Dhrubnarayan and Rambabu, 1983), so water conservation is not only essential to avoid the erosion but also to increase the availability to crop and men, particularly in arid and semi-arid regions where rainfall is predominantly erratic and inadequate both in quantity and distribution.

Rainfed agriculture is complex, diverse and risk prone and is characterized by low level of productivity, low input uses, fallowing, severe soil and water erosion with very poor social status of the communities. Variability in rainfall results in wide variation and instability in yields. Therefore, it is a challenge before Indian agriculture to avert rainfed farming into more sustainable and productive systems and to better support the population dependent upon it.

Broadly speaking, rainfed-farming areas in India can be grouped into following categories:

1. Arid and semi-arid areas of western India and Deccan plateau including the rain shadow areas,

2. High rainfall, high slope mountainous Himalayas ecosystem,
3. Waterlogged and coastal saline ecosystem
4. Inland waterlogged Himalayan foot hills and southern peninsular India, and
5. Cold desert ecosystem in the Himalayan range

In general, the crop production in rainfed areas experiences periodic drought during crop critical growth stages resulting into poor growth and yield. Therefore, conserving the rainwater during excess rainfall is the only option to provide life saving irrigations at crop needs for enhancing the crop production. The rain water harvesting techniques may vary with soil, slope, intensity of rain and requirements. Hence, the water harvesting techniques may be divided into two main categories;

Water harvesting techniques in Hills

Almost 60% population in hill region depends on natural water springs and streams for their needs. These water sources are diminishing due to increasing need of population for road, house, drinking and crop production purposes with environmental degradation mainly by deforestation and pollution. Almost 45% of existing water sources are depleting, drying and getting contaminated. In hilly regions, following water harvesting techniques are followed:

1. ***Harvesting of natural springs***: There are very small perennial/seasonal springs under hillocks with average flow rate of 0-5 liters per minute (LPM). They are used for drinking purposes and other household use. These springs may used for irrigation purpose by storing the water in small tanks and conveyed down to cultivated fields through pipelines. In case of high gravity head availability, a small portable sprinkler or drip irrigation system can be used for irrigation. A small area up to 65 ha can be irrigated particularly for growing vegetables etc.
2. ***Harvesting of small sub surface water streams:*** The small water streams flow in the valleys in the upper portion of the watersheds round the year but the flow gradually reduces from September to June. The average flow rate of these streams in the contributing valley portion varies between 0.01 to 0.1m^3. The small masonry ***check dam*** can be constructed across the stream to collect the sufficient water upstream site of the stream. If the stream is flowing through porous bed, the stream can be diverted to a storage structure by constructing a seepage trench and head wall with weir across the stream. The storage water can be either lifted for irrigation on the side bank cultivated area or conveyed through pipelines down stream fields.
3. ***Construction of Hydram (Hydraulic Dam):*** It is a water lifting device. It is self powered impulse pump using the energy of falling water for lifting water to higher altitudes. In hilly area, the hydrams may be constructed if the flow is 5 liters / minute and height of the waterfall is minimum one

meter from the site of hydram. In this process, the water can be lifted up to 150 meters otherwise normal lift rarely exceeds 50 metres.

4. ***Collection of roof water:*** In high hills, it is common practice to collect rainwater into cisterns or tanks constructed or fabricated below or above the ground surface. Thus cisterns may be used for community water supply if the total runoff is collected into a centralized tank. The individual tanks are more feasible in high hills due to topographical problems. These water storage structures may have the capacity of 25m^3 in the farmers' fields with top length ranges from 5 to 7m. width 3 to 5 m and depth 1-2 m. These tanks are used for both crop productions as well as drinking purposes (P.7).
5. ***Run-off water harvesting:*** In high rainfall areas, the excess runoff water is collected at central level into larger tanks of 150m^3 capacity and this water may be reused by two methods for irrigation purposes;
   a. ***Intra-terrace water harvesting:*** The runoff water is collected from a single terrace and recycled back to the same terrace.
   b. ***Micro-watershed water harvesting system:*** The runoff water is collected from a micro watershed having area < 1000m^2 and recycled to a community area in a down stream (P.2)

P. 2. Micro watershed water harvesting

6. ***Construction of small canals /Gules :*** It is a very common practice of irrigation in hilly areas. The concrete and cement made small channels are constructed parallel to large streams. The small flow of water from large streams/rivers is diverted to this channels/canals/gules and used for irrigating both sides of plain area. The canals/ gules finally end into small streams/rivers (P.3).

P.3. Canal system

7. ***Water spreading*:** In this simple technique, the flood water is deliberately diverted from its natural courses (rivers or larger streams) over adjacent plains through ditches, dikes, small dams and bush fences. These wet flood plains/valleys are thereby used to grow crops successfully. The fertility of the area is enriched due to silting of fine clay particles. Further, the water holding capacity of the area is also increased.
8. **Check Dams/Cement Plug/Nala Bund:** Check Dams/Nala Bund is made in high rainfall region. Check dams are normally constructed across small streams having gentle slope and are feasible both in hard rock as well as alluvial formations. The water stored in these structures is mostly confined to stream course and the height is normally less than 2 m. These are designed based on stream width and excess water is allowed to flow over the wall. In order to avoid scouring from excess run off, water cushions are provided at downstream side. To harness the maximum runoff in the stream, series of such check dams can be constructed to have recharge on regional scale. A series of small bunds or weirs are made across selected nala sections such that the flow of surface water in the stream channel is impeded and water is retained on pervious soil/ rock surface for longer body. Nala bunds are constructed across bigger streams of second order in areas having gentler slopes. A nala bund acts like a mini percolation tank.

Normally the catchment of the check dams/nala bund ranges from 40 to 100 ha. The dimension of the nala should be 5-15m wide and at least 1 m deep. The check dams are made in the area having annual rainfall of around 100cm. The check dams/nala bund were the small earthen dams but nowadays stone and cement are used to make these check dams much more durable and purposeful.

B. Water harvesting techniques in Arid and semi-arid areas

Number of water harvesting techniques has been used for long times back in India and abroad specially for drought proofing. The old techniques have been modified scientifically to increase its efficiency and reduce the cost. Some of the important water harvesting techniques for drought proofing are given below;

1. *Nadi /Pond system*

The Nadi is a small village pond either excavated or embanked at the lowest point of the village for collection of rain water from adjoining area or catchments. These Nadis are made under low precipitation areas. The collected water is used for both crop production and drinking purposes. The water availability may vary from two months to whole year depending on the characteristics of catchment, amount and intensity of rainfall and management. The capacity of a Nadi may very from 1200-1500 m^3 depend upon physiographic conditions and rainfall pattern (P.4).

It is considered one of the oldest systems of water collection for its conjunctive use. The first masonary Nadi was constructed in 1520 AD near Jodhpur (Rajasthan) during the regime of Rao Jodhaji. At present, most of the villages in the region have one or more Nadis. According to 1982 census, 1436, 592 and 1822 Nadis were found in Nagaur, Barmer and Jaisalmer districts, respectively. Generally the loss of water was quite high in traditional Nadis through evaporation, seepage and heavy sedimentation. The research findings indicate that the surface to volume ratio should be < 0.3 for reducing the evaporation loss. The silt traps may be used at inlet point to avoid siltation. The LDPE sheet or stone or concrete layer may be used for lining the bottom or side of the Nadi to avoid seepage losses. The Nadis/ponds may be square, rectangular or circular in shape but circular ponds have the highest storage capacity for a given surface area and side slopes but its curved area remains unused, hence, square or rectangular Nadis/ponds should be preferred (Table.1)

P. 4. Nadi / Pond system

Table 1 Surface area, perimeter and storage capacity of different shapes of ponds

Shape of pond	Side slope	Surface area (m^2)	Perimeter (m)	Storage capacity (m^3)
Circular	1.5:1	900	100	1499
Square	1.5:1	900	120	1464
Rectangular	1.5:1	900	122	1458

Source: Rudder *et al.* (1995)

2. *Tanka system*

The Tanka is also known as *Kund* or *Kundi (P.5).* Tanka system is an ancient water harvesting technique in dry regions and still very common in eastern Rajasthan. Tanka is a small circular or square underground tank made of lime mortar or cement plaster. Generally the Tankas are constructed at lower point of the catchments so that the runoff may be collected. In several places, there are *kuchcha Tankas* also-meaning underground wells without proper concrete and cover. *Tanka* generally caters to drinking needs of one family and is used both for harvesting and storing rain water. The most common size of a Tanka is 10-12 feet deep with a diameter of 10 feet. In the desert economy it is considered as an asset of a family. The small Tankas of 3 to 4.22 m diameter and about 21-59m^3 capacities are built by individual households and larger ones of 6m, diameter with 200m^3 capacities are built for village communities. The depth is kept equal to the diameter.

The individual Tankas of variable size but on an average of 21m^3 capacity are built in front of the house in an open area of about 10m x 10m size, while the community Tankas of 200 m^3 capacity require the catchment (open area) of 30m x 30m. The land surface of the catchment should be firm and sandy with gentle slope of about 3% with a fall of 3 cm in 1 m length. The catchment of the Tankas are treated with murrum, coal ash, gravel, pond silt, bentonite, soil cement mix, lime concrete, sodium carbonate etc. so that maximum rain water can be collected. The catchment of the Tankas should be cleared before arrival of rains and also it remains prohibited for grazing and illegal entry of animals. The Tankas should be covered by either wire net or cement/stone slabs with 1m x 1m opening for with drawl of water. The level of stone slabs should be 1m high from the land surface. The inlets and outlets are covered with iron bars and gratings to protect the entry of waste materials into Tanka. The silt traps are provided at inlets to avoid siltation.

In ancient system, leakage, seepage and siltation were the major problems, so CAZRI designed improved Tanka of capacity ranging from 10000 to 600000 litres. The vertical surfaces in improved Tanka are made of stone masonry with cement plaster and the base is made with 10-15 cm thick cement concrete layer. The stored water is used for drinking purpose and life saving irrigation to the crops.

(a)

(b)

P.5. Tanka System (a) Individual Tanka, and (b) Community Tanka

3. *Khadin system*

The *khadin* system is based on the principle of harvesting rainwater on farmland and subsequent use of this water-saturated land for crop production. A *khadin*, also called as *dhora*, is an indigenous construction designed to harvest surface runoff water for agriculture. Paliwal Brahmin of Jaisalmer, western Rajasthan was the first person who designed and developed the Khadin in the 15th century. This system has great similarity with the irrigation methods of the people of Ur (present Iraq) around 4500 BC and later of the Nabateans in the Middle East. A similar system is also reported to have been practiced 4,000 years ago in the Negev desert, and in southwestern Colorado 500 years ago. It is practiced generally at lower rainfall (100-225 mm) areas.

The soil of lower area develops from the silt load carried in runoff from adjoining uplands and becomes fine textured in the natural catchment. Generally the lower part of the catchments has a rocky base that restricts deep percolation. Such areas are developed into productive agricultural lands by constructing earthen embankments at lower periphery of field with suitable spillway arrangement even at as low as 150-350 mm rainfall. This type of productive land or basin is known as khadin (P.6). The size of the Khadin is designed on the basis of local rainfall patterns, catchment characteristics and soil type. On an average, the cultivated area under each Khadin is 10-14 ha with an average dam size between 1.2-1.7 m high x 1.0-1.5 m wide and 100-300 meters in length, depending upon catchment area and number of land holdings. The spillways and sluice gates are usually provided at a proper location for excess water during flood condition. For efficient agriculture, a minimum of 15:1 ratio of catchment area to crop area is required. A rainfall of 75-100 mm is sufficient to charge the Khadin soils with sufficient soil moisture content to raise a successful local crop. In the Khadin area, the collected runoff percolates into the ground with time recharges the subsoil. Of the total runoff collected only 50-60% of the water is utilized. The remainder is lost to evaporation, or percolates into the underground, recharging an aquifer. The scientific study reveals that the EC of khadin soil is 2-16 times lower than the soil of upper part of the khadin because leaching of salts through seepage water. The khadin soils store enough moisture and nutrients for successful crop cultivation. The Khadins also provide positive impact on the ecology by improving vegetation cover in the region.

This ancient practice of khadin cultivation and its development was neglected slowly during 19th century due to rivalry among the local inhabitants. This led poverty, scarcity of food grains and fodder and lead to adverse socio economic impacts. Hence, the government realized its importance and renewed the old khadins and so far more than 500 khadins have been revived. Presently, CAZRI is working on scientific planning and development of khadins in arid areas of Rajasthan.

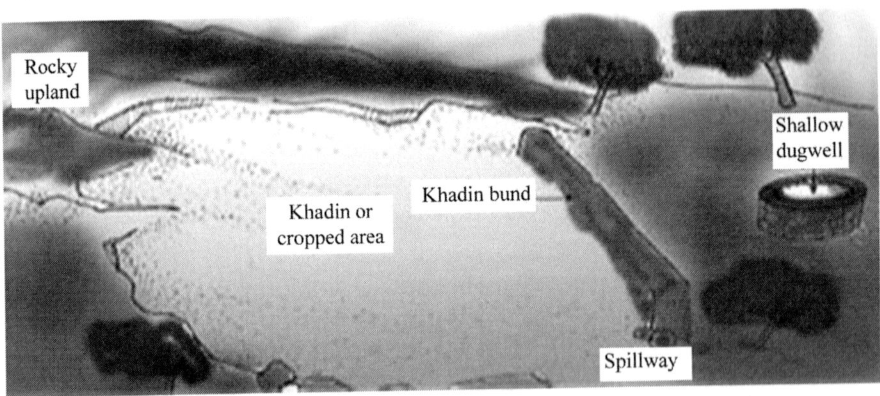

P.6. Khadin system

4. *Anicut*

In arid and semi-arid areas, the small streams carrying run-off water during excess rainfall are restricted with earthen fill structure with a proper spill way just to hold sufficient water upstream during rainfall. Such type of earthen structure is called Anicut. The main objectives of anicuts are to conserve the soil and water and recharge of groundwater table of the region. The studies indicate that the water levels in wells located down stream in Ujalion watershed areas increased from 1.8 to 2.2 as compared to increase only 0.5 m in wells located in adjoining areas. The moisture conservation upstream helps to establish silvipastoral systems that may fulfill the need of fuel and fodder of the region.

5. *Roof water collection*

Roof water harvesting is not a new practice of rainwater harvesting in India. In India this method is largely practical in north-eastern region and western Rajasthan. The rain water from the roofs of the buildings may be collected into underground cement or concrete made cisterns/tanks for household and kitchen garden purposes. Recently the importance of roof water harvesting has been realized greatly because of more demand of water for agriculture and household purposes for ever increasing population.

The yield of rain water from roof surface depends upon the type of roof and material used in its construction. It has been observed that the runoff efficiency of roof rainwater can be increased if the roofs are made of corrugated G-I-Sheet (85%) and stone slab and lime concrete (81%), while the traditional roofs made of clay soil and straw have as low as 40% run off efficiency (P.7).

6. *Percolation tanks*

Presently, the water use pattern has changed and total quantity increased many fold in most of the sectors including domestic and irrigation use. Further, the erratic and insufficient rainfall pattern has reduced the ground water recharge. The pressure of more food grain production and supply of drinking water has worsened the situation of ground water status as it is decreasing at very fast rates. For examples, summer rice cultivation in Punjab, Haryana and Tarai belt of Uttarakhand is very harmful for indiscrimination of ground water and number of reports has thrown light on decreasing trend of ground water level. Some of the states are very prone to reduce ground water level due to poor rain water harvesting like Maharashtra, Andhra Pradesh, Madhya Pradesh, Tamil Nadu, Karnataka, Rajasthan and Gujarat. The arid and semi arid areas, mountainous region and also hard rock areas experience most often water scarce situations due to inadequate recharge, indiscriminate withdrawal of ground water and mismanagement. Therefore, the percolation tanks are constructed exclusively to recharge the ground water. These percolation tanks may also be at the lowest point of the catchment or on the small streams or rivulets with adequate catchments for impounding surface runoff (P.8).

P.7. Roof water harvesting

Percolation tanks are artificially created surface water bodies, submerging a land area with adequate permeability to facilitate sufficient percolation of impounded surface run off to recharge the ground water. They are more or less similar to check dams or nala bund with a fairly large storage reservoirs. The location of the percolation tanks should be downstream of the runoff zone or in the upper part of the transition zone with a land slope of 3-5%. The size of the percolation tanks should not be more than 50% of the total quantum of utilizable runoff from the catchments. A percolation tank may be built either across the small streams by creating low elevation check dams or in uncultivated land adjoining streams through excavations and providing a delivery canal connecting the tanks and the streams. However, in hills, small tanks/ponds are constructed at suitable lower points for collecting the rain water for recharge purposes. The utility and viability of percolation tanks depend on quantity and pattern of rainfall, size and slope of a catchment, soil and topography of a catchment, location, percolating capacity and climatic parameters of the area.

The percolation rate in Noyili, Amravathy and Ponnani basin ranged from 10-50 mm/day (Table.1). It has been observed considerable reduction in percolation rate when the standing water persisted for a longer duration. Therefore, the desiltation of percolation tanks is required to remove the sedimentation regularly. In general, percolation tanks are very effective in deep and pervious soils.

Table 1 Percolation and evaporation in Noyil, Ponnani and Amaravathy basin

Name of tank	Name of basin	Tank capacity (m^3)	Av. annual rate of percolation (mm/day)	Percolation rate (%)	Evaporation rate
Chinnathadagam	Noyil	0.0354	20.00	89.85	10.15
Annur	do	2.579	10.00	66.63	33.37
Asanallip alayam	do	0.3800	30.00	92.29	7.71
Kuravallu	do	0.6428	30.00	81.82	18.18
Uiltamalai	Amaravathy	0.0671	42.00	85.49	14.51
Kothavedi	Ponnani	1.224	26.60	88.00	12.00

Source: Khan (2002)

P.8. Percolation Tank

7. *Sub-surface barriers*

The sub surface barriers are constructed across the streams particularly to recharge the ground water aquifer. These sub surface barriers are useful in arid and semi-arid areas where recharging capacity of wells and ground water is very poor because of scanty rainfall. The studies indicate the rate of replenishment of ground water table of wells situated in the nearby sub surface barriers increased and depletion rate decreased after construction of subsurface barriers across the streams. The recharge from 2.25 ha area and storage capacity of 15000 m^3 were observed nearly 1000 m^3 in artificial recharge study in alluvial soil during rainy season in Gujrat.

8. *Sand filled Dam/Tank*

Generally in arid and semi-arid areas, the sandy river beds have hard impervious subsurface and no surface flow except in rainy season. The sand filled dams are made across the streams as sub surface barriers to stop the runoff and store water in sand beds just for domestic water supply and recharge ground water in the vicinity. Initially the low head masonry dam is constructed across the stream and let the basin filled with sand and gravel and later the dam height is increased.

The basic principle of raising the dam height gradually to allow flood water to transport most of the fine sediments in suspension over the dam crest otherwise it gets deposited in the reservoir and leads to reduced storage capacity. The excess water of dam may be diverted for nearby Nadis or ponds for irrigation and domestic uses. In one study made in Jodhpur reveals that evaporation losses of sand filled dams was only 8% and reduction in storage capacity was nearly 50 % but quality of water was very high.

9. *Water Spreading*

Water spreading involves deliberately diversion of flood water from stream source or catchments area to micro catchments with the help of dykes and ditches and field leveling to improve the soil moisture regime and also to facilitate the regeneration of ground water. It helps to regenerate the natural and artificial vegetation in the water spreading area. It has been found that the excess rain water of 4-12 ha area can be spread in 1 ha area. Deep tillage also helps more infiltration of water. The water spreading practice is very common in Jalore district of Rajasthan and locally is called '*rela farming*'. The viability and economics are yet to be investigated for better acceptability of this system in arid & semi arid regions.

10. *Water harvesting dams*

The ephemeral streams carry heavy flood during rainy season. This flood water caused heavy soil erosion by making ravines and gullies. This excess water is collected by earthen or concrete dams across the streams and stored water is diverted through canals to larger areas to fulfill the irrigation and household water needs. This stored water also helps in regenerating the ground water table in adjoining areas.

11. *In situ water conservation*

The inter plot and inter row water harvesting techniques are now very common in arid and semi arid areas. The basic principle of *in-situ* water conservation is to store or use rainwater at its original falling or receiving place. Under inter plot water harvesting techniques, the rainwater is collected in the lower part of the field and recycled as per crop needs. One or two life saving irrigation may be given to the crops. This technique was initiated for first time at CAZRI, Jodhpur in 1909 by S.D. Singh and colleagues.

Inter row water harvesting techniques has also been found very useful in dry land areas. The 50-60 cm wide raised beds or ridges are made at 30-40 cm distance. The height of ridges should be 15-20 cm. The rainwater falling over both sides of ridge is collected in the furrows and improves the soil moisture states of the furrows where crops can be grown successfully. Tillage is restricted to furrows only, so energy can be substantially reduced.

12. *Increase run off*

Increase run-off and reduce infiltration rate are more important principles in arid and semi arid areas for collecting runoff as much as possible during rainy season for life saving irrigation to crops during dry spells. Therefore, different materials and chemicals are used to collect more run off water. The sheet metal, butyl rubber, fiber glass, asphalt roofing, bentonite paraffin, reinforced asphalt, plastic sheeting with gravel cover, Janta emulsion, sodium carbonate etc. may be used on catchments area to increase runoff volume for its further use. The Janta emulsion and sodium carbonate spray increased 68 and 66 % runoff, respectively over control at CAZRI, Jodhpur.

Traditional water harvesting structures in Uttarakhand Hills

1. Naula (little depression aquifer),
2. Dhara (springs),
3. Gadhera (small river tributaries),
4. Gul (traditional irrigation canals),
5. Chaal and Khaal (artificial ponds on hilltops and slopes)
6. Simar
7. Dhamm

These structures collect and supply water and still persist in the villages of Uttarakhand. Among these, *Naula* and *Dhara* are the most important and are still used as the prime source of drinking water in many hilly areas of Uttarakhand, especially in the Kumaon division.

A *Naula* is a naturally-occurring water aquifer. It is a stone-lined tank which catches dripping water from springs and streams. It looks like a temple from outside and is equally regarded as such. *Naula* and *Dhara* have been considered as sacred in

Uttarakhand's tradition and culture. Sculptures of Lord Vishnu or any local deity are usually carved onto the walls of the *Naula* or *Dhara* to declare it a sacred site and prevent it from being polluted. The entrance of the *Naula* was always made small. This was done to prevent animals from entering it and to ensure that only one person entered it at a time. On the other hand, *Dharas* are mostly protected by constructing boundary walls.

Simar is a fertile niche or a marshy track of land in an agricultural field nestling in small micro catchments in the hills characterized by high moisture availability, greater biomass productivity and rich biodiversity and it is created by ground water. *Dhaam* refers to water collected from small and big streams in shape of lake. This water is mainly used for domestic animals and irrigation purposes.

Naula Dhara

Other traditional water harvesting techniques in other parts of India

a. ***Bambu pipe line*** – It is very unique system of water harvesting and conservation system utilizing natural springs and streams through bamboo pipe lines. It is very popular in Mehgalaya for irrigating plantations of betel leaf and black pepper crops. The whole irrigation system is made up of different forms of bamboo pipes of varying cross sections which take the water from perennial springs on the tops of the hill. It is a brilliant drip irrigation system in which bamboos of various sizes and reduces the output to 20-80 drops per minute from 18-20 litres of water. The flow of water in controlled by changing pipe positions. The water is dropped at the base of the plant and there is no runoff and wastage of water.

b. ***Zabo system (Ruza system)*** – It is locally known as ruza system practiced in Nagaland.The rain water is collected on hilltops in to a pond like structures that are carved out on the hill sides. The water is then passed onto cattle yards below from where the water enters the paddy fields rich in manure. Zabo is used to deal with a lack of drinking water. So this system of water harvesting is used for drinking purposes and also for animal, forest and agriculture.

c. ***Roop top water harvesting*** –In Mizoram, Meghalya, Nagaland and Tripura, the farmers collect the rain water from roofs and collect in tanks. This water is used to irrigate their crops. This system is mainly used by tribal farmers of Meghalaya's Khasi and Jaintia hills for drip irrigation.

d. ***Haveli system*** – It is practiced in black soil of Bundelkhand region. The rainwater is collected from catchments by embarkments or bund in the field in rainy season and after receding of this water, the rabi crops are grown on residual soil moisture. This system also helps to recharge the ground water of the area. There is also provision to drain out the flooded water in September-October so that Rabi crops can be grown. The region has number of havelis but have defunct in due times, so recently Government is providing financial support to the farmers to renovate haveli system to avoid the water stress conditions in region.

Haveli system of Bundelkhand region

e. ***Jalkund (lined small ponds)*** – Maharashtra and Assam

f. ***Series of tanks connected through field channels*** – T N and Western Orissa,

g. **Virdas**- This system was developed by the nomadic Maldhari tribes of Rann of Kutch, Gujarat. Virdas are shallow wells dug within a natural depression (Lake/Jheel). Since the area around is very saline, when rainwater seeps down the soil, it collects over the saline groundwater due to the difference in density (rainwater being less dense). The tribesmen identify areas on basis of flow of the monsoon runoff and build these shallow wells. This smart method helps them separate freshwater from saltwater and provide water for a variety of purposes. The crops are also planted along virdas to protect them from water stress.

Virdas system

Zabo system

h. ***Surangam*** – It is practiced in Karnatka and Kerala. Surangam is a traditional water conservation system in which fine tunnels are made and the horizontal wells are dug out at specific distance to collect rainwater for drinking as well as irrigating crops like rice and coconut. The Surangam can be of varying length and can even go up to 300 metres and water is collected into a storage tank using gravitational force. Vertical shafts are provided for airflow in areas of Karnataka and Kerala. The vertical shafts are provided for airflow in the surangams. Water quality of surangam is of high quality.

i. ***Ahar Pynes***- It is traditional water conservation technique of South Bihar. ***Ahar*** is a catchment basin from which flood water flow in form of rivulet and ***Pynes*** are artificial channel through which rivulet water is directed to field for irrigation. The flood water collected by embankment of catchment area from three sides and end point of rivulet is linked with pynes. Although the system suffered under British rule, it has again been rejuvenated for agricultural purposes, especially in the district of Gaya.

Ahar Pynes system

Surangam system

j. ***Eri*** – Eri means tank and it is one of the oldest widely practiced for water conservation in Tamil Nadu, India. This tank water is used for irrigation purposes. Besides, it helps in preventing soil erosion, recharging groundwater and flood control. Eri can either be fed through channels that divert river water, or rain-fed ones. They are usually interconnected to balance the water in case of excess or lesser supply.

k. ***Johads*** – Johad is a small earthen check dam that captures & conserves rainwater, improving percolation & recharging groundwater. So normally the johad is a crescent shaped small check dam built from earth and rock to intercept and conserve rainwater and practiced in Alwar district of Rajasthan. After the drought of the 1980s, the villagers attempted to revive the traditional method of rain water harvesting in form of johad. This system of water harvesting not only helps to recharge the groundwater of the area but also make flow of river Arvari perennial. Johads also help to increase agricultural production of the area.

Johad system

Eri system

l. ***Kund*** **-** A kund is a saucer shaped catchment area that has gentle slope towards the central circular underground well. The main purpose of rainwater harvesting is for drinking purposes.The kunds are made in western Rajasthan and Gujarat. Traditinally these kunds were covered in disinfectant lime and ash. Nowadays the cement is used for construction of kunds. Raj Sur Singh is said to have built the earliest known kunds in village of 'Vadi Ka Melan' in year 1607 AD.

m. ***Baoli***- Baoli means a wide mouth well or reservoir with stairs used for reaching to the water level of the baoli. It is a secular structure from which every one can draw water.The step wells have beautiful arches, carved motifs and sometimes rooms on their sides. These baolis are used for social gatherings of the villages. They are exclusively used for agriculture. Baolis were made in the trade routes so that traders may rest in day and nights. These baolis are mainly found in UP, MP, Rajasthan and Gujarat states.

Kund system

Baoli system

n. ***Jhalaras*** **-** They are rectangular shaped step wells that have tiered steps on three or four sides. These step wells collect the subteranean seepage of an upstream reservoir or a lake. They were made in Jodhpur dates back to 1660 AD for regular supply of drinking water for religious rites, royal ceremonies and common use.

o. ***Bawari*** – It is a unique step well or man made tank built in cities of Rajasthan to reguair supply of drinking water in the cities. The rain water is collected in tanks from hilly outskirts of cities through canals.The series of layered steps were built around the tank/well narrow and deepen the wells.

Jhalara (Jodhpur)

Bawari

Questions

a. Short and Descriptive

1. Write short notes on following;
   a. Dhara
   b. Naula
   c. Hydram
   d. Gul
   e. Chal and Dhal
   f. Anicut
   g. Percolation tanl
   h. Khadin
   i. Tanka
2. What is the difference between following;
   a. Kund vs. Baoli
   b. Baoli vs. Bawari
   c. Jhalara vs. Tanka
   d. Haveli vs. Khadin
   e. Johad vs. Eri
   f. Ahar pynes vs. Surangam
3. Describe the important water harvesting structures used in hills.
4. Describe the important water harvesting structures used in arid and semi arid areas.
5. What do you mean by roof water collection? Also describe why it is important in hilly areas of North East and North West Himalaya?

6. What do you mean by *in-situ* water harvesting? Also discuss the in detail in-situ water harvesting.

b. Objective questions (MCQs)

1 Water harvesting techniques in hills are

a Harvesting of natural spring

b Harvesting of small sub surface water streams

c Construction of Hydram (Hydraulic Dam)

d All above

2 Water harvesting techniques in hills are

a Collection of roof water | b Construction of small canals

c Gules Water spreading | d All above

3 Run-off water harvesting is

a Intra-terrace water harvesting | b Micro-watershed water harvesting system

c Both a & b | d None above

4 Water harvesting structure in arid and semi arid areas are

a Nadi /Pond system | b Tanka system

c Khadin systen | d All above

5 Water harvesting structure in arid and semi arid areas are

a Anicut | b Roof water harvesting

c Percolation tanks | d All above

6 Water harvesting structure in arid and semi arid areas are

a Sand filled Dam/Tanks | b 'In situ' water conservation

c Increase runoff | d All above

7 Irrigation potential generated in India is

a 106.61 m/ ha | b 105.61m/ha

c 102.61m/ha | d All above

8 Which of the following is a rainfed-farming areas in India ?

a Arid and semi-arid areas of western India | b High rainfall

c Waterlogged and coastal saline ecosystem | d All above

9 The rain water harvesting techniques may vary with

a Soil
b Slope
c Intensity of rain
d All above

10 Water harvesting techniques in Hills are

a Harvesting of natural springs
b Harvesting of small sub surface water streams
c Construction of Hydram (Hydraulic Dam)
d All above

11 Average flow rate of water Harvesting of natural springs in hill region approximately

a 0-4 liters per minute
b 0-5 liters per minute
c 0-9liters per minute
d 0-10 liters per minute

12 The average flow rate of small sub surface water streams in valley portion varies

a Between 0.02to 0.3m^3
b Between 0.01 to 0.1m^3
c Between 0.04 to 0.1m^3
d None above

13 Hydraulic Dam is

a Water lifting device
b Check dam
c Both a & b
d None above

14 Hydram is also known as

a Check dam
b Hydraulic Dam
c Both a & b
d None above

15 In high hills areas, roof water is used for

a Crop productions
b Drinking purposes
c Both a & b
d None above

16 Canals is also called

a Channels
b Gules
c Both a & b
d None above

17 Canals is made up of

a Concrete
b Cement
c Both a & b
d None above

18 Check Dams is made in

a High rainfall region
b Low rainfall region
c Arid region
d None above

19 Check Dams is also known as

a Cement Plug b Nala Bund

c Both a & b d None above

20 The height of check Dam is approximately

a < 2 m b <3m

c < 4m d <5m

21 Catchment ranges of the check dams are approximately

a from 30 to 100 ha b from 40 to 100 ha

c from 60 to 100 ha d from 70 to 100 ha

22 The first masonary Nadi was constructed in

a Jodhpur (Rajasthan) b Karnataka

c Both a & b d None above

23 According to 1982 census, 1436, 592 and 1822 in which districts Nadis were found?

a Nagaur b Barmer

c Jaisalmer d All above

24 The capacity of a Nadi may vary from approximately

a 1100-1500 m^3 b 1200-1500 m^3

c 1400-1500 m^3 d 1200-1600 m^3

25 The shape of Nadis/pond is

a Square b Circular

c Rectangular d All above

26 Storage capacity of Circular pond is

a 1499 m^3 b 1458 m^3

c 1464 m^3 d 1455 m^3

27 Storage capacity of Rectangular pond is

a 1455 m^3 b 1464 m^3

c 1458 m^3 d 1456 m^3

28 Storage capacity of Square pond is

a 1464 m^3 b 1455 m^3

c 1458 m^3 d 1456 m^3

29 Tanka system also known as

a Kund b Kundi

c Both a & b d None above

30 Tanka system is commonly found in

a Dry region | b Low rainfall region
c High rainfall region | b None above

31 Khadin also called

a Dhora | b Kund
c Tanka | d None above

32 Who was first designed and developed the Khadin in the 15th century?

a Rao Jodhaji | b Paliwal Brahmin
c Dhrubnarayan | d Rambabu

33 The length of Khadin system is approximately

a 100-300 m | b 100-400 m
c 100-500 m | d None above

34 For efficient agriculture a minimum ratio of catchment area to crop area is required

a 15:1 | b 15:4
c 15:3 | d 14:1

35 EC of Khadin soil is

a 2-15 | b 2-18
c 2-16 | d None above

36 Roof water collection is found mainly in

a North-eastern | b Western Rajasthan
c Both a & b | d None above

37 Traditional roofs made of

a Clay soil | b Straw
c Both a & b | d None above

38 The evaporation rate of Noyil basin is

a 10.15 | b 14.51
c 33.37 | d 12.00

39 The evaporation rate of Amaravathy basin is

a 10.15 | b 14.51
c 33.37 | d 12.00

40 The evaporation rate of Ponnani basin is

a 33.37 | b 14.51
c 12.00 | d 10.15

41 Sub surface barriers are useful in

a Arid areas | b Semi-arid areas
c Both a & b | d None above

42 Storage capacity Sub-surface barriers is

a 15000 m^3 | b 17000 m^3
c 16000 m^3 | d 18000 m^3

43 Water Spreading locally known in Rajasthan

a Rela farming | b Nadis
c Both a & b | d None above

44 Water harvesting dams are carry

a Light flood | b Heavy flood
c Both a & b | d None above

45 *In-Situ* water conservation technique was initiated first time at

a Gujarat | b Punjab
c CAZRI, Jodhpur | d Haryana

46 Who initiated first *in-situ* water conservation technique?

a S.D. Singh | b Rao Jodhaji
c Dhrubnarayan | d None above

47 Which of the following chemicals are used to collect more runoff water?

a The sheet metal | b Butyl rubber
c Fiber glass | d All above

48 Which of the following increased 68 and 66 % runoff CAZRI, Jodhpur.

a Janta emulsion | b Sodium carbonate spray
c Both a & b | d None above

49 Most popular indigenous structures in arid and semi arid areas are

a Nadi, | b Tanka
c Khadins | d All above

50 Nadi, Tanka, Khadins water are used for

a Drinking water | b Crop production
c Both a & b | d None above

51 Water harvesting refers to

a Runoff collection and its reuse | b Rain water collection and its reuse
c Both a & b | d None above

52 Nadi is a

a Village pond
b Canal
c Underground tank
d All above

53 Khadin system is popular in

a Dry areas
b Wet areas
c Waterlogged areas
d None above

54 Nala bund acts like a

a Tanka
b Percolation tank
c Khadin
d None above

55 Anicut helps in

a Soil conservation
b Water conservation
c Recharge of ground water
d All above

56 Roof water is mainly used for

a Agriculture
b Household
c Both a & b
d None above

57 Percolation tanks are suitable in

a Hilly and rocky soils
b Sandy soils
c Deep and pervious soils
d Marshy soils

58 Anicut helps in

a Soil and water conservation
b Recharge of ground water
c Both a & b
d None above

59 Khadin help in

a Recharging ground water
b Crop production
c Reduce soil and water erosion
d All above

60 Which of the following is an indigenous water harvesting structure?

a Nadi
b Tanka
c Khadin
d All above

61 Which of the following water harvesting technique is employed in hilly region?

a Hydram
b Constrcution of gules/ canals
c Taping of natural springs
d All above

62 Which of the following water harvesting technique is not employed in hilly region?

a Khadin system
b Hydram
c Taping of natural springs
d All above

63 Which of the following water harvesting technique is employed in arid and semi arid region?

a Hydram
b Construction of gules/ canals
c Tanka
d All above

64 Which of the following water harvesting technique is not employed in arid and semi arid region?

a Hydram system
b Khadin system
c Tanka system
d Nadi system

65 Which of the following water harvesting technique is useful under low rainfall araes?

a Nadi system
b Tanka system
c Khadin system
d All above

66 Who is considered the originator of Nadi system?

a Paliwal Brahmin of Jaisalmer
b Rao Jodhaji of jodhpur
c Raja Udai Singh of Udaipur
d All above

67 Which type of Nadi has the highest water storage capacity for a given area?

a Circular
b Rectangular
c Square
d All above

68 Which is the main problem of poor acceptance of Nadi system in recent days?

a Silting
b High evaporation
c Poor quality of water
d Poor management

69 Tanka is a

a Underground Tank
b Canal
c Lake
d Pond

70 Which of the following is a major problem of Tanka system?

a Leakage
b Seepage
c Siltation
d All above

71 Khadins were developed for

a Drinking water
b Better crop production
c Better employment
d All above

72 Which of the following is a traditional water harvesting structure in Uttarakhand hills?

a Dhara
b Naula
c Chal and Khaal
d All above

73 What is true for Dhara ?

a Natural water springs
b Dhara is channeled through carved outlets in shape of cow, elephants, lion or Ganesha.
c It is common source of drinking water
d All above

74 What is true for Naula?

a It is percolation tank built across a stream or spring
b It is stone lined percolation tank/pond covered with pacca walls and roofs
c Water is used primarily for drinking purposes
d All above

75 What is Gadhera?

a Mountain stream and water is used for animals and agricultural production
b Percolation tank
c Pond
d None above

76 What is Khaal?

a Large depression on hillocks in which rain water is collected for recharging ground water
b Mountain stream
c Percolation tank
d None above

77 'Dhaam' refers to

a Water is collected from small and big streams gives the shape of lake
b Dhaam water is used for animals and irrigation
c Water harvesting structure in hills
d All above

78 'Simar' refers to

a Fertile niche or marshy track of land in agricultural field nestling is small micro catchments in hills

b It is characterized by high moisture availability

c It has high biodiversity

d All above

79 Haveli system of water harvesting technique is prevalent in

a Nagaland b Mizoram

c Bundelkhand region d Meghalaya

80 Zabo system of water harvesting technique is prevalent in

a Nagaland b Mizoram

c Bundelkhand region d Meghalaya

81 Roof top water harvesting technique is prevalent in

a Nagaland b Mizoram

c Bundelkhand region d Meghalaya

82 Bambu pipe line system of water harvesting technique is prevalent in

a Nagaland b Mizoram

c Bundelkhand region d Meghalaya

83 Series of tanks connected with field channels is popular in

a TN b Orissa

c Both a and b d None above

84 'Johad' system of water harvesting is popular in

a Alwar (Rajasthan) b TN

c South Bihar d Runn of Kutch

85 'Eri' system of water harvesting is popular in

a Alwar (Rajasthan) b TN

c South Bihar d Runn of Kutch

86 'Ahar Pynes' system of water harvesting is popular in

a Alwar (Rajasthan) b TN

c South Bihar d Runn of Kutch

87 'Virdas' system of water harvesting is popular in

a Alwar (Rajasthan) b TN

c South Bihar d Runn of Kutch

88 Surangam system of water harvesting is famous in

a Karnataka
b Kerala
c Both a and b
d None above

89 Which of the following is not a indigenous system of water harvesting structure?

a Baoli
b Kund
c Jalkund
d Canal

90 Which of the following water harvesting structure is used for both rain water harvesting and crop production?

a Haveli system
b Eri
c Johad
d Tanka

91 Which of the following water harvesting structure has stairs for reaching to the levels of stored water?

a Baoli
b Kund
c Bawari
d All above

92 Which of the following is true for bawaris?

a They are unique step wells
b Mostly found in cities of Rajasthan
c They help in recharging the ground water and aquifers
d All above

93 Which of the following is true about talab?

a Small pond of less than 5 bighas is called talai
b Mediun size lake/ reservoir is called bandhi
c Large size lake/reservoir is called sagar or samand
d All above

94 Which of the following is true for Jhalara, indigenous water harvesting structure for Rajasthan?

a It is rectangular shaped step well, first made in Jodhpur dates back to 1660 AD
b It collect the subterranean seepage of an upstream reservoir or a lake
c It is used for regular water supply for religious rites, rotyal ceremonies and common use
d All above

95 Panam Keni, an indigenous water havesting structure is popular among

a Native tribe of Wayanad
b Khasi
c Jaintia
d None above

96 Bhandara Phad, an indigenous water harvesting structure is popular in state

a Maharashtra
b T N
c Rajasthan
d Gujarat

97 Which of the following is true abour Zing?

a It is small tank that collect melting glacier water
b It is an indigenous water haresting structure in Ladakh
c It is used for irrigating the crops
d All above

98 Which of the following is true abour Kuhls?

a Kuhl is a surface water channel
b It is found in mountainous regions of Kangra valley in H. P.
c The channels carry glacial water from rivers to fields
d All above

99 Who made the grand anicut, 'Kallanai across the river Cauvery?

a King Karikala
b King Chandragupta Maurya
c King Kanishik
d None above

100 Water harvesting structures are useful for

a To store rain water for drinking and irrigation puposes
b To rechnarge the ground water
c To reduce the run off losses
d All above

Answers

1. **d**	2. **d**	3. **c**	4. **d**	5. **d**	6. **d**	7. **a**	8. **d**	9. **d**	10. **d**
11. **a**	12. **b**	13. **a**	14. **b**	15. **c**	16. **c**	17. **c**	18. **a**	19. **c**	20. **a**
21 **b**	22. **a**	23. **d**	24. **b**	25. **d**	26. **a**	27. **c**	28. **a**	29. **c**	30. **a**
31. **a**	32. **b**	33. **a**	34. **a**	35. **c**	36. **c**	37. **b**	38. **a**	39. **b**	40. **c**
41. **c**	42. **a**	43. **a**	44. **b**	45. **c**	46. **a**	47. **d**	48. **c**	49. **d**	50. **c**
51. **c**	52. **a**	53. **a**	54. **b**	55. **d**	56. **c**	57. **c**	58. **c**	59. **d**	60. **d**
61. **d**	62. **d**	63 **c**	64 **a**	65 **d**	66. **b**	67. **a**	68. **c**	69. **a**	70. **d**
71. **b**	72. **d**	73. **d**	74. **d**	75. **a**	76. **a**	77. **d**	78. **d**	79. **c**	80. **a**
81. **b**	82. **d**	83. **c**	84. **a**	85. **b**	86. **c**	87. **d**	88. **c**	89. **d**	90. **a**
91. **d**	92. **d**	93. **d**	94. **d**	95. **a**	96. **a**	97. **d**	98. **d**	99. **a**	100. **d**

8

Crop Adaptation and Mitigation Mechanisms to Drought

Outlines

"The capacity to endure without injury and intense loss of water (is) one of the most important properties of true drought-resistant plants. This property is connected with the capacity of the protoplasm to endure considerable fluctuations in its degree of swelling, without the loss of vital activity and without the appearance of irreversible coagulation"

(Maximov, 1929)

Introduction

Drought is a recurrent feature of climate or climatic anomaly or weather related natural disaster. It occurs when the precipitation remains deficient over an extended period of time, usually for a season or more and affects water shortage in water bodies, streams, soil moisture, ground water etc. resulting into low crop production and economic performance of large regions or entire country. Drought occurs in virtually all regions varying its severity, low and high from regions to regions. Other climatic factors like high temperature, high wind velocity, low relative humidity are associated with low precipitation and drought intensity.

Definition

Drought can be defined by two types of definitions;

1. ***Conceptual definition*:** Conceptual definitions are of the "dictionary" type, generally defining the boundaries of the concept of drought, and thus are generic in their description of the phenomenon. For example, the American

Heritage Dictionary (1976) defines drought as "a long period with no rain, especially during a planting season. Therefore it helps to understand the meaning of drought and its effects on agriculture, livestock, environment and society. Hence, drought is a protracted period of deficient precipitation that causes extensive damage to crops resulting in poor crop production.

2. ***Operation definition*:** It indicates the timing of drought i.e. time of beginning and end of drought as well as its severity. Therefore the operational definition quantifies the current departure of rainfall from average over some 10 or 30 years period. As per the Irrigation Commission, Govt. of India and Indian Meteorological Department of Govt. of India defines drought year when the rainfall is equal or less than 75% of an average normal. If an area experiences such type of condition for >20% have years, is categorized as 'drought prone area' provided the area does not have irrigation facilities.

   An operational definition in relation to agriculture, it refers to the relationship between precipitation and evapotranspiration, soil water balance and related drought impacts on crop production. The operational definition is also used to analyze drought frequency, severity and duration for a specified period. Therefore, the weather data and crop yield data are required to developing correlation and interpretation.

3. **Indian Meteorological Department (IMD)** has defined drought as 'a situation occurring in any area when the mean annual rainfall is less than 75% of the normal rainfall'. IMD has further classified droughts into two broad categories;

   ***a. Severe drought*:** when the deficiency of rainfall exceeds 50% of the normal rainfall and

   b. Moderate drought: when the deficiency of rainfall is between 25 and 50% of the normal rainfall.

4. **Ramdas (1960)** defined drought as , 'Drought is a situation when actual seasonal rainfall is deficient by more than twice the mean deviation'.

5. **American Meteorological Society** defined drought as, 'a period of abnormally dry weather, sufficiently prolonged for lack of water to cause a severe hydrological imbalance in the area affected'.

6. **Irrigation Commission of India** defined drought area and chronic drought area based on dryness in the region. *Drought zones* are areas with 25 per cent probability of rainfall departure from the normal. *Chronic drought zones* are areas with probability of rainfall departure $\geq$ 40 per cent from the normal.

Periodicity of drought

Drought prone areas in the country, classified on annual rainfall departure, fall either in arid, semiarid and dry sub-humid regions where droughts occur frequently. The periodicity of drought in different meteorological sub divisions is given in table.1. Historical rainfall data of the country suggests that the monsoon rainfall recorded in the country during drought year of 1918 was the lowest. The severe drought years in the past is listed in table.2.

Table 1. Periodicity of drought in different meteorological sub divisions of India

S.N	Meteorological sub divisions	Frequency of deficit rainfall
1	Assam	Very rare or once in 15 years
2	WB, MP, Konkan, Bihar and Odisha	Once in 5 years
3	Soth Karnataka, Eastern UP, Gujarat, Vidharbha and Eastern Rajasthan	Once in 4 years
4	WesternUP, TN, J&K, Telangana and Royalaseema	Once in 3 years
5	Western Rajasthan	Once in 2.5 years

Table 2. Severe drought years that occurred over the past 200 years (1801-2000)

S.N	Period	Drought years	S.N	Period	Drought years
1	1801-1825	1801, 4,6,12,19,25	6	1926-1950	1939,41
2	1826-1850	1832,33,37	7	1951-1975	1951,65,66,71,72,74
3	1851-1875	1853,60,62,66,68,73	8	1976-2000	1977,78,79,82,83,85,87, 88,92
4	1876-1900	1877,83,91,97,99	9	2001-2020	
5	1901-1925	1901,4,5,7,11,18,20			

Administrative districts frequently affected by drought are given in Table .3

Table 3. Administrative districts frequently affected by drought in India

S.N	State	Districts
1	Andhra Pradesh	Anantpur, Chittor, Kadapa, Mahabibnagar, Nalgonda, Prakasam and Rangareddy
2	Bihar	Munger, Nawadha, Palamau, Raphtas, Bhojpur and Gaya
3	Gujrat	Ahmedabad, Amrely, Banaskanta, Bhavanagar, Bharuch, Jamnagar, Kheda, Kutch, Mehsana, Panchmahal, Rajkot and Surendarnagar
4	Haryana	Bhewani, Gurgaon, Mahendragarh and Rohtak
5	J & K	Dosa and Udhampur
6	Karnataka	Bangalore, Belgaun, Bellary, Bijapur, Chitragurga, Chickmangalur, Dharwad, Gulbarga, Hassan, Kolar, Mandya, Mysore, Raichur and Tumkir
7	Madhya Pradesh	Betul, Datia, Dewas, Dhar, Jhabuva, Khandak, Khargaon, Shahdol, Shahjanpur, Sidhi and Ujjain
8	Maharashtra	Ahmednagar, Aurangabad, Beed, Nanded, Nashik, Osmanabad, Pune, Parbhani, Singli, Satara, Sholapur
9	Orissa	Phulbani, Kalakhandi, Bolangir and Kendrapada

Types of drought

Drought can be categorized based on precipitation, moisture index, aridity index and remote sensing data on vegetation index i.e. normalized differential vegetation index (NDVI). The detail is given below;

1. Based on precipitation

Climate scientists, hydrologists and disaster management specialists have traditionally distinguished between three different kinds of drought, meteorological, agricultural and hydrological droughts. Recently socio-economic drought, ecological drought, soil moisture drought and famine have also been added in the list of drought. The detail is given below;

a. Meteorological drought: It occurs when rainfall is less than average over a significant period, often a month. The degree of dryness or rainfall deficit and the length of the dry period are deciding factors to meteorological drought. It is region specific because precipitation varies region to region. The data of weather parameters like temperature, relative humidity, wind velocity, sunshine hours, evaporation, atmospheric pressure are required to quantify meteorological drought of a specific region or country. It is important to note that effectiveness of rainfall is more important than amount of rainfall. On an average India receives 118 cm annual rainfall that is highest in anywhere in the world but uncertain, unreliable and erratic nature of rainfall by south-west monsoons creates drought conditions in different parts of the country.

The major causes of meteorological drought may be summed up as under

i. Lean monsoon and below average rainfall due to absence of depressions over India.

ii. Late onset or early withdrawal of monsoons.

iii. Prolonged dry spells in monsoon.

iv. Re-establishment of southern branch of jet stream.

b. Agricultural drought: It refers to the conditions of rainfall deficits, soil water deficits, reduced ground water, or reservoir levels leading to reduced water availability for irrigation resulting into poor crop production. The data of soil texture, soil fertility, soil moisture, crop type and its acreage, crop water requirements, pest and climate are required to assess the agricultural drought. National Commission on Agriculture (1976) considered agricultural drought as an occasion when at least four consecutive weeks receive rainfall half of the normal (normal rainfall being 5 mm or more) during the crop season (mid-May to mid-October) or six such weeks during other period.

c. Hydrological drought: A hydrological drought occurs when a meteorological drought significantly reduces the availability of water resources in rivers, lakes and underground table. Hydrological drought is a natural phenomenon, but it may be exacerbated by human activities. Hydrological droughts are usually related to meteorological droughts and their recurrence interval varies accordingly. Changes in land use and land degradation can affect the magnitude and frequency of hydrological droughts. Data sets required to assess hydrological drought are surface-water area and volume, surface runoff, stream flow measurements, infiltration, water-table fluctuations, and aquifer parameters. Therefore are two types of hydrological droughts,

i. *Surface water drought: It* is related to drying up of surface water resources like rivers, streams, ponds, tanks, reservoirs etc. and deforestation and other human activities are considered main reasons of it.

ii. *Groundwater drought:* It is related to fall of groundwater level mainly because of exploitation of groundwater without compensatory replenishment.

c. Socio-economic drought: The Socio-economic drought occurs when the demand for an economic good exceeds supply as a result of a weather-related deficit in water supply. It has great impact on supply and demand of some economic goods such as fruits, vegetables, grains and meat. The data related to production, population, demand, supply, market rates/prices, trade and delivery system etc. are required for evaluating the impact of socio-economic drought.

d. Soil Moisture Drought: This is a situation of inadequate soil moisture particularly in rainfed areas which may not support crop growth. This happens in the event of a meteorological drought when the water supply to soil is less and water loss by evaporation is more. This type of situation is adverse to crop production and lead to sparse plant population and poor yield and its severity may cause even crop failure.

e. Ecological drought: The ecological drought takes place when the productivity of a natural ecosystem fails significantly as a consequence of distress induced environmental damage.

f. Famine: This is the peak of drought when the surface and ground water resources are decreased to the threshold level and no more water is available to crop production and livestock leading to large scale collapse of life and mass starvation in larger areas.

2. Based on rainfall and crop production

The rainfall, its quantity, distribution and intensity are the major component for agricultural production in rainfed areas because the rain water is the only source for fulfilling the water requirement of crops.

a. **Permanent drought:** It refers to the conditions where rainfall is always scanty and insufficient for crop production, so crop production is possible only after providing irrigation. It occurs in driest climate where sparse vegetation is found. It represents the extremely arid region.

b. **Seasonal drought:** It is defined as the condition where drought is experienced seasonally or during the crop growth period, is known as seasonal drought.

It occurs in arid and semi arid areas of the country where rainy and dry seasons are distinguished. The crops are grown in rainy season and dry season is used for conserving soil and water. The crop sowing is adjusted as per the arrival and retreat of monsoon and crops and varieties are selected on the basis of available length of growing period of the region.

c. **Unpredictable drought**: It involves deficient, erratic and uneven rainfall. It may occur almost anywhere but is most characteristic of humid and sub humid climates.It often affects only a relatively small area. However, ongoing large-scale droughts of this kind are possible, especially in drier regions with several subsequent years of inadequate rainfall or snowpack and affect adversely food, feed, fodder and fibre production.

d. **Contingent drought:** This involves an abnormal failure of rainfall. It may occur almost anywhere especially in most parts of humid or sub-humid climates. It is usually brief, irregular and generally affects only a small area.

e. **Invisible drought:** It occurs even when there is frequent rain in an area. The crop evapotranspiration is higher than precipitation and crop suffers borderline soil moisture deficiency resulting sub optimal crop yield. It occurs usually in humid regions.

f. **Atmospheric drought:** The situation where there is low relative humidity accompanied by high temperature and hot dry winds and crop suffers water shortage even if the crop is grown under favorable soil moisture conditions, is termed as atmospheric drought.

3. Based on Moisture AdequacyIndex (MAI)

The moisture adequacy index is the ratio of actual to potential evapotranspiration and computed for short period like weeks or 5 days in crop growing season to quantify the status of soil moisture availability for crops and intensity of drought in rainfall deficient periods.

MAI = (AE/PE) x 100

Where, AE represents actual evapotranspiration and PE, the potential evapotranspiration.

On the basis of MAI, drought can be grouped into

a. **Severe drought** : MAI = $<$ 25%

b. **Moderate drought**: MAI = 25 - 29.9%

c. **Crop growing season**: MAI = $\geq$ 50%. In this condition, sufficient soil moisture is available for crop production.

d. **Humid region** : MAI = $\geq$75%. Precipitation is more than ET of crop and drainage is essential for maintaining optimum soil moisture regime for crop production.

7. Based on Aridity Index (AI)

a. Thornthwaite used water balance approach for evaluating drought on the basis of aridity index as given in table.4 below;

Table 4. Classification of drought based on aridity index

S.N	Drought class	Aridity index
1	Moderate	< 0.5
2	Large	0.5 – 1.0
3	Severe	1.0 – 2.0
4	Disastrous	> 2.0

b. Palmer Index: The drought index proposed by Palmer is known as Palmer Index. According to him, the drought can be grouped in to four categories;

Table 5: Classification of Plamer's aridity index

S.N	Drought class	Values of Drought / Aridity index
1	Mild	1 – 1.99
2	Large	2 – 2.99
3	Severe	3 - 3.99
4	Disastrous	>4.00

8. Based on NDVI

9. Annawary system of classification

Under this system, the drought is estimated based on visual observation of crop conditions. Most of the states follow this system for declaring different types of drought in monsoon season. The detail is given below;

1. No drought : If the production is >75% of the normal,
2. Moderate drought : If production is in between 50 – 75% of the normal.
3. Severe drought : If production is in between 25 – 50% of the normal.
4. Disastrous drought : If production is <25% of the normal.

10. CAZRI classification

CAZRI, Jodhpur categorized drought based on moisture stress during crop growing season using following crieria;

1. **Drought free period (No drought):** When cumulative AE is >PE/2, AE is actual evapotranspiration and PE is potential evapotranspiration.
2. **Moderate drought period:** When cumulative AE lies in between PE/1 and PE/4.
3. **Severe drought:** When cumulative AE is <PE/4.

Table 6. Classification of drought based on AE/PE (CAZRI, Jodhpur)

S.N	AE/PE (%) during different phenophases	Drought intensity	Code		
			Seedling	Vegetative	Reproductive
1	76-100	Nil	S_0	V_0	R_0
2	51-75	Mild	S_1	V_1	R_1
3	26-50	Moderate	S_2	V_2	R_2
4	≥ 25	Severe	S_3	V_3	R_3

11. Climate scientists, hydrologists and disaster management specialists have traditionally distinguished between three different kinds of drought:

i. ***Meteorological drought***: It occurs when rainfall is less than average over a significant period, often a month. The degree of dryness or rainfall deficit and the length of the dry period are deciding factors to meteorological drought.

ii. ***Agricultural drought***: It refers to the conditions of rainfall deficits, soil water deficits, reduced ground water, or reservoir levels leading to reduced water availability for irrigation resulting in to poor crop production.

iii. ***Hydrological drought***: A hydrological drought occurs when a meteorological drought significantly reduces the availability of water resources in rivers, lakes and underground table.

iv. ***Socio-economic drought***: The Socio-economic drought occurs when the demand for an economic good exceeds supply as a result of a weather-related deficit in water supply. It has great impact on supply and demand of some economic goods such as fruits, vegetables, grains and meat.

Under drought stress, reductions in carbon assimilation result in an imbalance between electron excitation and utilization through photosynthesis, which results in the production of reactive oxygen species (ROS), primarily superoxide (O_2-) and hydrogen peroxide (H_2O_2). These ROS damage cell membranes, proteins, and nucleic acids, causing oxidative stress.

Similarly, under drought conditions, plants can alter water relations to maintain cellular functions. For example, plants exhibit osmotic adjustment by synthesizing and accumulating compatible solutes such as free amino acids, sugars, and proline. Osmotic adjustment allows the plant to maintain turgor pressure and cell volume at low water potential, which is important for maintaining metabolic functions (Akashi *et al.*, 2001. and Drennan, *et al.*, 1993). In addition, osmotic adjustment facilitates the recovery of metabolic activities after relief from stress (Abdul *et al.*, 2007).

Effects of drought on crop plants

Plants responses to water stress vary with the following factors.

1. Degree and duration of water stress.
2. Time of occurrence.

3. Kind of plant type.
4. Type of plant produce.

Water stress affects the growth, yield and quality of the produce in various ways. Different plant processes like root development, tiller formation, branching, flowering, grain filling etc. are affected most if the degree and duration of water stress is more. All plant physiological stages are not equally sensitive to water stress, so most water sensitive stages like crown root initiation in wheat, tasseling in maize, boll formation in cotton, flowering in oilseeds etc are affected most under even moderate water stress conditions. Succulent plants like agricultural crops could not tolerate water stress for longer period compared to forest or woody trees. Similarly, fodder and vegetable crops suffer water stress earlier than grain crops. Similarly, C_4 plants can tolerate more water deficit than C_3 plants. The green plants produce like fodder fresh vegetables and fruits need higher soil water potential than dry fruits and grain crops.

The growing plant organs are affected first than older organs by water stress conditions. It has been observed that cell growth and cell wall synthesis (cell elongation and division) are the first and foremost plant process affected adversely under water deficit condition followed by reduction in carbohydrate synthesis protein synthesis, nitrate reductase and accumulation of ABA and proline. In general, water deficit leads to reduce the synthetic processes and activate degradation processes. Imposition of mild to moderate water stress sometimes proves beneficial to improve water use efficiency in following manners;

1. Allow roots to develop into deep soil layers in search of water,
2. Allow plants to modify physiologically tolerant to water stress,
3. Use water effectively- higher WUE, and
4. Save water that may be used for other enterprises.

Following effects have been observed under water deficit conditions;

1. Stunted plant growth,
2. Reduced tillering and branching,
3. Reduced leaf area,
4. Reduced flowering and pollination,
5. Poor seed filling – shriveled seed and low test weight,
6. Forced maturity – early senescence of leaves,
7. Rolling and wilting of leaves,
8. Reduction in diameter of root and bulb crops,
9. Increase protein content in wheat and nicotine in Tobacco,
10. Closing of stomata – reduction of CO_2 intake,
11. Nutrient uptake reduced, and
12. Reduced photosynthesis → Low productivity.

The effects of drought on some of the important plant responses are given below;

1. ***Growth and Development*** – The seed germination is reduced and so the plant stands. The cell division, cell enlargement and differentiation are reduced resulted in to poor growth and development mainly due to interruption of water flow from xylem to the surrounding elongated cells. Therefore the impaired mitosis, cell elongation and expansion result in reduced plant height, leaf area and crop growth. The crop yield is also affected drastically under water stress conditions mainly due to poor plant stand and growth. Again the crop growth stage of water stress is important. The reduced flowering, pollination, barren seed, poor seed filling, small size of ear/spikes and test weight lead to poor yield under water stress conditions. Therefore, light intensity, sunshine hours, temperature and water stress of plant and soil are contributing factors to maturity period and crop yield.
2. ***Photosynthesis*** – Photosynthesis is reduced by moisture stress due to reduction in photosynthetic rate, chlorophyll content, leaf area and increase in assimilate saturation in leaves. Under water stress conditions, the stomata are closed, therefore, the CO_2 intake rate is reduced. Besides, the boundary layer, stomatal and mesophyll resistance are also increased resulting into reduction in CO_2 intake rate. Similarly the conductance of mesophyll cells is decreased by both age and water stress, hence CO_2 concentration in leaf chlorophyll is reduced and finally photosynthesis is reduced. The chlorophyll activity including enzymatic activity reduced under water stress condition that leads to poor photosynthesis. The other cause of reduced photosynthesis is the loss of balance between the production of reactive oxygen species and the antioxidant defense causing accumulation of reactive oxygen species which induces oxidative stress in proteins, membrane lipids and other cellular compounds. Plasmolysis of cells, wilting of leaves, shrinking of vacuoles, disintegration of cytoplasm and destruction of photosynthetic enzymes are the major symptoms of severe water stress that finally lead to death of plant. The detail of reduced photosynthesis can be described below;

i. Stomatal oscillation

The stomata are closed due to reduced cell turgidity of guard cells to reduce transpirational water loss in drought stress conditions. The closure of stomata affects the intake of CO_2 for photosynthesis and some spare electrons produce active oxygen species. It is observed that stomatal oscillation is often more closely governed by soil moisture than leaf water status. It suggests that stomata respond to chemical signals like ABA, pH and cytokinin level of leaf sap. The ABA is synthesized and control stomatal movement in water stress conditions. It was also found that the pH of leaf sap increased due to more transpiration loss and promotes ABA synthesis, while cytokinin in xylem cells promotes stomatal opening in presence of ABA, so it is a very complex phenomena due to inter specific variation in response of stomatal conductance, photosynthesis and leaf water potential. Therefore it is well accepted that stomata close progressively as drought progresses, followed by a parallel decline in net photosynthesis. However, stomatal conductance is not controlled by soil water availability alone, but by a complex interaction of intrinsic and extrinsic factors.

ii. Photosynthetic enzymes

Water stress decreases cell turgidity due to higher pull for transpiration leading to dehydration of cell that causes cell shrinkage and consequently a decline in cell volume making cytoplasm more viscous due to accumulation of more solutes. It affects adversely the photosynthetic machinery and declined Rubisco activity. A rapid decline in photosynthesis under drought was accompanied by decreased maximum velocity of ribulose-1, 5-bisphosphate carboxylation by Rubisco, speed of ribulose-1, 5-bisphosphate regeneration, Rubisco and stromal fructose bis-phosphatase activities, and the quantum efficiency of photosystem II in higher plants (Reddy *et al.*, 2004; Zhou *et al.*, 2007). Moreover, under severe drought, carboxylation efficiency by Rubisco was greatly declined, and it acted more as oxygenase than carboxylase. During water stress, activities of the phosphoenolpyruvate carboxylase, nicotinamide adenine dinucleotide phosphatemalic enzyme, Rubisco, fructose-1, 6-bisphosphatase and pyruvate orthophosphate dikinase decreased linearly with lowered leaf water potential. Du *et al.* (1996) reported that pyruvate orthophosphate dikinase activities were decreased 9.1 times during water stress, a much greater reduction than other enzymes, which were from 2 to 4 times, suggesting that pyruvate orthophosphate dikinase is very likely to be the limiting enzyme to photosynthesis under water stress.

iii. Adenosine triphosphate synthesis

The enzymes involved in carbon assimilation and in adenosine triphosphate synthesis are retarded and sometimes inhibited depending upon the extent of available moisture. The Rubisco, which shows dual functions, acts as oxygenase under water-limiting conditions; and therefore limited CO_2 fixation is noticed. Therefore, impaired photophosphorylation and adenosine triphosphate synthesis are the main culprits limiting photosynthesis even under mild drought (Tezara *et al.*, 1999).

3. *Plant Nutrition*

Water is a universal solvent and plant nutrients are dissolved in soil water that is absorbed against transpirational pull. The plants absorb most of the essential plant nutrients by mass flow. Under water deficit condition, nutrient solubility is decreased. Similarly, the root exchange is reduced due to break down of capillary continuity of moisture film around soil particles. Therefore, the nutrient availability and uptake is reduced. However, plant species and genotypes may differ in acquisition of mineral uptake and its effect on plant growth and development. In addition to reduced uptake of nutrients, the energy required for assimilation of NO^{-3} /NH^{+4} , PO_4^{3-} and SO_4^{2-} is also limited due to reduced production of ATPs through respiration. The specific nodule activity and leg hemoglobin in nodules are reduced under water deficit condition. The nitrogenous activity is also reduced. Hence nitrogen fixation in nodules is reduced. It has been observed that nitrogen deficiency increases stomatal resistance while phosphorus deficiency increases the sensitivity of newly developed leaves and finally leads to closure of stomata. In general, drought stress reduces the availability, uptake, translocation and metabolism of nutrients

4. *Respiration*

Photosynthesis and respiration are two important bio-chemical processes that ensure the productivity of crop plants. Respiration is destructive phenomena in which carbon is lost and ATPs are formed that govern the plant life activities. The rate of photosynthesis is often limits the plant growth and may have negative carbon balance under water stress conditions (Huang and Fu, 2000). Lambers *et al.* (1996) pointed out that root was a major consumer of carbon fixed in photosynthesis and used it for growth, dry matter production and maintenance. It was found that >50% of daily assimilates were transported to the roots and around 60% was respired in wheat under drought stress. Severe drought is more lethal and reduced shoot and root biomass, photosynthesis and root respiration rate. Therefore it is obvious to reduce respiration for higher productivity under drought stress. In respiration process, two mitochondrial electron transport pathways occur from Ubiqinoneto oxygen in plants. The alternative pathway branches from the cytochrome pathway and donates electrons to oxygen directly by alternative oxidase (Moore and Siedow, 1991) and it is not coupled with adenosine triphosphate synthesis, however it can be induced in response to stress or inhibition of the main electron transfer pathway. The plants produce reactive oxygen species and damage membrane components during drought stress, so alternative oxidase activity could be useful in maintaining normal levels of metabolites and reduce reactive oxygen species production during stress. Shugaeva *et al.* (2007) reported in sugar beet that at an early drought stage (10 days), a decrease in the activity of the cytochrome-mediated oxidation pathway was largely counterbalanced by the activation of mitochondrial alternative oxidase, whereas long-term dehydration of plants was accompanied by activation of additional oxidative systems insensitive to both potassium cyanide and salicylic hydroxamate. Normally, the mild stress enhances the respiration rate but severe water stress decreases the respiration because of degradation of enzymes responsible to respiration process. Finally it can be concluded that water deficit in the rhizosphere leads to an increased rate of root respiration, leading to an imbalance in the utilization of carbon resources, reduced production of adenosine triphosphate and enhanced generation of reactive oxygen species.

5. *Anatomical changes*

Water stress leads to following anatomical changes;

a) Decrease in size of the cells.

b) Decrease in intercellular spaces.

c) Increase in thickness of cell wall.

d) Greater development of mechanical tissues.

e) Increase number of stomata per unit leaf area.

6. *Protein synthesis*

RNA and DNA synthesis are hampered severely under water stress condition leading to reduction in protein synthesis. In contrast, the protein is degraded into *'Proline'* and ABA synthesis. Accumulation of sugars and amino acids take place because of reduced conversion of amino acids into protein synthesis.

7. *Hormonal Relationships*

Water deficit may alter hormonal balance in plant systems. It has been observed that the activity of growth promoting hormones like IAA (auxin), GA (Gibberalic acid), cytokinin etc. is decreased but the activity of growth regulating hormones like 'ABA' 'Betain' and Ethylene' etc. is increased under water stress conditions. The ABA content is inversely correlated to leaf water potential and also acts as water deficit sensor to minimize the loss of tissue water potential. High 'ABA' content checks the growth and saves water loss. High 'Ethylene' concentration promotes leaf and fruits drop under water stress conditions. 'Betain' is produced under moisture stress condition and is used as indicator of moisture stress.

8. *Protoplasmic Dehydration*

The following anomalies in protoplasm have been observed;

i. Protoplasm becomes dense with high viscosity,

ii. Dehydration of protoplasm, and

iii. Disintegration of protoplasmic organelle.

9. *Reproduction and Grain Growth*

Flowering and grain filling stages are very crucial for economic yield. Water deficit at flowering stages reduces the cell division and total number of fruits. Similarly, the water stress at grain filling stage is detrimental to test weight. Again stress at anthesis stage affects the pollens viability and drying of pollen and finally pollination is affected adversely. The economic yield is reduced due to reduction in leaf area and photosynthesis. In general vegetative and grain development stages are not very crucial for water stress conditions.

10. *Oxidative damage*

Reactive oxygen species are formed as by-products in the electron transport chains of chloroplasts mitochondria, peroxisomes and the plasma membrane.

The environmental stresses including water stress, in general, lead to production of reactive oxygen species, including superoxide anion radicals (O_2^-), hydroxyl radicals (OH^-), hydrogen peroxide (H_2O_2), alkoxy radicals (RO) and singlet oxygen (O_2^1) which react with proteins, lipids and deoxyribonucleic acid resulting into oxidative damage and impairing the normal functions of cells (Foyer and Fletcher, 2001). The chloroplasts are potentially important sources and produce reactive oxygen species because excited pigments in thylakoid membranes may interact with O_2 to form strong oxidants such as O_2^- or O_2^1 OH^- and H_2O_2. Similarly the interaction of O_2 with reduced components of the electron transport chain in mitochondria and peroxisomes during photorespiration produce different reactive oxygen species including H_2O_2. Both enzymatic and non enzymatic reactions are responsible for production of reactive oxygen species under O_2 deprivation stress. Among enzymatic reactions, xanthine oxidase, peroxidases, catalases and oxidases i.e. amino acid oxidases and glucose oxidase are involved into different biochemical reactions. In non-enzymatic reactions, electron O_2 reduction can occur at higher oxygen concentrations and even at very low O_2 concentrations, plant terminal oxidases and the formation of reactive oxygen species via the mitochondrial electron transport chain still remain functional. Overall, the production of reactive oxygen species is linear with the severity of drought stress, which leads to enhanced peroxidation of membrane lipids and degradation of nucleic acids, and both structural and functional proteins. Various organelles including chloroplasts, mitochondria and peroxisomes are the seats as well as first target of reactive oxygen species produced under drought stress.

11. *Yield*

The economic part is important because the crop are harvested not only for its grains but also for fodder, vegetables, fruits, leaf (tobacco and teas) etc., therefore, the green parts are affected even at mild stress at vegetative stage. The grain portion is very sensitive to water stress at flowing stage. The fibers or chemical products are not much affected by moderate water stress. In general, water deficit reduces the economic and biological yield of crop plants. Hsiao (1973) pointed out that cell growth and cell wall synthesis were the first and foremost affected plant processes under water stress condition. Later on, protein synthesis proto chlorophyll formation, nitrate reductions were affected. Some other plant processes like stomatal movement, CO_2 assimilation, respiration, proline formation and sugar accumulation were affected under moderate to severe water stress conditions (Table.7).

Table 7. Effect of water stress on different plant processes (Hsiao, 1973)

Plant processes	Sensitivity to stress			Remark
	Very sensitive Relatively insensitive			
	Tissues Ψ required to affect process			
	0 MPa	-1.0 MPa	-2.0 MPa	
Cell growth	--			Fast growing tissues
Wall synthesis	--			Fast growing tissues
Protein synthesis	--			-
Proto chlorophyll formation	--			-
Nitrate Reductase	--			-
ABA accumulation	--			-
Cytokinin level	--			-
Stomatal opening	--			Depends on species
Co_2 assimilation	--			Depends on species
Respiration	--			-
Proline accumulation	--			-
Sugar accumulation	--			-

Notes

1. Length of horizontal lines represents the range of stress levels within which a process first becomes affected. Dashed lines signify deductions based on more tenacious data, and
2. Ψ of well watered plants under mild evaporative demand as the reference point.

Drought resistance mechanism

The crop plants have differential ability to grow and survive in dry conditions. The morphological, biochemical and physiological changes in plant help them to adapt in different degree of drought stress. The crop plants can be grouped into three categories based on mechanism of mitigating drought conditions as given below;

1. Escaping drought
2. Drought resistance
3. Drought tolerance

1. Escaping drought (Ephemerals)

The plants that have very short life span and complete their life cycle within the availability of soil moisture are known as ephemerals or drought escaping plants. They have very fast growth from germination to maturity and complete its life cycle within 5-6 weeks before depletion of soil moisture. Generally they germinate at onset of monsoon and complete its life cycle at before monsoon or just after retreat of monsoon. Such plants do not have any drought mechanism to overcome soil moisture stress but they avoid drought or moisture stress conditions. Basically such plants are not categorized under drought resistance. Only their seeds persist during times of drought. Normally drought escaping plants are annuals and complete their life cycle in one season. In arid and semi arid conditions, the time of flowering is a major trait of a crop adaptation when crop is prone to terminal drought and high temperature. Therefore, in the regions where growing period is short, the development and selection of high yielding short duration crop varieties is effective strategy for minimizing the terminal drought (Kumar and Abbo,2001), however the yield is proportionally related to length of crop duration under favorable growing conditions (Turner *et al.*, 2001).

2. Drought resistance

The crop plants that modify their morphology and physiology suitable to either avoid or tolerate drought stress are known as drought resistant plants and the mechanism as drought resistance. These plants can avoid drought stress by conserving more water and allowing more water uptake. Those plants which have ability to conserve more water by morphological modification like closure of stomata, high photosynthetic efficiency, high water use efficiency, deposition of lipids on cuticle, fleshy stems, development of tendrils, spines and awns, reduced leaf area and leaf area index to reduce transpirational losses, are known as water savers or 'drought evading plants', while those have high root/shoot ratio, efficient deep root system, high osmotic potential may absorb more water to fulfill its requirement and avoid drought stress are known as 'water spenders' or 'drought enduring plants' because they absorb and transpire more water to adjust canopy temperature and avoid drought stress.

3. Drought tolerance

Drought tolerance or true drought resistance in plants can be developed either by mitigating the actual drought stress or by tolerating the drought stress. Plant can maintain high water potential to mitigate the actual stress by resistance to dehydration and preventing leaf collapse but the plant can tolerate actual drought conditions by developing metabolic and plastic strains. Succulence is the most important characteristic of drought resistant plants. The drought resistant plants have very shallow root system, stem heavily waxed, leaves turn into spines, stomata open in nights and they produced 'dehydrins' under dehydration and other late embryogenesis proteins (LEA) and both are induced by ABA systnesis under water stress conditions. Accumulation of compatible osmolytes, glycine, betaine, proline and other quaternary compounds as osmoprotectants help into developing drought tolerance in plants.

Porembski (2000) have classified the plants that can sustain its growth and yield under water deficit conditions in to 4 major categories;

i. ***Drought escaping plants***: The annuals that germinate and grow only during times of sufficient moisture to complete their life cycle.

ii. ***Drought-evading plants***: The non-succulent perennials which restrict their growth only to periods of moisture availability. Examples are Bergenia, *Cichorium intybus* (common chichory) and *Plantarum* spp.

iii. ***Drought-enduring plants***: They are also known as xerophytes. These plants are evergreen shrubs with extensive root systems along with morphological and physiological adaptations which enable them to maintain growth even in times of extreme drought conditions. Examples are *Cereus peruvianus* (Cactus), *Euphorbia virosa*, *Agave americana*, Accacia spp. and *Proposes cineraria*.

iv. ***Drought-resisting plants***: also known as succulent perennials, they have water stored in their leaves and stems for sparing uses. Examples are aloe and cactus.

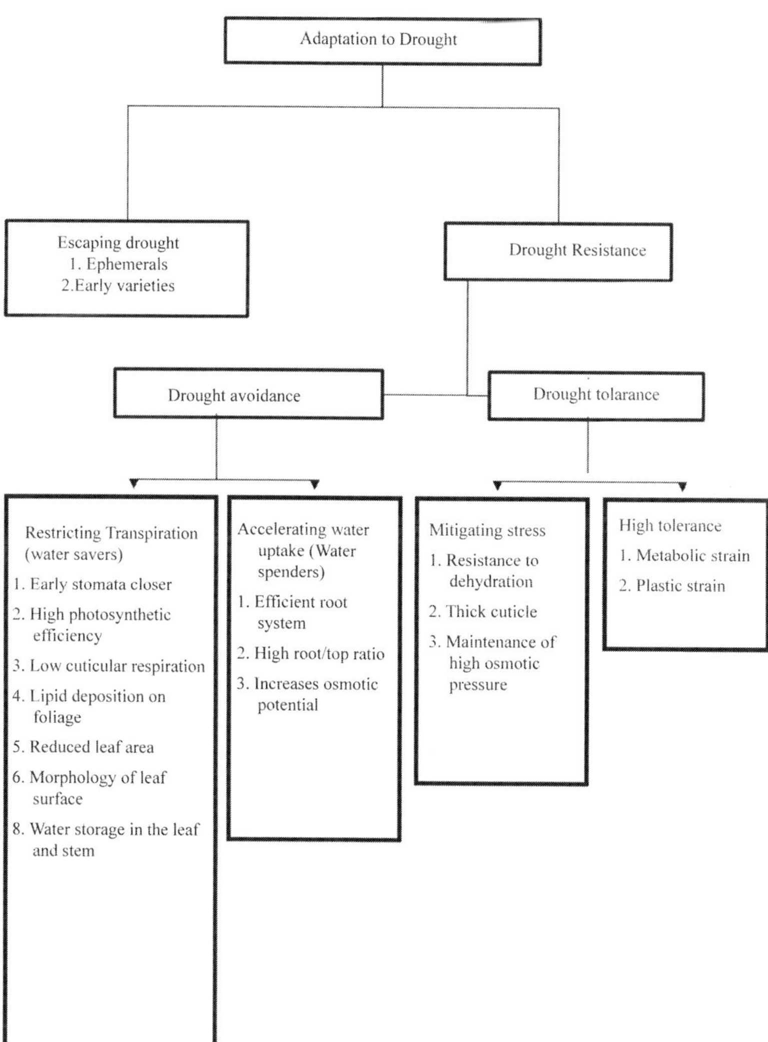

Fig. 1: Mitigation mechanism of drought adaptation

The crop plants can mitigate the drought resistance either by avoiding drought stress or tolerating drought stress. The detail is given below;

a. Drought avoidance

The plant can avoid drought stress by maintaining favorable water balance either by conserving water and reduce transpiration losses or by more water uptake from soil. Therefore morphological adaptation like early closure of stomata, high photosynthetic rate, reduced cuticular transpiration, deposition of lipids on leaf surface, reduced leaf area and leaf area index, water storage in plants, high root-shoot ratio and increased osmotic potential. Therefore two types of plant species i.e. plants which conserve more water known as 'water savers' and plants having high water uptake ability, termed as 'water spenders' are more suitable to sustain under moisture stress conditions.

Drought avoidance consists of mechanisms that reduce water loss from plants, due to stomatal control of transpiration, and also maintain water uptake through an extensive and prolific root system (Turner *et al.*, 2001; Kavar *et al.*, 2007). The root characters such as biomass, length, density and depth are the main drought avoidance traits that contribute to final yield under terminal drought environments (Subbarao *et al.*, 1995; Turner *et al.*, 2001). A deep and thick root system is helpful for extracting water from considerable depths (Kavar *et al.*, 2007).

The glaucousness or waxy bloom on leaves helps with maintenance of high tissue water potential, and is therefore considered as a desirable trait for drought tolerance (Richards *et al.*, 1986; Ludlow and Muchow, 1990). Varying degrees of glaucousness in wheat led to increased water-use efficiency, but did not affect total water use or harvest index. Determination of leaf temperature indicated that, compared with non-glaucous leaves, glaucous leaves were 0.7^{0}C cooler and had a lower rate of leaf senescence (Richards *et al.*, 1986). These authors suggested that a 0.5^{0}C reduction in leaf temperature for six hours per day was sufficient to extend the grain-filling period by more than three days. However, yield advantages are likely to be small as many varieties already show some degree of glaucousness.

Maintaining favorable water balance: Plants can maintain favorable water balance by two ways;

i. Conserving water (Water savers): plants conserve or save water within plant system by following mechanisms;

a. *Early closure of stomata*

Stomatal responses to drought stress are mainly controlled by chemical signals such as ABA that are produced in dehydrating roots. It is well documented that there is a direct correlation between conductance of stomata and the xylem ABA content. Changes in leaf to air vapor pressure deficit, xylem sap pH, status of plant nutritional, plant hydraulic conductance as well as farnesyltranferase activity, and decrease in relative water content are other factors that play roles in stomatal regulation in plants. The stomata are very sensitive to moisture stress and change in stomata morphology like shape, size and density have been reported in many crops even under short period of drought stress. The density of stomata is increased and size is decreased in moisture stress conditions resulting into reduced transpiration water loss. The average of stomata number per square of epidermis is called stomatal density. Drought stress effects on stomatal morphology may vary depending on plant species and plant varieties. Under long drought, the synthesis of ABA in roots, leaves and guard cells have recently been observed. The high ABA in roots increases hydraulic conductivity facilitating water uptake and transportation, while enhanced ABA levels in the leaves induce and regulate stomatal closure that helps to minimize water loss by reducing transpiration occurring through stomata.

b. *Increased photosynthetic efficiency*

The photosynthetic rate is decreased under drought stress due to closure of stomata that causes reduced CO_2 entry into the leaves, however the magnitude of decline in photosynthetic rate remains lower under severe drought tolerant plants than sensitive plants. It has also been observed imbalances between electron excitation and utilization in PS-II due to reduced carboxylation process resulting in to production of reactive oxygen species (ROS), primarily superoxide (O_2^-) and hydrogen peroxide (H_2O_2) and damages cell membranes, protein and nucleic acids, causing oxidative stress. The intercellular concentration of malondialdehyde (MDA) indicates the extent of oxidative stress. The plants have both enzymatic and non enzymatic mechanisms to detoxify ROS. The enzyme, 'superoxide dismutase' catalyzes the conversion of O_2^- to less reactive H_2O_2 and in presence of enzymes 'catalase' and 'ascorbate peroxidase', the H_2O_2 is further converted into O_2 and H_2O. Similarly the 'Glutathione (GSH)' and 'carotenoids' are among the non-enzymatic antioxidants involved in cellular defense. GSH protects the chloroplasts from ROS damage by increasing the ratio of reduced glutathione to oxidized glutathione (GSH/GSSG), whereas carotenoids safeguard the photosynthetic apparatus by dissipating excess excitation energy into heat. Therefore it is clear that drought tolerant plants have higher concentration of these enzymes and reduce oxidative stress and maintain higher photosynthetic rate than sensitive plants. The drought tolerant plants also recover its physiological and biochemical functions quicker than sensitive and less tolerant plants after re-watering. The adverse effect of drought on photosynthetic rate and stomatal conductance is higher at jointing stage onwards than vegetative stage i.e. tillering stage in wheat.

There were two tolerance mechanisms employed in response to drought stress; the first involved the upregulation of antioxidant enzyme (SOD, CAT, and APX) activity and production of non-enzymatic (GSH and carotenoids) antioxidants; whereas the second involved accumulation of soluble sugars, free amino acids, and proline to facilitate osmotic adjustment. These metabolic allow the wheat plant to withstand and survive water-deficit conditions. The more drought tolerant cultivar exhibited greater photosynthetic stability during drought and more rapid recovery following drought, primarily due to greater ability to scavenge ROS and to osmotically adjust. Moreover, plants were better able to recover from stress imposed during the tillering stage than the jointing stage, emphasizing the importance of drought timing in determining productivity. These results revealed that the plant's ability to maintain physiological functions during drought and recover after re-watering during vegetative periods are important for determining final productivity in wheat (Abid *et al.*, 2018).

c. *Low rate of cuticular transpiration*

There are two ways of transpirational water loss in plants, 1. Stomatal transpiration and 2. Cuticular transpiration. The stomatal transpiration is a major contributor in normal conditions but under drought conditions, the cuticular transpiration becomes major route of water loss and it may accounts 50 or more than 50% water loss. The presence of thick cuticle is a very common feature of xerophytes. The cuticular transpiration is associated with drought tolerance in plant species. The plant cuticle is a protective layer covering the outer epidermal surface of whole plants. It minimizes not only the incident of light, UV irradiance and pollutant entry but also defends against pathogens and herbivores. The major function of cuticle is to reduce water loss and maintain the optimum water status in plant systems. The presence of epicuticular waxes turns leaves in to bluish-white coloration termed glaucousness. The thickness of cuticle is more pronounced in drought tolerant than drought sensitive plants.The cuticular thickness is higher in plants like *Arabidopsis*, alfalfa, commelina and barley that are drought tolerant. It again indicates that presence of cuticular layer is safeguard against the drought stress conditions. It was observed in *Arabidopsis* that total wax amount per unit leaf area was increased by 75% and cuticle thickness by 49% under water deficit conditions. Similarly in tea leaves, total wax accumulation was increased by more than 150%. Therefore the cuticle layer reduces the permeability to water and helps to save water and improves water use efficiencies as observed in CAM plants.

d. *Deposition of lipids on leaf surface*

The leaf glaucousness or waxy bloom on leaves helps with maintenance of high tissue water potential and is therefore considered as a desirable trait for drought tolerance (Richards *et al.*, 1986; Ludlow and Muchow, 1990). Richards *et al.* (1986) reported that varying degrees of glaucousness in wheat led to increased water-use efficiency, but did not affect total water use or harvest index. The glossy leaves were 0.7^{0}C cooler than non-glossy leaves, and also had a lower rate of leaf senescence and higher crop yield. Some plants may secrete resins and waxes (epicuticular wax) on their surfaces, which reduce transpiration. Examples are the heavily scented and flammable resins (volatile organic compounds) of some chaparral plants, such as Malosma laurina, or the chalky wax of *Dudleya pulverulenta*.

It is noted that total lipids and galactolipids (monogalactosyldiacylglycerols-MGDGs and digalactosyldiacylglycerols-DGDGs) as well as phospholipids (phosphoryl glycerol (PG), phosphoryl ethanolamine (PE), phosphoric acid (PA) and phosphoryl serine (PS) decreased in stressed sensitive plants but in contrast, the galactolipids and unsaturated fatty acids i.e. oleic and linolenic acids (18:2 and 18:3) as well as antioxidant secondary metabolites i.e. flavonoids were increased in tolerant plants under severe drought conditions (Moradi *et al.*, 2017). In contrast, Thi *et al.* (1990) reported that the resistant crop varieties always had less polar lipids, expressed on a dry weight basis, than the sensitive varieties and MGDG clearly related to drought resistance. The resistant plants had less MGDG and less linolenic acid than the sensitive ones. Therefore it was summarized that cells of drought-resistant plants contained less membrane lipids and less monogalactosyl·diacylglycerol (MGDG) as the MGDG was less unsaturated, because of less linolenic acid, and therefore they were less susceptible to degradative processes. It is therefore stated that drought stimulates the biosynthesis of membrane structural lipids in tolerant plants to protect cell and chloroplast membranes from damage induced by severe drought and preserve their structure and function. Besides, the higher level of antioxidant lipids and flavonoids would help to scavenging ROS production during drought conditions. However, the recent molecular or marker induced approaches like transcriptomics and RNAseq are required to add new insights into plant adaptation mechanisms related to lipid synthesis under drought stress conditions.

e. ***Reduced leaf area and leaf area index***

Water is the most important factor in the initial phase of plant growth and development from germination to seedling formation and a reduction in water supply in this stage may lead to dehydration and even plant death. Therefore the plant growth and development is severely affected under soil moisture deficit conditions mainly because of reduced diffusion of $CO_{2,}$ water uptake and cell turgidity resulting into reduced photosynthesis, respiration and energy production. The plant architecture is altered, cell enlargement and cell division are reduced, leaf area and leaf area index are reduced due to reduced leaf numbers, its length and width and even more senescence of lower leaves. These are the adaptive mechanism of drought resistant plants just to save water for plant functions. Under moisture stress conditions, some osmoregulatory processes occur to protect cell membrane integrity and maintain inflow of water to the cell along with accumulation of organic solutes as sugars and quaternary ammonium compounds (glycine, betaine, alanine), hydrophilic proteins, soluble protein and amino acids (proline) that is formed from hydrolysis of proteins and plays an important role as an osmoprotectant in many cultivated species. Proline is considered to play important role in reducing cell water potential (Knipp and Honermeier, 2006), stabilizes membrane and proteins and removes free radicals. The increase of proline has also been linked to the reduction in leaf water potential in addition to its role as an osmoregulator, proline stabilizes membranes and proteins and contributes to the removal of free radicals (Ashraf and Foolad, 2007).

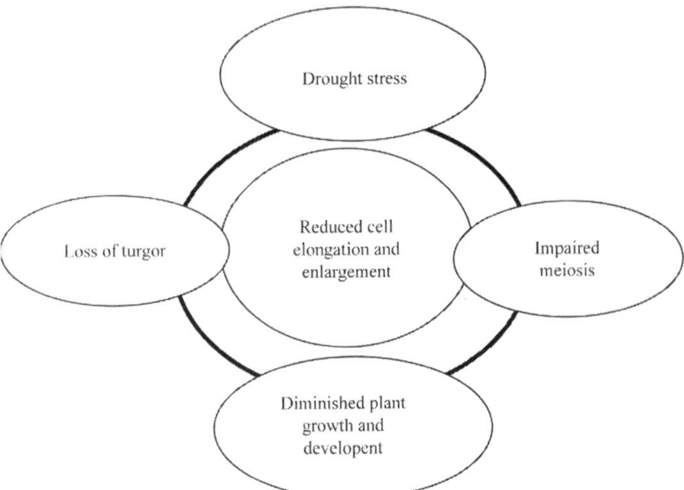

Fig. 2. Effect of drought on plant growth and development

f. *Change in leaf morphology*

The morphological traits like small leaves, leaf pubescence, shorter plants and deeper root system are xeromorphic traits that help plants to withstand drought very well. The small leaves have an ability to maintain their water status by developing microclimate surrounding leaf boundaries and survive though their growth rate and biomass are relatively low (Ball *et al.,* 1994). The leaf surface reduction in form of spines in cactus, compaction and reduction of branching in barrel cacti, leaf compaction at the base in basal rosette are the examples of surface area reduction. Similarly the leaf pubescence protects leaves from excessive heat load and reduces leaf temperature and transpiration (Sandquist and Ehleringer, 2003). The leaf hairiness increases light reflectance and minimizes water loss by increasing the boundary layer resistance to water vapor movement away from the leaf surface. In addition, the presence of tiny hairs and spines on leaf surface of some xerophytes reduces the direct contact of deep stomata to outer warm air and reduces transpiration. These leaves maintained more saturated localized environment in presence of spines and hairs mainly because of trapping of moist layer around the leaf surface and so reduced external water vapor pressure potential gradient near stomata resulting in to lower rate of transpiration. The plant surface covered with tiny hairs is called *tomentose.*

Some plant species like wheat also induces the production of trichomes on both sides of wheat leaves under drought stress but they do not have significant influence on boundary layer resistance. The shorter plants have lower number of leaves and total volume exposed to outer hot environment and so they save water by reducing transpiration loses. The dwarf wheat varieties have greater root biomass and develop thick root systems with more accumulation of water and photosynthates that protect plants under moisture stress conditions (Richards *et al.*, 1986; Ludlow and Muchow, 1990), however the plant species may have variable response of root distribution to dry soils ((Liu *et al.*, 2004). This adaptation is exhibited by some Agave and Eriogonum species, which can be found growing near Death Valley.

g. *Development of awns: awned varieties have 12% higher photosynthesis.*

The conversion of growing points of plant species into awns and tendrils reducing the surface area is also a mechanism of adaptation under moisture stress conditions because reduced surface area help to reduce transpiration. Experimental findings have reported that in wheat that awned wheat has nearly 12% higher photosynthetic rate than awnless varieties because more favorable water potential. Similarly the development of tendrils mainly in pulses like cowpea, beans etc. is a adaptive mechanism to drought.

h. *Leaf wilting and abscission*

The plants get completely wilted at permanent wilting point because sufficient water is not available for their consumptive use. It has been seen that plants start suffering water deficiency just after critical level of available water in soil. Under water stress conditions, plants start drying and falling of lower leaves slowly just to reduce transpiration loss and save soil moisture. At wilting points, the plant loses whole leaves or abscission and finally die. The xerophytic plants have ability to survive under water deficit conditions better than mesophytes as they shed their lower leaves up to 2/3rd of stem and maintain only 1/3rd upper leaves for reproduction and seed production. Only they die in extremely dry conditions.

Some of the plants change their morphology and store water in leaves, stems, trunks tuber, roots etc. under xerophytic conditions for their survivals. For example, fleshy tubers in Raphionacme and succulent leave and stem in Euphorbia and kalanchoe.

Table 8. Mechanism of drought adaptation in plant species

S.N	Mechanism	Adaptation	Plant specieis
1.	Water uptake	*Extensive root system*	*Acacia and Prosopsis*
2.	Water storage	Fleshy tuber	Raphionacme
		succulence	*Euphorbia and Kalanchoe*
3.	Reduced water loss	Surface area reduction	Barrel cactus, Basal rosette, *Eriogonum compositum*
		Sunken stomata and hairs	Pine, *Nassauvia falklandica*, Bromeliads etc.
		Waxy leaf surface	Prikly pear, Malosma laurina, Dudieya pulverulenta
		Nocturnal stomata	Tea plant, Alfalfa, *Quercus trojana*, *Brachychiton discolor*

		CAM photosynthesis	Cactus, Pineapple, *Agave americana*, Aeonium haworthii, Sansevieria trifasciata (Snake plant)
		Curled leaves	Esparto grass
4	Dormancy and reduced photosynthesis	Resurrection plants	Ramonda nathaliae, Ramonda myconi, Haberlea rhodopensis, Anastatica, Craterostigma pumilum
		Dormant seeds	Californian poppy
		Leaf abscission	Coastal sage scrub, Wiliwili, Geoffroea decorticans

Source: https://en.wikipedia.org/wiki/Xerophyte

ii. Improving water uptake (water spenders): Plants absorb more water by following mechanisms;

a. Efficient root development

The drought resistant plant must have well developed root system so that the roots have high volume and may penetrate to deeper layers so that they can extract water even from moisture deficient soils for their survival. Therefore the root characters such as biomass, length, density and depth are the main drought avoidance traits that contribute to final yield under terminal drought environments (Subbarao *et al*., 1995; Turner *et al*., 2001). A deep and thick root system is helpful for extracting water from considerable depths (Kavar *et al*., 2007). *Acacia and Prosopsis* species are the good examples of xerophytic nature with extensive root system that help to extract more water from deeper soil layers under moisture stress conditions.

b. High root/shoot ratio

The high root-shoot ratio is related to efficient root system. It is generally observed that drought resistant plants have lower volume of shoot than root mainly because of reduced size of plant height, number of leaves and branches in short of available water. It has already been reported that 55% of the total root weight was usually found between 0 and 25 cm for most crops, about 18% in between 25 and 50 cm, 15% of the root mass in between 50 and 75 cm, and below 12% of the root mass, It is particularly important under drought stress conditions. The influence of Rht genes, especially cereals on root shoot-ratio is often discussed and analyzed. Interestingly, the Rht genes reduce plant height i.e., have a significant effect on the above-ground part habitus, thus often also modifying the described ratio.

There are number of theories related to root and shoot growth and its ratio. In moisture stress conditions, the low absorption of nutrients is caused poor growth of root and shoot but shoot growth is affected more adversely than root. In addition, the synthesis of plant hormones like auxin and gibberellic acid are reduced resulting into poor shoot growth. Besides, upon exposure of plants to drought stress, ABA is the major hormone synthesized in roots and translocated to leaves to initiate adaptation of plants to drought stress through stomatal closure and reduced plant growth.The photosynthesis is reduced due to lower leaf area as well as reduced diffusion of CO_2 because of closing of stomata. The partitioning of photosynthates towards roots is channelized for developing vigorous root system so that roots may come to contact with larger volume of soil and extract more water. Therefore the root/shoot ratio is increased under moisture stress conditions and makes plants more tolerant to drought conditions.

c. High osmotic potential

The drought resistant plant can alter water relations to maintain cellular functions in moisture stress conditions. These plants exhibit osmotic adjustment by synthesizing and accumulating compatible solutes such as free amino acids, sugars and proline and improves water status that allows plants to maintain turgor pressure and cell volume at low water potential. It is required feature of plants to maintain physiological activity for extended period of drought (Kramer and Boyer, 1995). Osmotic adjustment not only delays dehydrative damage and also maintain cell turgor and physiological processes (Taiz and Zeiger, 2006) but also facilitates better translocation of pre anthesis carbohydrates for grain filling (Subbarao *et al*., 2000). High cell turgor also leads to higher photosynthetic rate and growth (Ludlow and Muchow, 1990 and Subbarao *et al*., 2000). Yokota *et al*., 2002) reported that wild water melon survive under severe drought without wilting of leaves mainly because of accumulation of 'citrulline', 'glutamate' and 'arginine' in leaves that maintain water status of leaves.

Tolerating stress: two types

Drought tolerance is the ability of plants maintaining its biomass production during arid or drought conditions. Some plants are naturally adapted to dry conditions, surviving with protection mechanisms such as desiccation tolerance, detoxification, or repair of xylem embolism. (Agrawal *et al*., 2008). Other plants, specifically crops like corn, wheat, and rice, have become increasingly tolerant to drought with new varieties created via genetic engineering. The mechanisms behind drought tolerance are complex and involve many pathways which allow plants to respond to specific sets of conditions at any given time. Some of these interactions include stomatal conductance, carotenoid degradation and anthocyanin accumulation, the intervention of osmoprotectants (such as sucrose, glycine, proline and ROS-scavenging enzymes. (Ahmadi and Baker, 2001). The molecular control of drought tolerance is also very complex and is influenced by other factors such as environment and the developmental stage of the plant. (Abdul *et al*., 2007). This control consists mainly of transcriptional factors, such as dehydration-responsive element-binding protein (DREB), abscisic acid (ABA)-responsive element-binding factor (AREB), and NAM (no apical meristem).

a. Mitigating drought stress

i. *Resistance to dehydration*

The crop plants, in general, suffer water stress just after critical level of available soil water and wilt at wilting point. Field capacity holds maximum and permanent wilting point almost nil available water to crop plants. As the moisture stress increases, the available water decreases and it has adverse effect on plant growth and development. Few crops and varieties differ in their tolerance capacity to drought stress. Hence the maintenance of optimal cell water or cell water potential is necessary to maintain cell turgidity for normal physiological functions. Pannu *et al.* (1993) described that determination of leaf water status in the morning and water content in leaves in the afternoon were potentially useful for screening drought tolerance in chickpea, while Riccardi *et al.*(2001) pointed out that water potential was useful for describing the drought stress in faba bean.

The drought tolerant plants have higher root/shoot ratio and greater osmotic adjustment under moisture stress conditions. The osmotic adjustment allows cell to decrease osmotic potential that increases the gradient of water influx, improves cell water status and also maintain the cell wall elasticity and turgidity. This is essential phenomena for maintaining physiological activity during extended drought stress conditions (Kramer and Boyer, 1995). Morgan (1990) concluded that osmotic adjustment helped to maintain cell water balance with the active accumulation of solutes in the cytoplasm, thereby minimizing the harmful effects of drought. Turner *et al.* (2001) reported that osmotic adjustment, abscisic acid (ABA) and induction of dehydrins might confer drought tolerance by maintaining high tissue water potential. The osmotic adjustment delays dehydrative damage by continued maintenance of cell turgor and physiological processes (Taiz and Zeiger, 2006), facilitates a better translocation of pre-anthesis carbohydrate partitioning during grain filling (Subbarao *et al.*, 2000) maintain high turgor that lead to higher photosynthetic rate and growth (Ludlow and Muchow, 1990; Subbarao *et al.*, 2000) in moisture stress environments.

ii. ***Prevention of cell collapse (leaf with thick cuticles)***

The cell membrane is the first target to be damaged under water stress conditions, so the drought tolerant plants must maintain its cell integrity and stability under water (Bajji *et al.*, 2002). Cell membrane stability *vis-a-vis* cell membrane injury as physiological index is widely used for the evaluation of drought tolerance (Premachandra *et al.*, 1991). It is a genetically controlled phenomenon since quantitative trait loci for this have been mapped in drought-stressed rice at different growth stages (Tripathy *et al.*, 2000). Dhanda *et al.* (2004) also reported that membrane stability of the leaf segment was the most important trait to screen the germplasm for drought tolerance.

Normally the cell membrane stability is declined rapidly in Kentucky bluegrass exposed to drought and heat stress simultaneously (Wang and Huang, 2004).

Gnanasiri *et al.* (1991I) reported improved drought tolerance in maize with K nutrition. Premachandra *et al.* (1991) found that seed hardening increased drought tolerance in holm oak (*Quercus ilex*) primarily by reducing osmotic potential and stomatal regulation, improved new root growth capacity and enhanced cell membrane stability. However, variation in cell membrane stability, stomatal regulation and root growth capacity were negatively related to osmotic adjustment (Villar-Salvador *et al.*, 2004).

The cell membranes are disintegrated under moisture stress mainly due to decrease in cellular volume and increase the viscosity of cytoplasmic components that increase the chances of molecular interactions leading to protein denaturation and membrane fusion. Another possibility of ion leakage from the cell may be due to thermal induced inhibition of membrane-bound enzymes responsible for maintaining chemical gradients in the cell (Reynolds *et al.*, 2001). A broad range of compounds like proline, glutamate, glycinebetaine, carnitine, mannitol, sorbitol, fructans, polyols, trehalose, sucrose and oligosaccharides, have been identified that prevent adverse molecular interactions and maintain the integrity and stability of cell membranes under drought stress (Folkert *et al.*, 2001). *Arabidopsis* leaf membranes are very resistant to water deficit as it maintains polar lipid contents and the stability of their composition under severe drought (Gigon *et al.*, 2004).

iii. *Maintain higher osmotic pressure by accumulating higher amount of solutes*

The drought tolerant plants produce number of compatible organic solutes in response of water deficit. These organic solutes are highly soluble compounds with low molecular weight and non toxic even at high cytosolic concentrations. They protect plants from stresses by osmotic adjustment, detoxification of reactive oxygen species, stabilization of membranes and maintaining of native structures of enzymes and proteins. Under moisture stress conditions, the osmotically active molecules/ions like soluble sugars, sugar alcohols, proline, glycinebetaine, organic acids, calcium, potassium, chloride ions, etc. are accumulated and lower the osmotic potential of cell, that attracts water into cell and maintain cell turgor and other physiological activities like growth, photosynthesis, respiration, partitioning of phtosynthates etc. take place at normal place. Therefore the mechanism of osmotic adjustment is an important trait to postpone the dehydration stress.

i. Proline is one of the most important organic solutes accumulated in cytoplasm and has wider response in higher plants, algae, animals, bacteria under moisture stress (Zhu, 2002; Wahid and Close, 2007). Proline helps in many ways like slow oxidation of mitochondria, stabilization of macromolecules, a sink for excess reductant, a store of carbon, nitrogen for use after relief of water deficit (Zhu, 2002), osmoprotectant and induced drought tolerance (Yamada *et al.*, 2005).

ii. Glycinebetaine (N, N, N-trimethyl glycine), one of the most extensively studied quaternary ammonium compounds and compatible solutes in plants, animals, bacteria and algae, plays an important role in enhancing plant tolerance under a range of abiotic stresses including drought (Quan *et al.*, 2004). It has a positive effects on enzyme and membrane integrity or as an osmoprotectant, also protect cells from environmental stresses indirectly by participating in signal transduction pathways (Subbarao *et al.*, 2000).

iii. Citrulline is an amino acid found in leaves of wild water melon under drought stress, so water melon adaptation of in severe drought stress is linked with accumulation of citrulline, glutamate and arginine (Akashi *et al.*, 2001). Citrulline is a novel and the most effective OH−scavenger among compatible solutes examined so far. Moreover, it can effectively protect DNA and enzymes from oxidative injuries (Akashi *et al.*, 2001; Bektasoglu *et al.*, 2006).

iv. Rapid accumulation of a non protein AA, γ-aminobutyric acid has been identified in plant tissues under drought stress. It acts as a zwitterion, exists in free form and has a flexible molecule, very similar to proline. The physiological roles of γ-aminobutyric acid in drought tolerance entail osmotic regulation (Shelp *et al.*, 1999), detoxication of reactive oxygen radicals, conversion of putrescine into proline, intracellular signal transduction (Kinnersley and Turano, 2000), pH regulation, nitrogen storage, plant development and defense, a compatible osmolyte and an alternative pathway for glutamate utilization (Shelp *et al.*, 1999 and Wahid *et al.*, 2007). Under drought stress, a signal transduction pathway is initiated with increased cytosolic Ca^{2+} activates Ca^{2+}/calmodulin dependent glutamate decarboxylase activity, leading to γ-aminobutyric acid synthesis (Shelp *et al.*, 1999). Elevated H+ and substrate levels can also stimulate glutamate decarboxylase activity resulting into to γ-aminobutyric acid accumulation.

v. Trehalose, a non-reducing disaccharide of glucose, is a compatible solute in the stabilization of biological structures under abiotic stress (Goddijn *et al.*, 1997). It is biosynthesized as a stress response by a variety of organisms including bacteria, fungi, algae, insects, invertebrates, lower plants (Wingler, 2002), drought tolerant fern and desiccation tolerant angiosperm ,*Myrothamnus flabellifolia* (Drennan *et al.*,1993), in tobacco (Goddijn *et al.*, 1997) and many higher plants (Kosmas *et al.*, 2006). It has many physiological roles including efficient stabilization of dehydrated enzymes, proteins and lipid membranes, as well as protection of biological structures under desiccation stress (Wingler, 2002) rather than regulating water potential (Lee *et al.*, 2004). Karim *et al.* (2007) reported that it enhanced drought tolerance by improved water status and extensive root development. Trehalose sysnthesis is controlled by genes as reported in transgenic plants having higher drought tolerance by overproducing trehalose-6-phosphate synthase trehalose (Karim *et al.*, 2007).

Therefore the osmatic adjustment, maintenance of antioxidant defense system for the scavenging of reactive oxygen species, stomatal regulation, plant growth regulators and polyamines, γ-aminobutyric acid, free amino acids and sugars like trehalose play a vital role in providing drought stress in plants by protection of macromolecules and cell water balance.

b. High degree tolerating stress

a. *Metabolic strain*

The metabolic strain refers to the response of drought occurred in plants under drought stress and it is reversible in absence of strain/drought. The number of genes is induced in response of drought at the transcriptional level and gene products confer drought tolerance in plants. The gene expression is triggered either directly or from secondary stresses/injury responses. The number of stress proteins is induced under drought stress as given below;

i. **Aquaporins**: Aquaporins, intrinsic membrane proteins, are synthesized in roots and leaves. They are abundantly found in plasma membrane and vascular membrane. It is believed that aquaporins have greater role in plant drought resistance by regulating hydraulic conductivity of membranes and potentiate a 10-20 fold increase in water permeability under drought stress. Phosphorylation, Ca and pH may influence the activity of aquaporins.

ii. **Dehydrins**: Dehydrins are the late embryogenesis associates (LEA) proteins, synthesized in response to drought and cold and involve in hydrophobic interactions leading to macromolecule stabilization by protecting cellular proteins from denaturation. They behave like molecular chaperones and inducing drought tolerance in plants.

iii. **Low molecular heat shock proteins**: They are produced under high temperature, drought, anaerobic conditions and low temperatures. They act as molecular chaperones that participate in adenosine triphosphate dependent proetiens and prevent protein denaturation under stress.

b. *Plastic strain*

The plastic strain refers to the response of drought occurred in plants under drought stress and it is irreversible in absence of strain/drought. The molecular responses to severe drought stress involve highly regulated genes and signal transduction processes that averse the stress conditions. C-repeat/dehydration-responsive element binding factors (CBF/DREB), mitogen-activated protein (MYB), cup-shaped cotyledon (CUC), no apical meristem (NAC) TFs and zinc-finger proteins (ZFPs) are recognized as significant moieties in conferring plant drought tolerance. The chemical signals i.e. ROS, Ca and plant hormones are involved in inducing stress tolerance by acting via transduction cascades and activate genomic re-programming. Mitogen activated protein kinase cascades are considered to have important roles in signaling various stresses including drought (Wrzaczek and

Hirt, 2001). Following genes have been identified providing drought tolerance in different plant species;

i. GsZFP1gene induced drought tolerance in *Medicago sativa, so this gene may be used effectively in breeding and* genetic engineering breeding to promote drought tolerance in crop plants.

ii. The over expression of SNAC1 in *Gossypium hirsutum* produced bigger roots and maintain water balance under drought stress.

iii. *GUDK* and *HYR* genes stabilize and improve grain yield of rice in normal as well as drought stress (Ambavaram *et al.*, 2014 and Ramegowda *et al.*, 2014).

iv. BdWRKY36 gene stimulated transcription of stress-related genes, reduced electrolyte leakage, decreased ROS level, elevated chlorophyll amount, plant water status and antioxidant enzyme activities to enhance the drought tolerance.

v. MpCYS4 gene boosted closing of stomata, triggered the transcription activity of abscisic acid (ABA) and water-deficit-associated genes to confer drought tolerance and was associated with ABA induced stress signal transduction.

vi. VaNAC26 gene is useful for synthesizing Jasmonic acid (JA) that enhances plants drought tolerance by stimulating root growth, decreasing level of ROS and promoting stomatal closure.

vii. ABA-induced genes encoding drought-related proteins such as dehydrins, ROS-detoxifying enzymes, regulatory proteins and phospholipid signaling enzymes can improve drought stress tolerance. Improved amount of ABA induced a signaling pathway in guard cells which results in outflow of guard cells K+ and reduced turgor pressure, ultimately causing stomata closure.

viii. Auxin regulates root development, functioning of ABA related genes and ROS metabolism to improve drought-tolerance.

ix. Ethylene mediates synthesis of guard cell antioxidant flavanols in an EIN2 dependent manner and adversely affects stomata closing by suppressing drought mediated ROS formation, thereby resulting in another possible target for genetically engineered plants tolerant to drought.

x. Increased level of cytokinin amount in xylem sap induced stomata opening by diminishing its sensitivity to ABA.

xi. Ca improved water stress tolerance in *Catharanthus roseus* by increasing γ-glutamyl kinase and reducing the proline oxidase activities (Abdul Jallel *et al.*, 2007). So far 29 Ca depended protein kinase genes have been identified.

xii. A number of phopholipid systems are activated by osmotic stress, generating an array of messenger molecules and may function upstream of osmotic stress activated protein kinases.

xiii. Over expression of Arabidopsis *EDT1/HDG11* was shown to increase drought tolerance of poplar and cotton through increased accumulation of solutes such as proline and soluble sugars and also increase the yield of cotton in the field (Du *et al.*, 2010).

Pineapple and cactus are typical examples of plastic strains. Pineapple leaves store enough water in special non-chlorophyllous plastic strain tissues which is utilized during drought. The xerophytes such as cacti are capable of withstanding extended periods of dry conditions as they have deep-spreading roots and capacity to store water. Their waxy, thorny leaves prevent loss of moisture. It is assumed that *saguaro cactus* can store water up to 4200 pounds of water. The water content in drought-treated mature stems declined by 4% and water potential by –0.25 MPa. It is observed that active phloem supply of assimilates and associated water reserves from mature stems was the mechanism that allowed developing stems of *Hylocereus undatus* to maintain growth under drought conditions (Nerd and Neumann, 2004). Moreover, girdling the phloem of growing stems rapidly inhibited stem elongation, but secretion of sucrose-containing nectar was maintained during drought. The water potential gradient was in the wrong direction for xylem transport from mature to young growing stems and axial hydraulic conductivity was low to negligible (Nerd and Neumann, 2004).

Therefore it is urgent need to identify drought stress genes and develop transgenic crops and varieties having higher drought stress. The classical breeding and genetic engineering approaches play a major role in developing drought stress and climate resilient future crops and varieties (Fig. 3).

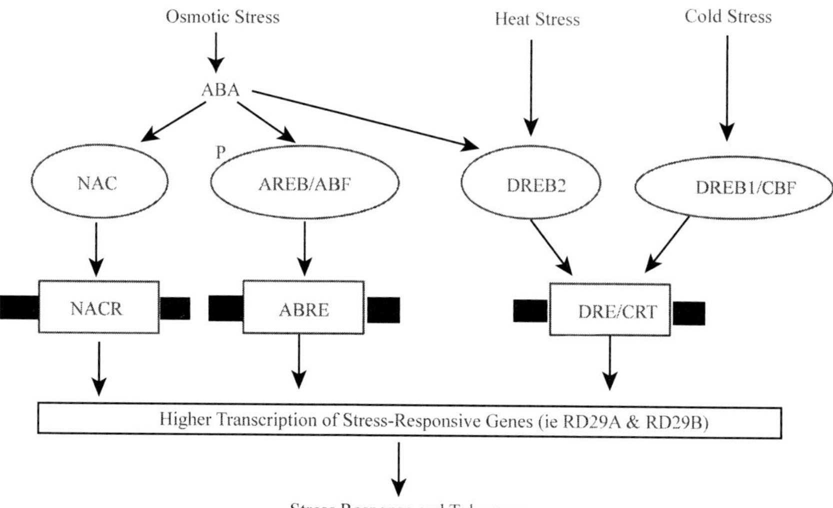

Fig. 3: Abiotic stresses (such as drought) induce the expression of the following transcription factors. They bind to cis-elements, resulting in a change in stress response and tolerance (Ref. Farooq et al., 2006) .

Molecular approach of water stress

The molecular approach of water stress and its effects on plant processes depends on a specific gene responsible for a unique plant process. The DNA/RNA synthesis followed by protein synthesis is governed by unique gene sequences. Under water stress conditions, it is essential to identify a specific gene for particular plant process .Further, the plants resist to water stress due to internal as well as external adaptations that are controlled by a unique gene or set of genes. Bohnert *et al.* (1995) described the range of molecular, biochemical and physiological responses of plants to water deficit as given in fig.4 below;

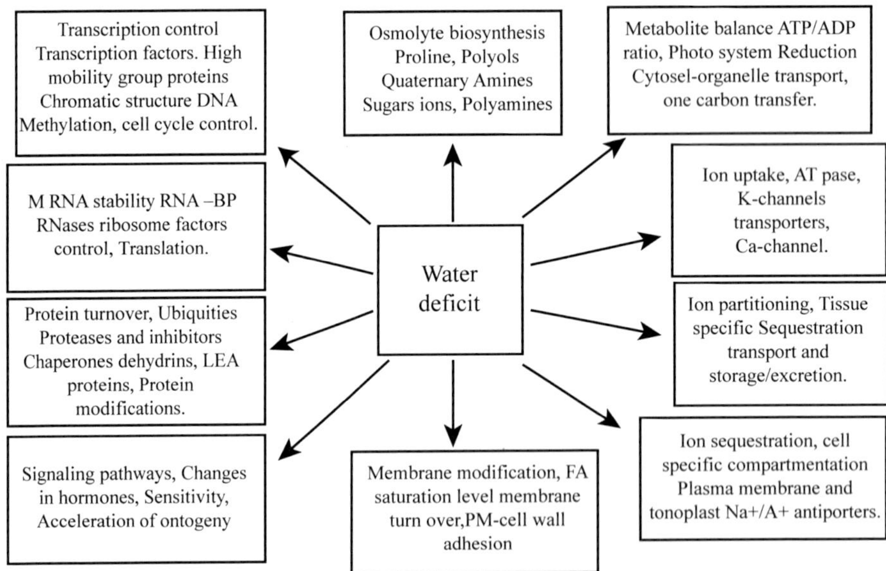

Fig. 4: Range of molecular, biochemical and physiological responses of plants to water deficit

Drought management

The risk involved in successful cultivation of crops depends on the nature of drought (chronic and contingent), its probable duration and periodicity of occurrence within the season. In the arid region where mean annual rainfall is less than 500 mm, drought is almost an inevitable phenomenon in most of the years (Ramakrishna, 1997). In semi-arid region (mean annual rainfall 500-750 mm), droughts occur in 40 to 60% of the years due to deficit seasonal rainfall or inadequate soil moisture availability between two successive rainfall events. Even in dry sub-humid regions (annual rainfall 750-1200 mm), contingent drought situations occur due to break in monsoon conditions. Therefore, drought management strategies need to be identified separately for each climatic region and for each major crop. Crop production in rainfed areas is generally affected by five distinct categories of drought, viz., early season, mid-season, late season, chronic and apparent drought (Rao, *et al.*, 1999).

i. ***Early season drought*** generally occurs either due to delayed onset of monsoon or due to prolonged dry spell soon after the onset, resulting in seedling mortality, need for resowing or poor crop stand and seedling growth. The early withdrawal of monsoon results in reduced water availability period for crop growth. Thus, the crops suffer from acute shortage of water during reproductive stage.

ii. ***Mid-season drought*** occurs due to inadequate soil moisture availability between two successive rainfall events during crop growth. Its impact depends on crop growth stage, duration and intensity of the drought spell.

iii. ***Late season or terminal drought*** occurs as a result of early cessation of monsoon, mainly during the years with late commencement or weak monsoon activity. Terminal droughts are more critical as the crop yield is strongly related to water availability during the reproductive stage. Rainfed rice in dry sub-humid regions is often subjected to terminal droughts due to failure of September rains with 40 to 50% of dry spells of over 5 days during that month.

iv. ***Chronic drought*** is common in arid areas where rainfall and stored soil moisture arc inadequate to meet crop water requirement during most of the years. Here, the assured growing period is hardly 6 to 7 weeks. These are characterized as chronic drought or highly drought prone areas.

v. ***Apparent drought*** conditions are observed in low to medium rainfall regions due to mis-matching of the cropping pattern in relation to rainfall / moisture availability.

Drought Management Strategies

Following strategies can be adopted for managing drought stress;

i. Selection and breeding strategies

ii. Molecular and functional genomics

iii. Induction of drought resistance

   a. Seed priming

   b. Use of plant growth regulators

   c. Use of osmoprotectants

   d. Application of Silicon

i. Selection and breeding strategies

It is an urgent need to breed new plant type having higher heritability and high yielding potential with drought tolerance because of increasing global warming. Conventional breeding and selection are normally used for breeding new crop varieties, however it is not very appropriate as the yield is a quantitative trait with low heritability. The expression of genotypic potential is also controlled by environmental conditions. The molecular and physiological understanding of crops and its yield contributing traits are important for breeders so that these traits may be transferred to newly evolved varieties. Further the screening under natural conditions in target environments is difficult because of the irregular and erratic drought response. Therefore the classical breeding is a good option for developing drought tolerance which relies upon multi-location testing of progenies in different agro-climatic conditions followed by selection of better plant type with higher drought tolerance. The genetic analysis of secondary traits like root system architecture, leaf water potential, panicle water potential, osmotic adjustment and relative water content and its related heritability with yield under drought may help to develop high yield varieties with greater drought tolerance. The water use efficiency of crop plants is differed and also reduced under drought stress conditions either by reducing leaf area or restricting stomtal opening or both resulting low biomass production. Condon *et al.* (2004) proposed three key processes to be looked in to breeding programs to improve water use efficiency in crop plants; 1. increasing water uptake, 2. improving biomass production, and 3. partitioning of produced biomass towards the harvested product.

ii. Molecular and functional genomics

The differential expression of gene and its related trait in different growing environment requires special attention to indentify the stress genes. Therefore the molecular and biochemical studies were employed to identify the abscisic acid and stress relative genes and few transcription factors (Buchanan *et al.*, 2005, Poroyko *et al.*, 2005) and its transgenic expression related to improvement in drought tolerance in crop plants. Presently these transgenic approaches are currently the mainstream method to bioengineer drought tolerance in crop plants (Behieldina *et al.*, 2005) but these drought enhancer genes are inversely related to growth and yield, so limit its practical applications. Therefore it is an urgent need to understand more and more molecular and genetic basis of drought resistance (Xiong *et al.*, 2006). Bioinformatic tools have great potentials to explore number of stress genes that could be translated in to newly breed high yielding crop varieties. One group of proteins that is consistently induced by dehydration are dehydrins, they are also part of late embryogenesis abundant proteins (LEA). The LEA and dehydrins can also be induced by ABA under water stress conditions. It has also observed that these proteins are also induced even in the absence of water stress in some plants. Accumulation of compatible osmolytes, glycine, betaine, proline and other quaternary compounds as osmoprotectants as a mechanism of drought resistance. Genes for their overproduction have been identified and they are being used for making transgenic plants resistant against drought stress. It is little difficult to identify regions in the genome of crop species which contribute to resistance to drought in a broad range of drought environments. The ultimate goal of the breeders is to identify QTLs that can increase crop yield under drought or at least increase yield stability under drought. Therefore, the genomics, proteomics, trascriptomic approaches and mapping of quantitative trait loci related technologies are presently displaying greater role in identifying stress genes. The transgenic and genetically modified organisms (GM crops) have shown greater tolerance in biotic and abiotic stresses and the future relies in them for developing more crop varieties with greater drought tolerance.

iii. Induction of drought resistance

The application of growth hormones, osmoprotectants, seed priming and chemicals have shown its worth in inducing the drought tolerance in crop plants. Therefore following strategies can be employed for increasing drought tolerance in crop plants;

1. Use of growth hormones

In general, the synthesis of auxin, gibberellic acid and cytokininis reduced under drought stress but in contrast, abscisic acid and ethylene is increased. It has been realized that combined application of two or more growth regulators enhanced the drought tolerance in crop plants mainly due to improving water status and chlorophyll content (Zhang *et al.*, 2004). Following observations indicate that proper exogenous application of growth regulators may help to induce drought stress in crop plants;

i. Taiz and Zeiger (2006) concluded that length and fresh weight of hypocotyle is reduced under drought stress but applying GA reversed the effect by improving the water status of the seedling and protein synthesis.
ii. Todd *et al.* (2004) proposed that application of '*1-aminocyclopropane-1-carboxylic acid*' improved drought tolerance by delaying senescence.
iii. Exogenously application of *uniconazole*, *brassinolide* and *abscisic acid* increased soybean yields both under well-watered and water deficit conditions.
iv. Application of *jasmonic acid* induced drought tolerance by increasing the betaine level in pear (Gao *et al.*, 2004). Similarly, application of *brassinolid*e, *uniconazol*e and *methyl jasmonate* improved drought tolerance in maize owing to increased activities of superoxide dismutase, catalase and ascorbate peroxidase, abscisic acid and total carotenoid contents (Li *et al.*, 1998). Jasmonates play an essential role in the signaling pathway, triggering the expression of plant defense genes under different stresses (Koda,1997).
v. Shang (2000) found that *benzyladenine* is an active cytokinin that helps in increasing drought resistance in plants.
vi. The synthesis of salicylic acid is increased (Shang, 2000) and improves plant growth under drought stress (Senaratna *et al.*, 2000). The application of salicylic acid improved drought tolerance in winter wheat due to increasing catalase activity. The seed soaking or foliar spray of combined salicylic acid and acetyl-salicylic acid (a derivative of salicylic acid), protected muskmelon (*Cucumis melo*) seedlings under drought stress conditions as both are related to signaling for gene expression rather than their direct

effects (Senaratna *et al.*, 2000). The salicylic acid is believed to increase ABA and proline content (Shakirova *et al.*, 2003).

2. Use of osmoprotectants

The osmoprotectants help in osmotic adjustment and protecting subcellular structures in stressed plants.The most common osmoprotectants are proline, trehalose, fructan, mannitol, glycinebetaine and others (Zhu, 2002) and they involve in signaling and regulating plant responses to multiple stresses including drought stress. The level of osmoprotetcnatns in plants can be increased by three ways (Ashraf and Foolad , 2007):

i. Use of traditional plant genetics and breeding to develop new cultivars with natural abilities to produce high levels of these compounds under stress conditions,

ii. Genetically engineering modified plants capable of producing sufficient amounts of these compounds in response to environmental stresses, and

iii. Exogenous use of these osmolytes on plants to enhance their stress tolerance ability.

Following observations have indicated that use of osmoprotectnats may help crop plants to avoid water stress conditions;

a. *G*lycinebetaine improves the growth and production of some plants under stress (Naidu *et al.*, 1998; Chen *et al.*, 2000; Hussain *et al.*, 2008) by maintaining leaf water status due to improved osmotic adjustment and enhanced photosynthesis, primarily due to a greater stomatal conductance and carboxylation efficiency of Rubisco (Sakamoto and Murata, 2002).

b. Application of proline enhanced the endogenous accumulation of free proline and improved the drought tolerance in petunia (Yamada *et al.*, 2005).

c. Exogenous spermidine application before the drought stress significantly improved the stress tolerance in barley (Kubi´s, 2003).

d. Application of polyamines improves drought stress in wheat mainly due to injury alleviation and growth promotion (Liu *et al.*, 2004).

e. Application of spermidine/free spermine and insoluble-conjugate putrescine improves drought stress in rice (Yang *et al.*, 2007).

3. Seed priming

Seed priming before sowing is important strategy to improve germination in dryland areas. It is a technique by which seeds are partially hydrated to a point where germination related metabolic processes begin but radicle emergence does not occur (Farooq *et al.*, 2006). Primed seeds usually exhibit increased germination rate, greater germination uniformity and sometimes greater total germination percentage (Kaya *et al.*, 2006 and Farooq *et al.*, 2007). This approach has been applied to overcome the drought stress effects in a range of crop species. Harris *et al.* (2002) reported that osmopriming of rice with 4% KCl solution and saturated $CaHPO_4$ solution was successful in improving the seedling emergence, crop stand establishment and yield under stress. In drought-prone areas primed rice seeds germinated well and seedlings emerged faster and more uniformly, leading to increased yield. In sunflower, osmopriming with KNO_3 and hydropriming improved the germination and stand establishment under stress conditions (Kaya *et al.*, 2006). Similarly the seed priming of wheat seeds improved germination and water-use efficiency of drought-stressed plants by 44% compared with unprimed seeds (Ajouri *et al.*, 2004. The beneficial effects of priming included faster emergence of crop seedlings, early flowering and higher grain yield even under drought stress (Kaur *et al.*, 2005).

4. Silicon

Silicon is considered to play an important role in drought tolerance in crop plants. It helps to improve water status of plants, root: shoot ratio, photosynthetic efficiency, stomatal conductance, antioxidant defense mechanism. The higher drought tolerance was observed in barley, wheat, rice, sorghum etc by different workers at application of silicon indicated its role in increasing drought tolerance in plants.

Summary

Water stress affect morphological and physiological responses of crop plants, however timing, duration, severity and speed of development undoubtedly have pivotal roles in determining how a plant responds to water stress. In general, seed germination is reduced and so plant stand. The growth and development of plants is greatly diminished due to reduced cell turgidity as a result of reduced cell enlargement, cell division and leaf area of the plants. The photosynthesis is also reduced due to stomatal closure and CO_2 intake. The stomata closure is not only governed by alone deficit soil moisture but also some other of intrinsic and extrinsic factors including ABA synthesis are increase, adenosine triphosphate synthesis is reduced and production of reactive oxygen species in chloroplast, mitochondira and peroxisomes like superoxide anion radicals (O_2^-), hydroxyl radicals (OH^-), hydrogen peroxide (H_2O_2), alkoxy radicals (RO) and singlet oxygen (O_2^1) which react with proteins, lipids and deoxyribonucleic acid resulting in to oxidative damage of cellular membrane lipids and degradation of enzyme proteins, nucleic acids and impairing the normal functions of cells. Finally the decreased activity RuBisCO, pyruvate orthophosphate and dikinase and assimilate partitioning gave poor crop yield and even crop failure in severe drought.

'Drought resistance' is a broader term applied to plant species with adaptive features that enable them to escape, avoid, or tolerate drought stress. Drought resistance mechanism is very complex and involves a number of physiological and biochemical processes at cell, tissue, organ and whole-plant levels at different stages of plant development. The drought-related traits i.e. root traits, leaf traits, osmotic adjustment capabilities, water potential, ABA content and stability of the cell membrane are used as indicators to evaluate the drought resistance of plants, therefore drought resistance includes reduction in water loss by increasing stomatal resistance, increased water uptake by developing large and deep root systems, accumulation of osmolytes and osmoprotectant synthesis. Amongst plant growth substances, salicylic acid, cytokinin and abscisic acid have been reported to play an important role in drought tolerance. Scavenging of reactive oxygen species by enzymatic and non-enzymatic systems, cell membrane stability, expression of aquaporins and stress proteins are also vital mechanisms of drought tolerance. Drought stress effects can be managed by production of most appropriate plant genotypes, seed priming, plant growth regulators, use of osmoprotectants, silicon and some other strategies.

Recently the scientists have investigated the genetic and molecular mechanisms of drought resistance to enhance the drought resistance of various crops and significant progress has been made with regard to drought avoidance and drought tolerance. With increasing knowledge to comprehensively decipher the complicated mechanisms of drought resistance in model plants, it still remains an enormous challenge to develop water-saving and drought-resistant crops to cope with the water shortage and increasing demand for food production in the future. Besides, the physiological and molecular basis of assimilate partitioning, high root:shoot ratio, production of osmoprotectants, production of drought resistant proteins like glycinebetaine, proline, betaine etc. and role of H_2O_2 and also its role as signaling molecule and identification of regulatory components need to be identified in future for increasing drought tolerance in crop plants. Therefore, the future research is imperative on genomics, proteomics and trascriptomic approaches for better understanding of the molecular basis of plant drought tolerance and improved water-use efficiency of crop plants. The new plant type or mutants with C_4 and CAM pathways or transgenic plants like genetically modified organisms (GMOs) with greater drought tolerance mechanisms or capabilities need to be identified at molecular level for developing drought tolerant crop varieties.

References

Abdul, Jaleel, C., Manivannan, P., Sankar, B., Kishorekumar, A., Gopi, R., Somasundaram, R. and Panneerselvam, R. 2007. Water deficit stress mitigation by calcium chloride in *Catharanthus roseus*: Effects on oxidative stress, proline metabolism and indole alkaloid accumulation. Colloid Surf. B 60: 110–116.

Abid, M., Ali, S., Qi, L.K., Zahoor, R., Tian, Z. W., Jiang, D. J., Snider, J. L. and Dai, T. 2018. Physiological and biochemical changes during drought and recovery periods at tillering and jointing stages in wheat (*Triticum aestivum* L.). Scientific Reports 8: 4615. (DOI:10.1038/s41598-018-21441-7).

Ajouri, A., Asgedom H. and Becker, M. 2004 Seed priming enhances germination and seedling growth of barley under conditions of P and Zn deficiency. J. Plant Nutr. Soil Sc. 167: 630–636.

Akashi, K., Miyake, C. and Yokota, A. 2001. Citrulline, a novel compatible solute in drought-tolerant wild watermelon leaves, is an efficient hydroxyl radical scavenger. FEBS Lett. 508: 438–442.

Ambavaram, M. M, Basu, S. and Krishnan, A. 2014. Coordinated regulation of photosynthesis in rice increases yield and tolerance to environmental stress. Nat Commun. 5: 5302. 10.1038/ncomms6302 [PMC free article] [PubMed] [CrossRef] [Google Scholar]

American Heritage Dictionary. 1976. Houghton Mifflin, Boston.

Ashraf, M. and Foolad, M. R. 2007. Roles to glycine betaine and proline in improving plant abiotic stress resistance. Environmental and Experimental Botany 59:206-216.

Bajji, M., Kinet, J. and Lutts, S. 2002. The use of the electrolyte leakage method for assessing cell membrane stability as a water stress tolerance test in durum wheat, Plant Growth Regul. 36:61–70.

Ball, R.A., Oosterhuis, D.M. and Mauromoustakos A. 1994 Growth dynamics of the cotton plant during water-deficit stress, Agron. J. 86:788–795.

Bektasoglu, B., Esin, C.S., Ozyürek, Mustafa, O., Kubilay, G. and Resat, A. 2006. Novel hydroxyl radical scavenging antioxidant activity assay for water-soluble antioxidants using a modified CUPRAC method. Biochem. Bioph. Res. Co. 345:1194–2000.

Bohnert, HJ., Nelson, DE. and Jensen, RG: 1995 Adaptations to Environmental Stresses. Plant Cell.;7(7):1099–111. 10.1105/tpc.7.7.1099 [PMC free article] [PubMed] [Cross Ref] [Google Scholar]

Buchanan, CD., Lim S., Salzman R.A., Kagiampakis, I., Morishige, D.T., Weers, B.D., Klein, R.R., Pratt L.H., Cordonnier-Pratt, M.M., Klein, P. and Mullet, J. 2005 Sorghum bicolor's transcriptome response to dehydration, high salinity and ABA, Plant Mol. Biol. 58: 699–720.

Chen, W.P., Li, P.H. and Chen, T.H.H. 2000. Glycinebetaine increases chilling tolerance and reduces chilling-induced lipid peroxidation in Zea mays L. Plant Cell Environ. 23: 609–618.

Condon, A.G., Richards R.A., Rebetzke, G.J. and Farquhar, G.D. 2004 Breeding for high water-use efficiency. J. Exp. Bot. 55: 2447–2460.

Dhanda, S.S., Sethi, G.S., Behl, R.K. 2004 Indices of drought tolerance in wheat genotypes at early stages of plant growth, J. Agron. Crop Sci. 190:6–12.

Drennan, P.M., Smith, M.T., Goldsworthy D. and van. Staden, J. 1993. The occurrence of trehalose in the leaves of the desiccation-tolerant angiosperm Myrothamnus flabellifolius Welw. J. Plant Physiol. 142: 493–496.

Du, H., Wang, N. and Cui, F. 2010. Characterization of the beta-carotene hydroxylase gene DSM2 conferring drought and oxidative stress resistance by increasing xanthophylls and abscisic acid synthesis in rice. Plant Physiol.154(3):1304–18. 10.1104/pp.110.163741 [PMC free article] [PubMed] [CrossRef] [Google Scholar]

Du,Y.C., Kawamitsu, Y., Nose A., Hiyane, S., Murayama S., Wasano, K. and Uchida, Y. 1996. Effects of water stress on carbon exchange rateand activities of photosynthetic enzymes in leaves of sugarcane (*Saccharum* Sp.), Aust. J. Plant Physiol. 23:719–726.

Farooq, M., Basra, S.M.A. and Wahid, A. 2006. Priming of field-sown rice seed enhances germination, seedling establishment, allometry and yield, Plant Growth Regul. 49:285–294.

Folkert, A.H., Elena A.G. and Buitink, J. 2001. Mechanisms of plant desiccation tolerance. Trends Plant Sci. 6:431–438.

Foyer, C.H. and Fletcher J.M. 2001. Plant antioxidants: colour me healthy. Biologist 48:115–120.

Gao, X.P., Wang, X.F., Lu, Y.F., Zhang, L.Y., Shen, Y.Y., Liang, Z. and Zhang, D.P. 2004. Jasmonic acid is involved in the water-stress-induced betaine accumulation in pear leaves. Plant Cell Environ 27:497–507.

Gigon, A., Matos, A., Laffray, D., Zuily-fodil,Y. and Pham-Thi A. 2004. Effect of drought stress on lipid metabolism in the leaves of *Arabidopsis thaliana* (Ecotype Columbia), Ann. Bot. 94: 345–351.

Gnanasiri, S.P., Saneoka, H. and Ogata ,S. 1991. Cell membrane stability and leaf water relations as affected by potassium nutrition of waterstressed maize, J. Exp. Bot. 42:739–745.

Goddijn, O.J.M., Verwoerd, T.C., Voogd, E., Krutwagen, P.W.H.H., Degraaf, P.T.H.M., Poels, J., Vandun, K., Ponstein, A.S. and Damm, B., Pen J. 1997 Inhibition of trehalase activity enhances trehalose accumulation in transgenic plants. *Plant Physiol.* 113:181–190.

Huang, B. and Fry, J.D. 1998. Root Anatomical, Physiological, and Morphological Responses to Drought Stress for Tall Fescue Cultivars. *Crop Sci.* 38(4): 1017–1022. (10.2135/cropsci1998.0011183X003800040022x [CrossRef] [Google Scholar]).

Hussain, M., Malik, M.A., Farooq M., Ashraf M.Y. and Cheema M.A. 2008. Improving Drought tolerance by exogenous application of glycinebetaine and salicylic acid in sunflower, J. Agron. Crop Sci. 194: 193–199.

Karim, S., Aronsson, H., Ericson, H., Pirhonen, M., Leyman, B., Welin, B., Mäntylä, E., Palva, E.T., Dijck, P.V. and Holmström, K. 2007. Improved drought tolerance without undesired side effects in transgenic plants producing trehalose. Plant Mol. Biol. 64:371–386.

Kaur, S., Gupta, A.K. and Kaur, N. 2005. Seed priming increases crop yield possibly by modulating enzymes of sucrose metabolism in chickpea, J. *Agron. Crop Sci.* 191: 81–87.

Kavar, T., Maras, M., Kidric, M., Sustar-Vozlic, J. and Meglic, V. 2007. Identification of genes involved in the response of leaves of *Phaseolus vulgaris* to drought stress, Mol. Breed. 21: 159–172.

Kaya, M.D., Okçub G., Ataka M., Çıkılıc Y., Kolsarıcıa, Ö. 2006. Seed treatments to overcome salt and drought stress during germination in sunflower (*Helianthus annuus* L.), Eur. J. Agron. 24:291–295.

Kinnersley, A.M. and Turano, F.J. 2000. Gama aminobutyric acid (GABA) and plant responses to stress. Crit. Rev. Plant Sci. 19: 479–509.

Knipp, G. and Honermeier, B. 2006. Effect of water stress on proline accumulation of genetically modified potatoes (*Solanum tuberosum* L.) generating fructans. Journal of Plant Physiology 163:392-397.

Koda, Y. 1997 Possible involvement of jasmonates in various morphogenic events, Physiol. Plant. 100, 639–646

Kosmas, S.A., Argyrokastritis, A., Loukas, M.G., Eliopoulos, E., Tsakas, S. and Kaltsikes, P.J. 2006. Isolation and characterization of drought related trehalose 6-phosphate-synthase gene from cultivated cotton (*Gossypium hirsutum* L.). Planta 223: 329–339.\

Kramer, P.J., Boyer, J.S. 1995. Water relations of Plants and Soils Academic Press, San Diego

Kubi´s, J. 2003. Polyamines and "scavenging system": influence of exogenous spermidine on catalase and guaiacol peroxidase activities, and free polyamine level in barley leaves under water deficit. Acta Physiol. Plant. 25: 337–343.

Kumar, J. and Abbo, S. 2001. Genetics of flowering time in chickpea and its bearing on productivity in the semi-arid environments. Adv. Agron.72:107–138.

Lambers, H., Atkin, O.K. and Scheureater, I. 1996. Respiratory patterns in roots in relation to their function, in: Waisel Y. (Ed.), Plant Roots,The Hidden Half. Marcel Dekker, NewYork.

Lee, S.B., Kwon H.B., Kwon S.J., Park S.C., Jeong M.J., Han S.E., Byun M.O. and Daniell H. 2004. Accumulation of trehalose within transgenic chloroplasts confers drought tolerance, Mol. Breed. 11: 1–13.

Li, L., Van Staden J. and Jager A.K.1998. Effects of plant growth regulators on the antioxidant system in seedlings of two maize cultivars subjected to water stress, Plant Growth Regul. 25, 81–87.

Liu, H.S., Li F.M. and Xu H. 2004. Deficiency of water can enhance root respiration rate of drought–sensitive but not drought-tolerant spring wheat, Agr. Water Manage. 64: 41–48.

Ludlow, M.M. and Muchow, R.C.1990. A critical evaluation of traits for improving crop yields in water-limited environments, Adv. Agron. 43:107–153.

Moore, A.L., Siedow, J.N. 1991. The regulation and nature of the cyanideresistant oxidase of plant mitochondria, Biochim. Biophys. Acta 1059: 121–140.

Moradi, Parviz , Mahdavi, Atiyeh , Khoshkam, Maryam and Iriti, Marcello. 2017. Lipidomics Unravels the Role of Leaf Lipids in Thyme Plant Response to Drought Stress. Int. J. Mol. Sci. 18(10): 2067 (doi: 10.3390/ijms18102067).

Morgan, P.W. 1990. Effects of abiotic stresses on plant hormone systems, in: Stress Responses in plants: adaptation and acclimation mechanisms, Wiley-Liss, Inc. pp. 113–146.

Naidu, B.P., Cameron, D.F. and Konduri, S.V. 1998 Improving drought tolerance

Nerd, A. and Neumann, P.M. 2004. Phloem water transport maintains stem growth in a drought-stressed crop cactus (Hylocereus undatus). J. Am. Soc. Hortic. Sci. 129: 486–490.

Pannu, R.K., Singh, D.P., Singh P., Chaudhary B.D. and Singh, V.P. 1993. Evaluation of various plant water indices for screening the genotypes of chickpea under limited water environment. Haryana J.Agron. 9:16–22.

Porembski, S. 2000. Granitic and gneissic outcrops (inselbergs) as centers of diversity for desiccation-tolerant vascular plants. Plant Ecol. 2000;151(1):19–28.10.1023/A: 1026565817218 [CrossRef] [Google Scholar

Poroyko, V., Hejlek, L.G., Spollen, W.G., Springer, G.K., Nguyen, H.T., Sharp, R.E. and Bohnert, H.J. 2005. The maize root transcriptome by serial analysis of gene expression Plant Physiol. 138:1700–1710.

Premachandra, G.S., Saneoka, H., Kanaya, M., Ogata S. 1991. Cell membrane stability and leaf surface wax content as affected by increasing water deficits in maize. J. Exp. Bot. 42: 167–171.

Quan, R.D., Shang, M., Zhang, H. and Zhang, J. 2004. Improved chilling tolerance by transformation with betA gene for the enhancement of glycinebetaine synthesis in maize. Plant Sci. 166:141–149.

Ramegowda, V. Basu, S., Krishnan, A. 2014. Rice Growth Under Drought Kinase is required for drought tolerance and grain yield under normal and drought stress conditions. Plant Physiol. 2014;166(3):1634–45. 10.1104/pp.114.248203 [PMC free article] [PubMed] [CrossRef] [Google Scholar]

Reddy, A.R., Chaitanya, K.V. and Vivekanandan, M. 2004. Drought-induced responses of photosynthesis and antioxidant metabolism in higher plants. J. Plant Physiol. 161:1189–1202.

Reynolds, M.P., Oritz-Monasterio, J.I. and Mc Nab, A. 2001. Application of physiology in wheat breeding, CIMMYT, Mexico.

Riccardi, L., Polignano, G.B. and de Giovanni, C. 2001. Genotypic response of faba bean to water stress, Euphytica 118:39–46.

Richards, R.A., Rawson, H.M. and Johnson, D.A. 1986. Glaucousness in wheat: its development, and effect on water-use efficiency, gas exchange and photosynthetic tissue temperatures, Aust. J. Plant Physiol. 13: 465–473.

Sakamoto, A. and Murata, N. (2002) The role of glycinebetaine in the protection of plants from stress: clues from transgenic plants. Plant Cell Environ. 25:163–171.

Sandquist, D.R. and Ehleringer, J.R. 2003. Population- and family-level variation of brittlebush (Encelia farinosa, Asteraceae) pubescence: its relation to drought and implications for selection in variable environments, Am. J. Bot. 90: 1481–1486.

Senaratna, T., Touchell, D., Bunn, E. and Dixon, K. 2000. Acetyl salicylic acid (aspirin) and salicylic acid induce multiple stress tolerance in bean and tomato plants, Plant Growth Regul. 30: 157–161.

Shakirova, F.M., Sakhabutdinova, A.R., Bezrukova, M.V., Fatkhutdinova, R.A. and Fatkhutdinova, D.R. 2003. Changes in the hormonal status of wheat seedlings induced by salicylic acid and salinity. Plant Sci.164: 317–322.

Shang Z. 2000 Effect of 6-BA and KT on photophosphorylation activity in wheat flag leaves under water stress, Acta Agr. Boreali-Sinica 15:34–38.

Shelp, B.L., Bown, A.W. and McLean M.D. 1999. Metabolism and functions of gammaaminobutyric acid. Trends Plant Sci.11: 446–452.

Shugaeva, N., Vyskrebentseva, E., Orekhova, S. and Shugaev, A. 2007 Effect of water deficit on respiration of conducting bundles in leaf petiole of sugar beet. Russ. J. Plant Physiol. 54: 329–335.

Subbarao, G.V., Johansen, C., Slinkard, A.E., Rao R.C.N. , Saxena, N.P. and Chauhan, Y.S. 1995 Strategies and scope for improving drought resistance in grain legumes, Crit. Rev. Plant Sci. 14: 469–523.

Subbarao, G.V., Nam, N.H., Chauhan, Y.S., Johansen, C. (2000) Osmotic adjustment, water relations and carbohydrate remobilization in pigeonpea under water deficits, J. Plant Physiol. 157, 651–659.

Taiz, L. and Zeiger, E. 2006. Plant Physiology, 4th Ed., Sinauer AssociatesInc. Publishers, Massachusetts

Tezara, W, Mitchell, VJ. And Driscoll, SD, 1999;: Water stress inhibits plant photosynthesis by decreasing coupling factor and ATP. Nature. 401:914–7. 10.1038/44842 [CrossRef] [Google Scholar]

Thi, Anh Thu Pham, Da Silva, Jorge Vieira and Mazliak, Paul. 1990. The role of membrane lipids in drought resistance of plants, Bulletin de la Société Botanique de France.Actualités Botaniques, 137:1, 99-114. (DOI: 10.1080/01811789.1990.10826991).

Todd, E.Y., Robert, B.M., and Daniel, R.G. 2004. ACC synthase expression regulates leaf performance and drought tolerance in maize. Plant J. 40:813–825.

Tripathy, J.N., Zhang, J., Robin, S., Nguyen, T.T. and Nguyen H.T. 2000. QTLs for cell-membrane stability mapped in rice (*Oryza sativa* L.) under drought stress, Theor. Appl. Genet. 100: 1197–1202.

Turner, N.C., Wright, G.C., and Siddique, K.H.M. 2001. Adaptation of grain legumes (pulses) to water-limited environments. Adv. Agron. 71: 123–231.

Villar-Salvador, P., Planelles, R., Oliet J., Peñuelas-Rubira, J.L., Jacobs, D.F., González M. 2004. Drought tolerance and transplanting performance of holm oak (Quercus ilex) seedlings after drought hardening in the nursery. Tree Physiol. 24: 1147–1155.

Wahid, A., Gelani, S., Ashraf, M. and Foolad, M.R. 2007 Heat tolerance in plants: an overview, Environ. Exp. Bot. 61: 199–223.

Wang, Z. and Huang, B. 2004. Physiological Recovery of Kentucky bluegrass from simultaneous drought and heat stress. Crop Sci. 44, 1729–1736.

Wingler, A. 2002. The function of trehalose biosynthesis in plants, Phytochemistry 60: 437–440.

Wrzaczek, M., Hirt, H. 2001. Plant MAP kinase pathways: how many and what for? Biol. Cell 93: 81–87.

Xiong, L.,Wang, R.,Mao G., and Koczan, J.M. 2006. Identification of drought tolerance determinants by genetic analysis of root response to drought stress and abscisic acid. Plant Physiol. 142: 1065–1074.

Yamada, M., Morishita, H., Urano K., Shiozaki, N., Yamaguchi-Shinozaki K., Shinozaki K. and Yoshiba Y. 2005. Effects of free proline accumulation in petunias under drought stress, J. Exp. Bot. 56: 1975–1981.

Yang, J., Zhang, J., Liu K., Wang, Z. and Liu L. 2007. Involvement of polyamines in the drought resistance of rice, J. Exp. Bot. 58: 1545–1555.

Yokota, A., Kawasaki S., Iwano M., Nakamura, C., Miyake C. and Akashi, K. 2002. Citrulline and DRIP-1 Protein (ArgE Homologue) in Drought Tolerance of Wild Watermelon. Ann. Bot. 89: 825–832.

Zhang, M., Duan, L., Zhai, Z., Li J., Tian, X., Wang, B., He Z. and Li Z. 2004. Effects of plant growth regulators on water deficit-induced yield loss in soybean, Proceedings of the 4th International Crop Science Congress Brisbane, Australia.

Zhou, Y., Lam, H.M., Zhang, J. 2007. Inhibition of photosynthesis andenergy dissipation induced by water and high light stresses in rice, J. Exp. Bot. 58: 1207–1217

Zhu, J.K. 2002. Salt and drought stress signal transduction in plants, Annu. Rev. Plant Biol. 53: 247–273.

Questions

a. Short and Descriptive

1. Write short notes on following;
   a. Drought escape
   b. Drought periodicity
   c. Water savers
   d. Water spenders
   e. Metabolic strains
   f. Plastic strains
   g. Aquaporines
   h. Seed priming
   i. Dehydrins
   j. Stress proteins
2. Describe the effect of drought stress on following;
   a. Photosynthesis
   b. Respiration
   c. Nutrient uptake
   d. Protein synthesis
   e. Hormonal relations
   f. Oxidative changes
3. Define drought. Also describe different types of drought.
4. What are the different effects of drought on crop production?
5. What do you mean by drought tolerance? Describe mitigating drought stress and high tolerating stress.
6. Write short note of molecular approach of drought tolerance.
7. Describe the draught stress management strategies for higher crop production in water deficit region.

b. Objective questions

1. Which of the following factor is responsible for drought ?
   - a Climate
   - b Climatic anomaly
   - c Weather related natural disaster
   - d All above

2 Which of the following climatic factor (s) is responsible for drought?
   - a High temperature
   - b High wind velocity
   - c Low relative humidity
   - d All above

3 A long period with no rain, especially during a planting season is considered as
   - a Conceptual definition of drought
   - b Operational definition of drought
   - c Both a & b
   - d None of above

4 The operational definition is also used to,
   - a Analyze drought frequency
   - b Analyze drought severity
   - c Both a & b
   - d None of above

5 Severe drought refers to
   - a When the deficiency of rainfall exceeds 50% of the normal rainfall
   - b When the deficiency of rainfall is between 25 and 50% of the normal rainfall
   - c Both a & b
   - d None of above

6 Moderate drought refers to
   - a When the deficiency of rainfall exceeds 50% of the normal rainfall
   - b When the deficiency of rainfall is between 25 and 50% of the normal rainfall.
   - c Both a & b
   - d None of above

7 Drought is a situation when actual seasonal rainfall is deficient by more than twice the mean deviation'. This definition of drought was proposed by
   - a Ramdas
   - b Irrigation Commission of India
   - c American Meteorological Society
   - d All above

8 The major cause of meteorological drought is
   - a Lean monsoon and below average rainfall due to absence of depressions over India.
   - b Late onset or early withdrawal of monsoons

c Prolonged dry spells in monsoon
d All above

9 Agricultural drought is related to
a Rainfall deficit
b Soil water deficit
c Both a & b
d None of above

10 Hydrological drought is related to
a Surface water drought
b Ground water drought
c Both a & b
d None of above

11 Soil moisture drought occurs in
a Rainfed areas
b Dry land area
c Both a & b
d None of above

12 Permanent drought occurs in
a Arid region
b Semi arid region
c Both a & b
d None of above

13 Seasonal drought occurs in
a Arid area
b Semi arid area
c Both a & b
d None of above

14 Unpredictable drought refers to
a Deficient rainfall
b Uneven rainfall
c Erratic rainfall
d All above

15 Unpredictable drought may occur in
a Humid climate
b Sub humid climate
c Both a & b
d None of above

16 Contingent drought occurs in
a Humid climate
b Sub humid climate
c Both a & b
d None above

17 Invisible drought may occur in
a Humid region
b Sub humid region
c Both a & b
d None above

18 Early season drought refers to
a Delayed onset of monsoon
b Prolonged dry spell
c Both a & b
d None of above

19 Ephemerals is also known as
a Escaping drought
b Drought resistance
c Drought tolerance
d All above

20 The plants that have very short life span and complete their life cycle is within the period of sufficient soil moisture, is called as

a Escaping drought
b Drought resistance
c Both a & b
d None of above

21 Which of the following is related to oxidative stress ?

a Primarily superoxide (O_2^-)
b Hydrogen peroxide (H_2O_2)
c Damages cell membranes
d All above

22 Which of the following mechanisms is responsible for water conservation in plants?

a Early closure of stomata
b Increased photosynthetic efficiency
c Low rate of cuticular transpiration
d All above

23 Which of the following mechanisms is response to drought stress?

a Accumulation of soluble sugars
b Upregulation of antioxidant enzymes
c Both a & b
d None above

24 Which of the following is responsible for transpirational water loss in plants?

a Stomatal transpiration
b Cuticular transpiration.
c Both a & b
d None above

25 The plant surface covered with tiny hairs is called as

a Tomentose
b Trichomes
c Both a & b
d None above

26 Which of the following hormone reduces shoot growth in plant?

a Auxin
b Gibberellic acid
c Both a & b
d None above

27 Aquaporins are synthesized in

a Roots
b Leaves
c Both a & b
d None above

28 Aquaporins are found in

a Plasma membrane
b Vascular membrane
c Both a & b
d None above

29 Which of the following gene induces drought tolerance in *Medicago sativa?*

a GsZFP1
b SNAC1
c MpCYS4
d All above

30 Jasmonic acid (JA) is synthesized from.

a VaNAC26 | b GsZFP1
c MpCYS4 | d All above

31 Auxin is responsible for

a Root development | b Functioning of ABA
c Improve drought-tolerance | d All above

32 Drought resistance of plants can be

a Drought avoidance, | b Drought tolerance,
c Drought escape | d All above

33 Drought-escaping plants are characterized by

a Germinate and grow only during times of sufficient times of moisture | b Restrict their growth only to periods of moisture availability.
c Both a & b | d None above

34 Xerophytes are also known as

a Drought-enduring plants | b Drought-escaping plants
c Drought-evading plants | d All above

35 Succulent perennials are also known as

a Drought-enduring plants | b Drought-resisting plants
c Drought-evading plants | d All above

36 Which of the following type of plants can store water in their leaves and stems?

a Drought-enduring plants | b Drought-resisting plants
c Drought-evading plants | d All above

37 C_3 pathway of photosynthesis is found in

a Rice | b Soybean
c Wheat | d All above

38 Which of the following solute (s) helps in protecting the plants from detrimental effects of drought stress ?

a Proline | b Glycine
c Both a & b | d None above

39 Which of the following enzyme (s) plays major role in OA?

a Betaine aldehyde dehydrogenase (BADH) | b Pyrroline -5-carboxylate reductase (P5CR)
c Ornithine δ-aminotransferase (OAT | d All above

40 Which of the following feature is found in drought-evading plants?

a Polymorphic leaves
b Stem photosynthesis
c Both a & b
d None above

41 C_4 pathway of photosynthesis is found in

a Sorghum
b Sugarcane
c Fingermillet
d All above

42 CAM pathway of photosynthesis is found in

a Cactus
b Pineapple
c Orchids
d All above

43 Which of the following morphological feature is found in CAM plants?

a Thick cuticle
b Reduced leaf size
c Sunken stomata
d All above

44 CAM plants are found in

a Humid region
b Arid region
c Temperate region
d None above

45 'Drought year receives the rainfall equal or less than 75% of an average normal'. This definition was given by

a Indian Meteorological Department of Govt. of India
b Ministry of Agriculture, Govt. of India
c Indian Institute of Remote Sensing, Dehradun
d None above

46 If an area receives rainfall equal or less than 75% of an average normal for > 20 years, is called as

a Arid area
b Drought prone area
c Humid area
d None above

47 As per Irrigation Commission of India, the drought zone area has rainfall departure equal to

a 25 per cent probability of rainfall departure from the normal
b 40 per cent or more probability of rainfall departure from the normal
c 50 per cent or more probability of rainfall departure from the normal
d 60 per cent or more probability of rainfall departure from the normal

48 As per Irrigation Commission of India, the chronic drought zone area has rainfall departure equal to

a 25 per cent probability of rainfall departure from the normal
40 per cent or more probability of rainfall departure from the normal

c 50 per cent or more probability of rainfall departure from the normal

d 60 per cent or more probability of rainfall departure from the normal

49 The condition in which the surface and ground water resources are decreased to the threshold level and no more water is available to crop production and livestock leading to large scale collapse of life and mass starvation in larger areas, is known as

a Meteorological drought b Agricultural drought

c Hydrological drought d Famine

50 Which of the following condition prevails in permanent drought?

a Rainfall is always scanty and insufficient for crop production

b Drought is experienced seasonally or during the crop growth period

c Abnormal failure of rainfall

d Crop failure even under frequent rainfall conditions

51 Which of the following condition prevails in seasonal drought?

a Rainfall is always scanty and insufficient for crop production

b Drought is experienced seasonally or during the crop growth period

c Abnormal failure of rainfall

d Crop failure even under frequent rainfall conditions

52 Which of the following condition prevails in contingent drought?

a Rainfall is always scanty and insufficient for crop production

b Drought is experienced seasonally or during the crop growth period

c Abnormal failure of rainfall

d Crop failure even under frequent rainfall conditions

53 Which of the following condition prevails in invisible drought?

a Rainfall is always scanty and insufficient for crop production

b Drought is experienced seasonally or during the crop growth period

c Abnormal failure of rainfall

d Crop failure even under frequent rainfall conditions

54 Permanent drought normally occurs in

a Arid region b Arid & semi arid regions

c Humid and sub humid region d Humid region

55 Seasonal drought normally occurs in

a Arid region b Arid & semi arid regions

c Humid and sub humid region d Humid region

56 Contingent drought normally occurs in

a Arid region
b Arid & semi arid regions
c Humid and sub humid region
d Humid region

57 Invisible drought normally occurs in

a Arid region
b Arid & semi arid regions
c Humid and sub humid region
d Humid region

58 Which of the following condition (s) is responsible for atmospheric drought?

a Low relative humidity
b Hot dry winds
c High temperature
d All above

59 Moisture adequacy index (MAI) is represented by

a (Actual ET / Potential ET) x 100
b (Actual rainfall / Potential ET) x 100
c (Actual E / Potential ET) x 100
d None above

60 On the basis of moisture adequacy index, severe drought is represented by

a MAI = < 25%
b MAI = 25 - 29.9%
c MAI = ≥ 50%.
d MAI = ≥75%.

61 On the basis of moisture adequacy index, moderate drought is represented by

a MAI = < 25%
b MAI = 25 - 29.9%
c MAI = ≥ 50%.
d MAI = ≥75%.

62 On the basis of moisture adequacy index, crop growing season is represented by

a MAI = < 25%
b MAI = 25 - 29.9%
c MAI = ≥ 50%.
d MAI = ≥75%.

63 On the basis of moisture adequacy index, humid region is represented by

a MAI = < 25%
b MAI = 25 - 29.9%
c MAI = ≥ 50%.
d MAI = ≥75%.

64 What is the aridity index in moderate drought as proposed by Thornthwaite ?

a < 0.5
0.5 – 1.0
c 1.0 – 2.0
>2.0

65 What is the aridity index in large drought as proposed by Thornthwaite?

a < 0.5
b 0.5 – 1.0
c 2.0 – 2.0
d >2.0

66 What is the aridity index in severe drought as proposed by Thornthwaite?

a < 0.5 b 0.5 – 1.0

c 1.0 – 2.0 d >2.0

67 What is the aridity index in disastrous drought as proposed by Thornthwaite?

a < 0.5 b 0.5 – 1.0

c 1.0 – 2.0 d >2.0

68 What is the drought index/aridity index in mild drought as proposed by Palmer?

a 1 – 1.99 b 2 – 2.99

c 3 - 3.99 d >4.00

69 What is the drought index/aridity index in large drought as proposed by Palmer?

a 1 – 1.99 b 2 – 2.99

c 3 - 3.99 d > 4.00

70 What is the drought index/aridity index in severe drought as proposed by Palmer?

a 1 – 1.99 b 2 – 2.99

c 3 - 3.99 d > 4.00

71 What is the drought index/aridity index in disastrous drought as proposed by Palmer?

a 1 – 1.99 b 2 – 2.99

c 3 - 3.99 d > 4.00

72 Annawary system classifies drought on the basis of crop condition and its production. If the production is >75% of average, the climatic condition is classified as

a No drought b Moderate drought

c Severe drought d Disastrous drought

73 Annawary system classifies drought on the basis of crop condition and its production. If the production is 50 -75 % of average, the climatic condition is classified as

a No drought b Moderate drought

c Severe drought d Disastrous drought

74 Annawary system classifies drought on the basis of crop condition and its production. If the production is 25-50% of average, the climatic condition is classified as

a No drought b Moderate drought

c Severe drought d Disastrous drought

75 Annawary system classifies drought on the basis of crop condition and its production. If the production is <25% of average, the climatic condition is classified as

a No drought b Moderate drought

c Severe drought d Disastrous drought

76 Crop growth and yield is reduced under drought mainly due to

a Poor plant stand b Reduced leaf area

c Reduced photosynthesis d All above

77 Photosynthesis is reduced under moisture deficit conditions due to

a Stomata closure b Low intake rate of CO_2

c Reduced Rubisco activity d All above

78 Stomata closure under drought condition is governed by

a Low cell turgidity b Synthesis of ABA

c High pH of cell sap d All above

79 Which of the photosynthetic enzyme activity is reduced under water deficit conditions?

a Rubisco b Phosphoenolpyruvate carboxylase

c Pyruvate orthophosphate dikinase d All above

80 Which of the following activity limits photosynthesis under mild water stress?

a Impaired phtophosphorylation b Low synthesis of adenosine triphosphate

c Both a & b d None above

81 Nutrient uptake is reduced under water stress mainly due to

a Low concentration of nutrients in soil solution b Break down of capillary continuity of moisture film around soil particles

c Reduced assimilation of nutrients d All above

82 Rubisco acts under water deficit conditions as

a Oxygenase
b It helps to increase respiration rate
c Both a & b
d None above

83 Protein synthesis under water stress is reduced due to

a Reduced activity of RNA and DNA
b Accumulation of sugars and amino acids
c Conversion of protein into proline, ABA and other by products of proteins.
d All above

84 Which of the following phytohormone synthesis is reduced in water deficit conditions?

a Auxin
b Gibberalic acid
c Cytokinin
d All above

85 Which of the following phytohormone synthesis is increased in water deficit conditions

a ABA
b Ethylene
c Betain
d All above

86 Which of the following is responsible for production of reactive oxygen species under drought stress?

a Chloroplast
b Peroxisomes
c Mitochondria
d All above

87 What is the role of reactive oxygen species in plants?

a Peroxidation of membrane lipids
b Degradation of structural and functional proteins
c Degradation of nucleic acids
d All above

88 Which of the following process is first affected under water deficit?

a Cell elongation
b Cell division
c Photosynthesis
d Protein synthesis

89 Which of the following is a water saving mechanism in plants?

a Closure of stomata
b Reduced leaf area
c Lipid deposition on foliage
d All above

90 Which of the following is a water spending mechanism in plants?

a Efficient root system
b High root-shoot ratio
c Increase osmotic potential
d All above

91 ABA is synthesized in

a Roots
b Leaves
c Guard cells
d All above

92 What is role of ABA synthesis in roots under water deficit conditions?

a Increases hydraulic activity of roots
b Improves water uptake from roots
c Maintain transpirational need
d All above

93 What is the role of ABA synthesis in leaves under drought stress?

a Induces stomata closure
b Reduces transpirational loss
c Regulates water balance in plant cells
d All above

94 Which of the following is an antioxidant that saves cell from reactive oxygen species?

a Super dismutase
b Ascorbate peroxidase and Catalase
c Glutathione
d All above

95 What is role of super dismutase in cell?

a Conversion of O_2^- to less reactive H_2O_2
b H_2O_2 is converted into O_2 and H_2O.
c Protect the chloroplasts from ROS damage by increasing the ratio of reduced glutathione to oxidized glutathione
d It safeguards the photosynthetic apparatus by dissipating excess excitation energy into heat

96 What is role of ascorbate peroxidase and catalase in cell?

a Conversion of O_2^- to less reactive H_2O_2
b H_2O_2 is converted into O_2 and H_2O.
c Protect the chloroplasts from ROS damage by increasing the ratio of reduced glutathione to oxidized glutathione
It safeguards the photosynthetic apparatus by dissipating excess excitation energy into heat

97 What is role of Glutathione in plant cell in drought stress?

a Conversion of O_2^- to less reactive H_2O_2
b H_2O_2 is converted into O_2 and H_2O.
c Protect the chloroplasts from ROS damage by increasing the ratio of reduced glutathione to oxidized glutathione
d It safeguards the photosynthetic apparatus by dissipating excess excitation energy into heat

98 What is role of carotenoids in plant cell under water deficit?

a Conversion of O_2^- to less reactive H_2O_2

b H_2O_2 is converted into O_2 and H_2O.

c Protect the chloroplasts from ROS damage by increasing the ratio of reduced glutathione to oxidized glutathione

d It safeguards the photosynthetic apparatus by dissipating excess excitation energy into heat

99 Which of the following type of water loss is more important under drought stress?

a Leaf transpiration
b Cuticular transpiration
c Both a & b
d All above

100 Which of the following fatty acid and lipids are synthesized under drought tolerant plants?

a Galactolipids
b Oleic and linolenic acids
c Flavonoids
d All above

101 What is the role of unsaturated fatty acids synthesized under water deficit conditions?

a Protect cell and chloroplast membranes from damage induced by severe drought

b Preserve their structure and function

c Protect from reactive oxygen species

d All above

102 Which of the following is a osmoprotectant?

a Organic solutes like sugars

b Quaternary ammonium compounds (glycine, betaine, alanine)

c Hydrophilic and soluble proteins like proline

d All above

103 What is the role of proline in cell?

a Reduces cell water potential
b Stablises membrane and proteins
c Removes free radicals
d All above

104 Which of the following is a xeromorphic trait ?

a Smaller leaves
b Leaf pubescence
c Deep root system
d All above

105 Sunken stomata and hairs are found in

a Pine | b *Nassauvia falklandica*
c Bromeliads | d All above

106 Nocturnal stomata are found in

a Tea plant | b Alfalfa
c *Quercus trojana* | d All above

107 Waxy leaf surface is found in

a Prikly pear | b Malosma laurina
c Dudieya pulverulenta | d All above

108 Extensive root system is found in

a Acccia | b Prosopsis
c Subabul | d All above

109 Water melon survives under severe drought mainly due to leaf accumulation of

a Citrulline' | b Glutamate
c Arginine | d All above

110 Drought tolerant plants have

a High root/shoot ratio | b Greater osmotic adustment
c Maintain high cell turgidity | d All above

111 Which of the following is a major role of osmotic adjustment in drought tolerating plants?

a Maintain cell water balance | b Maintain of cell turgor and physiological processes
c Delays dehydrative damage | d All above

112 Which of the following compounds maintain the cell integrity under drought?

a Proline | b Glycine and Betaine
c Polar lipids | d All above

113 Which of the following compounds are responsible for maintaining higher osmotic pressure as mitigating drought stress mechanism in crop plants?

a Proline | b Glycinebetaine
c Cirulline | d All above

114 What is the physiological role of non protein AA, 'γ-aminobutyric acid' synthesized in drought tolerating plants?

a Regulate osmotic pressure
b Detoxify the reactive oxygen radicals
c Convert putrescine into proline
d All above

115 What is the physiological role of Trehalose in crop plants under desiccation stress?

a Stabilization of dehydrated enzymes, proteins and lipid membranes
b Protect cell organelles
c Improved water status of plants
d All above

116 What is metabolic strain?

a Strain that is reversible after removal of strain
b Strain that is not reversible after removal of strain
c Both a & b
d None above

117 What is plastic strain?

a Strain that is reversible after removal of strain
b Strain that is not reversible after removal of strain
c Both a & b
d None above

118 What is true about 'aquaporin'?

a It is an intrinsic membrane proteins
b It is synthesized in roots and leaves
c It is found in plasma membrane and vascular membrane
d All above

119 The role of aquaporins in plants is

a It increases drought tolerance in plants
b It increases 10-20 fold water permeability
c It regulates hydraulic conductivity of cell mmbrane
d All above

120 Which of the following affects the activity of aquaporins?

a Phosphorylation
b Ca ions
c pH of cell sap
d All above

121 Which of the following statement is true for dehydrins?

a They are late embryogenesis associates (LEA) proteins
b They are synthesized in response to both drought and cold
c They support protecting cellular proteins from denaturation
d All above

122 What is molecular chaperonses?

a They are proteins that assist the conformational folding or unfolding and the assembly or disassembly of other macromolecular structures.

b They are four types i.e.Hsp60s, Hsp70s, Hsp90s and sHsps.

c They assist in the folding of unfolded and misfolded polypeptides by stabilization of folding intermediates and prevention of protein misfolding and aggregation.

d All above

123 Molecular chaperones are found in

a Prokaryotes

b Cytosol of eukaryotes

c Mitochondria

d All above

124 Low molecular heat shock proteins are synthesized under

a High and low temperature

b Drought

c Anaerobic conditions

d All above

125 Low molecular heat shock proteins are important for

a They participate in adenosine triphosphate dependent proteins.

b They prevent protein denaturation under stress.

c They increase drought tolerance in plants

c All above

126 Which of the following are considered moieties in conferring drought tolerance in plans?

a Mitogen-activated protein (MYB),

b Zinc-finger proteins (ZFPs)

c C-repeat/dehydration-responsive element binding factors (CBF/DREB)

d All above

127 Which of the following chemical signal (s) is responsible for inducing drought tolerance in crop plants?

a Reactive oxygen species (ROS)

b Ca

c Plant hormones

d All above

128 Which of the following crop specific gene is responsible for drought tolerance in plants?

a GsZFP1gene in *Medicago sativa*

b SNAC1 in *Gossypium hirsutum*

c *GUDK* and *HYR* genes in rice

d All above

129 Which of the following gene is responsible for synthesis of Jasmonic acid responsible for drought tolerance in plants?

a VaNAC26 | b MpCYS4 gene
c BdWRKY36 | d All above

130 Which of the following gene is related to ABA synthesis and closing of stomata in crop plants?

a VaNAC26 | b MpCYS4 gene
c BdWRKY36 | d *EDT1/HDG11*

131 Over expression of EDT1/HDG11gene in popular and cotton increased drought tolerance mainly due to

a Increased accumulation of proline | b Increased accumulation of sugars
c Maintain osmotic pressure | d All above

132 Which of the following strategy is responsible for improving water use efficiency of plants under water deficit conditions?

a Increasing water uptake | b Improving biomass production
c Partitioning of produced biomass towards the harvested product | d All above

133 Which of the following hormones/chemicals can be applied for improving soybean yield under drought stress?

a Uniconazole | b *Abscisic acid*
c Brassinolide | d All above

134 Which of the following hormones/chemicals can be applied for improving maize yield under drought stress?

a Uniconazole | b Methyl Jasmonate
c Brassinolide | d All above

135 Benzyladenine acts as a

a Auxin | b GA
c Cytokinin | d ABA

136 Salicylic acid improves drought tolerance in

a Maize | b Rice
c Winter wheat | d Sugarcane

137 Which of the following is correct combination of osmoprotectant used for increasing drought tolerance in plants?

a Spermidine in rice and barley | b Polyamines in wheat
c Proline in petunia | d All above

138 Which of the following is used for seed priming for increasing drought resistance

a KCl
b $CaHPO_4$
c KNO_3
d All above

139 What is role of silicon in crop production?

a Improves drought tolerance
b Improve water balance
c Antioxidant defense mechanism
d All above

140 Which of the following is a drought enduing plant?

a *Cereus peruvianus*
b *Euphorbia virosa*
c *Agave americana*
d All above

141 Which of the following is a drought evading plant?

a Bergenia
b *Cichorium intybus*
c *Plantarum* spp.
d All above

142 Which of the following is a drought resistant plant?

a Aloe
b Cactus
c *Dudleya brittonii*
d All above

143 Which of the following feature belongs to drought resistant plants?

a Thick cuticle
b Fleshy leaves
c Sunken stomata
d All above

144 Which of the following can be used as an indicator of water stress in plants?

a Relative water content
b Leaf water potential
c Leaf temperatre
d All above

145 Which of the following is linked with hydraulic and chemical signaling of water storage in drought resistant plants?

a Aquaporins
b Guard cell turgor
c ABA
d All above

146 Stomata open at night and close during days time in

a Xerophytes
b Succulents
c Mesophytes
d Hydrophytes

147 Active uptake of minerals by roots mainly depends on the

a Temperature
b Availability of O_2
c Light
d Availability of CO_2

148 Which of the following is a water absorbent used in dryland agriculture for increasing water availability to crops?

a Polyacrylate
b Hyaluronic acid
c Polyacrylamide derivatives
d All above

149 What is *Alcaligenes latus* that absorb water from soil?

a Soil bacterium
b Chemical compound
c Virus
d None above

150 Which of the following plant type has higher drought tolerance ?

a C_3 plants
b C_4 plants
c CAM plants
d All above

Answers

1. d	2. d	3. a	4 .c	5. a	6. b	7. a	8. d	9. c	10. c	11. a	12. a	13. c
14. d	15. c	16. c	17 .a	18. c	19. a	20. a	21. d	22. d	23. c	24. c	25. a	26. c
27. c	28. c	29. a	30 .a	31. d	32. d	33. a	34. a	35. b	36. b	37. d	38. c	39. d
40. c	41. d	42. d	43 .d	44. b	45. a	46. b	47. a	48. b	49. d	50. a	51. b	52. c
53. d	54. a	55. b	56 .c	57. d	58. d	59. a	60. a	61. b	62. c	63. d	64. a	65. b
66. c	67. d	68. a	69 .b	70. c	71. d	72. a	73. b	74. c	75. d	76. d	77. d	78. d
79. d	80. c	81. d	82. c	83. d	84. d	85. d	86. d	87. d	88. a	89. d	90. d	91. d
92. d	93. d	94. d	95. a	96. b	97. c	98. d	99. b	100. d	101. d	102. d	103. d	104. d
105d	106. d	107. d	108. d	109. d	110. d	111. d	112. d	113. d	114. d	115. d	116. a	117. b
118.d	119. d	120. d	121. d	122. d	123. d	124. d	125. d	126. d	127. d	128. d	129. a	130. b
131. d	132. d	133. d	134. d	135. c	136. c	137. d	138. d	139. d	140. d	141. d	142. d	143. d
144. d	145. d	146. b	147. b	148. d	149. a	150. c						

9

Sustainable Crop Production in Rainfed Areas

Outlines

Introduction

Rainfed agriculture occupies about 1.22 billion hectare (Bha) area worldwide, which is about 82% of total area (1.5 Bha) under agriculture (Yadav *et al.*, 2006) and contributes nearly 58% to the global food basket. It is practiced under a wide variety of agro-climatic, edaphic and socio-economic conditions and globally, rainfed agriculture accounts for more than 95% of farmed land in sub-Saharan Africa, 90% in Latin America, 75% in the Near East and North Africa; 65% in East Asia and 60% in South Asia (Singh *et al.*, 2018). According to UNO (2017) report, a number of hungry people are rising rapidly and 11% of the global population suffered from hunger in 2016. Asia region has the largest number of hungry people (63.8%) in the world followed by 29.82% in Africa and 5.15% in Latin America and the Caribbean regions in 2016. The world population is increasing at an alarming rate and it is expected that global population will increase up to 8.6 billion (B) in 2030, 9.8 B in 2050 and 11.2 B in 2100.

Rainfed agriculture, in general, specifies the field conditions in which the agricultural production depends on rainfall. India ranks first in the world in terms of area (74 mha) under rainfed agriculture and value of the produce. Rainfed agriculture is a pre-dominating production system occupying nearly 52% of total net cultivated area (140.0 mha) and contributes nearly 44% to the total food grain production of the country. Nearly 60% farmers cultivate crops without irrigation. Between 55-60% of gross domestic product of the agriculture and allied activities (agri-GDP) comes from rainfed agriculture. Besides it contributes nearly 85% of coarse cereals, 83% of pulses, 70% of oilseeds, 42% of rice and 65% of cotton area (Dass, *et al.*, 2017). However, average yields of rainfed crops are less than a t/ha, as low input-low risk-low yield concept dominates in the rainfed regions. The total food grain production of India has increased from 50.82 mt in 1950-51 to 292 mt in 2019-19 mainly because of improved irrigation facilities and new technologies (Table.1). The poverty ratio, land productivity, social development index and per capita food availability is lower in rainfed area than irrigated mainly because of poor development of rained area (Table.2).

Table 1. Population growth, net cultivated area and food grain production

S.N	Year	Population (millions)	Net cultivated area (mha)	Net irrigated area (mha)	Food grain production (mt)
1.	1950-51	361.1	118.75	20.9	50.82
2.	1960-61	439.2	133.20	24.7	8402.02
3.	1970-71	548.2	140.27	31.1	108.42
4.	1980-81	683.3	140.00	38.7	129.59
5.	1990-91	846.4	143.00	48.0	176.32
6.	2000-01	1028.7	141.34	55.2	196.81
7.	2010-11	1210.9	141.56	63.7	244.49
8.	2014-15	-	140.13	-	252.02
9.	2016-17	1338.7	-	-	175.11
10.	2018-19	1352.6	-	-	285.01
11.	2019-20	1366.4	-	-	292.0
12.	2020-21	1380.00	-	-	

Source: Agricultural Statistics (2018)

Table 2. Comparison of important characteristics of predominantly rainfed and irrigated regions of India

S.N	Parameters	Rainfed region	Irrigated regions	All regions
1.	Poverty ratio, Head count (%)	37	33	35
2.	Land productivity (INR/ha)	5716	807	6867
3.	Labor productivity (INR/ha)	6842	9830	8336
4.	Per capita food grain consumption (kg/year)	260	471	365
5.	Infrastructure development index	0.30	0.40	0.35
6.	Social development index	0.43	0.44	0.43

Source: Sharma *et al* :

2. Major constraints of rainfed agriculture

i. Fragmented land

The size of operation holdings declined to 1.08 ha in 2015-16 compared to 2.28 ha in 1970-71. Similarly the per capita land availability declined from 0.91 ha in 1951 to 0.32 ha in 2001 which further estimated to decline to 0.19 ha in 2050.

ii. Shrinking cropped area

The net sown area was reduced from 140.96 mha in 1970-71 to 140.13 mha in 2014-15. It will further reduce in near future because of infrastructure development of the country.

iii. Land degradation

Nearly, 94 mha land is suffering from water erosion, 16 m ha from acidity, 14 m ha from floods, 9 mha from wind erosion, 6 m ha from salinity and 7 m ha by combination of factors. The climate change and its effect on extremes of heavy rainfall, cyclones, flood enhanced the degree of land degradation. In summer, the temperature is shooting very high and increased the events of forest fire leading naked soil cover, prone to land degradation.

iv. Poor soil fertility

Imbalanced, inadequate and over exploitation of land resources leave soils deficient in multi nutrients including essential macro and micro plant nutrients. In India, about 59, 36 and 5% area are low, medium and high in available N, respectively. Similarly, soils of about 49, 45 and 6% area are low, medium and high in available P, respectively and soils of around 9, 39 and 52% area are low, medium and high in available K, respectively (Chaudhari *et al.*, 2015). Furthermore, nearly 49, 15, 6, 8, 11 and 33% soils are suffering by zinc, iron, manganese, copper, molybdenum and boron deficiency, respectively. The deficiencies of macro-and micro-nutrients have become major constraints to productivity, stability and sustainability of soil.

Rainfed soils are fragile and marginal lands, poor in organic carbon, soil moisture, soil depth, fertility and prone to soil degradation. Besides, uneven topography like sand dunes in arid and semi arid areas while landslides in hills are very common features of rainfed agriculture. Similarly, soils are either saline-alkaline in arid areas or acidic hills. Alfisols (30%), Vertisols (35%) and Entisols (10%), Aridisols, Oxisols and Mollisols are the major soil groups of rainfed agriculture. Poor quality of ground water, high air and soil temperature, light soil texture and structure, low water holding capacity of soil, high wind velocity coupled with deficient rainfall make rainfed agriculture unproductive and less profitable.

v. Decreasing factor productivity

The total factor productivity is continuously decreasing. A recession has started in Haryana and Punjab, but the total wheat productivity is still increasing in these two states. In a study, Tripathi (2010) sub-divided the period from 1969-70 to

1979-80, 1980-81 to 1990-91, and 1991-92 to 2005-06 in to three parts and found that contribution of total factors productivity was in negative at -0.11% during the entire period.

vi. Poor irrigation facility

Presently 48% net cropped area is under irrigation and 52% under rainfed farming. The crop productivity is lower (0.1 t/ha) in rainfed areas than irrigated areas. Agriculture is using 81% of total water resources available in India and it is expected that only 73% of total available water will be available for agriculture in 2035 mainly because of competition with other developmental activities like industries, drinking purposes, wild life etc.

vii. Climatic constraints

Rained agriculture faces harsh climatic conditions characterized by inadequate, uneven and erratic rainfall, high temperature, high solar incidence, high wind velocity, low relative humidity, weather aberrations, drought, flood, frost, gale, cyclones, heat waves and high potential evapo-transpiration that limit crop production. Most often the region suffers either heavy rains causing flood and landslides in hilly tracts or long dry spells or drought may cause partial or complete crop failure. It is a major concern that the paradigm shift in rainfall pattern of country in last few years indicates that total annual rainfall as well as number of rain days have reduced and so the growing period has reduced. The farmers have little choice for selecting crops and varieties. The monsoon aberrations are common now-a-days in form of late arrival, early retreat of monsoon and mid way dry spells causing area vulnerable, poor crop growth and productivity.

Climate change is again a true fact with extremes events like rainfall, flood, heat and cold waves cause more drastic effect on rainfed than irrigated areas. The Global Climate Risk Index-2020 published by 'International Environmental Think Tank Germanwatch' puts India fifth in the list of 181 countries worst hit by extreme events. IPCC has also revealed that climate change may cause 1-2°C rise by 2050 and up to 3°C by end of century. This is something extraordinary rise of temperature that may decrease productivity of most of the crops by 10-35% varying with crops and agro-climatic conditions, so food and nutritional security of Indian continent is on sever risk.

viii. Technological constraints

The results of frontline demonstrations of different crops conducted in different agro-climatic regions have demonstrated that there is a wide yield gap between potential and actual crop yields. The gap may vary from 15-70 percent in different crops mainly due to poor adaptation of improved technologies. The percolation of improved technologies from lab to land is still very low around 20-25% that limits the realization of potential of crop varieties. This gap is wider in rainfed areas because of poverty, illiteracy, mindset to bear risk, unawareness and also poor extension infrastructure. Governments and other private institutions playing a major role for technology transfer including launching of government programs like 'Krishi Vigyna Kendra (KVKs)', 'Technology Missions' on different crops like oilseeds, pulses, maize, millets etc., 'Sansad Adarsh Gram Yojana', 'Digital India', 'ICT programs' etc. to enhance the process of technology transfer in rural India. Further, new development in conservation agriculture, organic farming, nano fertilizers, use of remote sensing and GIS in precision agriculture need to be propagated in rural India for better result in future.

ix. Market and storage facilities

The size of produce is very small because of larger share of small and marginal farmers as well as poor crop yield. The marketable surplus is very less in both quantity and quality. In addition, the area is producing diversified produce because of prevalence of mixed and intercropping. Further the marketable surplus is available only in one season i.e. *Kharif* season. Therefore, the middlemen and Mandi's are not available at village and Taluka levels. The mandi's are very far from the villages, so farmers have to face many problems related to sale of their produce. There is no scope of minimum or procurement support prices as middlemen always purchase produce at lower prices. Similarly, the storage facility of cereals, perishable crops like vegetables and fruits are not up to mark in the region, so farmers have to sell their produce just after crop harvest at very low prices. The farmers are poor and they need money for their inputs and implements but they have to take money from village land lords and middlemen at high interest rate, so most of farmers cannot come out from vicious cycle and live under heavy load of debt that is the main reason of suicides.

x. Pest outbreaks

The crop losses may range from 10-30% varying with crop and environment and complete crop failure may occur in case of serious pest outbreak. Improper and heavy use of pesticides in rainfed crops may cause harmful side effects such as direct toxic effect on applicator or consumer, development of strains or pests resistant to pesticides, resurgence of pest species, outbreak of secondary pesticides and destruction of non-target organisms such as parasites and predators and accumulation of harmful residues of food products. Cotton is the important crop of the rainfed agriculture and provides livelihood more than 42 million people in India and its yield loss is 5-15% due to pests and diseases annually and may shoot up to 50% if not managed properly. Recent attack of locust in 2020 summer season destroyed cotton crop in districts Sri Ganganagar, Hanumangarh, Nagaur and Bikaner of Rajasthan. Similarly it has already been reported that outbreak of sugarcane woolly aphid (*Ceratovacuna lanigera* Zehntner) in sugarcane belt of Karnataka and Maharashtra states during 2002-03 resulted in 30% yield losses.

The large and sporadic populations of insect pests are usually called outbreaks. This situation typically occur when the pest population rises significantly above its general equilibrium level and becomes a threat to human interests or endeavors.

The climate has strong correlation with outbreak of pest and diseases in crops. Though weather forecasting services in India are available to the farmers, which help with weather-based planning of farm operations, yet proper weather-based pest and disease forewarning services are not available to rainfed region. Therefore, pest forewarning, integrated pest management, machine learning algorithms etc. are required for control of pest and diseases.

xi. Poor mechanization

The rainfed agriculture faces many challenges and poor mechanization is one of them as dominance of small and marginal holdings and poor status of the farmers deny to purchase heavy machinery like tractors, planters, threshers, levelers, fertilizer applicators etc. that limit the mechanization in region. The sowing period is very short in rainfed regions. If the sowing is delayed, the crop productivity is decreased due to reduced growing period and terminal drought. Similarly the depth of sowing and placement of fertilizers are important aspect of rainfed agriculture and both may be facilitated with the help of proper implements, if available. These implements may also reduce the labor requirement and cost of cultivation.

xii. Socio-economic constraints

The farming community of rainfed region is poor and live under scarce resources mainly because of small and marginal holdings, low adoption of innovative technologies, low crop yield, low annual income, unemployment, poor marketing infrastructure, illiteracy, indebtedness and unawareness. Hence, 'Low Input- Low Risk-Low Yield' concept is prevailing in the rainfed region.

3. Sustainable management of rainfed agriculture

Following aspects are to be dealt with efficiently in the regions

1. *Adoption of soil and water conservation measures to avoid the soil degradation*

Numerous soil and water conservation measures based on topography, soil, capital, labour and technology may be adopted. There are two types of measures like *temporary measures* for arable lands like basin listing, sub soiling, bunding, terracing, contour trenching, diversion channels, gully or nala structures and percolation tanks (*mechanical measures*), contour farming, strip cropping, crop rotation, mulching, minimum tillage/zero tillage and cover cropping (*agronomic measures*) and agroforestry systems like agri-silviculture, agri-horticulture, horti-silvi-pastoral, silvi-pasture, agri-silvi-pasture, agri-hort-silviculture, ley farming and afforestation (*vegetative/agrostological measures*) and permanent measures for non arable lands like diversion drains, trenches, geojute, retaining walls (gabions) and kutta-crate structures (*slope stabilization*), temporary check dams, gabion check dams (*Gully/Channel stabilization structures*) and spurs, retards and gabions (torrent control measures). Besides, drop spillways, chute spillways and drop inlet spillways are of permanent nature to control the runoff and conserve land resources.

2. *Rain water harvesting*

Rain water is the only source of water for crop production in rainfed areas, so every drop of rainfall is to be conserved either '*in-situ*' or '*ex-situ*'. Among the *in-situ*' rain water harvesting structures- summer ploughing, deep tillage, field bunding, contour bunding, pitting, strip bunding, terracing, land leveling, inter row or inter plot water harvesting, vegetative barriers and mini or micro watersheds/catchments and among the '*ex-situ*' structures –naula, gadera, chal and dhal, hydrams, gules/channels and diversion channels (hills) and nadi/pond, tanka, khadin, anicut, roof eater collection, percolation tanks, sub surface barriers, sand filled dams, water spreading, dams, increase run off (dryland areas) are used to reduce the run off velocity but conserve it and also help to recharge the ground water. Number of research findings were found encouraging on rainwater harvesting and its recycling in crop production. Saxena *et al.* (2011) revealed in his study conducted at CAZRI, Jodhpur that the yield of pearlmillet was obtained almost 3 times with water harvesting compared to conventional practice in low rainfall condition and an increase of about 100% water use efficiency (WUE) was recorded in the study (Table.3). They also reported that in-situ water harvesting helped in efficient use of N-fertilizer in pearlmillet. Application of 40 kg N//ha with in-situ water harvesting improved pearlmillet grain yield by 27% with 69 mm of additional runoff compared to rainfed condition. Kumar *et al.* (2011) reported that summer ploughing enhanced soil moisture retention by 20%, reduced weed by 40% and increased crop yield. Oswal (2001) also reported higher seed yield of *Kharif* crops like pearlmillet, green gram, castor etc, under deep ploughing compared to conventional tillage. Khan and Singh (2009) reported reduced runoff under silvi-pasture system. Khan and Singh (2005) recorded the highest seed yield of chickpea (1803 kg/ha) grown in Khadin under conserved moisture conditions. Bhandarkar (2010) reported that recycling of the runoff water at different stages significantly increased grain yield of chickpea, linseed and safflower.

Table 3. Water use, yield and WUE of pearlmillet under water harvesting

System	Good rainfall			Low rainfall		
	Water use	Yield (kg/ha)	WUE (kg/ha/mm)	Water use	Yield (kg/ha)	WUE (kg/ha/mm)
Conventional	440	2320	5.3	143	400	2.9
In-situ water harvesting	630	2425	3.9	212	1240	5.9

Source: Saxena *et al.* (2014)

3. Adopt innovative and improved crop management practices

Rainfed agriculture is a home of traditional farming with poor access of new technologies. Recently agricultural science has progressed with many new innovations and technologies but they have not reached to major section of rainfed communities. Therefore following technologies/management practices may be taken care of for improving the crop production in rainfed farming;

i. Land preparation

The land preparation is the pre-requisite of crop sowing, so the tillage is part and parcel to make field ready for sowing. Tillage may be defined as the practice of modifying the state of the soil in order to provide conditions favorable to crop growth (Cuplin, 1986). Tillage has following main objectives;

i. To conserve soil moisture,
ii. Control weeds,
iii. Incorporate crop residues, manures and fertilizers, agrochemicals into soil,

The goal of proper tillage in rainfed agriculture may be listed as following;

i. Provide a suitable environment for seed placement and its proper germination,
ii. To provide suitable environment for root growth,
iii. To control weeds,
iv. To control soil erosion, and
v. To conserve soil moisture by holding maximum rainwater.

Therefore the success of tillage operations in dryland areas should be evaluated on the basis following points;

i. Moisture management,
ii. Erosion control,
iii. Weed control,
iv. Management of crop residues,
v. Improvement of soil tilth,
vi. Soil aeration,
vii. Provide good seed-soil contact,
viii. Fine soil surface for seeding operation,
ix. Incorporation of crop residues and manures-fertilizers,
x. Insect-pest control,
xi. Temperature control for seed germination.

The tillage operation requires primary and secondary tillage for proper seedbed preparation. Primary tillage is an initial major soil working operation and normally designed to reduce soil strength, cover plant materials and rearrange the soil aggregates.

It cuts and shatters soil and may bury trash by inversion, mix it into the tilled layer or leave it basically undisturbed.It is more aggressive, relatively deeper operation and usually leave the surface rough. The implements like country plough, Mould board plough, disc plough, chisel plough and sub soiler for primary tillage, on the other hand, the secondary tillage is important to make soil more pulverized with good soil tilth and main objectives of secondary tillage are;

i. Improve seedbed by greater pulverization of soil,

ii. To cut up and mix crop residues and trash in soil,

iii. To break clods from the top soil and put it in better tilth for seeding and germination of seed.

ii. Selection of suitable crops and varieties

The public and private institutions have bred many crop varieties suitable to different biotic and abiotic factors. In rainfed areas, the crops and varieties should be drought tolerant and suitable to the regions, therefore the C_4 plants or genetically modified (GM) crops are better suited to the regions. The crop varieties need to be developed having C4 and CAM characteristics for higher yield. In addition, the crop and its varieties are selected based on crop maturity period and available length of growing period because the rainfall is always deficient and uneven, so growing period differs with agro-climatic regions. In general, short duration crop and varieties are preferred so that they may mature within available growing period.

In general, short duration, photo insensitive, initial fast growth habit, low water requirement and deep rooted crops are preferred in rainfed areas. Therefore millets like maize, sorghum, bajra, ragi, madira, kodo, kakun (cereal), green gram, black gram, pigeon pea, garm, lentil, horsegram, lathyyrus (pulses), groundnut, sesame, safflower, soybean, castor, linseed, niger, sunflower (oilseeds), cotton and tobacco are grown in rainfed areas.

The important crop varieties suitable to dryland agriculture are given in table .4 below;

Table 4. Recently released high yielding crop varieties suitable for drought stress

S.N	Crop	Area and State	Drought stress varieties
1	Rice	Rajasthan (Banswara area)	Ashoka-200F
		Gujarat (Heavy rainfall zone	Gujarat Nagli-4,
		WB (New alluvial & coastal saline zone)	Kalinga, Sahabhagi, IR-36
		Uttarakhand & Chhattisgarh (North-East Plain zone)	Pant Dhan 16, Barani, Dhan-1, Aditya, Kalinga-3,
		Chhattisgarh (Bastar Plateau)	Pant Dhan 16, Barani, Dhan-1, Aditya, Kalinga-3, Pradhan, Poornima
		Bihar	Sahbhagi, Sushak, Samrat
		HP (Mid hills)	Sukara Dhan-I, VL-221

S.N	Crop	Area and State	Drought stress varieties
2	Wheat	Bihar, Eastern UP, West Bengal	HD-2888, K-8027
		Chhattisgarh (North Hills)	Sujata, C-306, JWS-17, HI-8627, HI-1531
		HP i. High Hills ii. North Hills	HPW-155, HPW-236 VL-421, HS-277, VL Gehun-829, HPW-249, VL-907, HS-420, HPW-236
		Punjab, Haryana, UP, Rajasthan (North West Plain Zone)	PBW-644, PBW-527, PBW-175, WH-1080, HD-3043, PBW-396
		Rajasthan (Luni basin & Jaipur area)	HI-1500, Raj-3077, Raj-3765, KRL-1-4
		Uttarakhand (Hill zone)	UP-1109, UP-2572
		MP & Rajasthan (Central zone)	HI-1531 and HI-8627, DBW-110, MP-3288, MP-3173, HI-1531, HI-1500
		Karnataka & Maharashtra (Peninsular zone)	NIAW-1415, HD-2987, HD-2781
3	Maize	Bihar (North East Plain zone)	Suwan
		Chhattisgarh (North Hills)	HQPM-5, HQPM-1, Vivek-21, Vivek-9
		Karnataka	Nithyashri, Hema (NAH-1147)
		Punjab	PMH-2, Parkash
		Rajasthan	Mahi Dhawal, PEHM-1, Pratap,Makka-5, Pratap Makka-3, Hybrid Makka-1
		Haryana, UP, AP, Maharashtra	HM 4
4	Pearl millet	Gujarat	GHB-538 and GHB-719
		Karnataka	WCC-75,
		Rajasthan, Gujarat & Haryana	Raj-171, GHB-538, RHB-177,HHB-67(I), RHB-177, RBH-154, RBH-173, HHB-234, HHB-226, HHB-216, Bio-70, GHB-757, GHB-719, CZP- 9802
5	Barley	Punjab	PL-419
		UP, Bihar	K-603
		Rajasthan, Haryana & UP	RD-2660
6	Foxtail millet	Karnataka	RS-118, K-211-1, PS-4, SIA-326
		MP	MPMH-17
7	Finger millet	Chhattisgarh	VR-708 (Padmavati), HR-374
		Karnataka	MR-1, MR-6, GPU-66, GPU-28, KMR-301, ML-365
		Maharashtra	Phule Nachani

S.N	Crop	Area and State	Drought stress varieties
8	Sorghum	Karnataka	CSH-5, CSH-9, CSV-4, DSV-2
		Maharashtra & Gujarat	Parbhani Moti, GJ-39
		Rajasthan	CSV-17
		Maharashtra (Rabi season-Rabi growing areas)	M-35-1, Phule Chitra, Phule Vasudha, Phule Panchami, CSH-19R, CSV-18
		Uttarakhand	Pant Chari-5, Pant Chari-7
9	Sugarcane	Tamil Nadu, Karnataka, Maharashtra UP, Haryana, Punjab & Bihar	Co-86032, Co-85019, Co-94008, Co-94012, Co-99004, Co- 2001-13, Co-0218, Co-0403 and Co-06027 Co-2001-15, Co-98014, Co-0238, Co-0118, Co-0124, Co-0239
10	Blackgram	Maharashtra	BDU-1
11	Chickpea	Chhattisgarh	JG-14, Indira Chana, JG-315, JG-11
		Karnataka	JG-11, C-101, ICCV-10
		Madhya Pradesh	JG-16, JAKI-9218, JG-6, JG-11, RVG-201, RVG-202, RVG-203
		Maharashtra	BDN-9-3
		Punjab	PDG-3, PDG-4
		Rajasthan	Pratap Chana-1, RSG-973 (Aabha), RSG-991, RSG-888, GNG-663, RSG-14, RSG-44
		MP, Maharashtra, Gujarat, Rajasthan & UP	Raj Vijay Gram-202, JSC-55, JSC-56
12	Field bean	Karnataka	HA-4
13	Green gram	Rajasthan	RMG-268, RMG-62, SML-668
14	Horse gram	AP, Karnataka, TN and Kerala	CRIDA-18R, CRHG-4, CRHG-19, PHG-9, HG-2 & 5
		Maharashtra	Dapoli Kulthi-1
		Rajasthan	AK-21, Pratap Kulthi-1 (AK-42), GC-3
15	Lathyrus	Chhattisgarh	Mahateoda, Prateek
16	Lentil	MP	JL-3, RVL-31, Vamban
17	Moth bean	Rajasthan	RMO-40, RMO-225, RMO-425, RMO-297,CZM-1, CZM-2, CZM-3
18	Pigeonpea	Chhattisgarh	Rajeev Lochan, Asha, UPAS-120, Paras, Number-148, C-11
		Karnataka & AP	TTB-7, BRG-1, BRG-2
		Maharashtra	Paras
		Rajasthan	BDN-711, BDN-708, ICPL-87, ICPL-151, Paras

S.N	Crop	Area and State	Drought stress varieties
19	Brassica (Brown Sarson)	Himachal Pradesh	KBS-3
20	Brassica (Rapeseed & Mustard)	Rajasthan	Aravali, RGN-48, RB-50, RGN-48
		Uttarakhand	Pant Rai 20
21	Brassica (Raya)	Himachal Pradesh	RCC-4
22	Castor	Karnataka	DCS-9 (Jyothi), Jwala
23	Groundnut	Gujarat	GG-2 and GG-5, ICCV-9114
		Karnataka	JL-24, TMV-2, KCG-2,GPBD-4, ICGV-91114, GG-5, IGN-3
		Madhya Pradesh	GG-5
		Maharashtra	LGN-1
		Andhra Pradesh	Kadiri-5, Kadiri-9, Abhaya, Greeshma, Narayani , Dharani
		Andhra Pradesh & Tamil Nadu	K-134
		TN	TMV-13
		UP, Bihar, Haryana, Punjab & Rajasthan	ICGS-5
		MP	JGN-3, JGN-23
		UP, Rajasthan & Haryana	CSMG-84-1
24	Karan Rai	Himachal Pradesh	Jayanti
25	Linseed	Chhattisgarh	R-552, RLC-78, RLC-81, RLC-92
		Himachal Pradesh	Nagarkot, Himani
26	Safflower	Maharashtra	PBNS-40, PBNS-12, PhuleKusum
27	Sesame	Gujarat	G.Til-4, G Til-10, RT-46, RT-125, RT-146, Rama, TKG-22
28	Soybean	Maharashtra	JS-80-21, Indira Soya-9, Ahilya, JS-71-05, NRC -7
29	Sunflower	Karnataka	KBSH-44, KBSH-53
		Maharashtra	LSPH-35
30	Toria	HP	Bhawani
31	Cotton	Gujarat	G. Cot-18, AKA-8, BN bt.
		Maharashtra	Phule Anmol (RAC 024), SVPR-4
		Tamil Nadu	SVPR-4
32	Jute (*C. capsularis*)	Assam, North Bengal, UP	JRC-80, JRC-7447, JRC-212
		West Bengal	JRO-524, JRO-204
		Assam, North Bengal, UP	JRC-80, JRC-7447, JRC-212
33	Tossa jute (*C. olitorius*)	All jute growing states	JRO-524, JRO-204,

S.N	Crop	Area and State	Drought stress varieties
34	Mesta (HS)/ Roselle (*Hibiscus. sabdariffa)*	Odisha, AP, Maharashtra, Tamil Nadu	GR27, HS-4288, HS-7910
		Odisha, AP, Maharashtra, Tamil Nadu, Karnataka	AMV-5, AMV-7
35	Mesta (HC) / Kenaf (Hibiscus cannabinus)	Odisha, AP, Maharashtra, Tamil Nadu	MT-129, MT-150, AMC-108
		Odisha, AP, Maharashtra, Tamil Nadu, Karnataka	MT-129, MT-150 (for paper pulp production), HC-583, AMC-108
	Fodder crop		
37	Anjan grass/ Buffalo grass	-	Marwar Anjan (CAZRI-75)
		All over India	Bundel Anjan-1, Bundel Anjan-3
		Tamil Nadu	Kolukkattai grass CO-1
38	Sorghum	Haryana & Punjab	Haryana Chari-171, Haryana Chari-308
		Tamil Nadu	CO-27, CO(FS)-29
		Two cut in Gujarat	Pusa Chari-9,
		Maharashtra & Gujarat	Pusa Chari-23
39	Rice bean	WB, Orissa, Jharkhand, Assam, Sikkim, Mizoram, Manipur, Nagaland & Bihar	Bidhan-1, Bidhan-2, Bidhan-3
40	Bajara-Napier (Fodder)	Tamil Nadu	BN Hybrid CO-1
		Himachal Pradesh	NB-37
		Maharashtra	Yeshwant
		TN	KKM-1
		Punjab, Haryana, & HP	PBN-83
41	Cowpea (Forage)	All India	Type -2 , Bundel Lobia-1, GFC-2, EC-4216
		South zone	MFC-09-1
		Haryana	HFC -42-1, C-88
42	Dhaman grass / Bird wood grass	Rajasthan	Marwar Dhaman (CAZRI-76)
		MP, Chhattisgarh	Jawahar Pennisetum-12
		All over India	Bundel-1, Bundel, Dinanath-2
		Tamil Nadu	CO-1
43	Forage Cowpea	Uttarakhand	UPC 9202, UPC 628
44	Guinea grass	All over India	Bundel Guinea-2, JHGG-08-1, RSDGG-1
		Kerala	Haritha, Makueni
		Rainfed condition all over country	Bundel Guinea-2
		Kerala, Tamil Nadu & AP	Riversdale

S.N	Crop	Area and State	Drought stress varieties
45	Niger	Karnataka	KBN-1, JNC-1, JNC-6, JNC-9
		Maharashtra	PNS-6, JNC-1, JNC-6,JNC9
46	Setaria grass	HP, Uttarakhand	Setaria-92, S-18,PSS-1, Nandi
47	Tall Fescue	HP & Uttarakhand	Hima-4, EC-178182
G	**Fruit**		
48	Apple	J & K	York Imperial
49	Apricot	J & K	Badami, Inzhirnyl, Rannil
50	Banana	Southern A.P., Northern Karnataka	Karpuravalli (ABB)
51	Ber	Rajasthan	Sev, Gola, Umran
		J & K	Seb, Mudia, Jogia, Gola
52	Citrus	J & K	Mosambi
53	Water Melon	Rajasthan	Durgapura, Madhu, Durgapura Kesher
54	Guava	J & K	Allahabad Safeda, Lucknow-49
55	Mango	Orissa and A.P.	Arka, Neelachal, Kesari
		Rajasthan	Sinduri, Jalore Seedless /Sindhuri
H	**Vegetable**		
56	Ash gourd		Kashi Dhawal
56	Bottle gourd	Rajasthan	Thar Samridhi
		UP	Kashi Ganga
57	Brinjal	Tamil Nadu	4 PKM-1
58	Carrot	Tamil Nadu	Ooty-1
59	Cassava	Chhattisgarh	Shri Sahya
60	Chillies	Karnataka	Samrudhi
		Uttar Pradesh, Bihar, M.P. Karnataka	Kashi Anmol
61	Cluster bean	Rajasthan	RGC-936, RGC-1017, RGC-1003, RGC-1066, Thar Bhadavi
62	Colocasia	Chhattisgarh	13 White Goriya, Haloo Kesoo
63	Cowpea	Karnataka	IT-38956-1, TVX -944-02E,
		Rajasthan	RC-19, RC-101
64	Cumin	Rajasthan	GC-4
65	Drumstick	Karnataka	Bhagya-KDM-01
66	Elephant Foot Yam	Chhattisgarh	NDA-5, Gajendra
67	Onion	Chhattisgarh	Agrifound Dark Red, Arka Kalyan
		Rajasthan	Raseedpura local
68	Taramira	Rajasthan	RTM-314
69	Tomato	Chhattisgarh	Arka Meghal, Arka Vikas

iii. Seed and sowing

Broadcasting is a common sowing method in rainfed areas, therefore the seed remains on soil surface and it does not come in proper content of moist soil resulting in to poor germination and low plant population. Besides, the seed is subjected to various losses like bird and ant damages. The root growth is poor and prone to severe damage to moisture stress and finally low crop productivity. Therefore, it is recommended that seed rate should be increased by 20-25% for achieving optimal plant population.

The sowing time is very crucial in rainfed areas because it is mainly decided by onset of monsoon in *Kharif* season, so the crop should be planted as early as possible after onset of monsoon. Early onset of monsoon and early withdrawal of monsoon are common in rainfed areas, so crop suffers soil moisture stress either in early, middle and terminal water stress that limit crop productivity. Early sowing not only ensures avoiding water stress but also offers an opportunity to utilize residual soil moisture for *Rabi* crops.

The crops should be sown at optimum crop specific geometry to avoid low or high plant stand, hence optimum seed rate should be used. The sowing depth is again very important in dryland areas as the soil moisture may not be sufficient at sowing time. It is also recommended that field should be leveled and compacted by 'leveler' or 'pata' to ensure better seed-soil contact for better germination. On the basis of number of field experimentation, it has been concluded that sowing depth should be 5 to 7.5 cm deep for better germination. The seed priming either with water soaking or water absorbing chemicals like sodium/potassium polyacrylate, hydrogel etc. before sowing has been found beneficial for germination as well as crop yield. It is always desirable to treat seed before sowing with suitable fungicides/ Rhizobium culture/biofertilizers to avoid seed and soil born diseases. Ridge and furrow method of sowing is preferred under limited moisture conditions. In hills, planting should be done across the slopes. Rabi crops should be sown just after harvest of *Kharif* crops so that residual soil moisture may be utilized efficiently.

iv. Precision nutrient management

Balanced nutrition is the pre requisite for better crop yield. 4R's i.e. right source of nutrient, right dose of nutrients, right application time, and right place of application are to be decided for precision nutrient management. The remote sensing and GIS is important tool for precision nutrition. Soil test values, soil health cards, previous crop history, variable rate technology (VRT) are important strategies for precision plant nutrition. Presently, drone operated imageries, 'artificial intelligent'(AI), Agribotics etc may be futuristic approach for plant nutrition. In general, the crop specific fertilizer recommendations are there for rainfed areas, however as a thumb rule, the fertilizer dose should be half in rainfed areas compared to irrigated areas and full dose should be applied at the time of sowing. The top dressing is recommended provided rainfall is received at required time of fertilizer application. The top dressing of urea should be made in afternoon around 4-5 PM under poor soil moisture in anticipation of some dews in night hours. The fertilizers should be placed at 6-8cm deep for higher WUE.

v. Weed control

Weeds affect plant growth by competing for moisture, nutrients, light and space, resulting in severe yield loss. The weed problems in the rainfed regions is more complex compared to irrigated conditions owing to several factors such as edaphic, climatic, social (small farm holdings, illiteracy and limited resources) and technologies. The factors like climatic, geographical, cropping pattern, crop diversification and management practices determine the composition of weed species. Weeds also interfere with agricultural operations, increase cost of cultivation and are potentially responsible for about 34% yield loss worldwide (Oerke, 2006). The weeds grow fast and utilize soil moisture leaving crop plants to suffer water stress. Therefore, precision weed control is mandatory to reduce the yield loss occurred due to severe crop-weed competition. Presently number of high grade crop specific herbicides is available in market, so right dose and time of application are of much significance for better weed control. The tractor operated VRT sprayers are available for the job. Drones are used for uniform herbicide spray for weed control.

vi. Mulching

Poor soil moisture and its maintenance in rainfed farming is a major hurdle for crop production and it is aggravated high temperature, wind velocity, low organic matter, low water holding capacity and poor soil structure and texture. Water is required for consumptive use including evaporation, transpiration and metabolic activities. Mulching is one of the effective techniques to avoid soil moisture through evaporation. The mulch not only conserves the soil moisture but also prevent soil loss by intercepting runoff and acting as insulation barrier for evaporation. Besides, mulching improves the soil fertility, soil structure, minimize the weed infestation, moderate the soil temperature and thereby increase the crop yield (Choudhary *et al*., 2003). The materials like crop residues, straw, grasses, weeds, tree leaves (live mulch), polyethylene, gunny bags, soils, etc. can be used for mulch purpose. Khan and Singh (2005) reported higher grain yield and WUE of chickpea under polythene mulch (Table.5). In one study conducted at Dehradun concludes that application of organic mulch @ 4t/ha effectively reduced runoff from 37 to 15% and soil loss from 18 to 5.4 t/ha in maize crop planted at 8% slope and grain yield was increased from 1.9 to 2.4t/ha. Application of mulch @ 2/ha also reduced 50% erosion. Similarly in Chandigarh, a field study reveals that wheat yield was increased 58% by using grass mulch @ 7.5 t/ha (Table.6). The live mulches like leaf twigs, weeds, green manuring crops like *Leucaenia leucocephaala* and Sesbania sesban, cowpea, green gram etc can also be used as a mulch materials. Some of the important weeds of hilly region like *Lantana camera*, *Urtica dioica* (scorpian grass or Bichhu ghas), eupatorium (*Eupatorium adenophorum*) may be used for mulching. Similarly crop weeds and other intercrops/alley crops like sesbania sesban, cowpea, subabul etc. can be used as a live mulch during Rabi/ summer season to protect main crop from excessive heating.

Table 5. Effect of mulching on seed yield and WUE of chickpea

S.N	Treatment	Seed yield (kg//ha)	WUE (kg/ha-mm)
1.	No mulch	1465	5.78
2.	Weed straw mulch	1632	6.78
3.	Polythene mulch	1757	7.40
	CD (P = 0.05)	152	-

Source: Khan and Singh (2005)

Table 6. Effect of grass mulch on wheat yield at shiwalik region (Chandigarh)

S.N	Treatment	Grain yield of wheat (t/ha)	Increase over control (%)
1.	Control	1.70	-
2.	Dust mulch	1.87	8.2
3.	Grass mulch @ 5t/ha after maize harvest	2.13	25.4
4.	Grass mulch @ 5t/ha after sowing of wheat	2.30	35.6
5.	Grass mulch @ 7.5t/ha after maize harvest	2.50	47.2
6.	Grass mulch @ 7.55t/ha after sowing of wheat	2.68	57.8

vii. Use of antitranspirants

The crop plants require water to fulfill the consumptive use of water. The water is lost in evaporation and transpiration. The transpirational loss can be minimized either by increasing leaf resistance to water vapor transfer, reducing amount of energy absorbed by leaf surface, reduce leaf temperature or reduce plant growth and leaf area index. Therefore the leaf transpiration can be reduced by application of antitranspirants.

Antitranspirant refers to any material that reduces transpiration on application. Normally they are used as a foliar spray on leaf surfaces. They are also known as 'transpirant retardant', 'transpirant suppressant' or 'antidessicant'. The antitranspirants have better result in arid and semi arid areas.

Types of antitranspirants

1. ***Film forming type***: plastic & waxy material which form thin film on leaf surface

   Examples: Waxol (mobleaf), silicon oils, Ethyl alcohol, S-789 foliate, S-800, Hexadecanol etc.

2. ***Stomata closing type: They close the stomata on application.*** Examples: PMA (Phenyl mercuric Acetate), Atrazine (fungicide & herbicide @ 250g/ha), Phosphon, ABA (Natural antitranspirant), CCC (2 Chloroethyl-Trimethyl Ammonium Chloride), Ascorbic acid, Alkenyl succinic acid, 8-Hydroxydecanol sulphonate etc.

3. ***Reflectant type:*** They form coating on leaves and increase the leaf reflectance (albedo). They are white in color. Examples are Kaoline @5% (Diatomaceous earth), chalk, ciliate, Lime water etc.
4. ***Growth retardant* type:** The reduce shoot growth and increase root growth. In addition they interfere in biosynthesis of GA that helps in movement of stomata and also induce stomata closure. Examples are CCC (cycoceal), ABA, Phosphan-D, low concentration of 2,4-D.

viii. Adoption of suitable cropping systems

The pre-requisite of selecting cropping systems for a region is either higher crop productivity, yield equivalents of the system, net profit and B:C ratio. The cropping systems i.e. intercropping, mixed cropping and sequential cropping should be well fitted into available soil types and length of growing period. The intercropping and mixed cropping are more popular in the rainfed areas as they are treated as insurance because atleast of the system will survive even in adverse weather conditions. Number of remunerative cropping systems has been identified for different ecology of rainfed areas. Nanwal *et al.* (2012) proposed cropping systems based on rainfall, soil type and length of growing period (Table.7). The cropping systems also differ with agro-climatic regions with available growing period (Table.8). AICRPDA (2010) proposed some of the remunerative cropping systems for different dryland regions of India (Table.9). In addition, some important cropping systems of Uttarakhand are also mentioned in the above table. Some of the prevalent cropping systems of different regions are given in table.10. Singh (1995) proposed intercropping and double cropping systems for different rainfed regions based on soil types and growing period and moisture availability (Table.11).

Table 7. Potential cropping systems in relation to rainfall and soil type

S.N	Rainfall (cm)	Soil type	Growing period (weeks)	Cropping system
1.	320-600	Red & shallow black soils	20	Single *Kharif* cropping
2.	350-600	Deep Sierozems and alluviums	20	Single cropping in either *Kharif* or *Rabi*
3.	350-600	Deep black soils	20	Single *Rabi* cropping
4.	600-750	Red, black and alluvial soils	20-30	Intercropping
5.	750-900	Alluviums, deep black, deep red and sub mountainous soils	>30	Double cropping with monitoring
6.	>900	Alluviums, deep black, red and sub mountainous soils	<30	Double cropping

Source: Narwal *et al.* (2012)

Table 8. Length of effective cropping season in different areas

S.N	Growing period (weeks)	Effective cropping season (weeks) in different dryland areas	Proposed cropping systems
1	<20	Bellary (8), Jodhpur (11), Anantpur (13), Hisar, Rajkot & Bijapur (17)	Sole cropping
2	20-30	Jhansi & Koralpati (21), Hyderabad, Udaipur (22), Sholapur (23), Agra (24), Anand (25) & Akola (27)	Intercropping
3	>30	Bhuveneshwar & Hishiarpur (35), Varanasi (32), Benagaluru, Indore & Rewa (36), Dehradun & Samba (44) & Ranchi (45)	Sequential cropping

Source: Jat *et al*. (2013)

Table 9. Remunerative cropping systems in different rainfed regions of India

S.N	Remunerative cropping systems	Rainfed regions
1.	Groundnut+Pigeonpea intercropping	Scarce rainfall zone of AP
2.	Fodder pearlmillet-Green chilli & Pearlmillet + Pigeionpea (10:2)	Southern dry zone of Karnataka
3.	Maize+Blackgram(2:2) and Groundnut+Sesame (6:2)	Southern Rajasthan
4.	Soybean+Pigeonpea (3:1) intercropping	Budelkhand of MP
5.	Cotton+Sorghum (6:1) and Sorghum+Pigeonpea(2:1)	Vidharba Region of Maharashtra
6.	Pearlmillet+Pigeonpea (2:1) and Sunflower+Pigeonpea (2:1)	Scarce rainfall region of Maharashtra
7.	Groundnut+Castor (3:1)	Northern Saursatra of Gujarat
8.	Castor+Cowpea(1:2) for aberrant weather situation and Relay cropping of Castor in Greengram	Northern Gujarat
9.	Pigeonpea+Rice (1:3) and Pigeonpea +Maize (1:1)	Rainfed upland of Jharkhand
10.	Pigeonpea+Sesame (2:1)	Eastern UP
11.	Wheat+Mustard (2:1)	Kandi region of Punjab
12.	Maize+Pigeon pea	Eastern Ghat of Odisha
13.	Ragi+Blackgram, Soybean+Barnyard millet Ragi+Soybean (Black) mixed cropping and Fingermillet-Pea, Fingermillet-Lentil, Fingermillet-Fallow and Barnyard Millet-Wheat	Uttarakhand valleys and mid hills
	Finger millet+Amaranthus/pulses mixed cropping	High hills of Uttarakhand

Source: AICRPDA (2010)

Table 10. Important cropping systems of different regions

S.N	Cropping systems
A	**Northern India**
1	Sorghum – Safflower / Mustard
2	Greengram/Blackgram/Cowpea – Gram/Wheat
3	Sorghum – Wheat / Gram

4	Rice – Gram (Low lying areas)
5	Bajra – Gram+Linseed
6	Bajra – Wheat/Barley/Gram/Mustard
7	Soybean – Gram/Wheat/Barley/Mustard
8	Maize – Gram /Safflower
B	**Central India**
1	Greengram – Rabi Sorghum
2	Greengram - Safflower
3	Cotton/Sorghum - Gram
4	Pearlmillet-Gram
5	Sorghum -Gram
C	**Southern India**
1	Maize – Sorghum+Pigenpea
2	Sorghum - Gram
3	Pearlmillet - Horsegram
4	Greengram/Blackgram - Safflower
5	Greengram - Pigenpea
6	Rice –Blackgram/Gram/
7	Sugracane + Blackgram

Table 11. Intercrooping and double cropping systems for different regions

S.N	Soil and region	Growing period (days)	Water availability	
			Double cropping system	Intercropping system
A	**Vertisols and related soil zones**			
1.	Malva plateau (MP)	191-210	Sorghum- Safflower/ Chickpea	Sorghum+Pigeonpea
		210-230	Maize-Safflower/ Chickpea	Maize+Soybean
2.	Bundelkhand (UP)	190-220	Sorghum-Chickpea Blackgram-Mustard/ Safflower/Pulses	Sorghum+Pigeonpea
3.	Vidharbha (Maharashtra)	170-190	Greengram-Sorghum/ Safflower	Pearlmillet+Greengram Sorghum+Greengram
		190-210	Groundnut-Safflower Sorghum-Safflower	Sorghum+Pigeonpea Cotton+Pigeonpea
4.	Southern Maharashtra	160-180	Greengram-Sorghum/ Safflower/sunflower	Pearlmillet+Pigeonpea Sunflower+Pigeonpea Groundnut+Pigeonpea
5.	Southern Rajasthan	160-180	Greengram-Safflower	Maize+Pigeonpea Groundnut+Pigeonpea Chickpea+Mustard

6.	Northern and Southern Karnataka	100-120	-	Sorghum+Corainder Safflower+ Coriander
		130-150	Cowpea-Sorghum	Pearlmillet+Pigeonpea Greengram+Safflower Groundnut+Pigeonpea
7.	Saurastra (Gujrat)	130-140	-	Groundnut+Pigeonpea Pearlmillet+Pigeonpea
8.	Southern TN	120-130	-	Sorghum+Blackgram Cotton+Pigeonpea
B	**Alfisols and related soil zones**			
	Southern Karnataka	190-220	Cowpea-Fingermillet Soybean-Fingermillet	Fingermillet+Pigeonpea Groundnut+Pigeonpea Fingermillet+Soybean
	Telangana region	140-160	-	Sorghum+Pigeonpea Castor+Clusterbean
	Rayalseema (AP)	110-130	-	Groundnut+Pigeonpea Groundnut+Pearlmillet
C	**Aridisols**			
	North Western Gujarat	75-90	-	Pearlmillet+Clusterbean Clusterbean+Blackgram Pigeonpea+Blackgram
		100-120	-	Pearlmillet+Blackgram Pearlmillet+Clusterbean Clusterbean+Greengram

Source: Singh (1995)

4. Crop diversification

According to Niti Aayog, doubling real income of farmers till 2022-23 over the base year of 2015-16 requires annual growth of 10.41% in farmers' income. Therefore, strong measures like improvement in productivity, resource use efficiency or saving in cost of production, increase in cropping intensity, diversification towards high value crops etc. were suggested. In the due course, many useful strategies were proposed like Irrigation ('Per Drop More Crop'), Quality seeds (Improving seed replacement rate), Soil-test based nutrient management (Distribution of soil-health cards), Post-harvest crop losses (Large investments in warehousing and cold chains), value addition, creation of a national agricultural market, removing distortions and e-platform, new crop insurance scheme (Pradhan Mantri Fasal Bima Yojana), promotion of ancillary activities (poultry, sericulture, beekeeping and fisheries) etc. to support in doubling farmers' income by 2022. Three prolonged strategies focused on (i) development initiatives, (ii) technology and (iii) policy reforms in agriculture are needed to double farmers' income. The expansion of cropped area is difficult but crop productivity may certainly be increased by 2 to 3 folds in rainfed areas by adopting suitable technological innovation. Crop diversification and integrated farming system approach are another options to improve farmers' income. Crop diversification refers to the addition of new crops or cropping systems to agricultural production on a particular farm taking into account the different returns from value-added crops with complementary marketing opportunities. The aim of crop diversification is to increase crop portfolio so that farmers are not dependent on a single crop to generate their income. The inclusion of new high yielding crop and varieties will not only improve crop productivity, farm profit, health and nutritional value, biodiversity, crop resilience to biotic and abiotic stresses, conserve natural resources and food security but also reduce dependency on off farm inputs, risk and environmental pollution. Therefore crop diversification is a paradigm shift in crop production approach for larger benefit of farming communities particularly in dryland areas where scarce resourced farmers have least scope to invest in improvement of their agricultural production. Crop diversification is easy to adopt with least investment. The prerequisite of crop diversification is to improve awareness about new technology like high yielding varieties for different climatic region, precision farming, conservation farming, organic farming, new farm implements like happy seeder, bed and ridge planter, laser land leveler, drip and micro irrigation, biofertilizers etc. so training, technology extension and availability of inputs need to ensured for efficient farming and sustainable income enhancement.

Gulati (2021) reported that the growth rate of dairy, fisheries, horticultural crops were higher than cereals, pulses and oilseeds from 2000-01 to 2018-19 and dairy has the highest share of agriculture commodities in gross value of output (2018-19). It has also been reported that the annual consumption of cereals per person had decreased, while the consumption of dairy products, fisheries, poultry, fruits and vegetables have increased during last decade. It also favors for crop diversification and require much more acreage and investment in production of dairy, poultry, fisheries, fruits and vegetables than food crops.

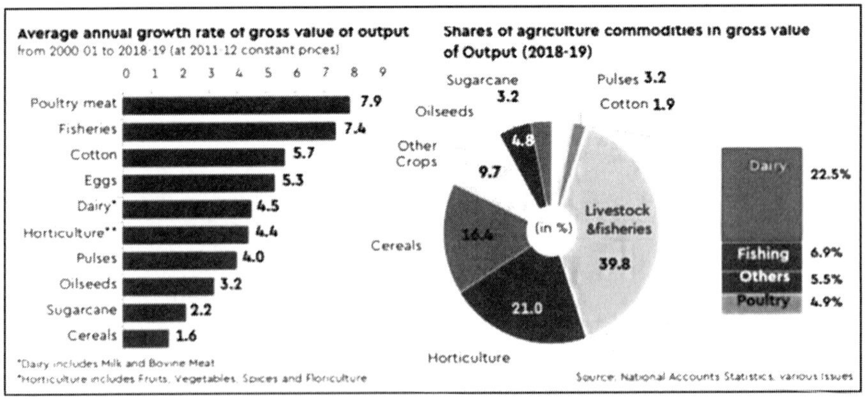

Source: Gulati (2021)

A study carried out at IARI, New Delhi by Singh *et al.* (2019) revealed that the introduction of babycorn-mustard cropping systems in rainfed condition fetched

Rs. 122629/ha as net returns compared to Rs.17271/ha with pearl millet-lentil cropping systems (Table .12).

Table 12. Crop productivity and economics of crop diversification in rainfed condition

S.N	Cropping system	Pearlmillet equivalent yield (t/ha)	Net returns (Rs/ha)	B:C ratio	Income (Rs/ha/ Day)	Increase net returns over Pearlmillet-Lentil system (Rs)
1.	Baby corn-Mustard	15.61	122629	1.42	573.0	105358
2.	Cowpea (veg)-Barley	9.03	52305	0.72	251.71	35034
3.	Greengram-Chickpea	8.94	48118	0.66	227.31	30847
4.	Pearlmillet-Lentil	5.50	17271	0.31	82.64	-

Source: Singh *et al.* (2019)

5. Conservation Agriculture

The term "conservation agriculture" was coined in the 1990s, but the idea to minimize soil disturbance has its origins in the 1930s, during the Dust Bowl in the United States of America. Later CIMMYT pioneered no-till training programs and trials in the 1970s, in maize and wheat systems in Latin America. In the 1980s this technique was also used in agronomy projects in South Asia. Farmers worldwide are increasingly adopting conservation agriculture. In the 2015/16 season, conservation agriculture was practiced on about 180 mega hectares of cropland globally, about 12.5% of the total global cropland, 69% more than in the 2008/2009 season. FAO defined conservation Agriculture as 'a farming system that promotes minimum soil disturbance (i.e. no tillage), maintenance of a permanent soil cover, and diversification of plant species. It enhances biodiversity and natural biological processes above and below the ground surface, which contribute to increased water and nutrient use efficiency and to improved and sustained crop production. There are 3 major pillars of conservation agriculture i.e. i. Minimum mechanical soil disturbance (no tillage), ii. Permanent soil organic cover with crop residues and live mulches (at least 30%) and iii. Species diversification (crop rotation and intercropping at least three different crops). The main purpose of conservation agriculture is to maintain a permanent soil cover to assure its protection, avoids soil tillage and cultivates a diverse range of plant species to improve soil conditions, reduce land degradation and increase water and nutrient use efficiency.

Rainfed agriculture is characterized with low agricultural productivity due to low soil moisture, poor soil fertility, prone to wind and water erosion and climate change. So the conservation agriculture is the safeguard of all above problems. Kumar *et al.* (2018) reported 8.45% higher grain yield of maize with 72% water saving and 281% increase in water productivity under zero tillage with mulch compared to tradition method. Singh *et al.* (2019) reported that 14.48, 0.41 and 14.7% increase in the yield, net returns and water use efficiency, respectively from the crops grown during winter season crops with the use of residue mulch in zero tillage compared to conventional tillage under rainfed condition. Jat *et al* (2012) reported that rice-wheat (RW) system integrated with mungbean improved the system productivity by 10%, profitability by 20–30% using 15–30% less irrigation water and 20–25% less energy input compared to conventional RW system in the IGP.However, the replacement of rice with maize improved the productivity by 10-15%and profitability by 40–50% using 70% less irrigation water. However suitable solution is required for certain constraints related to adoption of conservation agriculture like lack of appropriate seeders or planters especially for small and medium scale farmers, competition of crop residues between CA use and livestock feeding, burning of crop residues, availability of skilled and scientific manpower and overcoming the bias or mindset about tillage.

6. Follow alternate land use systems (ALUS)

Rainfed agriculture faces numerous challenges related to climate change, aberrant monsoon rainfall, soil degradation with multiple nutrient and water deficiencies, declining the ground water table, poor resource base of the farmers and finally unstable crop yields but marginal and sub marginal lands of rainfed areas are brought under cultivation forcefully to fulfill the increasing demand of food, fodder, feed and fibre without considering its capability and suitability to crop production leading to imbalances in the ecosystem. Therefore it is necessary to think and evaluate about present land use pattern based on productivity and profitability and shift or change the present land use pattern based on land capability and suitability classes so that production system must become system more profitable and sustainable on long term. This change in present land use pattern to new land use pattern is known as alternate land use system. Therefore it may combine crop production with other remunerative enterprises like dairy, poultry, sericulture, mushroom production, agroforestry, honey bee production, processing and value addition activities etc. Normally the land capability class I to IV are good for crop production with some limitations but class V to VIII are good for other activities like agroforestry, landscaping, pasture and grasslands, industries, dairying, poultry etc. Hence alternate land use system is a way of maximizing land productivity, profitability and sustained ecology. In rainfed areas, agroforestry is a better option as it is a profitable enterprise with minimum investment in poor and infertile lands. The important agroforestry systems are detailed below (Table.13).

7. Adoption of alley Cropping

Alley cropping is the planting of trees or shrubs in two or more sets of single or multiple rows with agronomic, horticultural, or forage crops cultivated in the alleys between the rows of woody plants. It is an intercropping system of crops and woody trees. The food crops are grown in the alleys formed by hedgerows of planted trees and shrubs, preferably leguminous fodder or multipurpose trees. At planting of crops, the hedgerows are cut and pruned periodically before cropping season to prevent shading and reduce competition with the food crops and browse is used as fodder/green manures and mulch to save soil moisture. In addition, these hedge rows prevent soil and water erosion in summer and rainy season. The hedgerows are allowed to grow freely to cover the land when there are no crops. Alley cropping is the basic feature of the traditional fallow system in arid and semi arid regions. One major advantage of alley cropping is that the cropping and fallow phases take place concurrently on the same land, thus allowing the farmer to crop the land for an extended period without a break. The International Livestock Centre for Africa has proposed alley farming as livestock component has been added and the hedgerow foliage is used as fodder for animals.

The selection of woody plants should be suitable to soil and climate. The trees like *Leucaena leucocephala, Acacia auriculiformis, Gliricidia sepium, Cajanus cajan* for non- acid soils and *Acioa baterii, Calliandra calothyrsus, Flemingia macrophylla, Paraserianthes falcataria* for acid soils may be preferred for hedge rows. The selected trees should be easy to plant, coppice, deep rooted, leguminous and withstand for repeated pruning.

8. Integrated farming system

Mixed farming system has been the culture of Indian agriculture since time immemorial. It combines crop production along with livestock including two oxes, one or two milch animals, few goats and sheep, few fruit and multipurpose trees on their farms. Few other farmers may also have pigs, fisheries mainly in West Bengal and Odisha. Presently this age old mixed farming system has been coined freshly as integrated farming system (IFS) with wider scope and application as 'new wine in old bottle'.

Definition

1. Therefore IFS is defined as biologically integrated systems combining two or more enterprises with the objective of best use of farm waste or resources interchangeably for achieving maximum replacement of off farm inputs and sustain farm income.
2. Lal and Millar (1990) defines as, 'Farming system designates a set of agricultural activities organized while preserving land productivity, environmental quality and maintaining desirable level of biological diversity and ecological stability'.
3. Biswas and Singh (2003) defined IFS as , 'it involves integration of two or more enterprises with best use of available resources to satisfy maximum needs of the owner which leads to increase in productivity per unit area, efficient recycling of farm wastes, better utilization of resources, generate employment, reduce the risks and ensure sustainability.
4. Kumar *et al.* (2015) pointed out the IFS improved yield per unit time and area by virtue of intensification and diversification of crops with integration of allied enterprises. It also ensures nutrient recycling within the system

Table 13. List of important agroforestry systems

S.N	Agroforesty system	Components		
		I	II	III
1	Agri-silvi culture	Crops-cereals, oilseeds, pulses, fodder, sugar crops etc.	Fodder tress like Leucaena leucocephala, Acacia auricullformis, Albizia lebbek, Populus spp.	-
2	Agri-horti culture	Crops-cereals, oilseeds, pulses, fodder, sugar crops etc.	Fruit plants like Ziziphus mauritiana, Psidium guajava, Mangifera indica etc.	-
3	Silvi-pasture	Fodder tress like *Leucaena leucocephala, Acacia auricullformis, Albizia lebbek, Populus* spp.	Grasses like Cencharus ciliarus, Cenchrus setigerus, Lasiurus sindicus stylosanthes hamata,	
4	Horti-pasture	Fruits- Citrus spp, *Psidium guajava* etc.Emblica officinalis, *Aegele marmelos*,	Grasses like Cencharus ciliarus, Cenchrus setigerus, stylosanthes hamata	-
5	Agri-silvi-pasture	Crops like maize, bajra, sorghum, cowpea, greengram, mustard, barley, safflower, niger etc.	Fodder tress like Leucaena *leucocephala, Acacia auricullformis*, Albizia lebbek, Prosopis cineraria, Populus spp.	Grasses like *Cencharus ciliarus*, Cenchrus setigerus, Lasiurus sindicus stylosanthes hamata
6	Agri-horti-pasture	Crops like maize, bajra, sorghum, cowpea, greengram, mustard, barley, safflower, niger etc.	Fruits- Citrus spp, Psidium guajava etc.Emblica officinalis, Aegele marmelos,	Grasses like Cencharus ciliarus, *Cenchrus setigerus*, Lasiurus sindicus stylosanthes hamata
7	Boundary plantation/ shelter belts/wind breaks/live fence rows	*Acacia catechu*, Acacia nilotica , Acacia tortilis, Aegel marmelos *Codia myxa*, *Azadirchta* indica, Dalbergia sisso, Eucalyptus camaldulensis, Tecomella undulate etc.		

to economize and sustain the system and minimizes the dependence on chemical fertilizers for crop production to earn more profit and employment generation round the year.

Therefore, the IFS combine crops and cropping systems, livestock, agroforestry, aquaculture, duckery, poultry, horticulture, bee keeping, sericulture, piggery, mushroom cultivation, food processing, other agro based industries and other allied enterprises in an expanded symbiotic or synergistic system in which the wastes of one process become the valuable input for other processes making IFS more self reliant, climatic resilient, profitable and sustainable and leading to nation food and nutritional secured.

Concept of IFS

IFS is based on the concept that there is no waste on agricultural farm house rather the so called agricultural waste or by-products of one enterprise like crop residues/straw from crop production and gobar-urine from livestock etc., are the valuable input of another enterprise like crop residues is a feed for livestock and gobar-urine is a nutrient source for crop production and convert farm from chemical mode to organic mode with diversified quality produce and sustained farm income. It also reduces dependency of a farm on off –farm inputs like chemical fertilizer and pesticides and labor requirements, so the cost of production is decreased and profit increased. Besides, it also reduces risk, generating more employment, increasing farm diversification and biodiversity, effective utilization of family labour and over all making farm more profitable and self reliant.

Objectives

The objectives of integrated farming systems (IFS) may be listed as below;

i. To reduce dependency of farm on 'off- farm' inputs and labor requirement,
ii. To utilize waste of one activity as a valuable input for other activity and convert farm into organic mode,
iii. To utilize family labor effectively round the year,
iv. To conserve and utilize natural resources rationally,
v. To diversify farm with inclusion of all possible agro based enterprises with synergy,
vi. To generate employment round the year,
vii. To produce diversified quality products,
viii. To make farm productive and profitable, and
ix. To increase farm income and
x. To improve living standards of the farming communities.

Integrated Farming system models in rainfed areas

Some of the important integrated farming systems of different regions are listed below based on conclusion of numerous studies conducted in different agro climatic regions (Table 14 & Table.15).

Table 14: Comparison of conventional production system and integrated farming system

S.N	Region/State	Conventional Production System	Integrated Farming Systems	Increase over conventional farming (%)	Reference
A	**Jodhpur (Rajasthan)**				
i		Sole Hardwikia binata/ grass	*Harwickia binata + Cenchrus ciliaris*	+ 25// 65	Harsh and Tewari (2007)
ii		Sole pearlmillet	Pearlmillet + *Acacia Senegal*	+ 61	
iii		Alone arable crops	Crops+Trees+Fruits +Grass +Livestock	+ 193	Tanwar *et al.*(2016)
B	**Bikaner (Rajasthan)**				
i		Moth bean – Cumin	Aonla +Mothbean – Cumin	+ 57	Awasthi *et al.* (2009)
ii		Moth bean – Chickpea	Aonla +Mothbean – Chickpea	+54	
C	**Hisar (Haryana)**				
i		Alone cropping	Cropping + Crossbred cattle in place of	+346	Singh *et al.* (1993)
ii		Alone cropping.	Cropping + Baffalo	+35	Singh *et al.* (1993)
	Bawal (Haryana				
i		Clusterbean – Barley	Khejri + Guava+Clusterbean	+ 381	Kaushik *et al.* (2017)
ii		Clusterbean – Barley	Khejri + Aonla + Clusterbean - Barley	+ 327	
	Uttar Pradesh				
		Alone crops	Crops + Dairy	+ 56	Gill *et al.* (2009
		Alone crops	Crop + Dairy+ Horticulture	+ 62	
		Alone crops	Crop + Dairy + Apiculture	+ 102	
		Alone crops	Crop+ Dairy + Vermicompost	+ 110	

S.N	Region/State	Conventional Production System	Integrated Farming Systems	Increase over conventional farming (%)	Reference
	Punjab				
		Alone crops (1.0 ha)	Crop (0.6 ha)+ Dairy (0.4 ha)	+ 22	Gill *et al.* (2009)
		Alone crops (1.0 ha)	Crop (0.6 ha)+ Dairy (0.4 ha) + Poultry (100 birds)	+ 25	
	Northern dry zone (Rainfed), Karnataka				
		Sole Bengal gram	Ber + Bengal gram	+ 190	Yaragattikar and Itnal (2003)
		Sole safflower	Ber + Safflower	+ 171	
	Tamil Nadu				
		Sole Rice	Rice-Fish-Poultry	-	Srivastava (2018)
	Assam				
		Sole Rice	Rice-Fish	-	Srivastava (2018)
		Sole Rice	Pig/poultry -Fish-Vegetables	-	
	Patna				
		Rice-Fish	Rice-Fish+Water Chestnut' (Makhana)	-	Srivastava (2018)
	Odisha				
		Sole cropping	Crops+Vegetables+Poultry+Mushroom +Vermicompost	+ 82.5	-

Table 15. Suitable IFS models for arid and semi-arid regions of India

S.N	Region	IFS model	Reference
1	Semi-arid vertic Inceptisols of Tamil Nadu (1.0 ha)	Crop + poultry (20) + goat (4) + sheep (6) + dairy (1)	Solaiappan *et al.* (2007)
2	Tungbhadra project area of Karnataka (1.0 ha)	Crop (0.74 ha)+ horticulture (0.18 ha) + fodder crop (0.02 ha) + livestock (2 cow, 1 buffalo,14 goat)+ vermicompost units (4) + compost unit (1) + Azolla unit (1)	Basavanneppa and Gaddi (2017)
3	Dryland western zone of Tamil Nadu	Sorghum + cowpea (grain purpose), sorghum + cowpea (fodder) and C. glaucuseach in 0.33 ha intercropped with *Emblica officinalis* + goat (5+1	Radhamani (2001)
4	Semi-arid Gujarat	Crops:pearl millet – wheat (0.44 ha), mustard – pearl millet (0.22 ha), cotton – fodder sorghum (1 ha) + horticulture: papaya (0.04 ha) + dairy buffalo (6)	Patel *et al.* (2007)
5	Semi-arid irrigated Punjab (2.0 ha)	Crop (1.14ha) + dairy (0.22) + fishery (0.56) + piggery (0.24)	Gill *et al.* (2009)
6	Semi-arid Haryana	Cropping + buffalo or cropping + crossbreed cattle	Singh *et al.* (1993)
7	Arid Rajasthan (7.0 ha land)	Agroforestry (*Prosopis cineraria* + crop), agroforestry (*Hardwickia binata* + crop), agri-horti (*Zizyphus mauritiana* + crops), hortipasture (Zizyphus mauritiana + grass), silvipasture (*Colophospermum mopane* + grass) with six adult cattle unit (4 cow, 8 bucks and 4 rams)	Tanwar *et al.* (2016)

IFS adoption has not scaled up in our country despite many economic and ecological advantages in rainfed areas mainly because of small and marginal farmers, poor transport, processing and marketing facilities, illiteracy, unawareness and poor managerial skill among farming communities, lack of diversified programs and no government policies for IFS. Therefore, it is most desirable that farming communities should be trained for IFS systems, its benefits and related technologies supported by institutional policies. The facilities like transport, processing and market need be improved in rural areas. In addition, the arid and semi arid areas, the cropping season is limited to 3-4 months in monsoon season and remaining period of the year, the fields remain without crops and used for grazing of animals and they graze and destroy the newly planted saplings and other horticultural plants if not protected causing devastating effect on adoption of IFS in the region. The farmers' participation is also envisaged in designing, evaluating and refinement of farming systems research for better adoption of IFS.

It is very clear from the research findings of rainfed areas that sole crop production is neither remunerative not fulfill the demand of food and nutritional requirement of the masses, therefore the integrated farming system is required to be adopted for larger benefits of farming communities and nation for sustained life and environment of rainfed areas.

9. Organic farming

In India, nearly 51% of country's net sown area is under rainfed agriculture that contributes 40% to the food basket of the country. It is very complex, fragile, diverse and risk prone due to vagaries of unstable monsoon leading to low input usage and poor productivity. Therefore, it is urgently required to explore the possibilities of alternate land use pattern coupled with traditional wisdom and new innovations to make rainfed areas more productive and sustainable so that it can improve the living standards of the region. Organic farming is the answer for rainfed areas where fertilizer consumption is very low i.e. <25kg/ha/year (Mitra and Devi, 2016) particularly in hilly areas, northeastern states and also other parts of the country where rainfed farming is practiced. The rainfed areas are organic by default as very low chemical inputs are used but the poor soil health *vs*. low crop yield may be improved by organic farming that emphasizes on use of organic manures, biofertilizres, green manuring, legumes in crop rotations etc. Hence, organic farming has great significance in rainfed areas. The rational use of abundance of organic resources like crop residues, livestock products like gobar and urine in form of compost, FYM, vermicompost, oil cakes, poultry manures, weeds, forest litter, rural and urban solid wastes, bi-products of agro industries like molasses and bagasse, fruit and vegetable wastes, biofertilizers and also organic bio-enhancers like Kunapjal Jivamritu, Amritpani, Panchagavya, Beejamrita etc. is not only supports for adoption of organic farming but also fulfill the nutritional requirement of the crops and cropping systems of rainfed areas. The bio preparations like BD-500, BD-501, Cow Pat Pit, Biodynamic liquid manures

(Biodynamic agriculture), Agnihotra ash enriched water and Biosol (Homa organic farming) have been used successfully for improving soil fertility and plant health of many crops and horticultural plants.

The rainfed areas are more suited to organic farming mainly because of;

1. Soils are poor by both fertility and soil moisture,
2. Low application of chemical inputs,
3. Low crop yield,
4. Risk prone area due to unstable weather conditions,
5. Increasing rate of desertification due to over exploitation of canal and ground water,
6. Abundance of livestock and its products like gobal, urine, skin etc.,
7. Neem (*Azadirachta indica*), Karanji (*Pongimia pinnata*), and *Calatropis* spp, are abundantly available in drylands as they are used for preparation of biopesticides,
8. Rock phosphate, gypsum and lime are available naturally in large quantity in Rajasthan. These minerals help improve soil conditions and supply plant nutrients.
9. Abundance of crop residues, grasses, forest litter etc.
10. Unemployment is a serious problem, and
11. Low income.

Recently the organic agriculture is gaining popularity due to increasing demand of organic products and its higher market rate globally. Organic farming in India is in a nascent stage and accounts about 2.78 mha of farmland (March 2020) sharing 2% of 140.1 m ha net sown area of the country. At present, India ranks 9th in area and 1st in organic producers in the world. Sikkim is the 1st organic state followed by Tripura and Uttarakhand. FiBm-IFOAM survey 2020 indicated that the total area under organic farming was 71.5 mha in 2018 leading by Australia (35.7mha), Argentina (3.6mha), China (3.1mha), Spain (2.6mha), Uruguay (2.2mha), France (2.0mha), USA (2.0mha), Italy (1.96mha)m India (1.94mha) and Germany (1.5mha). In India, MP is leading in acreage with 0.76mha followed by Rajasthan and Maharashtra with 4.9, 2.0 and 1.6% of their net sown area, respectively under organic agriculture. States like Meghalya, Mizoram and Uttarakhand (hilly states) and Goa, Sikkim, Delhi, Dadra and Nagar Haveli and Daman and Diu (plain states and UTS) have 10% or more of their net sown area under organic farming.

In present era of chemical farming and its residues in food, feed, fodder, fibre, fruits and vegetables and dairy products has increased demand of organic products. Some niche crops like basmati rice, nutricereals including ragi, organic fodder, fruits and vegetable, Assam lemon, Joha rice, medicinal rice, passion fruits etc have high value and are demanded crops in the world and fetch high price in local market too. The organically produced dryland crops like ragi, sesame, groundnut, clusterbean, safflower, niger, lentil, soybean, cotton etc. have high demand in international market, so these crops should be promoted under organic farming in rainfed areas for high net profit. Besides, organic farming will not only enrich soil health and biodiversity, reduce off farm inputs, increase crop yield as and net profit but also produce safe, clean and nutritional products.

Climate resilient integrated organic farming system model: Climate resilient integrated organic farming system model was developed in ICAR for semi-arid region on one ha area for small and marginal farmers facing problems due to various kinds of abiotic stresses. The model includes crops and cropping systems, vegetables, fruits, livestock and boundary plantation, hence, 2500 m^2 area has been assigned to vegetable crops, 2500 m^2 for sugarcane-based cropping sequence, 1000 m^2 for fodder crops, 1000 m^2 for cereals and pulse crops, 2000 m^2 for orchard crops (pomegranate, sapota, custard apple, mango), 100 m^2 pond for storage of water to provide irrigation to crops through drip and 500 m^2 for poultry (Vanaraja, Kadaknath and ornamental breed), goatry (Osmanabadi breed), cattle (ghir) and compost unit. One solar-operated water pump (3 KV) was installed near water storage pond for supporting irrigation to the crops. The fruit plants like Mango, lemon, dragon fruit and papaya were planted on borders of all the experimental plots. The chickpea was intercropped in orchard crops for revenue generation from orchards during initial years, while drumstick with Napier grass for getting maximum return. Among the vegetables, bhindi, cowpea, cabbage, tomato, bottle gourd, pumpkin, onion, garlic, coriander and sweet corn crops were planted in an area of 2500 m^2 for regular income from the farm. It is a successful model for regions and others are adopting the models.

The central government has started new scheme, 'Paramparagat Krishi Vikas Yojana' in 2015 as an extended component of soil health management under centrally sponsored scheme, National Mission on Sustainable agriculture (NMSA) for promotion of organic farming and improvement in soil health. It will encourage farmers to adopt eco-friendly concept of cultivation and reduce their dependence on fertilizers and agricultural chemicals to improve yields. Government has made budgetary allocation of Rs. 300 crores for the scheme. However the adoption of organic farming is very slow due to doubts among producers and scientists for its sanctity and benefits because the food security based on major cereals like rice, wheat, sorghum and maize which require heavy fertilization due to decreasing soil health and how this heavy demand of nutrition can be met out. The supply of organic manures like FYM, compost, vermicompost and other organic fertilizers is not sufficient to fulfill the demand. Further, the yield is reduced in beginning with use of organic fertilizers and buffering may take 4-5 years and country cannot afford this incubation period. Another aspect is related to the quality of product as told of high quality under organic farming has not yet ensured. Organic certification, organic market, storage and trade are also important aspects to be developed for better support and promotion of organic farming in years to come.

It is therefore concluded that organic farming must be propagated in rainfed areas to overcome most of the prevalent problems of rainfed agriculture. However, organic certification, market and storage facilities, availability of subsidized organic inputs, capacity building and awareness, research and monitoring and government support with suitable programs and policies are pre requisites for promotion of organic farming in the region.

10. Adoption of watershed approach

Total of 312 districts i.e. 46.4% out of 672 districts constitute important rainfed systems and 197 districts are drought prone districts in the country. Another 115 districts, though having 40% irrigated area, have > 1.0 one lakh ha under rainfed agriculture. It shows that rainfed agriculture is still important for national food security because 47.5% food grains, 80.2% of pulses, 73% of oilseeds, over 90% of millets and 40% of rice come from rainfed areas but none of the Ministries have mandate to care for water resources management in rainfed areas. However, National Watershed Development Project for Rainfed Areas (NWDPRA) was launched in 1990-91 and National Rainfed Area Authority (NRAA) in 2006-07 for development of rainfed areas. NRAA listed 24 districts as most drought prone in India including 16 from Karnataka, 4 from Andhra Pradesh and 4 from Rajasthan. Earlier in 1990-91, National Watershed Development Project for Rainfed Areas (NWDPRA) was launched in 25 States and 2 Union Territories and later in IX Plan, the scheme was extended to 3 newly formed States of Uttarakhand, Jharkahand and Chhattisgarh. In 2000-2001, the scheme of NWDPRA has been subsumed under the Scheme for Macro Management of Agriculture (MMA). At present, this scheme is being implemented as a program of Centrally Sponsored Scheme of Macro Management of Agriculture in 28 States and 2 UTs. Presently watershed programs are under control of NRAA.

The watershed approach is based on the principles of 1. Conservation, development and sustainable management of natural resources, 2. Enhancement of agricultural production and productivity in a sustainable manner, 3. Restoration of ecological balance in the degraded and fragile rainfed eco-systems by greening these areas through appropriate mix of trees, shrubs and grasses, 4. Reduction in regional disparity between irrigated and rainfed areas, and 5. Creation of sustained employment opportunities for the rural community including the landless. Hence it is a concept of all round development of the area. Since the farmers involvement has been ensured in selection of watershed area, priority areas of development, action plan, execution and monitoring, the watershed mode of area development is excellent and also impact analysis of most of the watershed program are encouraging. It does not mean it is a foolproof approach for rainfed development because of few lacunae like insufficient fund allocation by government, involvement of party-politics in watershed institutions, frequent changes in policies, delay in work, improper evaluation, monitoring and reporting , so these hurdles need to be tackled timely for better result of watershed development programs.

Summary

Rainfed agriculture is equally important to ensure the food and national security of India, so it draws special attention of state and union governments to implement area specific programs based on watershed mode for all round development of area particularly in arid and semi arid region of the country where water is a most scarce input with due supported by sufficient funding and monitoring.

References

Agricultural Statistics. 2018. Pocket Book of Agricultural Statistics. Ministry of Agriculture & Farmers Welfare Department of Agriculture, Cooperation & Farmers Welfare, Directorate of Economics & Statistics, Govt. Of India, New Delhi.

AICRPDA. 2010. AICRPDA at a glance. AICRPDA, Hyderabad. 6p.

Awasthi O P, Singh I S and More T A. 2009. Performance of intercrops during establishment phase of aonla (Emblica officinalis) orchard. Indian Journal of Agricultural Sciences 79 (8): 587–91.

Bhandarkar, D.M.2010. Harvesting and Recycling Technology for Sustainable Agriculture in Vertisols with high Rainfall. Central Institute of Agricultural Engineering (CIAE), Bhopal (MP).

Biswas, C and Singh, R. 2003. Integrated farming system: An intensive approach. Intensive Agriculture 41 (7-8): 22-23.

Cropping systems in arid zone of western Rajasthan. Indian Journal of Arid Horticulture 1 (1):20-22.

Dass, Anchal and Bhattacharyya, R. 2017. Wheat residue mulch and anti-transpirants. ield Crops Research 2010:9-19.

Gill M S, Singh J P and Gangwar K S. 2009. Integrated farming system and agriculture sustainability. Indian Journal of Agronomy 54: 128–139.

Gulati, A. 2021. https://indianexpress.com/article/opinion/columns/crop-diversification-policy-agriculture-sector-farmers-hunger-index-children-malnutrition-7150654/

Harsh L N and Tewari J C. 2007. Agroforestry system in arid region (Rajasthan) of India. (In): Agroforestry Systems and Practices, pp 175–90. Puri S and Panwar P (Eds).New India Publishing Agency, New Delhi.

Jat, H. S., Datta, A., Choudhary,M, Sharma, P. C. and Jat, M. L. 2021. Conservation Agriculture: factors and drivers of adoption and scalable innovative practices in Indo- Gangetic plains of India –a review. International Journal of Agricultural Sustainability 19 (1): 40-45.

Jat, M. L., Bhakar, S. R., Sharma, S. K. and Kothari,A. K. 2013. Dryland technology. Scientific Publishers (India). 805p.

Kaushik N, Tikkoo A, Yadav P K, Deswal R P S and Singh S. 2017. Agri-silvi-horti systems for semi-arid regions of North-west India. Agricultural Research 62: 150–158.

Khan, M and Singh, 2000. Application of remote sensing data in characterization and mapping of soil resources for watershed planing in arid Western Rajasthan. *J. Indian Remote Sensing* 28(http://look.springer.com/article/10.1.1007/BF02989895)

Khan, M.A. and Singh Raj. 2005. Effect of cultivars, nutrient management and mulching on the production potentiality of chickpea under water harvesting condition of arid zone. J. Arid Legumes 2(1):10-14.41.

Kumar Shalander, Sharma, K. L., Kareemulla, K., CharyRavindra, G., Ramarao, C, A., Rao Srinivasa Ch. And Venkateswarlu. B. 2011. Techno-economic feasibility of conservation agriculture in rainfed regions of India.Current Science, 101(9):1171-81

Kumar, S., Bhatt, B. P., Dwivedi, S.K. and Shivan. 2015. Family farming-a mechanism to contribute towards global food security. Indian Farming 64 (12): 13-18.

Kumar, V., Jat, H. S., Sharma, P. C., Balwinder-Singh, Gathala, M. K.,Malik, R. K., Kamboj, B. R., Yadav, A. K., Ladha, J. K., AnithaRaman, Sharma, D. K., & McDonald, A. 2018. Can productivityand profitability be enhanced in intensively managed cereal systems while reducing the environmental footprint of production? Assessing sustainable intensification options in the bread basket of India. Agriculture, ecosystems and.Environment, 252, 132–147. (https://doi.org/10.1016/j.agee.2017.10.006)

Lal, R. and Miller, F. P. 1990. Sustainable farming for tropics. Sustainable Agriculture: Issues and Prospective, Vol 1 pp 69-89. Singh R P (Ed). Indian Society of Agronomy, IARI, New Delhi.

Oswal M.C. Characterizing physical environment of rainfed lands and their management for sustainable production. J. Agric. Phys.. 2001;1:71–75.

Ramakrishna, Y.S. 1997. Climate Features of the Indian Arid Zone. (In): Desertification Control in the Arid Eco-system of India for Sustainable Development (Eds. by Singh, S. and Kar, A.), Agro-Botanical Publishers, Bikaner. pp. 27-35.

Saxena Anurag, Goyal, R.K., Singh, R.K. and Roy, M.M. 2014. Water Management for Climate Resilience in Arid Region. Central Arid Zone Research Institute, Jodhpur, India. 52p.

Singh C B, Renkema J A, Dhaka J P, Singh K P and Schiere J B. 1993. Income and employment on small farmers. (In) Proceedings of International Workshop on Feeding of Ruminants on Fibrous Crop residues: Aspects of Treatment, Feeding, Nutrient evaluation, Research and Extension, Karnal, Haryana, 4-8 February, 1991, pp 67–76.

Singh, R., Das, A. and Singh, V K. 2020. Sustaining crop production in rainfed areas in India: issues and strategies. Current Journal of Applied Science and Technology 39 (25): 154-173.

Singh Raj, Dogra Atul, Sarkar Ashutosh, Saxena Anurag and Singh Bhagwan. 2018. Technology gap, constraints analysis and improved production technologies for yield enhancement of barley (*Hordeum vulgare*) and chickpea (*Cicer arietinum*) under arid conditions of Rajasthan. Indian Journal of Agricultural Sciences 88(2):93-100.

Singh Raj. 2011. Production potential, moisture useefficiency and monetary benefit as influenced by intercropping pattern of mothbean (*Vigna aconitifolia*) and pearlmillet (*Pennisetum glaucum*) in arid region. J. Arid Legumes 8 (1):35-38.

Srivastava, A. P. 2018. Selected integrated farming system models for enhanced income. Indian Farming 68 (01): 13-16.

Tanwar S P S, Akath Singh, Patidar M, Sirohi A S and Roy M M. 2014. Performance of integrated farming system in arid region diversified with perennials under delayed monsoon situations. (In) Extended summaries of volunteer paper. National Symposium of Agricultural Diversification for sustainable livelihood and environmental security, held at PAU Ludhiana, November 18-20, 2014. pp 564–565.

Tripathi, A. 2010. Total factor productivity growth in Indian agriculture. J. Global Economy 6(4): 286-298.

United Nations Organization. 2017. Department of Economics and social welfare, World Population Prospects.

Yadav, N.D., Beniwal, R.K. and Soni, M.L. 2006. Crop diversification under fruit based improve productivity and quality of rainfed soybean in semi-arid north-Indian plains. Rainfed Agriculture. Deptt.Of Agronomy, CCSHAU, Hisar. 170p.

Questions

a. Short and Descriptive

1. Write short notes on following;
   i. Conservation agriculture
   ii. Integrated farming system
   iii. Organic farming
   iv. Ley farming
   v. Alternate land use system
   vi. Alley cropping
   vii. Mulching
   viii. Antitranspirants
   ix. Soil and water conservation
   x. Watershed approach
2. Enlist the different crop management strategies for sustainable management of crop production in rainfed areas. Also describe the significance of tillage in rainfed agriculture.
3. Define antitranspirants and also enlist types of antitranspirant and its role in rainfed agriculture.
4. What do you mean by alternate land use system? Also describe the significance in sustainable crop production in rainfed areas.
5. Describe following in relation to crop production in rainfed areas;
   i. Seed and sowing
   ii. Nutrient management
   iii. Crops and varieties
   iv. Cropping systems
   v. Weed control
   vi. Pest and disease management
6. What do you mean by watershed? Justify with suitable reasons that watershed is helpful in development of rainfed agriculture.
7. What is rainfed agriculture and how does it differ with irrigated agriculture? Also enumerate the important constraints of rainfed agriculture.
8. What is the difference between conservation agriculture and organic farming. Describe the role in rainfed area development of the country.

b. Objective questions

1 What is the contribution of rainfed agriculture in food production in world?

a 25% b 40%

c 58% d 62%

2 What is the contribution of rainfed agriculture in food production in India?

a 25% b 40%

c 58% d 62%

3 What is percent area of total farm land under rainfed agriculture in South Asia?

a 60% b 65%

c 75% d 95%

4 What is percent area of total farm land under rainfed agriculture in East Asia?

a 60% b 65%

c 75% d 95%

5 What is percent area of total farm land under rainfed agriculture in Latin America?

a 60% b 65%

c 75% d 95%

6 What is percent area of total farm land under rainfed agriculture in Sub Saharan Africa?

a 60% b 65%

c 75% d 95%

7 What is percentage of hungry people of total population in the world?

a 5% b 11%

c 30% d 64%

8 What is percentage of hungry people of total population in Latin America?

a 5% b 11%

c 30% d 64%

9 What is percentage of hungry people of total population in Africa?

a 5% b 11%

c 30% d 64%

10 What is percentage of hungry people of total population in Asia?

a 5% b 11%

c 30% d 64%

11 What was the food production of India in 2019-20

a 245mt b 262mt

c 292mt d None above

12 What is the poverty ratio of India?

a 33 b 35

c 37 d None above

13 What is the poverty ratio of rainfed region of India ?

a 33 b 35

c 37 d None above

14 What is the poverty ratio of irrigated region of India?

a 33 b 35

c 37 d None above

15 What is the per capita food availability in rainfed areas of India ?

a 260 kg/yr b 365 kg/yr

c 417 kg/yr d None above

16 What is the per capita food availability in irrigated areas of India ?

a 260 kg/yr b 365 kg/yr

c 417 kg/yr d None above

17 What is the per capita food availability in India ?

a 260 kg/yr b 365 kg/yr

c 417 kg/yr d None above

18 What is the percentage of area low in nitrogen content?

a 5 b 36

c 59 d None above

19 What is the percentage of area medium in nitrogen content?

a 5 b 36

c 59 d None above

20 What is the percentage of area high in nitrogen content?

a 5 b 36

c 59 d None above

21 What is the percentage of area low in phosphorus content?

a 6 b 45

c 49 d None above

22 What is the percentage of area medium in phosphorus content?

a 6 b 45

c 49 d None above

23 What is the percentage of area high in phosphorus content?

a 6 b 45

c 49 d None above

24 What is the percentage of area low in K content?

a 9 b 39

c 52 d None above

25 What is the percentage of area medium in K content?

a 9 b 39

c 52 d None above

26 What is the percentage of area high in K content?

a 9 b 39

c 52 d None above

27 How much soils of India are deficient in Fe ?

a 6% b 8%

c 11% d 15%

28 How much soils of India are deficient in Mn ?

a 6% b 8%

c 11% d 15%

29 How much soils of India are deficient in Cu ?

a 6% b 8%

c 11% d 15%

30 How much soils of India are deficient in Mo?

a 6% b 8%

c 11% d 15%

31 How much soils of India are deficient in B?

a 15% b 33%

c 49% d None above

32 How much soils of India are deficient in Zn ?

a 15% b 33%

c 49% d None above

33 Rainfed soils are unproductive mainly due to

a Poor texture and structure b Low water holding capacity

c Deficient rainfall d All above

34 The Global Climate Risk Index-2020 published by 'International Environmental Think Tank Germanwatch' puts India at……. in the list of 181 countries worst hit by extreme events;

a 1st place b 3rd place

c 5th place d 8th place

35 How much crop yield will be decreased up to 2050 as per the estimation by IPCC due to global warming?

a 5-10% b 10-30%

c 35-50% d 52-60%

36 What is the general crop yield gap between potential and actual realization based on field demonstration conducted in whole country?

a 15-20% b 15-45%

c 15-70% d 40-80%

37 How much new technologies have been percolated from lab to land in India?

a 5-10% b 10-15%

c 15-20% d 20-25%

38 How much average crop losses have been observed due to pest infestation?

a 5-10% b 10-15%

c 10-30% d 35-50%

39 In which state the outbreak of sugarcane wooly aphid was observed in 2002-03?

a UP and Punjab b Karnataka & Maharashtra

c TN and Kerala d Bihar and West Bengal

40 Which of the following is a temporary measures of soil and water conservation?

a Basin listing b Sub soiling

c Bunding d All above

41 Which of the following is a temporary measure of soil and water conservation?

a Terracing b Trenching

c Diversion channel d All above

42 Which of the following is a agronomical measures of soil and water conservation?

a Contour farming b Strip cropping

c Cover cropping d All above

43 Which of the following is an agronomical measure of soil and water conservation?

a Mulching b Crop rotation

c Conservation tillage d All above

44 Which of the following is an agrostological measure of soil and water conservation?

a Agri-silviculture b Agri-horticulture

c Horti-silvi-pastoral d All above

45 Which of the following is used for slope stabilization as a permanent measure of soil and water conservation in non arable lands?

a Geojute b Gabions

c Kutta -crate structures d All above

46 Which of the following is used for gully stabilization as a permanent measure of soil and water conservation in non arable lands?

a Temporary check dams b Gabion check dams

c Both' a' & 'b' d None above

47 Which of the following is used for torrent control as a permanent measure of soil and water conservation in non arable lands?

a Spurs b Retards

c Gabions d All above

48 Which of the following water harvesting structures are prevalent in Uttarakhand hilss?

a Chal & Khal b Naula

c Gules d All above

49 Which of the following crop (s) have the highest area under rainfed agriculture?

a Pulses b Oilseeds

c Rice and wheat d Nutricereals

50 Which of the following rice variety is suitable for rainfed farming?

a Pant Dhan-16 b Gujarat Nagli-4

c Kalinga-3 d All above

51 Which of the following wheat variety is suitable for rainfed farming?

a HD-2888 b Raj-3765

c PBW 175 d All above

52 Which of the following wheat variety is suitable for rainfed farming?

a UP-1109 b UP-2572

c HI-8627 d All above

53 Which of the following maize variety is suitable for rainfed farming?

a PMH-2 b Prakash

c Vivek-21 d All above

54 Which of the following maize wheat variety is suitable for Rajasthaan and Gujarat under rainfed conditions?

a Pratap Makka-5 b Mahi Dhawal

c PEHM-1 d All above

55 Which of the following pearlmillet variety is suitable for Rajasthaan and Gujarat under rainfed farming?

a RHB-177 b GHB-757

c HHB-226 d All above

56 Which of the following barley variety is suitable for rainfed conditions?

a PL-419 b RD-2660

c K-603 d All above

57 Which of the following fingermillet wheat variety is suitable for rainfed conditions?

a Phule Nachni b GPU-28

c VR-708 d All above

58 Which of the following sorghum variety is suitable for rainfed farming?

a Prabhani Moti b CSV-17

c CSH-19R d All above

59 Which of the following chickpea variety is suitable for Rajasthan under rainfed farming?

a Abha b Pratap Chana-1

c RSG-888 d All above

60 Which of the following chickpea variety is suitable for MP under rainfed farming?

a JG-16 | b RVG-203
c JG-11 | d All above

61 Which of the following Horse gram variety is suitable for rainfed?

a Pratap Kulthi-1 | b CRIDA-18R
c Dapoli Kulthi-1 | d All above

62 Which of the following Mothbcan variety is suitable for rainfed?

a RMO-225 | b RMO-297
c CZM-3 | d All above

63 Which of the following Lentil variety is suitable for rainfed?

a JL-3 | b Vamban
c RVL-31 | d All above

64 Which of the following Pigeonpea variety is suitable for rainfed?

a UPAS-120 | b Paras
c Rajeev Lochan | d All above

65 Which of the following groundnut variety is suitable for rainfed?

a GG-5 | b Kadiri-9
c ICGS-5 | d All above

66 Which of the following Linseed variety is suitable for rainfed?

a Himani | b Nagarkot
c RLC-92 | d All above

67 Which of the following Safflower variety is suitable for rainfed?

a Phule Kusum | b PBNS-40
c PBNS-12 | d All above

68 Which of the following Sesame variety is suitable for rainfed?

a Guarat Til-4 | b TKG-22
c Rama | d All above

69 Which of the following soybean variety is suitable for rainfed?

a NRC-7 | b Ahilya
c Indira Soya-9 | d All above

70 Which of the following sunflower variety is suitable for rainfed?

a KBSH-44 | b KBSH-53
c LSPH-35 | d All above

71 Which of the following jute (*C. capsularis*) variety is suitable for rainfed condition?

a JRC-7447 b JRC-212

c JRC-80 d All above

72 Which of the following Jute (Tossa) variety is suitable for rainfed condition?

a JRO-524 b JRO-204

c JRO-632 d All above

73 Which of the following Mesta (HS) variety is suitable for rainfed condition?

a HS-7910 b AMV-7

c HS-4288 d All above

74 Which of the following Mesta (HC) variety is suitable for rainfed condition?

a MT-150 b MT-129

c AMC-108 d All above

75 Which of the following Mesta variety is suitable for paper pulp production?

a MT-150 b HS-7910

c AMC-108 d All above

76 Which of the following Sorghum fodder variety is suitable for rainfed condition?

a HC-308 b CO (FS)-29

c Pusa Chari-23 d All above

77 Which of the following Ricebean variety is suitable for rainfed condition?

a Bidhan-1 b Bidhan-3

c Bidhan-2 d All above

78 Which of the following Pearlmillet fodder variety is suitable for rainfed condition?

a Raj Bajra-1 b BAIF Bajra-1

c GFB-1 d All above

79 Which of the following Cowpea variety is suitable for rainfed condition?

a GFC-2 b Bundel Lobhia-1

c EC-4216 d All above

80 Which of the following cotton variety is suitable for rainfed condition?

a Phule Anmol | b SVPR-4
c BN bt. | d All above

81 Which of the following cluster bean variety is suitable for rainfed conditions?

a RGC-936 | b RCG-1017
c Thar Bhadavi | d All above

82 Which of the following chemical is used for seed priming in rainfed areas?

a Sodium Polyacrylate | b Potassium Polyacrylate
c Hydrogel | d All above

83 Which of the following is used for precision nutrient management in crops?

a RS & GIS | b Artificial intelligent
c Agribotics | d All above

84 Which of the following mulch type is most effective to conserve soil moisture?

a Dust mulch | b Straw mulch
c Green leaves mulch | d Polythene mulch

85 What is the role of antitranspirants in crop production?

a Check evaporation losses | b Check transpirational losses
c Check ET losses | d All above

86 Which of the following is a film forming antitranspitant?

a Mobleaf | b Silicon oil
c Hexadecanol | d All above

87 Which of the following is a stomata closing type antitranspitant?

a PMA | b CCC
c Atrazine | d All above

88 Which of the following is a stomata closing type antitranspitant?

a Ammonium chloride | b Ascorbic acid
c Alkenylsuccinic acid | d All above

89 Which of the following is a reflectant type antitranspitant?

a Kaaoline | b Lime water
c Chalk | d All above

90 Which of the following is a growth retardant type antitranspit

a Cycoceal
b ABA
c Phosphan-D
d All above

91 Which of the following type of cropping system can be adopted if the area receives rainfall <600mm and growing period is 20 weeks?

a Monocropping
b Intercropping
c Double cropping
d All above

92 Which of the following type of cropping system can be adopted if the area receives rainfall 600-750mm and growing period is 20-30 weeks?

a Monocropping
b Intercropping
c Double cropping
d All above

93 Which of the following type of cropping system can be adopted if the area receives rainfall 750-900mm and growing period is >30 weeks?

a Monocropping
b Intercropping
c Double cropping
d All above

94 Crop diversification refers to

a Addition of new crops and cropping systems
b Addition to animals and sericulture
c Addition of pulses and oilseeds
d Growing of trees with crops

95 Diversification helps in

a Income generation
b Employment generation
c Improving soil health & biodiversity
d All above

96 Which of the activity has the highest share in agriculture GDP?

a Dairy
b Crops
c Fisheries
d Agroforesrty

97 Conservation agriculture refers to

a Minimum soil disturbance
b Maintenance of a permanent soil cover
c Diversification of plant species
d All above

98 Which of the following institute is considred to start work on conservation agriculture?

a CIMMYT, Mexico
b IRRI, Philippines
c ICAR, India
d All above

99 Alternate land use system refers to

a To switch over from non profitable activity to more profitable one or more activities

b To include dairying

c To grow more crops

d All above

100 Integrated farming system refers to

a To add dairying, poultry, fisheries etc with crop production

b To add mushroom cultivation and poultry with crop production

c To include sericulture, bee keeping with crop production

d All above

101 Which of the following is an objective of IFS?

a To produce diversified quality products

To reduce dependency of farm on 'off- farm' inputs and labour requirement

c To increase farm income

d All above

102 Organic farming refers to

a Crop production without chemical inputs

b Crop production with organic products

c Both 'a' &'b'

d None above

103 Which of the following formulation (s) is used in Biodynamic agriculture?

a BD-500

b Jivamritu

c Amritpani

d Kunapjal

104 The rainfed areas are more suitable for organic farming mainly because of

a Low application of chemical inputs

b Soils are poor by both fertility and soil moisture

c Abundance of natural resources like crop residues, livestock and its products like gobal, urine, skin etc.

d All above

105 Which of the following is used for preparation of biopesticides?

a *Azadirachta indica*

b *Pongimia pinnata*

c *Calatropis* spp.

d All above

106 Which of the following is a first organic state in India?

a Uttarakhand b Sikkim

c Assam d Meghalaya

107 Which of the country has the highest cropped area under organic farming?

a USA b Australia

c India d Brazil

108 Which of the state in India is leading in acreage under organic farming?

a UP b MP

c Rajasthan d Maharashtra

109 When was the National Rainfed Area Authority set up in India?

a 1985-86 b 2006-07

c 2014-15 d 2018-19

110 How many districts have been listed under the list of most drought prone category in India?

a 24 b 48

c 156 d 221

111 Which of the following is an organic farming practice that helps maintain soil health?

a Monocropping b Synthetic fertiliers

c Crop rotation d Sewage and sluge

112 Which of the following is not allowed in organic farming?

a Crop rotation b Monocropping

c Sewage sludge d Buffer zone

113 How many years must the field/land be treated as organic without use of chemical inputs to be legible for organic certification?

a 2 yr b 3 yr

c 4 yr d 5yr

114 IFOAM is related to

a Organic farming b Chemical farming

c Mixed framing d Intensive farming

115 Which of the country has the highest organic producers in the world?

a Australia b Argentina

c India d USA

116 What is the name of a scheme to support organic farming endeavors?

a Prayogik Krishi Vikas Yojana b Paramparik Krishikaran Yojana

c Khushal Kheti Yojana d Paramparagat Krishi Vikas Yojana

117 Paramparagat Krishi Vikas Yojana was launched in year

a 2012 b 2015

c 2016 d 2018

118 The main objective of Paramparagat Krishi Vikas Yojana is

a Promotion of organic farming b Improvement of soil health

c Both 'a' & 'b' d None baove

119 National Center of Organic Farming (NCOF) is situated at

a Ghaziabad b New Delhi

c Bengaluru d Hyderabad

120 Organic farming is important mainly due to

a Free from chemical hazards b Improves soil health

c Protect environment d All above

Answers

1. **c**	2. **b**	3. **a**	4. **b**	5. **c**	6. **d**	7. **b**	8. **a**	9. **c**	10. **d**
11. **c**	12. **b**	13. **a**	14. **c**	15. **a**	16. **c**	17. **b**	18. **c**	19. **b**	20. **a**
21. **c**	22. **b**	23. **a**	24. **a**	25. **b**	26. **c**	27. **d**	28. **a**	29. **b**	30. **c**
31. **b**	32. **c**	33. **d**	34. **c**	35 **b**	36. **c**	37. **d**	38. **c**	39. **b**	40. **d**
41. **d**	42. **d**	43. **d**	44. **d**	45. **d**	46. **c**	47. **d**	48. **d**	49. **d**	50. **d**
51.**d**	52. **d**	53. **d**	54. **d**	55. **d**	56. **d**	57. **d**	58. **d**	59. **d**	60. **d**
61. **d**	62. **d**	63. **d**	64. **d**	65. **d**	66. **d**	67 **d**	68 **d**	69. **d**	70. **d**
71. **d**	72. **d**	73. **d**	74. **d**	75. **a**	76 **d**	77. **d**	78. **d**	79. **d**	80. **d**
81. **d**	82. **d**	83. **d**	84. **d**	85. **b**	86. **d**	87. **d**	88. **d**	89. **d**	90. **d**
91. **a**	92. **b**	93. **c**	94. **a**	95. **d**	96. **a**	97. **d**	98. **a**	99. **a**	100. **d**
101. **d**	102. **c**	103. **a**	104. **d**	105. **d**	106. **b**	107. **b**	108. **b**	109. **b**	110. **a**
111. **c**	112. **c**	113. **b**	114. **a**	115. **c**	116. **a**	117. **b**	118. **c**	119. **a**	120. **d**

10

Principles of Watershed Management

Outlines

Introduction

Due to continuous pressure on our food security even after independence, the cultivation has been expanded to marginal and sub marginal lands that are prone to different types of erosion. Though the rainfed farming is based on the erratic and uneven rainfall, it contributes nearly 50% food grain, 90% oilseeds and pulses and 65% cotton production. The watershed approach was realized to conserve the natural resources like soil, water and vegetation for food and environmental security. The watershed development approach was initiated by central government during 7th and 8th five year plan on pilot basis and its encouraging results gave new direction to watershed development particularly in rainfed areas.

The ministry of Agriculture have been actively involved in planning, formulation, monitoring and reviewing of watershed based developmental project activities which includes National Watershed Development Project for Rainfed Areas (NWDPRA) and sizeable number of internationally/bilaterally aided projects.

In consonance with overall developmental objectives and strategy for holistic and integrated development of rainfed areas, a centrally sponsored scheme, NWDPRA was launched in 1990-91 and implemented during 8th FYP that developed 4.23 m ha area in 2554 watersheds against the target of 2.8 m ha. During 9th FYP (1997-98-2000-01), it was targeted to treat 2.25m ha of rainfed areas at a cost of Rs 1020 crore. A watershed Development Fund (WDF) was established at NABARD from 1999-2000 onwards with objective of integrated watershed Development in 100 priority districts through participatory approach. The WDF enables States to access credit for treatment of larger areas under watershed development. Till the Eighth Plan period, a total of 16.5 million hectares of land had been treated through the watershed approach. According to 2025-year perspective plan prepared by the Planning Commission, it was proposed to treat 10 and 12 million hectares, respectively, during the IX and X Plan period.

Definition

A watershed is a natural hydrological unit that covers a specific land surface area from which runoff passes through a well defined outlet or an area that drains water and to a common outlet. It is also knows as drainage-based catchments or contributing area (sub watershed). Soil and water are the most important natural resources within our ecosystem and watersheds have been identified as planning units for administrative purpose to conserve these precious resources (FAO, 1985 and Honore 1999). The concept of watershed management recognizes inter-relationships among land use, soil and water and the linkage between uplands and down streams areas.

The shape and size of a watershed vary with places like plain, undulated and mountains. In watershed approach, major efforts are made to find solutions of watershed problems from the watershed area itself. The delineation, runoff rate, flood volume, soil and water conservation measures, crop diversification and land productivity are the major components of watershed management. The purpose of integrated watershed management is to practice improved and scientific land management, for improving the hydrologic behavior of watershed by reducing runoff and soil loss, and increased land productivity, resulting a sustainable all round development that maintains ecological balance between land, water plant and animal Kingdom in the area. Therefore, watershed management particularly in dry lands is regarded as holistic approach towards optimizing the integrated use of land water and vegetation and also the answer to natural resource conservation, prevention of flood and runoff, improve food, fuel, fodder and fibre and open new alternate land use practices and other allied agricultural activities like livestock production, sericulture, fisheries, mushrooming etc., for improving the health and economic status of the society.

Watershed Management : A Holistic Approach

The case studies of Operational Research Project (ORP) on watersheds at Fakot (Himalayan region) and 'Sukhomajiri' (Shivalik region) have shown that integrated approach is economically beneficial.

These success stories paved the way for launching country wide watershed management in different agro-ecological regions for scientific management of resources in an integrated manner with technical back up from National Agricultural Research Systems (NARS) and large scale involvement of the local people and voluntary agencies in their implementation.

Thus, watershed based development has now been accepted as a strategy for natural resource conservation for improving the crop production on a sustainable basis and restore ecological balance particularly in rainfed and mountainous areas.

Concept of Watershed Management

Watershed management in fact is not a new concept rather it is a new strategy of land and water management considering watershed as a unit for proper planning and execution of watershed management programs for sustainable development of the area and their residents. It is a dynamic system where the activities are integrated for synergy and multiple effects. Normally participatory approach is followed for most of the activities. The efficient management of natural resources at watershed mode is relevant both to low rainfall or rainfed areas as well as high rainfall areas like hills. Under low rainfall areas, main emphasis is given to control of wind erosion and conservation of water, where as, control of water erosion, land slides and safe disposal of run off are the major threats in high rainfall areas.

The ultimate aim of watershed management is to improve the quality of local residents through efficient natural resource management. Therefore, the sustainable development of the area or a watershed, meeting people's needs and arresting environmental degradation are the major thrust of watershed management programs.

Growth of Watershed Management Program in India

The brief history of watershed development is given below;

i. **In 1880**: Famine commission was constituted in India after series of famines during 1960s to 1970s with objectives to suggest the measures to avoid the famines. On the basis of its report, few irrigation projects were taken up particularly in Madras, Bombay and Mysore states,

ii. **In 1923:** Dr. H. H. Mann, Director of Agriculture, Bombay state and Mr. C. V. Mehta, Minister of Agriculture, Govt. of India made a strategy for development of rainfed agriculture. Finally, the work on dry farming was started under the leadership of Mr. V. A. Tamhane, soil physicist at 'Manjri farm'. Later this work was passed on to Mr. N. V. Kanitkar in 1926.

iii. **In 1928**:The Royal Commission on Agriculture highlighted importance of moisture conservation for rainfed/dryland areas.

iv. ***In between 1933 to 1935***: Five research centers on dry farming research were opened including *Bijapur* and *Solapur* in 1933, *Hagari* and *Raichur* in 1934 and *Rohtak* in 1935.

v. **1943:** Bombay Land Improvement Act was passed and it served as a model for country when field bunding/contour bunding works were taken up at large scale in Bombay Province for rural development during drought relief operations.

vi. **In 1949-50:** A multi-disciplinary soil conservation department was set up at Hazaribagh under Damodar Valley Corporation for watershed Management and land reclaimation.

vii. **In 1953:** Central Soil Conservation Board was set up and the emphasis was shifted from improving the crop yield to soil conservation and management. Later 8 soil conservation research centres were established at *Dehradun, Agra, Chandigarh, Kota, Vasad, Bellary, Ibraheempatnam* and *Ooty*. The reports of these centres emphasized on contour bunding and 42 small watersheds were identified for monitoring. The impact of conservation measures and impact of biotic and abiotic interferences on surface hydrology and soil erosion were discussed.

viii. **In 1960-62**: Centrally sponsored Watershed Protection Program in catchments of River Valley Projects was launched.

ix. **In 1980-81:** 'Flood Prone Rivers Projects' was launched.

x. **In 1974:** Four ORPs on watershed management were adopted by CSWCRTI, Dehradun at;

1 Sukhomajri (Foot hills of Shiwalik hills Haryana)

2 Fakot in middle Himalaya (U.P.)

3 Chitradurga in semi-arid region (Karnataka) &

4 Mahendragarh in Aravalis region (Haryana)

The case studies of Operational Research Project (ORP) on watersheds at Fakot (Himalayan region) and 'Sukhomajiri' (Shivalik region) have shown that integrated approach is economically beneficial. These success stories paved the way for launching country wide watershed management in different agro-ecological regions for scientific management of resources in an integrated manner with technical back up from National Agricultural Research Systems (NARS) and large scale involvement of the local people and voluntary agencies in their implementation. Thus, watershed based development has now been accepted as a strategy for natural resource conservation for improving the crop production on a sustainable basis and restore ecological balance particularly in rainfed and mountainous areas.

xi. **In 1983**: another 47 ORPs on watershed management were started jointly by CSWCRTI, Dehradun & CRIDA (Central Research Institute for Dryland Agriculture), Hyderabad.

xii. **In 1986**: National Watershed Development Project on Rainfed Agriculture (NWDPRA) was launched in 16 states with following broad objectives;

1 Conservation, development and sustainable management of national resources including their use,

2 Enhancement of agricultural productivity and production in a sustainable manner,

3 Restoration of ecological balance in the degraded and fragile rainfed ecosystems by greening these areas through appropriate mix of trees, shrubs and grasses,

4 Reduction in regional disparity between irrigated and rainfed areas,

5 Creation of sustained employment opportunities for the rural community including land less, and

6. Uplift the living standards of masses by increasing net income/annum

xiii. **In 1987**: Drought proofing potential of watershed management approach was realized.

xiv. **In 1990-91**: NWSPRA was launched in 25 states and 2 UT's and will continue in 9th and 10th FYP.

xv. Recently, the Watershed programs like 'Rolegaon Sidhi (Ahmednagar)' and 'Tarun Bharat Sangh (Alwar)' led by Shri Anna Hazare and Shri Rajendra Singh, respectively came in to light for good work.

Principles of watershed management

The watershed is a unit for not only planning but also for all round development of the area. Therefore, the basic principles of watershed management are based on resource conservation, resource generation and resource utilization as given below;

- To utilize the land resources based on its capacity,
- To protect fertile top layer of the soil,
- To minimize silting up of tanks, reservoirs and lower fertile lands,
- To maintain the vegetative cover throughout the year,

- To conserve '*in situ*' rain water,
- To divert surface runoff safely to storage structures through grassed water ways,
- To stabilize the gullies and construct check dams for increasing ground water recharge,
- To increase cropping intensity through inter and sequence cropping,
- To adopt alternate land use systems for efficient use of land resources,
- To encourage water harvesting for supplemental irrigation,
- To ensure sustainability of the ecosystem,
- To maximize farm income through agricultural related activities such as dairy, poultry, sheep & goat farming,
- To improve infrastructural facilities of storage, transport and agricultural marketing,
- To set up of small scale agro-industries, and
- To improve socio-economic status of farmers.

Objectives of watershed management

The main objectives of watershed development program are listed below;

1 To control damaging runoff,

2 To manage and utilize run off for useful purposes,

3 To control erosion and effect reduction in the sediment production,

4 To moderate floods in the downstream areas,

5 To enhance ground water storage wherever applicable, and

6 Appropriate use of the land resources in the watershed and thus developing forest and fodder resources.

Singh (1995) described the objectives of watershed in symbolic form, 'POWER'. The letters symbolize as follows;

Where,

P = Production of food-fodder-fuel-fruit-fibres-fish-milk combine on a sustained basis

- Pollution control, and
- Prevention of floods,

O = Over exploitation of resources to be minimized by controlling excessive biotic interferences like over grazing, and

- Operational practicability of all on-farm operations and follow-up programs including easy approachability to different locations in watershed.

W = Water storage at convenient location for different purposes, and
- Wild life and indigenous plant life conservation at selected places.

E = Erosion control Ecosystems,
- Eco system safety,
- Economic stability, and
- Employment generation.

R = Recharge of ground water,
- Reduction of drought hazards,
- Recreation, and
- Reduction of siltation in multi-purpose reservoirs.

Benefits of watershed management

1 Improved soil health,
2 Better drainage,
3 Efficient use of rainwater – improve WUE
4 Sustainable improvement in crop and animal production,
5 Protect down stream farm lands and human habitation from floods,
6 Reduces siltation of expensive irrigation structures
7 Protects natural resources,
8 Reduces water & wind erosion,
9 Increases farm profits, and
10 Improves living standard of the local people.

Delineation of watershed

The commonly used terms like primary river, secondary river, tertiary river and similarly catchment, watershed, sub-watershed, micro watershed etc. are required to define for better understanding. Normally, the watersheds are delineated from the ridge lines that join the highest points of the watersheds and the drainage lines join the streams. There is no land common between two watersheds. The procedure of delineating watershed is very flexible and the area of a watershed, sub watershed or even micro watershed may vary from place to place. The watershed projects sanctioned under Drought Prone Area Program (DPAP) and Desert Development Program (DDP) have area of 500 ha each, while the projects sanctioned under Integrated Watershed Development Program (IWDP) cover an area of 5000 to 6000ha (NPCM Team, 2009). The above terms have been defined and also delineated (Panda, 2003) as given below;

1. ***Primary River*** - *It represents a big river with a catchments of several lakh acres (several 40,000 ha). It has number of tributaries of* 1st, 2nd, 3rd, 4th, 5th *etc order. It discharges its load to sea.*

2. ***Secondary River*** – *It is a tributary of a primary river having a drainage area of more than a lakh acre or 40,000 ha.*
3. ***Tertiary River/Stream*** – *It is tributary of secondary river having an area of 10,000 to 1,00,000 acre (4000 - 40,000 ha).*
4. ***Quaternary River/Stream*** – *It is tributary of a tertiary river having a drainage area 5 to 10,000 acre (2,000 – 4,000 ha).*
5. ***Pentad River/Stream*** – *It is a tributary of a quaternary river with an area of 5,000 to 10,000 acre or 400 – 2,000 ha.*
6. ***Hexad River/Stream*** – It is a tributary of a pentad stream having a *drainage* less than 1000 acre or 400 ha.

Therefore, on the basis of above classification of rivers, the terminology related to the watershed may be correlated as follows;

a. ***Catchment*** – It represents the drainage area of a primary river covering an area of few lakh acres or more.
b. ***Sub-catchment*** – It is the drainage area of a secondary river having an area of more than a lakh acre or 40,000 ha.
c. ***Watershed*** – It has an area of 10,000 to 1, 00,000 acre or 4,000 to 40,000 ha and commonly represent a drainage area of a tertiary river/stream.
d. ***Sub watershed-*** It is drainage area of a quaternary river/stream with an area of 5,000 to 10,000 acre or 2,000 to 4000 ha.
e. ***Mini watershed-*** It represents a drainage area of a pentad river/strean with an area of 1,000 to 5,000 acre or 400 to 2000 ha.
f. ***Micro watershed-*** It is a drainage area of a hexad covering an area of less than 1000 acre or 400 ha.

However, the delineation of a watershed based on fixed area is difficult because of the topography of the terrains and natural boundaries. For the convenience, the watershed may be grouped as given below;

i. Macro watershed - 25,000-30,000 ha
ii. Medium watershed or sub watershed - 5,000-6,000 ha
iii. Micro watershed - 500 ha

The sub watersheds are divided based on sediment yield, run off potential index, degree of land degradation, underground water status etc.

Major components of watershed development

A. *Water resource improvement*

Most of the dry lands suffer water stress due to erratic and uneven rainfall pattern over the cropping period. The availability of water in a particular watershed varies according to area, topography, soil structure and texture, rainfall pattern and existing land use pattern. The rainwater is very limited and so its proper use is to be ensured during rainy season through following techniques;

1. **Water harvesting structures:** Depending on soil topography and rainfall pattern of a watershed, the water harvesting methods are used to collect the surface-runoff water during rainfall. The surface water ponds across the flow of runoff as embankments in uneven and dig outs in flat topography are made to check run-off and also store excess water for reuse during dry spells.
2. **Judicious use of stored rainwater:** The water stress is a common feature of dry land agriculture. The stored water is used judiciously for life saving irrigation to crops during severe dry spells.

B. *Soil and water conservation measures*

Soil and water conservation measures help to improve soil moisture. The contour bunds, graded bunds, bench, zing and broad base terracing, ditches etc. (*hardcore treatments*) and compartmental bunding, broad bed and furrow system, dead furrows, contour farming, strip cropping across the slope, mulching etc (*software treatments*) are important soil water conservation measures that are adopted according the soil type, topography, rainfall pattern and also socio-economic status of a particular watershed area.

C. *Alternate Land use systems*

Efficient land use for generation of income at minimum risk is a prime goal in dry land areas. The marginal and sub marginal, eroded and uneven land that is not fit for crop production, may be used for alternate systems like agro-forestry, agro-silvipasture, agro-horticulture systems etc. for better land use. The tree species should be chosen for its suitability to a the particular region/watershed conditions.

Watershed Development Planning

The areas situated beyond the reach of conventional irrigation and subjected to soil and water erosion and moisture stress should be developed on watershed basis. In high rainfall areas, the main objective of watershed development is to control soil and water erosion, while in dry region, to conserve the rain water for crop and drinking purpose. It is an integrated program, so proper help and coordination are required not only from different faculties but also from local farmers. The following developmental objectives should be considered while planning for a watershed development program:

1. To improve land productivity through modern technologies,
2. To ensure proper soil and water conservation measures for achieving sustained productivity,
3. To adopt water harvesting techniques for life saving irrigation in dry spells,
4. To ensure the availability of safe and drinking water for the villagers,

5. To increase the crop productivity per unit land per unit water or per unit time.
6. To generate employment to local inhabitants,
7. To increase income/head/day, and
8. To improve socio-economic status of the farming communities of the watershed area.

The agro-physical and also hydrological conditions may differ with agro-climatic regions, so the planning is to be made well in advance considering the needs and the capability of the participants. After deciding the area and its priority, an action plan is prepared as well as harness maximum natural resources on watershed basis. The following points need to be evaluated critically before implementing the action plan;

1. The soil movement should be minimized during mechanical soil conservation measures,
2. Cost of each component should be minimized without sacrificing the developmental need,
3. The infrastructure like, roads, sheds etc need to be created,
4. The farm ponds and other water harvesting measures should be adopted depending upon the situation,
5. The suitable cropping plan should be chalked out depending on local needs and also on land capability classes, and
6. The availability of budget should be worked out for maintenance work of soil and water conservation works.

Elements of Watershed Development Program

Watershed management essentially consists of 3 elements of operation;

(a) Basic land treatment to conserve soil moisture control, land erosion and harvest runoff,

(b) Application of crop and animal management practices, and

(c) Alternate land use such as agro-forestry, social and silvi-pasture system.

(1) Land treatment

a. Soil moisture conservation,

b Control erosion,

c Harvest run-off water into form ponds, small tanks or small dams, and

d Recharge ground water down stream.

In fact, these are soil – engineering measures for safe disposal of water through diversion channels and water ways.

A. For arable lands

- Leveling,
- Trenching (contour, staggered, vert. mulch),
- Terracing,
- Vegetative field boundaries (Live bunds), and
- Vegetative barriers.

B. Conservation work for non-arable lands

a. Gully control (cut down run off velocity)

i) Vegetative treatment (Soil binding grasses/shrubs).
ii) Vegetative check dams.
iii) Dry stone check dams.
iv) Woven wire check dams.
v) Crate wire check dams. and
vi) Cross barrier/weir

b. River/stream bank protection

i) Spurs, and
ii) Side wall (Masonry/crate wires)

c. River Training

i. Structures to deflect stream, and
ii. Wing wall

d. Road side erosion control – Retaining wall

e. Land slide treatment

i. Check dam, and
ii. Retaining wall.

In other words, following ***technological options*** are available for soil and water conservation in different types of soil is given below:

(a) Near flat or valley lands

i) Rainwater harvesting from mini-watersheds, broad & furrow systems with grassed waterways in black soils. In flat black soils with < 0-5% slope, drainage can be practices through narrow ridge and furrow and raised beds during rainy months. It requires community participation on whole village basis.

ii) Off season tillage, dry seeding and fertilizer placement are accomplished by animal drawn farm implements or tractor drawn, if possible.

ii) Double cropping is possible by harvesting of rain water. Crops can be raised in the furrows with stored moisture during the post-monsoon season.

(b) Foot hills – Submontane tracts with light textured soils

i) Contour farming and bench terracing at suitable intervals supplemented with organic mulches,

ii) In situ water harvesting – reduce erosion,

iii) Gully plugging or putting across check dams in upper parts of seasonal cracks increases the recharge of ground water in the down stream basis, and

iv) Horticultural & silvi-pastoral systems are recommended for higher sections.

(c) Steep slopes

(i) Contour bunding, bench terracing or a combination of both may help in retaining 90 to 100 % rainfall in the treated area of steeper sloped. The vertical intervals between contour bunds or bench terraces are kept within 0.5 - 1.0 m for operational convenience. Contour benches along the bunds or terraces carry excess flow through grassed waterways into a common pond.

(ii) A live hedge planted just above the bench acts as a soil filter preventing soil erosion and with time makes the terraced bed less sloppy.

(iii) For still steeper slopes with soil depth up to 1m, horticulture is the most appropriate form of agricultural activity. Hill tops are to be used for forestry purposes. Such a system allows retention of almost 90% of rainwater within the watershed reducing soil loss to less than 1.0 t/ha/year.

(2) Crop & Animal husbandry practices

- Contour tillage,
- Selection of quick growing crops,
- Inter cropping and cropping systems,
- Flat sowing followed by ridging up is a good practice to prevent soil erosion in rainy season.
- Bed and furrow system of planting is more appropriate in black soils,
- Use of organic mulches in widely spaced crops is often desirable and
- Cattle rearing, fisheries in reservoirs, poultry etc are beneficial enterprises.

(3) Alternate land use systems

- Afforestation,
- Agro-forestry,
- Silvi-pastoral system,
- Agro-silvi-pastoral and horti system,
- Energy plantation (Fuel wood) and
- Ley farming /pasture.

Watershed Work Plans

The work plan of any project refers to the written information related to the project area, its problems, the strategies to follow to solve the problems, materials and methods to be adopted, estimated cost, possible outcomes, evaluation and its indicators, modifications in strategies and technologies if required and post project evaluation. Therefore, the work plan of a watershed program describes about the watershed, its problems and the sequence of events to be followed during the implementation of the program. Since the problems of one watershed differ with other one, therefore, each watershed has its unique work plan. In general, the work plan may answer following questions;

1. What is the name of watershed?
2. What are the problems of watershed?
3. When to implement watershed?
4. What are the pre requisites for implementation of watershed management program?
5. What are the strategies/technologies to be adopted in the watershed management program?
6. Who will implement the watershed?
7. Who are the stake holders of the watershed program?
8. What is the cost of watershed program?
9. Who will evaluate the progress of the watershed project?
10. What should be the indicators for evaluating the progress of the watershed project?
11. What will be the impact on soil and water conservation, water harvesting structures, crop production, people needs, income, employment, environment etc.?

Tamthane (1967) described following contents in work plan of a watershed project;

1. Description of watershed

i. Location,

ii. Size and shape,

iii. Climate,

iv. Geology,

v. Slopes,

vi. Surface drainage,

  a. Nature of flow

  b. Drainage net

c. Morphology of streams

d. Soils

e. Physiology

f. Water used and needs

vii. Land use and cover conditions and

viii. Economic data,

2. Watershed problems

i. Flood water damage,

ii. Sediment damage,

iii. Erosion damage,

iv. Water management problems,

v. Financial problems,

vi. Political problems and

vii. Socio economic problems

3. Proposed Management Program: different treatment measures are described in this section.

a. ***Agricultural lands***: recommendations may be given on land capability classes.

i. *Agronomic practices*: crops, rotations, fertilizers, chemicals, seeds, implements, plant protections measures, yield and economics.

ii. *Engineering measures*: proposed measures, its justifications, specifications, with plans and design, calculations and cost.

b. ***Conservation irrigation:*** Selection of irrigation methods, their specification, water use efficiency, duration, water requirement and cost.

c. ***Protection against flood***: Channel improvements, flood retarding structures, their locations, design and cost.

d. ***Forest land:*** Protection of existing forests, degraded forests, shrubs, range lands and farm forestry. Engineering measures may be adopted if required.

e. ***Grazing lands***

i. ***Pastures:*** Species, growing techniques, techniques for maintenance of pastures.

ii. ***Grazing lands:*** Improved grazing systems and their management.

f. ***Road Construction:*** Layout of roads, protection of roadside from erosion etc.

g. ***Special problems:*** Land slides, stream bank erosion, gullies and ravines control etc.

4. Enhanced benefits or Effect of Works of Improvement

Under this section, the enhanced benefits of watershed management programs are discussed. The benefits may be in form of

i. Increased crop production,
ii. Flood control,
iii. High water use efficiency,
iv. Reduction of sediment loads,
v. Increased forest produce,
vi. High income/capita,
vii. High living standards of residents,

5. Comparison of benefits and costs*:* The benefit: cost ratio, gross income, net profit etc. are calculated.

Integrated Watershed Management

Integrated watershed management refers to the integration of all possible advanced technologies for sustainable development of the watershed. The sustainable development of an area at watershed mode may help to improve the socio-economic and political status of the local residents with increased income/head. Therefore, the basic objectives of integrated watershed management are listed below;

1. To conserve the basic resources of soil, rain water and vegetation,
2. To achieve higher biomass production both in arable and non arable lands,
3. To stabilize the crop yield through efficient use of rain water,
4. To reduce the dependency on monsoon,
5. To reduce risk in crop production,
6. To enhance income/head,
7. To restore ecological balance,
8. To improve socio-economic status of the people.

The recent advances in science and technology like use of *pressurized irrigation, zero tillage, new farming systems, application of remote sensing and GIS*, release of *new hybrids, genetically modified crops, laser leveling, new farm implements* etc. may improve the sustainability of the watershed. Similarly, the assistance rendered by World Bank, Asian Bank, UNDP, IWMI, Govt's of Netherland, Germany, Canada etc. would boost up the watershed development programs in India.

The conventional approach of watershed development has changed under integrated watershed development program (Dube, 1999). Under the new approach, the watershed development program is based mainly on demand of the local people and is executed on rural participatory appraisal mode. The watershed development program are executed either by Government departments or NGOs.

Internationally Aided watershed Projects

(A) World Bank aided watershed

1. **Integrated watershed Development (Hills-II) Project (IWDP-Hills-II)-** It was launched in 1999 with objective to restore the productive potential of Shivalak hills in Haryana, H.P, J&K., Punjab and Uttarakhand using cost effective measures and community participation approach.
2. **Assam Rural Infrastructure and Agriculture Project** – It was implemented in 1995 with objective to alleviate poverty and accelerate agriculture growth through improved use of resources.
3. **Diversified Agriculture Support Project-** It was launched in 1998 in UP with objective to increase agricultural productivity, promote private sector development and improve infrastructure. Presently, it is operational in U.P. & Uttarakhand states.
4. **Comprehensive watershed development Project (Karnataka)-** It was implemented in 2001.

(B) German Assisted Project

Watershed Development Project in Maharashtra- It was launched in December 1991.

(C) U.K. aided Project

Karnatika watershed Development Project (KAWAD) It was started in 1998 in three districts of Karnatika ie. Bellary, Bijapur and Chitradurga supported by Department for International Development (DFID).

Table 1. Comparison of conventional and new approaches to watershed development (Dube, 1999)

S.No.	Conventional approach	New approach
1	Executing driven agency	Participatory, people driven
2	Target based	Demand driven/process based
3	Aimed only at soil water and vegetation conservation	Aimed at poverty alleviation & over all human development
4	Transfer of technology extension method	People first approach, dovetailed to TOT approach
5	Based on imported technology & ideas	Based on indigenous technology traditions & culture and cosmic vision of people
6	Top-down planning monitoring a evaluation	Bottom-up (Participatory) planning, monitoring & evaluation
7	Land use based on land capability	Based on land suitability and people's needs and preferences

8	Empowered the agents of technology transfer officials	Empowered people
9	Attended to select, generally better-off farmers	Aimed at marginal, small a poor farmers with special emphasis on equity
10	Involvement of single department	Multi-departmental & multi-disciplinary
11	Villagers not empowered	Villagers empowered
12	Based on large watersheds	Based on small watershed with people participation

Guidelines for Watershed Development

The Ministry of Agriculture and Ministry of Rural Development have jointly formulated and adopted common approach for watershed development, incorporating strengths of their earlier watershed based programs. In the guidelines, due emphasis has been given to following;

1. Conservation and integrated development of natural resources,
2. Moisture conservation,
3. Sustainable farming systems,
4. Adoption of ridge to valley approach,
5. Demographical decentralization in decision making,
6. Livelihood support system for landless families,
7. Community organization and its empowerment, and
8. Production enhancement activities for land owners.

Watershed Development Activities may be grouped into various phases as given below;

a. *Project activities-Preparatory phase,*
b. *Capacity building phase,*
c. *Community organization phase,*
d. *Planning phase,*
e. *Implementation phase,*
f. *Monitoring and evaluation phase,*
g. *Success criteria, and*
h. *Post project activities-Withdrawal phase.*

Preparatory phase

Before implement of the watershed project, following activities are to be carried out in this preparatory phase;

i. *Identification of watershed implementing institutions*: A suitable Project Implementing Agency (PIA) and a District Nodal Agency (DNA) are identified by promoting healthy competition among various agencies. Apart from the above agencies, District Watershed Committee (DWC), Watershed Development team (WDT), Watershed Association (WA), Watershed Committee (WC), Self Help Groups (SHGs), User Groups (UG) etc. are also constituted.

ii. *Identification and prioritization of watersheds*: The watershed are identified and demarcated on map with the help of topography maps. The macro watersheds cover the area of 25000 to 30000 ha. These macro watershed are again divided in to sub watershed with 5000 to 6000 ha each. The prioritization of sub watersheds is based mainly on sediment yield, run off potential index, degree of land degradation, underground water status etc. The sub watersheds are again divided in to micro watersheds having an area of 500 ha each and its basis of prioritization remains same as in case of sub watersheds.

iii. *Identification of Villages having Prioritized Watersheds*: The villages are identified for implementing the watershed programs. The identified village should have ≥50% arable land under private owners, pre ponderance of resource poor SC/ST, severity of land degradation and willingness of villagers to participate and contribute in the program.

iv. *Project Commencement date*: Normally a date 3 months from the date of nomination of PIA is considered the date of commencement of the watershed project.

v. *Demarcation of Watershed Boundaries at the Field Level*: After selection of villages, the boundaries of micro watershed are demarcated.

Capacity Building Phase

In fact, the various personnel engaged in the watershed project are trained about the technical know how of the watershed program particularly for proper planning, execution and maintenance of the watershed projects. Therefore, following activities are important in capacity building phase;

1. *Different categories of personnel engaged in watershed development program*

Capacity building is a very important component of watershed development program for all categories of persons associated with it. There are three categories of personnel associated with the watershed programs namely 1. Administrators and managers, 2.Implementors, and 3. Trainers. They are located and functioning at different levels namely national, state and district and micro watershed.

a. **Administrators and Managers**: They are the officers from various organizations like Ministry of Agriculture at national level, Department of Agriculture, Soil and Water Conservation at State and district levels, heads of PIA and heads of the registered societies at micro watershed levels.

b. **Implementers**: They are the members of WDT belonging to respective PIAs. Office bearers of registered societies as well as para-professionals at the micro watershed level, other participating members, land owners and land less families in the micro watersheds.

c. **Trainers**: The trainers are full time faculty members of the identified training institutes in the public and NGO sector and also part time resource persons who are identified as trainers.

2. Decide courses for various personnel,

3. Preparation of short and long term training plan

In phase I, two types of courses are arranged 1.short term course is arranged for nodal officers at national and state levels and 2. long term courses are arranged for skill enhancement of internal and external trainers. These courses are organized in regional and state levels reputed institutes. In phase II, the short term course are organized for nodal officers from district levels officers of different departments and heads of PIAs from concerned micro watersheds but long term skill enhancement courses are carried out for members of WDT. In Phase III, the short term course are arranged for honorary members of the registered societies of the watershed level and long term course for paid members of the societies, secretary and volunteers of watershed committee, book keeper of SHGs, User Groups and para- professionals of the same and nearby villages.

4. Identification of training organizations: The training organizations are identified for capacity building.

5. Establishment/Identification of support organizations, and

6. Establish the net working of training organizations for capacity building.

Community Organization Phase

In this phase, the awareness program about the watershed project is organized in the watershed area so that the villagers may have idea about the benefits of watershed development project. Different organizations are formed and also the training program for different organizations is also organized. Therefore, following activities are performed under community organization phase;

1. *Mobilization of community and awareness building,*
2. *Release of 3% fund for early point activities (EPA) to WA/WC,*
3. *Identification of village level community organizations,*
4. *Organization of different groups at village level like SHG, UG, WA and WC,*
5. *Constitution of watershed association and watershed committee,*
6. *Identification of watershed secretary and volunteers,*
7. *Management of watershed committee office,*
8. *Training of watershed secretary, volunteers and members of watershed committee,*

9. *Balanced empowerment of WC and WA,*
10. *Making linkage of watershed institutions with Panchayat Raj, and*
11. *Integration of social resource management with natural resource management.*

Planning Phase

Following activities are carried out in the planning phase;

1. *Collection of first hand general information related to soil, water,* conservation measures, socio-economic status, production, marketing etc. with the help of participatory rural appraisal exercises,
2. Allocation of watershed budget for different activities,
3. Utilization pattern of development fund,
4. Selection of technological options available i.e. for soil and water conservation and production programs,
5. Testing and demonstration of new and sustainable technologies,
6. Adoption of proven and sustainable technologies,
7. Convergence of on going government production programs,
8. Technology management aspects for improving the income and livelihood of rural landless families,
9. Preparation of strategic plan for watershed development,
10. Approval of strategic plan year wise, and
11. Final approval of annual action plan by the District Nodal Agency.

Implementation Phase

The implementation phase includes following major activities;

1. Opening of watershed project account,
2. Setting up of watershed development (corpus) fund,
3. Decide the financial powers of different organizations,
4. Release of watershed budget,
5. Provision for flexibility in re allocation of funds for different activities,
6. Adoption of participatory approach in implementation phase by watershed committee,
7. Maintenance of records and accounts, and
8. Introducing management flexibility during implementation.

Monitoring and Evaluation Phase

Time to time monitoring and evaluation of watershed development program is necessary to achieve desired objectives and it should be continued from implementation phase to the termination phase. It is also required for some corrective measures during the course of project because of some specific problems of the watershed. In general, following activities are performed during this phase;

1. *Formation of monitoring and evaluation committees*

Following four committees review and monitor the activities of watershed development programs;

   1. National Watershed Committee,
   2. State Watershed Committee,
   3. District Watershed Committee, and
   4. Watershed Committee.

2. Monitoring of progress at district level,
3. Monitoring of physical and financial progress,
4. Concurrent evaluation by internal as well as external agencies,
5. Application of information technology in watershed program,
6. Auditing of accounts on regular basis, and
7. Evaluation of the project and documentation of the process.

Success Criteria Phase

In this phase, the criteria of success for different categories of work are chalked out. The criteria may be of either measurable and quantitatively or qualitatively for evaluation of their success. The criteria should be minimal but they must indicate the performance standards. The state Govt. may supplement these with more success criteria if found desirable or prescribe higher performance standards. The proposed criteria may be of 5 categories as given below;

1. Success criteria for physical development,
2. Success criteria for financial management,
3. Success criteria for human capital development,
4. Success criteria for social capital development, and
5. Success criteria for post project sustainability

Besides, the format of criteria for maintenance of records and supervision of quality work of watershed committee are also need to be developed for better management of different records.

Post project Activities: Withdrawal Phase

1. After the termination of the project, most of the activities are supposed to be completed. If not, extra time may be given with prior permission to complete the activities,

2. The left out project amount should be returned to the funding agency or it can be utilized for development activities not for administrative purposes,
3. All community assets created during the project period should be transferred to watershed association (WA) that have the right to maintain all the assets after project work,
4. If watershed association is no more functional, the community assets are to be transferred to village panchayat,
5. The watershed development fund (corpus) should be utilized for sustainable development,
6. Redeployment of experienced WDTs and office bearers of WC, if required for taking up new micro watersheds in the area.

Impact Evaluation

The evaluation of watershed management programs in 10 major agro-ecological regions of the country indicates that 10-80% soil loss and 2-42% run-off reduced and productivity of arable lands increased by 4.2-15.4 q/ha due to better in site moisture conservation in different regions. The casual employment opportunities also increased @ 215 man days/ha/year in post intervention period. A recent review (Samra, 1997) indicated that integrated development of watersheds moderated flooding of down streams areas and improved in site moisture conservation for increased biomass production (Table.2). Besides, the groundwater recharge capacity and water table increased significantly in different region of the country (Table.3). The integrated approach was observed economically viable with tremendous environmental externalities.

Table 2. Impact of integrated management practices on run-off (flood moderation) and soil loss in different watersheds.

Water sheds	Region	Rainfall (mm)	Run-off (%)		Soil loss(t/ha)	
			Pre treatment	Post treatment	Pre treatment	Post Treatment
Fakot, Uttaranchal	Himalaya	1900	42.0	14.2	11.9	2.5
Bajar-Ganiyar, Haryana	Aravati	495	7.3	3.5	-	-
Behdala, H.P.	Shivaliks	1820	30.0	15.0	12.0	8.0
Una, H.P.	Shivaliks	800	30.0	20.0	12.0	10.0

Source: Samra (1997), Samra and Narain (1998)

Table 3. Effect of watershed management strategies on ground water recharge in different regions.

Watershed	**Surface storage capacity created (ha m)**	**Observed rise in ground water table (m)**
Bajar-Ganiyar Haryana	79.0	2.0
Behdula, H.P.	18.0	1.0
Bungu, Haryana	60.0	1.8
Sika, Haryana	42.2	2.0

Source: Samra (1998).

Watershed Implementing Institutions/Agencies/Units

a. **District Watershed Committee (DWC):** This committee may be chaired by District Collector or Chairman of District Panchayat/Zila Parishad. The committee is supposed to review the progress of the watershed program, resolve administrative and management problems, guide in implementation, identify policy issues etc. The head of District Nodal Agency (DNA) shall be the convener of this meeting. This committee approves the selection of PIA and watersheds.

b. **District Nodal Agency (DNA):** It is consisted of members from different departments at district level like agriculture, soil conservation, watershed development, land Development Corporation etc. Some other institution like 'Agricultural Technology Management Agency (ATMA)' or some other agencies may also be given responsibility of handling watershed project in the given area. For this purpose, a committee shall be constituted under the chairman of District Head of the Nodal Department. The committee will consist of the following members;

   a. 4 representatives from major line departments in the district,
   b. 1 representative from the NGO (to be rotated on annual basis),
   c. 1 representative from District Rural Development Agency (DRDA),
   d. 1 representative from nearby research organization,
   e. 1 representative from rural engineering department/Public Works Department (PWD)/Irrigation department etc.

   This committee is a sub committee of the District Watershed Committee and will be known as District Nodal Agency. This committee will carry out over all management and supervision of the program including activities like identification of suitable PIAs, selection of new watersheds, office bearers of watershed committee, Watershed Association, training part etc.

c. **Project Implementing Agencies (PIA):** The watershed projects are implemented by District Rural Development Agencies/Zila Parishads (DRDA/ZPs) through Project Implementation Committee (PIA). The PIAs could be a Line Department (of the state government), Panchayati Raj or a reputed NGO. The PIA has a technical team of 4 experts called, '*Watershed Development Team (WDT*). The PIAs provide necessary technical guidance to the Gram Panchayat for preparation of Development plans for the watershed through Participatory Rural Appraisal exercise, undertake community organization and training for the village communities, supervise watershed development activities, inspect and authenticate project accounts.

d. **Watershed Association (WA):** This association refers to the local people living in the watershed area. The watershed association planned and execute the program.

e. **Watershed Committee (WC);** The watershed association elects the executive committee that has sole power to execute the program.

f. **Watershed Development Team (WDT**): It has at least one woman member, one sociologist, three graduate members from agricultural/civil engineering, agriculture, forestry, animal sciences etc. the retired government officers may also be considered for membership.

g. **Self Help groups (SHGs):** The group of 15-20 is selected from landless or underprivileged community. This group is trained for different activities of the program like management of credits, thrift activity, preparation of proceedings, maintenance of records, etc.

h. **User Groups (UG):** The user groups include the members who are land owners within the identified watershed area. These groups are also trained for different watershed activities like SHGs.

Major steps involved in planning and management of watershed

Following steps are involved in efficient planning and management of watershed program;

- Characterization of soil,
- Characterization of climate,
- Land capability class,
- Land use system (LUS),
- Vegetative cover,
- Cropping system,
- Extent of erosion,
- Vegetative and engineering measures being taken for Soil and water conservation,
- Vegetative and Engineering measures required,
- Land use survey and cropping systems,
- Suggested crop management technology,
- Water harvesting measures needed,

- Suggested Agro-forestry, Horticulture, fisheries and animal husbandry programs,
- Scope of income generating activities,
- Socio-economic survey,
- Resource inventory,
- People's perceptions, and
- Fixation of priorities (Matrix score).

Questions

a. Short and Descriptive

1. i. Define watershed.

   ii. What is the concept of watershed management?

   iii. What is the main principle of watershed management?
2. Discuss the holistic approach of watershed management. Also describe the objectives of the watershed management?
3. What do you mean by delineation of watershed? Discuss the criteria of delineation of watersheds?
4. Discuss the growth of watershed management program in India. Also throw light on National Watershed Development Program on Rainfed areas?
5. Briefly describe the major components of watershed development program?
6. What are the major problems in watershed management?
7. What are the technological options available for watershed management?
8. Discuss about the watershed implementing institutions/agencies and their major functions?
9. Write brief notes on following;
   a. Project Implementing Agency
   b. Watershed Committee
   c. Watershed Association
   d. IWDP
   e. DRDA
   f. NWDPRA
   g. Tarun Bharti Sangh
   h. Rolegaon Sidhi
   i. Alternate land use systems
   j. Sukho Majiri
   k. Micro watershed

10. Justify the statement that watershed approach is the only option for sustainable development of India ?
11. Describe the major steps to be followed in planning and management of a watershed project in new area?
12. Briefly describe the software and hardcore treatments of soil and water conservation. Also elaborate the plus and minus points of both type of treatments.

a. Objective Questions

1 Watershed is a

a Natural hydrological unit | b Natural metabolic unit
c Artificial hydrological unit | d Artificial metabolic unit

2 How many drainage points are found in a Watershed?

a One | b Two
c Three | d Many

3 Central Soil Conservation Board was set up in year

a 1947 | b 1951
c 1953 | d 1963

4 Model watershed program was initiated in year

a 1973 | b 1983
c 1993 | d 2003

5 National Watershed Development Program on Rainfed Agriculture was started in year

a 1976 | b 1986
c 1996 | d 2006

6 Which of the following is an objective of watershed program?

a Control run off & Reduce sedimentation | b Appropriate use of land resources
c Moderate floods | d All above

7 Word 'POWER' symbolizes

a Objectives of watershed | b Principles of watershed
c Concept of watershed | d None above

8 Which of the following is a benefit of watershed program?

a Improved soil health | b Better water conservation
c Sustainable development | d All above

9 Which of the following is a main principle of watershed program?

a Land use based on capability class
b Conservation of 'in situ' rain water
c Sustainable development of ecosystem
d All above

10 Normally macro watershed covers an area of

a 500 ha
b 5,000-6,000 ha
c 25,000-30,000 ha
d > 30,000 ha

11 Normally medium watershed covers an area of

a 500 ha
b 5,000-6,000 ha
c 25,000-30,000 ha
d > 30,000 ha

12 Normally micro watershed covers an area of

a 500 ha
b 5,000-6,000 ha
c 25,000-30,000 ha
d > 30,000 ha

13 Which of the following is the name of a watershed?

a Sukho manjri
b Fakot
c Rolegaon Sidhi
d All above

14 Tarun Bharat Sangh is a name of a

a Watershed
b NGO
c Govt. institution
d Cooperative bank

15 The watershed projects governed by Desert Development program has an area of

a 500 ha
b 5,000-6,000 ha
c 25,000-30,000 ha
d > 30,000 ha

16 The watershed projects governed by IWDP has an area of

a 500 ha
b 5,000-6,000 ha
c 25,000-30,000 ha
d > 30,000 ha

17 IWDP refers to

a Integrated Watershed Development Program
b Integrated Wasteland Development Program
c Intensive watershed Development Program
d Intensive Water Development program

18 Major component of water shed development is

a Soil and water conservation
b Water resource improvement
c Alternate land use systems
d All above

19 Fakot watershed is situated at

a Uttarakhand b U. P.

c Haryana d A. P

20 Karnatika Watershed Development Project was aided by

a USDA b DFID

c World Bank d Planning Commission of India

21 Maharashtra Watershed Development Project was aided by

a USDA b DFID

c World Bank d German Govt.

22 PIA refers to

a Project Implementing Agency b Project Initiation Agency

c Project Identification Agency d Project Interaction Agency

23 WDT refers to

a Watershed development Team b Watershed Development Treaty

c Water Deficit Team d All above

24 Watershed Association refers to

a Local people living in the watershed area b Local people living outside watershed area

c Working people in the watershed d Watershed executing team

25 Main function of the District Nodal Agency is

a Identification of suitable PIAs b Selection of new watersheds

c Training part d All above

26 The chairman of the District Watershed Committee is

a District Magistrate or Chairman of Zila Parishad b Local M. L. A.

c Local M. P. d Gram Pradhan

27 Watershed Committee is elected by

a District Magistrate b Gram Panchayat

c Watershed Association d B.D.O

28 Which of the following record is maintained by Watershed Development Team?

a Cash book | b Stock register
c Ledger | d All above

29 Which of the following record is maintained by Watershed Committee?

a Cash book | b Stock register
c Ledger (WSC) | d All above

30 The short courses organized for administrators and managers should cover the items/modules;

a Rationale for sustainable development of watershed | b Focused exposure visit to successful watersheds
c Orientation about common guidelines of watershed | d All above

31 Major role of administrators and managers is

a Timely selection of PIA and micro watershed | b Preparation of training program
c Facilitation of timely fund flow | d All above

32 Which of the following is a responsibility of Watershed association?

a Approval of watershed development plan | b Organization of monthly meeting to review the progress
c Annual auditing of accounts | d All above

33 Which of the following is a responsibility of Watershed Committee?

a Certification and authorization of payment to labourers | b Organization of monthly meeting to office bearers
c Ensure quality works | d All above

34 Watershed Development Team is constituted by

a District watershed Committee | b District Nodal Agency
c Project Implementing Agency | d Watershed Association

35 Watershed Development Committee has

a 2 members | b 3 members
c 4 members | d 5 members

36 The members of WDC are

a Permanent | b Temporary
c Both a & b | d Neither a nor b

37 District Nodal Agency has

a 4 members
b 6 members
c 8 members
d 10 members

38 Watershed committee is constituted by

a W A
b W D C
c PIA
d DNA

39 Which of the following is not correct?

a WC has 4 members from SHGs
b WC has 5 members from UGs
c WC has one representative from Gram Panchayat and WTC each
d All above

40 Who is the chairperson of WC?

a District Magistrate
b Gram Pradhan
c B. D. O.
d President of WA

41 Watershed Development (Corpus) Fund is utilized for maintenance and management of watershed

a During watershed project work
b After watershed project life
c Both a and b
d Neither a nor b

42 Which of the following is correct for Watershed Development Fund (WDF)?

a 1% of the project cost is earmarked as a WDF
b 1% is contributed by state Govt. and beneficiaries
c Some contribution by sale of common output
d All above

43 WDF is operated by

a President of Watershed Association
b Project Leader of WDT
c Both a & b jointly
d Neither a nor b

44 National Watershed Committee is chaired by

a Secretary (A &C)
b Secretary (RD)
c Both a & b jointly on rotation
d P. M.

45 State Watershed Committee is chaired by

a C .M.
b Chief Secretary or A.P.C
c Agriculture Minister
d None above

46 Which of the following technique is helpful in delineation of watershed?

a RS & GPS
b Laser technology
c Biotechnology
d Nanotechnology

47 The new approach of watershed development is based on

a People driven b Demand driven

c Village empowered d All above

48 Watershed projects are implemented by

a National Watershed Agency b State Development Agency

c Project Implementation Agency d All above

49 WARASA refers to

a Watershed Areas' Rainfed Agricultural Systems Approach

b Watershed and Rainfed Agricultural Systems Approach

c Water and rainfall for Semi arid areas

d Watershed Appraisal for Rainfed Agricultural Systems

50 Which of the following is a participatory technique for development of watershed?

a Mapping and modeling b Transect walk

c Timeline d All above

51 The purpose of participatory watershed development is

a Achieve enhanced production b Bring about socio economic equity

c Bring about social balance d All above

52 How much fund is available for Entry Point Activity

a 1% of total watershed project fund b 2% of total watershed project fund

c 3% of total watershed project fund d 5% of total watershed project fund

53 Which of the following is an advantage of participatory approach of watershed development?

a Availability of first hand information b Availability of map at the spot

c Baseline information from farmers d All above

54 Matrix ranking helps in

a Availability of first hand information b Availability of map at the spot

c Baseline information from farmers d All above

55 Which of the following is correct in relation to codification system of watershed?

a Catchment is co notated by capital alphabet usually first letter of catchment name

b Subcatchment is indicated by small alphabet succeeding the catchmnet code

c Watershed is co notated by numerals from 1-9, succeeding subcatchment code

d All above

56 The stream order of a watershed is determined by

a Horton's method b Jensen's method

c Karale's method d None above

57 '3' in Bc3b subwatershed represents

a 3rd watershed of Bc3 subcatchment b Sub watershed of Bc3 watershed

c Name of Catchment d Name of Subcatchment

58 'B' in Bc3b subwatershed represents

a 3rd watershed of Bc3 subcatchment b Sub watershed of Bc3 watershed

c Name of Catchment d Name of Subcatchment

59 'c' in Bc3b subwatershed represents

a 3rd watershed of Bc3 subcatchment b Sub watershed of Bc3 watershed

c Name of Catchment d Name of Subcatchment

60 'b' in Bc3b subwatershed represents

a 3rd watershed of Bc3 subcatchment b Sub watershed of Bc3 watershed

c Name of Catchment d Name of Subcatchment

61 Which of the following is a hardcore treatment?

a Contour bunds b Terracing

c Ditches d All above

62 Which of the following is a software treatment?

a Contour farming b Strip cropping

c Mulching d All above

63 Which of the following is correct from the reports of impact evaluation of watershed programs?

a Soil loss is reduced | b Crop productivity increased
c Sedimentation reduced | d All above

64 Conservation tillage includes

a Zero tillage | b Reduce tillage
c Minimum tillage | d All above

65 Watershed program can be implemented in

a Rainfed areas | b Hilly areas
c Both a and b | d None above

66 Which of the following is the most important natural resources within our ecosystem and watersheds?

a Soil | b Water
c Both a and b | d None above

67 The shape and size of a watershed vary with

a Plain | b Undulated
c Mountains | d All above

68 Which of the following is the major component (s) of watershed management?

a Delineation | b Runoff rate
c Flood volume | d All above

69 Which of the following is the major component (s) of watershed management?

a Crop diversification | b Land productivity
c Both a and b | d None above

70 Basic purpose of integrated watershed management is

a Scientific land management | b Reducing runoff
c Increased land productivity | d All above

71 Bombay Land Improvement Act was passed in year?

a 1943 | b 1945
c 1960 | d 1962

72 NWSPRA was launched in year

a 1990-91 | b 1990-92
c 1990-94 | d 1990-95

73 Which of the following is the basic principle (s) of watershed management?

a Utilize the land resources based on its capacity
b To protect fertile top layer of the soil,
c To minimize silting up of tanks, reservoirs and lower fertile lands
d All above

74 Which of the following is the basic principle (s) of watershed management?

a Increase cropping intensity through inter and sequence cropping,
b To adopt alternate land use systems for efficient use of land resources
c To encourage water harvesting for supplemental irrigation,
d All above

75 Main objectives of watershed development program is

a To control damaging runoff
b To manage and utilize run off for useful purposes
c To control erosion and effect reduction in the sediment production,
d All above

76 Dry lands suffer water stress due to

a Uneven rainfall
b Erratic rainfall
c Both a and b
d None above

77 Availability of water in a particular watershed varies according to

a Area
b Topography
c Soil structure
d All above

78 Which of the following is the element (s) of watershed development program?

a Basic land treatment to conserve soil moisture control land erosion and harvest runoff.
b Application of crop and animal management practices, and
c Alternate land use such as agro-forestry, social and silvi-pasture system
d All above

79 What is the main component of watershed management?

a Land management
b Water management
c Biomass management
d All above

80 Which of the following do affect watershed ?

a Climate
b Geology
c Topography
d All above

81 Which of the following is the objective(s) of integrated watershed management?

a To conserve the basic resources of soil, rain water and vegetation,
b To achieve higher biomass production both in arable and non arable lands
c To reduce the dependency on monsoon
d All above

82 Which of the following is the objective (s) of integrated watershed management?

a To reduce risk in crop production,
b To restore ecological balance
c To improve socio-economic status of the people.
d All above

83 The main function of watershed is

a Water capture
b Water storage
c Water release
d All above

84 Common feature of watershed management is

a Water quality
b Water stress
c Drainage
d All above

85 Which of the following is the most common channel characteristic in estimating the watershed hydrological processes?

a Channel order
b Channel length
c Channel slope
d All above

86 Which of the following is an aim of watershed management program?

a Develop primary resource of land
b Produce secondary resource of plant
c Improve the quality of local residents
d All above

87 Which among the following scheme initiated by Indian Government has the primary objective of using the available land and water efficiently in a sustainable manner to improve the farmer's livelihood?

a WDC-PMKSY
b PMRY
c RKVY
d MGNREGA

88 The heated effluents from thermal plants discharged into rivers and other water bodies have

a Reduced level of dissolved oxygen b Increased level of CO_2

c Reduced level of nitrogen d Reduced level of CO_2

89 Material deposited by flowing water is known as

a Alluvium b Albedo

c Alluvial fans d Bhangar

90 A geologic formation which contains water and transmits it from one point to another in quantities sufficient to permit economic development is called

a Aquifer b Aquiclude

c Alluvial fans d Albedo

91 Which of the following is a concentrate water saving measure?

a Check dams b Diversion channel

c Engineering structures d All above

92 Hydrograph means

a When discharge is plotted over time in minutes, hours b Days resulting curve is known as hydrograph

c Both a & b d None above

93 Mini watershed covers area of

a 0-100 ha b 100-1000 ha

c 1000-10,000ha d 10,000 – 50,000 ha

94 Micro watershed covers area of

a 0-100 ha b 100-1000 ha

c 1000-10,000ha d 10,000 – 50,000 ha

95 Milli watershed covers area of

a 0-100 ha b 100-1000 ha

c 1000-10,000ha d 10,000 – 50,000 ha

96 Sub watershed covers area of

a 0-100 ha b 100-1000 ha

c 1000-10,000ha d 10,000 – 50,000 ha

97 Macro watershed covers area of

a 100-1000 ha b 1000-10,000ha

c 10,000 – 50,000 ha d >50,000 ha

98 Which of the following is not considered the main function of watershed management?

a Water capture b Water storage

c Water release d Sewage water storage

99 What is dark watershed area?

a Surface water b Ground water

c Sea water d Pond water

100 Which of the following is a synonymous of watershed?

a Basin b Catchment area

c Catchment basin d All above

101 Which of the following is the world's largest watershed ?

a Amazon River Basin b Ganga River Basin

c Godavari River Basisn d Chambal River Basin

102 National Water Policy came into existence in year

a 2000 b 2002

c 2005 d 2010

103 In which year the watershed development guidelines were published?

a 1994 b 1995

c 1996 d 1998

104 Which of the following is the largest watershed of India?

a Ganga River Basin b Chambal River Basin

c Narmada River Basin d Gandak River Basin

105 Haryali guidelines were published in year

a 2003 b 2005

c 2007 d 2011

106 Ralegaon Siddhi watershed started by Anna Hazare in year

a 1975 b 1979

c 1980 d 1985

107 Which of the following committee related to watershed management is most powerful?

a Watershed Development Team b Watershed Association

c Project Implementation Agency d Watershed Group

108 As per the Watershed Atlas of India, river resource regions are

a 06 b 35

c 112 d 550

109 As per the Watershed Atlas of India, total number of basins are

a 06 b 35

c 112 d 550

110 As per the Watershed Atlas of India, total catchments are

a 06 b 35

c 112 d 550

111 As per the Watershed Atlas of India, total sub catchments are

a 06 b 35

c 112 d 550

112 As per the Watershed Atlas of India, total watershed in country are

a 112 b 550

c 3257 d 5000

113 Which of the following is the largest dam of India?

a Tehri dam, Uttarakhand b Lakhwar dam, Uttarakhand

c Bhakra dam HP d Idduki dam, Kerala

114 Which river of India is known as red river?

a Ganga b Brahmaputra

c Chambal d Krishna

115 Which of the following river is the deepest in India?

a Bagirathi & Alaknanda b Narmada

c Godavari d Tapti

116 Project Implementation Agency (PIA) of a watershed management program has a technical team of 4 experts, known as

a Watershed Development Team b Watershed Association

c Watershed Committee d District Rural Development Agency

117 Watershed projects are implemented by

a Watershed Development Team b Watershed Association

c Watershed Committee d District Rural Development Agency/Zila Parishad

118 Who is the chairman of District Watershed Committee?

a District Magistrate/ Chairman of Zila Parishad
b Local MP
c District Agriculture Officer
d Local MLA

119 What is the role of District Nodal Agency of watershed development program?

a Selection of new watershed
b Management of watershed
c Conducting training for watershed development program
d All above

120 What is true about the Watershed Association?

a Local people living in a watershed area
b Planning and execution of watershed program
c Selection of executive committee for execution of watershed program
d All above

Answer

1. **a**	2. **a**	3. **c**	4. **b**	5. **b**	6. **d**	7. **a**	8. **d**	9. **d**	10. **c**
11. **b**	12. **a**	13. **d**	14. **b**	15. **a**	16. **b**	17. **a**	18. **d**	19. **a**	20. **b**
21. **d**	22. **a**	23. **a**	24. **a**	25. **d**	26. **a**	27. **c**	28. **d**	29. **d**	30. **d**
31. **d**	32. **d**	33. **d**	34. **c**	35. **c**	36. **b**	37. **c**	38. **a**	39. **d**	40. **d**
41. **b**	42. **d**	43. **c**	44. **c**	45. **b**	46. **a**	47. **d**	48. **c**	49. **a**	50. **d**
51. **d**	52. **c**	53. **d**	54. **a**	55. **d**	56. **a**	57. **a**	58. **c**	59. **d**	60. **b**
61. **d**	62. **d**	63. **d**	64. **d**	65. **c**	66. **c**	67. **d**	68. **d**	69. **c**	70. **d**
71. **a**	72. **a**	73. **d**	74. **d**	75. **d**	76. **c**	77. **d**	78. **d**	79. **d**	80. **d**
81. **d**	82. **d**	83. **d**	84. **d**	85. **d**	86. **d**	87. **a**	88. **a**	89. **a**	90. **a**
91. **d**	92. **c**	93. **a**	94. **b**	95. **c**	96. **d**	97. **d**	98. **d**	99. **b**	100. **d**
101. **a**	102. **b**	103. **a**	104. **a**	105. **a**	106. **a**	107. **c**	108. **a**	108. **b**	110. **c**
111. **d**	112. **c**	113. **a**	114. **b**	115. **a**	116. **a**	117. **d**	118. **a**	119. **d**	120. **d**

11

Water Management for Sustainable Agriculture

Outlines

Introduction

Increasing population, declining natural resources, changing climate and political stability are the major threat to our food, nutritional as well as environmental security. It is expected that India's population will boom to 1.5 billion by 2030 and to feed them we need approximately 341mt food grain including 315 mt cereals and 26mt pulses) (Kumar *et al.*, 2016). As per the estimate, the total food grain production in 2017-18 will be around 277 mt, If so, again 64 mt extra food grain need to be produced with 5 mt increase every year during next 12 years. At present we have approximately 45% area under irrigation and remaining 55% area is still under rainfed conditions. The availability of water to agriculture will be reduced from present 84.6% to 83.2% from 2010 to 2015 (CGWB, 2011). Further changing scenario of climate change had shown drastic shift in monsoon and its pattern, therefore shrinking glaciers reducing rainfall and surface water in last few years have resulted in overexploitation of ground water resources.

Water in 'soil–plant–atmosphere continuum' is considered to be one of the most important factors affecting crop productivity (Boyer, 1982). The globally averaged combined land and ocean surface temperature data show a warming of 0.85^0 (0.65 to 1.06)°C over the period 1880 to 2012 (IPCC, 2014), so the evapotranspirational need of crops will increase further in years to come. It means that irrigation water requirement will be increased in contrast to reduced share of water to agriculture. Therefore the option of increasing water use efficiency is left. Several approaches based on plant, soil water and climatological factors are in practices for precision water management in different crops. Recently remote sensing and GIS and irrigation models are very much in practices for precision water management. Jin *et al.* (1999) reported that excessive irrigation can reduce crop WUE, while effective deficit irrigation may result in higher production and WUE. Generally the responses of grain yield and WUE to irrigation were strongly influenced by soil water contents and irrigation schedules (Kang *et al.*, 2002).

Status of water resources

Total utilizable water resource in the country has been estimated to be about 1123 billion cubic metres (BCM) including 690 BCM from surface and 433 BCM from ground water which is just 28% of the water derived from precipitation (Table.1). About 80% of the water (688 BCM) is being diverted for irrigation, which may increase to 910 BCM by 2025. On the basis of the available water resources, the total irrigation potential from surface and ground water resources is estimated to be 139.9 Mha. The major source for irrigation is groundwater. Annual groundwater recharge is about 431 BCM of which 212.5 BCM is used for irrigation and 18.1 BCM for domestic and industrial use (CGWB 2011). By 2025, demand for domestic and industrial water usage may increase to 29.2 BCM. The sector-wise estimated demand is given in table 2. A cumulative storage capacity of about 225 BCM from major and medium irrigation projects has made it possible to provide assured irrigation in command areas, extend the supply of drinking water to remote areas and ensure water supply to hydro and thermal power plants and to meet other requirements. The minor irrigation potential of about 100 BCM, created through check dams, small structures, ponds, etc. Finally, groundwater potential of about 243 BCM was available. All this has made it possible to irrigate about 66.10 million hectares (2012-13) of the 139.9 million hectares of net sown area.

Table 1. Status of water resources in India

S.N	Water resources	Units
1.	Estimated Annual Precipitation (including snow fall)	4000 BCM
2.	Average precipitation during monsoon (June-September)	3000 BCM
3.	Average annual runoff in rivers	1869 BCM
4.	Estimated Utilizable Water	
a	Surface water potential	690 BCM*
b	Ground water potential	433 BCM
c	Available Ground Water Resource for irrigation	361 BCM
d	Total Utilizable Water	1121 BCM
e	Net Utilizable Ground Water Resources for irrigation	325 BCM
f	Net utilizable water potential	605 BCM (44% of total)
4.	Irrigation Potential	
a	*Major and Medium (Surface Water)	
	Ultimate	58.5 Mha
	Created (upto 2009-10)	44.4 Mha
	Utilized (upto 2009-10)	34.4 Mha
b	*Minor Irrigation	
	Ultimate	81.4 Mha
	Created (upto 2009-10)	62.8 Mha
	Utilized (upto 2009-10)	52.5 Mha
c	Total (Major and Medium+Minor)	
5.	Ultimate irrigation potential	139.9 Mha
a	Created (upto 2009-10)	107.2 Mha
b	Utilized (upto 2009-10)	96.9 Mha
6.	Area Irrigated (2015-16)	
	Gross	96.46 Mha
	Net	68.38 Mha
7.	Annual per capita water availability (year 2000)	
	World	7000 m^3
	India	1869 m^3
8.	Per capita water availability	
	1951	5177m^3
	2011	1544 m^3
	2014	1504 m^3
	2025	1465 m^3
	2050	1100 m^3

*BCM=Billion Cubic Metre = 1 km^3

Source: CWC (2013)

Table 2: Projected water demand in India by different sectors

S.N.	Sector	Standing Sub-Committee of MOWR			NCIWRD					
		2010	2025	2050	2010		2025		2050	
					Low	High	Low	High	Low	High
1.	Irrigation	688	910	1072	543	557	561	611	628	807
2.	Drinking water	56	73	102	42	43	55	62	90	111
3.	Industry	12	23	63	37	37	67	67	81	81
4.	Energy	5	15	130	18	19	31	33	63	70
5.	Other	52	72	80	54	54	70	70	111	111
	Total	813	1093	1447	694	710	784	843	973	1180

Note : NCIWRD: National Commission on Integrated Water Resources Development; MOWR: Ministry of Water Resource, BCM: Billion cubic metre
Source: CGWB. (2011)

Water availability is declining with each passing day. If the international yardstick of 1700 m^3 per capita availability of water is taken as a criterion, the whole country is "Water stressed" as the per head water availability was 1545 m^3 in 2011 (Figure.1). It is estimated that the per capita availability of water is likely to reach a level of around 1100 m^3 in 2050 but in reality we may reach this threshold much earlier considering the wasteful ways of water usage. It may be noted that per capita availability of <1000 m^3 indicates "Water scarcity". If the per capita availability is scrutinized basin-wise, a significant number of basins in the country were either in the "water scarcity zone" or worse in 2010. (Figure.2). The data available indicate that the per capita storage in India at 220 m^3 is one fifth that of China (1110 m^3).

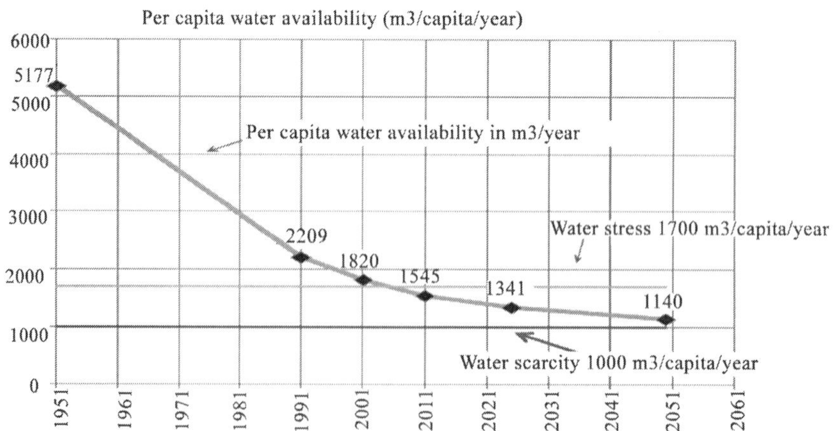

Fig. 1: Per capita availability of water in India

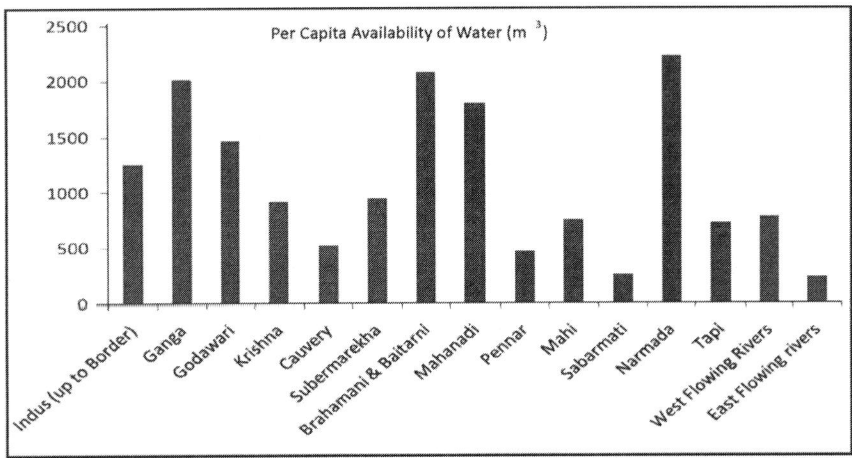

Fig. 2: The per capita water availability of water in the river basin of India in 2010

Ground water resources

Presently groundwater contributes more than 60% of the irrigation demand and playing a critical role in ensuring food security to the millions in the country. However, its over exploitation is now a very serious cause of concern and the potential of judicious exploitation of ground water resources is now confined only to some states in the eastern part of the country. Out of the 5842 assessed blocks, 4277 are safe while the rest are in the semi-critical (523), critical (169) and a large number of over-exploited (802) blocks (Figure.3). Considering the availability of the ground water resources, Government of India (GOI) has rightly decided to focus on Eastern India and has implemented a program aptly titled "Bringing Green Revolution to Eastern India (BGREI)". In view of the seriousness of the problem, the Government is, therefore, putting a lot of emphasis on artificial recharge of ground water. A project for setting up 225 thousand recharge structures at an estimated cost of approximately Rs. 20,000 crores has been launched with this objective and a clear defined goal that it has to be implemented in a participatory mode with involvement of local communities, locally elected bodies like *Panchayats,* and Non Government Organizations (NGOs).

The National Water Policy had clearly enunciated that to meet the required food grain production of 2050, the efficiency of surface water use has to increase to 60% and of ground water to 75%. The National Water Mission, one of the eight missions launched as a part of the National Action Plan on Climate Change of the Government of India, envisages an increase in water use efficiency by 20 % by the year 2017.

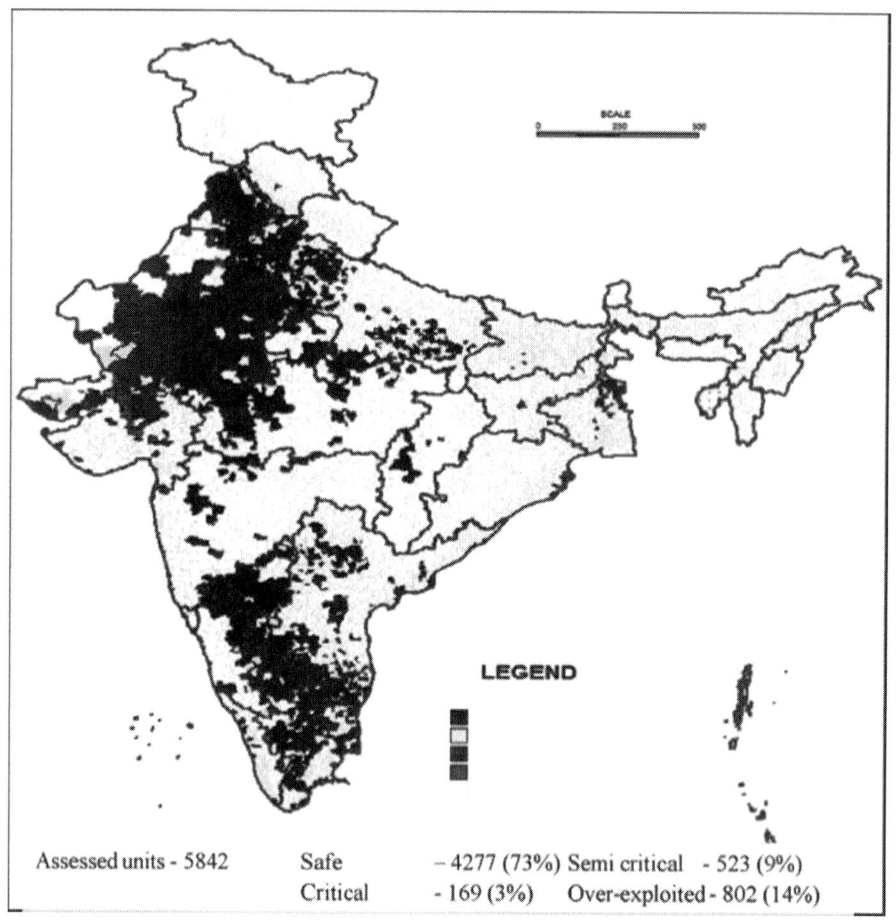

Fig. 3: Block–wise status of ground water exploitation

India has the largest irrigated area in the world, it has a very sizeable area under rainfed agriculture (nearly 80 Mha) which caters largely to pulses, oilseeds, minor millets, cotton and even rice (~45%) but the productivity of which is very low (<1 t ha^{-1})). The average annual rainfall in India is 117 cm which is higher than the global average of 110 cm but there is a wide spatial and temporal variation in its distribution and intensity. The ratio of the minimum, average and maximum is of the order of 1:10:100 with the maximum being around 11000 mm in Cherrapunji (Meghalaya) to 100 mm in the Thar desert. The number of rainy days can vary from 4 days to over 300 days a year and the intensity can vary anywhere from <1 cm hr^{-1} up to 15 cm hr^{-1}. It has also been estimated that almost 80% of the rain occurs in only 100 hr in a year. Soil and nutrients loss due to run off is of the order of 5.34 billion tons and 6 million tons, respectively, resulting in production losses estimated to be in excess of Rs. 115 billion.

Sharma *et al.* (2010) revealed on the basis of district level assessment on water harvesting that the area under rainfed can be increased with higher productivity if extra water is recycled for supplemental irrigation at critical growth stages of crops. The study identified about 28.5 Mha of potential rainfed area covering a large number of districts in central and eastern India that can generate sufficient runoff (114 BCM) for harvesting and reutilization. It is possible to raise the rainfed crop production by a total of 28-36mt from an area of 20-25 Mha during normal monsoon years which could result in about 12% increase over the present production. Watershed approach for sustainable development of the area is again a wonderful idea and whole area can be developed with available in-situ resources including water harvesting.

Irrigation Potential in India

The expansion of irrigation facilities along with consolidation of the existing systems have been the main strategy for increasing agricultural production. The irrigation potential of the country has been increased from 22.6 m ha in 1951 to 102.77 m ha at the end of 10th FYP (2002-07) including 42.35m ha from major and medium and 60.42 m ha from minor irrigation projects. The utilizable irrigation potential is only 87.23 m ha covering 31.01 and 49.05 m ha under major & medium and minor projects, respectively. The ultimate irrigation potential (UIP) of the country has been estimated as 139.9 (~140 m ha) million hectares (Table.5). The summary is given below;

a. Ultimate Irrigation Potential : 139.9 m ha ≈ 140 m ha
b. Major & Medium Projects : 58.5 m ha
c. Minor Projects : 81.4 m ha
    i. Ground Water :64.1 m ha
    ii. Surface Water : 17.3 m ha

Area under irrigation

As per the record of Ministry of Agriculture, Govt. of India, the net irrigated and gross irrigated area during 2014-15 was 68.38 (48.8% net) and 96.46 (48.6% gross) mha, respectively. The net and gross area under rainfed was 71.75 mha (51.2%) and 101.9 mha (51.4%), respectively.

The percentage of gross irrigated area to total gross cropped area (194.4 mha) was 48.6%. Among the states, Punjab has the highest irrigated area, 95.2% followed by Delhi, 89.3%, Haryana, 88.8 and U.P. 77.4%. Among the hilly states, Uttarakhand has the highest area, 44.6% followed by Meghalaya, 26.5%, Manipur, 22.7% and H. P., 19.4%.Among the zones, north zone has the highest irrigated area, 77.9%, followed by East, 37.5%, South, 34.7% and West, 30.6% during 2004-05.

Plan wise position of irrigation

S.N	Plan Period	Potential (Cumulative) created (m ha)			Potential (Cumulative) utilized (m ha)	Percent utilization	Gross area irrigated (m ha)
		Major & Medium projects	Minor projects	Total			
1	Pre-Plan period	9.70	12..90	22.60	22.60	100	22.56
2	1st Five Year Plan (1951-56)	12.20	14.06	26.26	25.04	95.41	NA
3	2nd Five Year Plan (1956-61)	14.33	14.75	29.08	27.80	95.59	27.98
4	3rd Five Year Plan (1961-66)	16.57	17.00	33.57	32.17	95.86	NA
5	Annual Plans (1966-69)	18.10	19.00	37.10	35.75	96.38	35.43
6	4th Five Year Plan (1969-74)	20.70	23.50	44.10	42.09	95.43	40.28
7	5th Five Year Plan (1974-78)	24.72	27.30	52.02	48.46	93.17	46.08
8	Annual Plans (1978-80)	26.61	30.00	56.61	52.64	92.99	49.21
9	6th Five Year Plan	27.70	37.52	65.22	58.82	87.10	54.53
10	7th Five Year Plan (1985-90)	29.92	46.61	76.44	68.59	89.63	61.85
11	Annual Plans (1990-92)	30.74	50.30	81.09	72.85	89.84	65.10
12	8th Five Year Plan (1992-97)	32.96	53.31	86.26	77.21	89.51	73.25

S.N	Plan Period	Potential (Cumulative) created (m ha)			Potential (Cumulative) utilized (m ha)	Percent utilization	Gross area irrigated (m ha)
		Major & Medium projects	Minor projects	Total			
13	9th Five Year Plan (1997-2002)	37.06	56.90	93.95	81.00	86.22	NA
14	10th Five Year Plan (2002-2007)	43.56	63.71	102.77	87.23	84.88	NA
15	10th Five Year Plan (2002-2007)	43.56	63.71	102.77	87.23	84.88	NA
16	11th FYP (2007-12)	9.00	7.00	16.00	-	-	-
17	2015-16						96.46
	Net irrigated area				68.38		

Source of Irrigation

Canal, tube wells, tanks, other wells and other sources are generally considered the other source of irrigation. At present in 2012-13, the highest area i.e. 62% area is irrigated by wells. Among the wells, the tube wells accounts the highest share of irrigated area, 46% followed by other wells, 16% and other sources, 11%. The irrigated area under canal is 24% and tanks 3% (Fig.4). The share of tube wells has increased recently due to greater exploitation of ground water.

Irrigated area Contribution by source in 2014-15

a. **Canal** : 16.18 mha area (23.66% =24%)

   Leading states = U.P.> Rajasthan > M.P. > Karnataka> Punjab

b. **Tube wells**: 31.61 mha area (46.23% =46%)

   Leading states = U.P. > Rajasthan> M.P.> Punjab>Maharashtra

c. **Tanks** : 1.72 m ha area (2.52% =3 %)

   Leading states = T.N.> M.P.>Karnatika>Telangana>Rajasthan

d. **Other wells**: 11.35 mha area (17%)

   Leading states = M.P.> Gujrat >Rajasthan>U.P.>T.N.

e. **Other source** : 7.52 mha (11%)

   Leading states = M. P.> Orissa> Karnataka>Rajasthan>Bihar

(Source: : Irrigation-Statistical Year Book of India 2018)

The irrigated area before independence was only 19.1%. The area under canal irrigation was higher than the well's irrigated area. After independence, the major emphasis was given to develop irrigation work under major and medium projects and minor irrigated projects. The extraction of ground water increased slowly and the area under tube wells increased much more than area under canal irrigation.

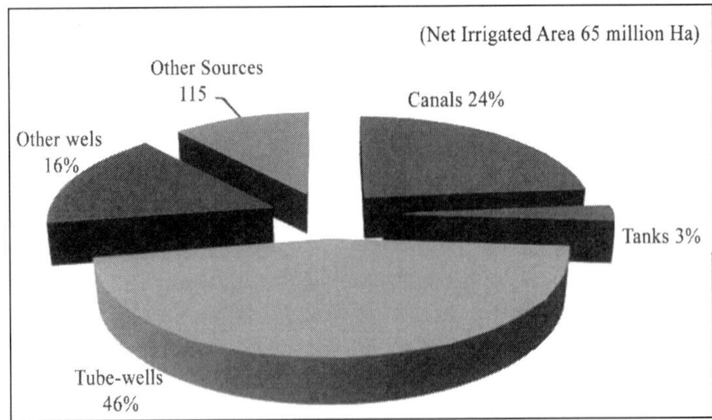

Fig. 4: Various Sources of Irrigation in 2012-13

Source: Land Use Statistics (2012-13), Ministry of Agriculture & Farmers Welfare

National Water Grid

1. The National Water Grid is a large scale engineering proposal for the management of water resources in India by the interlinking of the Indian rivers through networks of reservoir and canals.
2. The idea was given by Dr. K L Rao in 1972 to the inter-linking of rivers for irrigation cum hydro project.
3. The idea was to interlinking of the Perennial River to the non-perennial because perennial rivers, witness devastating floods whereas peninsular states suffer from scvcrc droughts.
4. But this idea was rejected by the Central Water Commission due to fund allocation issues and technical feasibility. In 1980, Government of India formulated the perspective plan for water resource development and set up the National Water Development Agency (NWDA) in July 1982. After long years of consideration, this idea was revived again in 2003 but the process of interlinking is very slow till date.

Significance of National Water Grid

Advantages of the National Water Grid Project

1. **Redistribution of water:** Indian rivers are two types-perennial and non-perennial. The perennial rivers, witness devastating floods whereas peninsular states suffer from severe droughts. This project controls the water distribution through the linking of different river which will not only control the flood like situation but also to the drought like situation.
2. **Hydropower generation:** If rivers in India are interlinked then there will be a majority of dams and reservoirs which can be used for the electricity generation. Therefore, we can say interlinking of the river has potential to fulfil the energy requirement of the industrial, agricultural as well as rural households.
3. **Irrigation Facilities:** As we know that India is a land of agriculture, but its dependency on the monsoon becomes a barrier to the agricultural growth.
4. **Commercial facilities:** Interlinking of the river will open the gate for new perspectives of the inland waterways. It will help in faster mobilization goods from one place to another. It will also open the new perspective of the fish farming.

Disadvantages of the National Water Grid Project

1. **Ecological Impact**: Some of the environmentalist states that the interlinking of the river will affects the ecosystem of the rivers.
2. **Social Impact**: If rivers will inter-linked then there will dams and reservoirs which causes the large number of displacement of the people. Therefore, rehabilitation will be major issues.

Soil Moisture Constants

The concept of irrigation cannot be understood unless we understand different soil moisture constants that play major role for deciding irrigation. Therefore soil moisture constants are listed below;

1. Field capacity (FC)

The soil moisture content that retains in soil pores after drainage of gravitational water. This stage comes after 48 – 72 hours after either irrigation or heavy rains in the field. It is the upper limit of available water that is held in soil at 0.1 – 0.3 bar or 0.01 – 0.03 MPa or an average of 0.3 bar.

2. Permanent wilting point (PWP)

The soil moisture level or points at which plants are wilted and can not recoop its original state is called permanent wilting point. At this stage, soil water is not available sufficiently to meet out the evapotranspiration (ET) demand of the field. The force at which plants get wilted is -15 bar or -1.5 MPa. This is also called as *'ultimate wilting point'* as the plants cannot survive but die even if the field is irrigated. It is the lowest limit of available water.

3. Available Soil Moisture (ASM)

The available soil moisture refers to the soil water between FC and PWP or -0.3 bar to -15 bar. The available soil water is generally expressed in terms of soil depth. The research findings indicate that most of the crops utilize 50 – 75 percent of available soil water. The course soils have lower amount of available water than heavy soils (Table.3). The available soil moisture decreases after FC towards PWP.

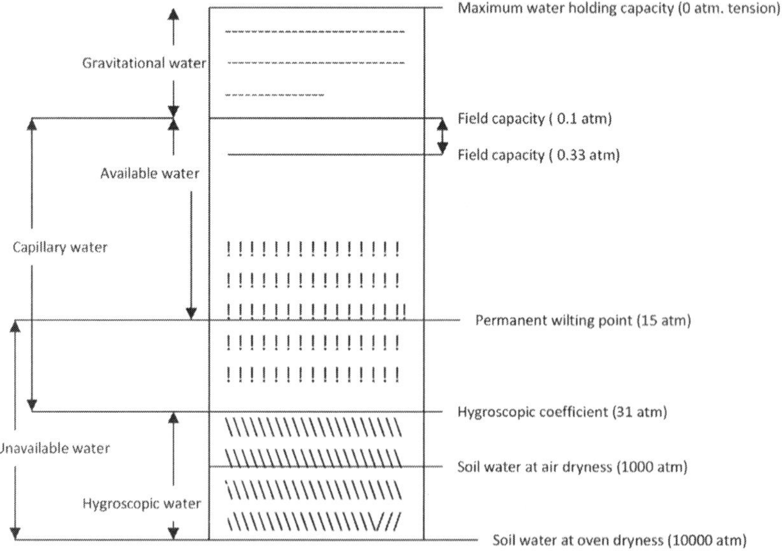

Fig. 5: Classification of soil water and soil water constants

Table 3. Available water holding capacity of different soil classes.

S.N	Soil Texture	Available water holding capacity (cm/m)
1	Course, fine and loamy sands	6 -10
2	Sandy and fine sandy loam	10 - 14
3	Very fine sandy and silt loam	12 - 19
4	Sandy and silty clay soil	13 - 20
5	Clay and silty clay loam	14 - 20

Source: (Sivanappan and Karai Gowder, 1977)

The plant growth and development is drastically decreased just below the point of available water in between FC and PWP is called *'critical soil water level'*. Therefore, the soil moisture is to be maintained just above to this critical soil water level for optimum plant growth and development as well as higher water use efficiency (WUE). However, the climate, plant characteristics and soil water conditions affect the available water pattern to crop plants. So,

$$ASM = E - PWP$$

4. Moisture Equivalent (ME)

The moisture equivalent refers to the amount of water retained by an initially saturated soil sample against centrifugal force of 1000 times of gravity force for half an hour. It is a laboratory analogue of FC and may be used as equal to FC. The moisture equivalent also denotes the moisture content in the inner capillary or microcapillary pores. The values of FC and ME are same for medium textured soils, slightly less than FC for sandy soils and slightly more than FC for very clayey soils.

5. Water Holding Capacity (WHC)

It shows the capacity of soil to hold water under fully saturated conditions. The soil pores, macro and micro, are fully filled with water. It is also referred as *'saturation capacity'* or *maximum water holding capacity* of the soil. The soil water is retained with almost zero force or the water remains at free state condition. The free or surplus or excess or gravitational water is drained out. The WHC of soil varies with soil texture, structure, organic carbon and colloidal matter. The plants are unable to utilize water between saturation point and field capacity of soil. The Keen – Raczkowski Box (20 gauze brass with 5 – 6 cm, internal diameter and 1.6 cm height and perforated at 4 mm apart at the bottom) is used to determine the saturation capacity of soil.

Concept of available water

Available water refers to the amount of water available between FC and PWP to plant communities as given below:

$$ASW = FC – PWP$$

It is also known as *'available water capacity of soil'*. The available water can be computed from different soil depths and later it can be summed up for total available water as given below:

$$ASW = \sum_{i=0}^{n} (FCi - PWPi).di$$

Where,

d_i = depth of ith soil layer

n = number of soil layer

The soil texture, structure, organic matter, colloidal substances and cations affect the availability of soil water to plants. The highest available water is found in clay soils and lowest in sandy soils (Table.4 & 5).

There are three concepts or theories related to available soil water as given below;

a. **Veihmeyer and Hendrickson (1950) theory:** They concluded that the soil water is equally available to plants in the range of FC and PWP. The plants functions also remain unaffected upto PWP stage but curtails abruptly at PWP.

b. **Richards and Wadleigh (1952):** Theory proposes the water availability to plants is maximum at FC and decreased towards PWP and plants suffer water stress before PWP.

c. **Recent Theory:** The available water is divided into groups i.e. readily available water and critical level of available water. It was said that sufficient water is available to plants during the range of readily available water till the critical level of soil available water and there is no adverse effect on plant functions. The plant will experience water stress just below the critical level of available water and plants will be wilted at PWP stage. Normally, irrigation is scheduled to maintain the readily available water status of soil or at critical moisture level of soil.

Table 4. Soil moisture constants and available water of different soils

Soil Texture	FC		PWP		Available water	
	%	cm/m	%	cm/m	%	cm/m
Sandy	9 (6 – 12)*	15 (10 – 20)	4 (2 – 6)	7 (3 – 10)	5 (4 – 6)	8 (7 – 10)
Sandy loam	14 (10 – 18)	21 (14 – 22)	6 (4 – 8)	9 (6 – 12)	8 (6 – 10)	12 (9 – 15)
Loam	22 (18 – 36)	31 (25 – 36)	10 (8 – 12)	14 (11 – 13)	12 (10 – 14)	17 (14 – 19)
Clay loam	27 (23 – 31)	36 (30 – 43)	13 (11 – 15)	18 (14 – 21)	14 (12 – 16)	19 (17 – 22)
Silt clay	31 (27-35)	40 (34-36)	15 (13-17)	20 (16-23)	16 (14-18)	21 (18-23)
Clay	35 (31-39)	44 (37-50)	17 (15-19)	21 (18-25)	18 16-20)	23 (19-26)

Figures in parenthesis are ranges.

Table 5. Representative available water holding capacities of different soils

S.N	Soil Class	Range/Average (in/ft)	Average(cm)
1.	Very coarse to coarse sand	0.50-1.25/.90	18
2.	Moderately Coarse Sandy Loams	1.25-1.75/1.50	45
3.	Medium Loams	1.50-2.30/1.90	57
4.	Clays and Clay Loams	1.60-2.50/2.10	63
5.	Peats and Mucks	2.00-3.00/2.50	75

Soil Water Measurement

Soil water can be measured by many methods but thermo gravimetric and volumetric methods are traditional and still used for soil water measurement. However, different instruments like tensiometer, resistance blocks, neutron moisture meter, Time Domain Reflectometer (TDR), pressure plate/membrane apparatus and also automatic soil moisture sensors are available and can measure soil water accurately.

a. ***Laboratory methods***
   i. Thermo gravimetric methods
   ii. Volumetric methods
   iii. H_2SO_4 method
   iv. Sprit burning method
   v. Immersion method

b. ***Field method***
   i. Feel and appearance method
   ii. Tensiometer and Eqitensiometer
   iii. Electrical resistance method
   iv. Neutron moisture meter
   v. Microwave radiation method (TDR)
   vi. Pressure plate/pressure membrane apparatus
   vii. Automatic moisture meter (sensors)

Consumptive use of water (CUW)

The consumptive use of water is the total quantity of water required for fulfilling the plant ET and metabolic requirements. It is also expressed as CUW in millimeters and centimeters. The peak consumptive water use differs among crop plant species (Table.6).

CUW = ET + water required for metabolic activities (Wm)

Table 6. Peak consumptive use of crops in different climatic conditions (Michael,1978)

S.N	Climatic conditions	Peak consumptive use (mm/day)
1	Cool and humid	3
2	Cool and dry	4
3	Moderate humid	4
4	Moderate dry	5
5	Hot and humid	5
6	Hot and dry	8

Potential evapotranspiration (PET)

The potential evapotranspiration of a plant or crop is the highest rate of ET obtained by a short and actively growing crop or vegetation with abundant foliage completely shading the ground surface with optimum water supply in a given climate. Under specific climatic conditions if any plant or crop achieves maximum ET under optimum growth conditions shading the soil surface, the value of ET is called PET. However, Gates and Hanks (1967) stated that ET reaches its maximum when crop foliage covers 50% of the ground, therefore it does not require full coverage of ground surface. The PET can be expressed as;

$$PET = Kc \times Epan$$

Where,

PET = Potential evapotranspiration

Kc = Crop factor (coefficient)

Epan = Evaporation rate of pan evaporimeter

Actual Crop Evapotranspiration (ET_{crop})

It refers to the rate of ET of a particular crop or plant in a specific period under prevailing soil and atmosphere conditions. In other words it is rate of ET or total ET of a plant or crop in prevailing field conditions therefore it varies with crop, soil season and place. Further it also differs with crop stages as mentioned earlier. It can be calculated either reference ET (ET_0) or evaporation rate of evaporimeter as given below;

$$ET_{crop} = K_c \times ET_0$$

Where,

ET_0 = reference crop ET

ET_{crop} = actual crop ET

K_c = crop coefficient or crop factor

If the evaporation rate of pan evaporimeter is Epan, the actual evaporation (E_0) under field conditions is calculated as given below;

$$E_0 = K_p \times E_{pan}$$

Where,

E_0 = actual evaporation

E_{pan} = evaporation rate of pan evaporimeter

K_p = pan evaporimeter coefficient

Reference Crop Evapotranspiration (ET_0)

The term 'reference crop evapotranspiration' was given by Doorenbos and Pruitt (1973) and it may be defined as "the rate of evapotranspiration from an extended surface of 8 to 15 cm tall green grass cover of uniform height actively growing completely shading the ground and not short of water". This is a unadjusted reference crop evapotranspiration and actual reference crop evapotranspiration is calculated by the multiplying correction factors (C) with unadjusted reference crop evapotranspiration as given below;

$$ET_0 = C \times ET_0^*$$

Where,

ET_0 = actual reference crop ET

ET_0^* = unadjusted reference crop ET

C = correction factor (adjustment factor)

The correction factor (C) varies under different weather condition including wind velocity for day and night (Appendix table) therefore. ET may be defined as

"The evapotranspiration of a disease free crop in a large field under optimal soil conditions with adequate water and fertility and having the full production potential of the crop under the prevailing environmental conditions."

Determination of Crop Water Requirement

The crop water requirement is estimated either in form of crop ET or consumptive use of water. Though number of methods are applied for estimating crop water requirement but following methods are commonly used;

1. Lysimeter method
2. Soil moisture depletion method
3. Field experimentation
4. Water balance method (inflow outflow)
5. Pan evaporimeter method
   i. USWB Class A Pan Evaporimeter
   ii. Sunken Screen Pan Evaporimeter
   iii. Pitch Evaporimeter (Atmometer)
6. Climatological approaches
   i. Blaney-criddle method (1950)

ii. Thornthwaite method (1948)
iii. Penman method (1948)
iv. Modified Penman method (1975)
v. Radiation method
vi. Bensen and Haise method

1. Lysimeter method

Lysimeter is a tank or container filled of soil is installed in the field to grow the crops with similar field conditions. The lysimeter may be weighing or non weighing types. The weighing type lysimeters are useful for short period studies (hourly and daily studies). They are costly and ET measurement is complex. The non weighing type lysimeters are cheap and ET measurement is comparatively simple.

Lysimeters not only measure the crop ET but also runoff, deep percolation, evaporation, transpiration etc. are also measured. The relationship may be expressed as;

$$ET = P + IR_n + \Delta S - (R + DP)$$

$$ET \text{ or } CUW = ER + IR_n + \Delta S$$

Where,

P = Precipitation (cm)

IR_n = Net irrigation requirement of the crop (cm)

ΔS = Soil water contribution (cm)

R = Surface runoff (cm)

DP = Deep percolation (cm)

Lysimeters may be of many types (Pelton, 1961, Slatyer and Mcllroy, 1958, Mcllroy and Angus, 1963). Harrold and Dreibelbis (1958)) described weighing and non weighing type lysimeters. Pruitt and Angus (1960) used weighing type lysimeter for estimation of crop ET. Gilbert and Van Bavel (1954) used non weighing oil drum type lysimeter

Installation of lysimeters in the field is again important. It has been observed that rectangular units of 4 m^2 are sufficient for most of the crops. The depth of lysimeter should be 100 cm for shallow and 150 cm for deep rooted crops. The available soil moisture in the lysimeter should not be more than 50%. Lysimeter, in general, should meet following conditions;

1. Lysimeters should be filled by soil in layer wise as found in the surround field,
2. The depth of the lysimeters should be maintained in such a way so that the root zone must maintain moisture regime similar to the surrounding field,

3. Lysimeters should be erected 0.5 cm above the soil surface to avoid inflow from the field,
4. Lysimeters should have similar management like the surrounding field, and
5. The ratio of the wall surface to the enclosed lysimeter area should be small to avoid small scale advection from non cropped area.

In weighing type lysimeters, weight of lysimeters is measured daily and change in weight is considered equal to ET loss but in non weighing or drainage type lysimeters, runoff, percolation, effective rainfall etc are measured but soil moisture content or soil water contribution is measured by either neutron meter or gravimetric method.

2. Soil moisture depletion method

The consumptive use of crop can be estimated by measuring soil moisture depletion before and after each irrigation or precipitation. This method is applicable only when the water table is fairly deep (at least 3 m deep) so that it does not influence the soil water fluctuations in the root zone. The soil water content is measured in each soil layer. The total soil moisture depletion is summed up after harvest of crop or calculation of seasonal consumptive use of water of a crop. The short period consumptive use is calculated from the soil between two successive soil sampling by the formula as given below;

$$CU = \sum_{i=1}^{n} \frac{M_{1i} - M_{2i}}{100} xAsi\ x\ Di$$

Where,

CU = consumptive use of water between two successive sampling periods (cm)

M_{1i} = soil moisture at first sampling in i^{th} layer (%)

M_{2i} = soil moisture at second sampling in i^{th} layer (%)

Asi = apparent specific gravity of the ith soil layer

Di = depth of the ith layer of soil (cm)

N = number of soil layers in the root zone

The seasonal consumptive use (SCU) may be estimated by adding CU values of each sampling intervals and potential ET losses during the intervening period between irrigation and the time of next soil sampling after irrigation or rainfall as given below;

SCU = Σ Cu + PET

Where,

SCU = seasonal consumptive use (cm)

PET = potential ET (= K_{crop} x E_{pan}) (cm)

E_{pan} = pan evaporation (cm)

K_{crop} = crop coefficient (dimension loss)

3. Field Experimentation

The field experiments are conducted with varying levels of irrigation. The measured water is applied at each irrigation. The soil water before irrigation and after irrigation is measured in each treatment. The contribution from effective rainfall and soil water is also measured. The water table should be well below the root zone (3m below). The total water consumed by each treatment is correlated with crop yield. The total water used in producing the lightest or most profitable yield is taken as the consumptive use of crop total water requirement of the crop. It is calculated as given below;

$$WR\,(CU) = ER + IR + \Delta S$$

$$= ER + IR + \sum_{i=1}^{n} \times \frac{Msi - Mhi}{100} \times Asi \times Di$$

Where,

WR = water requirement of the crop (cm)

CU = consumptive use of water by the crop (cm)

ER = effective rainfall (cm)

IR = irrigation water applied (cm)

Msi = soil water content at sowing time in i^{th} the soil layer (%)

Mhi = soil water content at harvesting time in i^{th} soil layer (%)

Asi = apparent specific gravity of soil

Di = depth of ith layer of root zone.

This method is simple but has limitations in estimating deep percolation, peak consumptive rate of crop actual soil water status before and after irrigation or rainfall. The effective rainfall and its contribution may be estimated by the any of the procedures discussed earlier.

4. Water Balance Method

This method is based on the principal of bank account having three points debit, credit and balance. Similarly, in this method total water inflow and outflow are estimated therefore, the method is also known as 'inflow – outflow' method the total water requirement is calculated as;

a. Inflow of water = $P + IR + \Delta S$
b. Outflow of water = $E + T + DP + R$

Therefore,

$E + T + DP + R = P + IR + \Delta S$

$\therefore ET = P + IR + S - (DP + R)$

Where,

P = precipitation (cm)

IR = amount of irrigation water applied (cm)

ΔS = soil water contribution (cm)

DP = deep percolation (cm)

E = evaporation

T = transpiration

R = surface runoff

5. Pan Evaporimeter Method

Estimation of crop ET (ET_{crop}) by pan evaprimeter method is simple and more realistic than using empirical formula. The evaporation value by pan evaporimeter is mainly affected by climatic factors of a location and accurately can be used for estimating ET_{crop}, while empirical formulae use number of weather parameters for calculating the crop ET. This method is most suitable for estimating crop ET for short period. The crop ET for most of the crops can be estimated by this method. The evaporation value of pan evaporimeter (E_{pan}) is multiplied by crop factor or coefficient (K_{crop}) as given below;

$$ET_{crop} = K_{crop} \text{ x } E_{pam}$$

$$Kcrop = \frac{ETcrop}{Epan}$$

The crop factor varies with crops, crop age, growth factors, location and ground cover therefore, the crop factor is estimated for each crop for its different growth stages for different locations. The lysimeter values of crop ET is taken for calculating the crop factor as given above. The crop factor is low at vegetative stage of the crops and it increases with crop age and reaches at its maximum value at flowing stage and then starts declining. Therefore, crop factor for different crops and locations have been estimated (table.8). If the crop factor is not available, the value 0.6 and 0.8 can be used for summer and winter crops, respectively for estimating ET_{crop} from E_{pan}.

Pan evaporation method requires measuring pan evaporation, meaning RH and winding velocity (km/day at 2 m ht). The pan coefficient (Kp) varies with wind velocity and relative humidity (table.7). It is the simplest inexpensive and reliable method for estimating the $ET_{0.}$

Table 7. Pan coefficient (Kp) for open pan for different wind velocity and relative humidity.

Wind velocity	**Wind ward side distance of green crop (m)**	**RH mean**		
		Low (< 40%)	**Medium (40-70%)**	**High(> 70%)**
Light (< 175)	01	0.55	0.65	0.75
	10	0.65	0.75	0.85
	100	0.70	0.80	0.85
	1000	0.75	0.85	0.85
Moderate (175-425)	01	0.50	0.60	0.65
	10	0.60	0.70	0.75
	100	0.65	0.75	0.80
	1000	0.70	0.80	0.80
Strong (425-700)	01	0.45	0.50	0.60
	10	0.55	0.60	0.65
	100	0.65	0.70	0.75
	1000	0.65	0.70	0.75
Very strong	01	0.40	0.45	0.50
	10	0.45	0.55	0.60
	100	0.50	0.60	0.65
	1000	0.55	0.60	0.65

Further the crop coefficient Kc is mainly affected by wind velocity and relative humidity. It varies with the development stage of the crop and also type of crops. Rice, banana and cocoa have higher and citrus, grapes, sisal and pineapple have higher Kc values (table.9). In general, Kc is higher in hot windy and dry climates than in cool calm and humid climates. The values vary among crops due to differences in reflectivity crop height, roughness, degree of ground cover and canopy resistance to transpiration. If the actual and maximum ET of a crop is known, the crop response factor can be calculated as;

$$1 - \frac{ETa}{ETm} = Ky\,(1 - \frac{Ya}{Ym})$$

Where,

ETa = actual ET of a crop

ETm = maximum ET of a crop

Yn = actual crop yield

Ym = maximum crop yield

$\frac{ETa}{ETm}$ = relative ET of crop

Ya/Ym = relative crop yield.

Table 8. Crop coefficient (Kc) for major crops at their different growth stages

Crop	Initial	Development	Mid season	Late season	Harvest	Average
Wheat	0.30-0.40	0.7-0.8	1.05-1.20	0.65-0.75	0.20-0.25	0.80-0.90
Rice	1.10-1.15	1.10-1.10	1.10-1.30	0.95-1.05	0.95-1.05	1.65-1.20
Maize	0.30-0.50	0.7-0.85	1.05-1.20	0.80-0.95	0.55-0.60	0.75-0.90
Sorghum	0.30-0.40	0.7-0.75	1.00-1.15	0.75-0.80	0.50-0.55	0.75-0.85
Ground nut	0.40-0.50	0.7-0.8	0.95-1.10	0.75-0.85	0.55-0.60	0.75-0.80
Sunflower	0.30 0.40	0.7 0.8	1.05-1.20	0.75-080	0.35-0.45	0.75-0.85
Soybean	0.30-0.40	0.7-0.8	1.00-1.15	0.75-0.80	0.40-0.50	0.75-0.90
Sugarcane	0.40-0.50	0.7-0.8	1.00-1.30	0.75-0.80	0.50-0.60	0.85-1.05
Cotton	0.40-0.50	0.7-0.8	1.05-1.25	0.80-0.90	0.65-0.70	0.80-0.90
Tabacoo	0.30-0.40	0.7-0.8	1.00-1.10	0.90-1.00	0.75-0.85	0.85-0.95
Cabbage	0.40-0.50	0.7-0.8	0.95-1.10	0.90-1.00	0.80-0.95	0.70-0.80
Onion	0.40-0.60	0.6-0.75	0.95-1.05	0.95-1.05	0.95-1.05	0.65-0.80
Tomato	0.30-0.40	0.7-0.8	1.05-1.05	0.80-0.95	0.60-0.65	0.75-0.95
Potato	0.40-0.50	0.7-0.8	1.05-1.20	0.85-0.95	0.70-0.75	0.75-0.90
Banana	0.40-0.50	0.7-0.85	1.00-1.10	0.90-1.00	0.75-0.85	0.70-0.80
Grape	0.35-0.55	0.7-0.8	.70-0.90	0.60-0.80	0.55-0.70	0.55-0.75

Source: Doorenbos and Kassons (1979)

Note: First figure is under high humidity (RHmin>75%) and low wind velocity(<5 m/sec), and Second figure is under low humidity (RHmin<20%) and strong wind velocity (>5 m/sec).

Measurement of evaporation

USWB Class A Pan Evaporimeter, *sunken screen pan evaporimeter* and *pitch atmometer* can be used for estimating evaporation but the evaporation value of USWB Class-A Pan Evaporimeter has been found most satisfactory and highly correlated with crop ET. It is calculated as given below;

$$E_0 = K_p \text{ x } E_{pan}$$

$$Kp = \frac{E_0}{E_{pan}}$$

Where,

E_0 = evaporation rate over the large free water surface

K_p = Pan evaporimeter coefficient (0.7)

E_{pan} = evaporation value from pan evoporimeter

i. ***USWB Class A Pan Evaporimeter:*** This is a standard open pan evaporimeter and commonly used for estimation of evaporation (Fig.6). It is made of 22 gauge galvanized iron, 120 cm in diameter and 25 cm in depth. It is normally pained white and stand on a wooden frame having 15-20 cm above the ground surface so that air may be circulated beneath the pan. The water is filled in the pan up to 20 cm depth. The daily water loss is measured with the help of hook gauge in a stilling well. The water loss is maintained by adding water to maintain the fixed point in the stilling well. The evaporation is computed as the difference between observed levels adjusted for any precipitation measured in a standard rain gauge. Since the pan evaporimeter has higher evaporation rate than large open water surface, therefore, the evaporation value is corrected by multiplying it with factor 0.7. This factor is known as pan coefficient.

Fig. 6: Open Pan Evaporimeter (USWB Class-A Pan Evaporimeter)
(*Source*: Michael, AM-Irrigation-Theory and Practices, 2001)

ii. Sunken Screen Pan Evaporimeter: The sunken screen evaporimeter was developed by Sharma and Dastane (1968) to measure the evaporation and it gives ET values more close to crop ET than USWB Class A pan evaporation. The pan coefficient of this evaporimeter was found in the range of 0.95 to 1.05 varying with climatic and crop conditions. The evaporimeter has a open pan of a 60 cm diameter and 45 cm deep made of 20 gauge galvanized iron sheet and a stilling well of 15 cm diameter and 45 cm depth attached to a pan with a tube (Fig.7). Both pan and stilling well are buried in the soil with 10 cm edge over the soil surface. The open pan is covered with 6/20 mesh wire net. The pan and stilling well are painted with white paints so that it may reflect most of the incoming radiation. The stilling well has the pointer inside the center to record the level of water. The water loss in the stilling well in 24 hours is recorded and the value is considered the evaporation rate per day. The water level of the stilling well is maintained by adding the fresh water after recording the values. The level of water in stilling well and so the tip of the pointer should be at the level of soil surface while installing the evaporimeter. The evaporimeters are installed in the field in such a way so that the wind circulation may not be obstructed over the pan. The evaporation value from this evaporimeter is reasonably good, however, it requires extensive evaluation for estimating the crop coefficient for different crops and locations.

iii. Pitche Atmometer: It is also called ' pitche evaporimeter' that consists of a 30 m long graduated tube with open end covered by a drier paper held in place by a metallic device. The tube is filled with water and turned upside down. The water slowly wets the paper and evaporates from the paper. The rate of water loss from the graduated tube is considered as a evaporation rate. The evaporation rate was found too high to use. It gives higher evaporation than USWB Class A pan evaporimeter. It is also poorly correlated with crop ET. It is mainly influenced by wind velocity but least by radiation. Hence, it gives very crude measurement of evaporation.

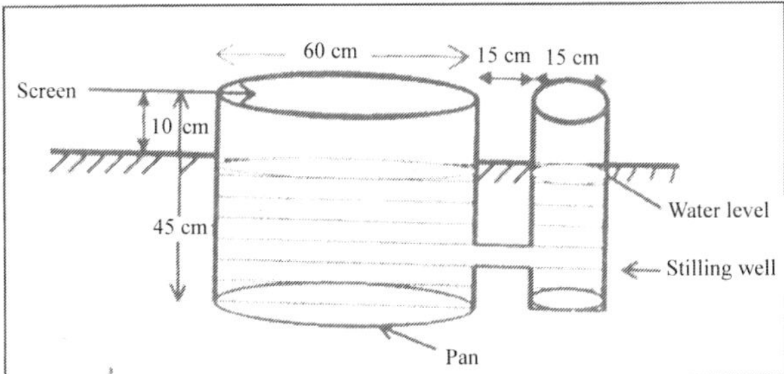

Fig. 7: Sunken Screen Evaporimeter (Sourc: Sharma and Dastane, 1968)

Methods of measurement

Irrigation water measurement is again important for efficient utilization of irrigation water. Therefore number of approaches is applied for measurement of irrigation water as listed below. The current meter, float meter, water meter and pipe method are used to measure small flow, while high water flow is measured by open channels, orifice, meter gates, weirs and flumes.

Volumetric measurements,

a. Velocity-area methods,

  i. Current meter,
  ii. Float method,
  iii. Water meter,
  iv. Pipe methods, and
  v. Open channels.

b. Measuring special structures i.e. orifices, weirs and flumes

  i. Orifice,
  ii. Meter gate,

iii. Weir, and

iv. Flumes

c. Tracer methods

Irrigation methods

An adequate water supply is not only important for plant growth but also for greater water use efficiency. Irrigation water is applied to crops under insufficient rainfall conditions to avoid water stress to crops. The manner in which irrigation water is applied to the land of crop is commonly referred to as method of irrigation. The selection of irrigation methods depends on landscape, amount of water, soil types, climatic conditions, season, equipment available and type of crop and its water requirement. The main objective of any irrigation method is to store water in the effective root zone uniformly and in maximum quantity ensuring water loss and higher crop productivity. Each method has its advantages and disadvantages, hence, these considerations are taken into account when choosing the method which is best suited to the local circumstances. **Normally following four main type of irrigation methods are used to irrigating crops.**

Types of Irrigation Methods

1. Surface irrigation
   - i. *Wild flooding,*
   - ii. *Border or border strip method,*
   - iii. *Check basin method,*
   - iv. *Furrow method,*
   - v. *Corrugation,*
   - vi. *Surge irrigation,*
   - vii. *Basin or ring method,*
   - viii. *Cablegation*
2. Sub-surface irrigation,
3. Sprinkler (over head) irrigation, and
4. Drip (tickle) irrigation.
5. Pitcher method: old method for very special and valuable plants.

Surface irrigation is preferred once there is no shortage of water, however furrow, check basin, border, basin and surge methods are used to not only to save irrigation water but also for uniform irrigation. The furrow method of irrigation is generally used to irrigate row crops and vegetables, and is suited to soils in which the infiltration rates are between 0.5 and 2.5 $cmhr^{-1}$. It is ideal for slopes varying from 0.2 to 0.5% and a stream size of 1-2 l/sec. Sub surface method is applied under conditions like high water, high infiltration rate and low water holding capacity of soil. It is also useful to shallow rooted crops like vegetables, cereals, fodder, oilseeds pulses, potato and beet. Sprinkler irrigation is well suited in arid and semi arid conditions as well as in undulated field conditions. All type of crops and tree plants planted either in rows or broadcasted may be irrigated by sprinkler method. Similarly, drip irrigation is usually considered good for widely spaced crops like trees, vegetables and fruit crops but presently field crops like sugarcane, pigeonpea etc. may be irrigated by drip method. Both drip and sprinkler methods are very efficient and have higher water use efficiency 70 and 85%, respectively but are costly and require constant check up and care. Pitcher method is an old practice and nowadays is in vague and not used for irrigation of field crops rather used only for very-2 special and valuable plants like medicinal and aromatic plants.

Scheduling irrigation

Proper irrigation schedule is determined for different crops under varied soil environments and agro-climatic conditions for higher water use efficiency. Therefore scheduling irrigation is the process of determining when, how and how much water is to be applied in the field or crop or in other words, the scheduling irrigation answers following questions?

a. When to irrigate?
b. How much to irrigate?
c. How to irrigate?
d. Where to irrigate?

Hence, following important facts are related to irrigation scheduling;

- Irrigation scheduling is the decision of when and how much water to apply to a field.
- Its purpose is to maximize irrigation efficiencies by applying the exact amount of water needed to replenish the soil moisture to the desired level.
- Irrigation scheduling saves water and energy.
- All irrigation scheduling procedures consist of monitoring indicators that determine the need for irrigation.

The purpose of irrigation scheduling is to determine the exact amount of water to apply to the field and the exact timing for application. The amount of water applied is determined by using a criterion to determine irrigation need and a strategy to prescribe how much water to apply in any situation

a. When to irrigate?

This is the first and foremost important problem of farming communities to decide when to irrigate. Most often, two types of farming situations arise;

a. When sufficient water is available, and
b. When limited water is available.

In first case, the main aim should be to produce maximum per unit of land without considering much about the total irrigation requirement, however, optimum amount of water should be applied at required crop stage or soil moisture regime so that crop may not experience water stress during its growing period. But in second case, every drop of water is applied for higher production or maximum production per unit of irrigation water. The limited water is used or distributed over the available land and irrigation is applied only at most sensitive crop stages and withholding irrigation at other non sensitive crops stages. Sometimes, deficit irrigation is also preferred to improve WUE and net returns.

The decision related to time of irrigation is very difficult as varied with number of factors related plant, soil, climate and crop management. However, irrigation should be applied when there is still adequate moisture to meet ET demand of the crop and atmosphere or irrigation should be applied before the soil moisture reaches to the level of critical soil moisture level. Again, the question may arise how to decide critical soil moisture level or critical period or critical water sensitive crop stages or optimum soil moisture regime for irrigation? The field experiments are carried out on different crops with application of irrigation at different soil moisture level, suction soil water potential, crop growth stages etc. growth and yield reduction are correlated with different soil water levels, crop stages, soil water potential or suction and the highest reduction in yield is considered most. Once the critical soil moisture level, crop growth, critical period of water need, soil suction or soil water potential etc. are decided, the irrigation schedules for different crops are chalked out.

b. How much to irrigate?

The amount of irrigation water for applying irrigation is to be decided based on crop and its stage, soil type, irrigation infrastructure, irrigation method and irrigation efficiency. The net irrigation depth is determined. The depth of irrigation is influenced by effective root zone, water retention capacity of soil, soil water depletion and apparent specific gravity of soil. The water content of the soil at the time of irrigation is estimated by gravimetric method and it will help to decide how much irrigation water is required or irrigation as given below;

$$\text{NIR}= \sum_{iri}^{n} \frac{Fci - Mbi}{100} \times Asi \times Di$$

Where

NIR = Net irrigation or depth of irrigation (cm)

Fci = Water content of ith layer of soil at field capacity (% by weight)

Mbi = Water content of ith layer of soil just before irrigation (% by weight)

Asi = Apparent specific gravity of ith layer of soil

n = Number of soil layers in the root zone.

The value of net irrigation requirement is the actual value of irrigation water to apply in the field to bring the soil moisture to the level of field capacity in the root zone soil. It does not include contribution by effective rainfall, ground water, seepage, dew/ hails/ snow etc. Similarly, gross irrigation requirement may be estimated based on the efficiency of irrigation system as given below;

$$\text{Gross Irrigation Requirement (GIR)} = \frac{NIR}{\textit{Efficiency of irrigation system}}$$

Example: Calculate the NIR and GIR for a wheat crop to be irrigated at 50% allowable soil moisture level if the FC of soil is 16 cm, apparent specific gravity1.2 g /cm^3 and efficiency of irrigation system 80 %

Answers

FC of soil = 16cm.

50% allowable soil moisture = 16/2 =8 cm

Apparent specific gravity =1.2 g /cm^3

$$NIR = \frac{16-8}{100} \times 1.2 \times 100$$

$$= 8 \times 1.2 = 9.6\ cm$$

$$GIR = \frac{NIR}{\textit{Efficiency of irrigation system}}$$

$$= \frac{9.6 \times 100}{80} = \frac{96}{8} = \underline{12\ cm}$$

c. How to irrigate?

This question is related to the irrigation methods so that higher WUE may be achieved. The irrigation method is decided on the basis of water availability, topography, soil type, plant type, socio- economic conditions of the farmers and also climatic factors. When water is in plenty, any irrigation method from surface, sub surface, sprinkler or drip may be chosen but sprinkler or drip methods may be preferred under limited water availability. Topography is again important factor as uneven and undulated topography may require sprinkler or drip irrigation. The drip and sprinkler methods of irrigation cost much, so high value crops should be irrigated through these costly irrigation methods. Proper management has always edge over other factors; hence, proper crop management practices may improve water use efficiency with higher net returns.

d. Where to irrigate?

The amount of available irrigation water, crop values and its growth stage, priority of crop and finally net return help to answer the above question. Further, law of an opportunity cost is to be applied for deciding where to irrigate with maximum benefit and water use efficiency. The amount of available irrigation water is very crucial factor because all grown crops may be irrigated at its critical stages provided sufficient irrigation water is available. If irrigation water is available only for limited irrigation, the most critical stage of a crop should be irrigated first followed by second most critical stage and so on. For example, if the irrigation water is available for only one irrigation to maize crop, the crop should be irrigated at tasseling stage. If two irrigations are possible, crop should be irrigated at tasseling and silking stages, otherwise, crop should be irrigated at all water sensitive growth stages, i.e. tassling, silking, grain filling and knee high stages.

For example

If a farmer is growing two crops, wheat and sunflower and has irrigation water equal to 5 irrigations. Normally, wheat and spring sunflower require 5 irrigations each during its growing period for maximum yield. Therefore, the farmers have to decide to irrigate both crops in such a way to get maximum yield or net profit as well as high water use efficiency. The detail is given below in table.9.

Table 9: Effect of irrigation levels on expected gross returns from wheat and Sunflower crops

Irrigation levels	Returns (Rs.)	
	Wheat	Sunflower
I	2000 I	1500
II	1800 II	1500
III	1400	1700 III
IV	1450	1650 IV
V	1600 V	1200
	3 irrigation	2 irrigation

Table.9 indicates that 1st irrigation provides higher returns from wheat than sunflower, so 1st irrigation should be applied to wheat crop and this irrigation may coincide to critical root initiation (CRI) stage. The second irrigation is again to be applied to wheat because of higher returns than sunflower and this stage may coincide to tillering to late jointing stage of wheat. The 3rd and 4th irrigation should go to sunflower for higher returns than wheat crop. This irrigation may coincide to bud initiation and head filling stages of sunflower. The 5th irrigation should be given to wheat crop and it may coincide to grain filling stage of wheat. Therefore, 3 irrigation to wheat and 2 irrigation to sunflower should be applied for higher returns.

Sometimes, the priority and choice of farmers may alter the whole irrigation schedules for different crops without considering the gross returns. The crops like wheat, rice etc. are preferred over vegetables, pulses and oilseeds as these are essential part of their livelihood even though these crops may give lower gross returns than vegetables per unit water used.

Critical period of water need

The crop plants constitute more than 90% water of their total body weight, so water is always required by crop plants for completing their life cycle. The crop plants experience number of phenophages or physiological stages and some of them are more sensitive to water stress. The growth or physiological stages may be broadly grouped in four stages;

a. Initial growth stage: Sowing to 10% ground cover
b. Crop development : Stage- 10-70% ground cover
c. Mid season stage : Flowering to grain filling stage
d. Late season stage : Ripening to harvest.

Some of the above physiological stages need more water and significant reduction in growth and yield observed under water stress condition. These stages are known as *'water sensitive critical crop stages'*. Generally, these stages do not coincide with the period of peak consumptive water use by crops. The water requirement at different crop stages may be correlated with the typical sigmoid growth curve or S shaped growth curve (Fig.7). As mentioned in the figure, the period between the 'point of inflection' and 'Point of deflection' is the most critical for water need. At point of inflection, there is sudden spurt in vegetative growth and at 'point of deflection' there is a initiation of reproductive stage or vegetative phase is changed to reproductive phase. At these points crops demand adequate water and can not afford water stress. Therefore, above both points are considered most critical stages of water demand. Apart from these points, bloom and grain filling stages also need sufficient water. However, the water requirements as well as the critical water sensitive stages differ with crops and agro-climatic conditions. The water sensitive crop stages are listed in its descending order in table 10.

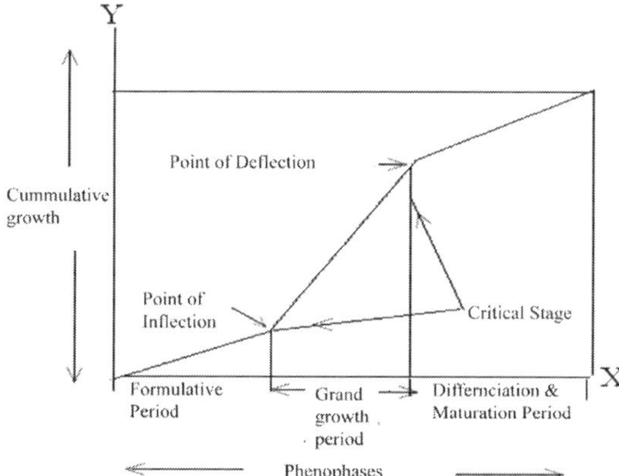

Fig. 7: Typical sigmoid (S Shaped) growth curve and related water sensitive period in crop plants.

Approaches for scheduling irrigation

Either; over or under irrigation leads to poor management of irrigation water. Proper and efficient management of irrigation water is always beneficial as the unnecessary water losses may be avoided under excess irrigation conditions and on contrary, water stress at critical growth stages may be avoided under deficit irrigation conditions. Therefore, numbers of approaches or criteria have been used to decide the exact time of irrigation for different crops. These approaches are related to either plant factors, soil factors or climatic factors. Therefore, all approaches have been grouped mainly into following three categories;

(1)Approaches based on plant factors (Plant indicators)

i. Plant appearance
ii. Plant growth and development
iii. Critical growth stages
iv. Plant or leaf water potential
v. Indicator plant
vi. Leaf temperature
vii. Leaf diffusion resistance
viii. Stomatal aperture

(2) Approaches based on soil factors (soil indicators)

i. Feel and appearance: the detail is given in Table .11.
ii. Soil water content (Table. 12)
iii. Critical level of soil water
iv. Depth and interval of irrigation
v. Soil water potential (Table.12)
vi. Digital soil moisture probes

(3) Approaches based on climatic factors (Climatological Approaches)

i. Empirical formulae
   a. Penman (1948) formula
   b. Thornthwaite (1948) formula
   c. Blaney – Criddle (1950) formula
   d. Penman modified formula (Doorenbos and Pruitt, 1975)
   e. Israelson and Hensen (1962) formula
ii. Irrigation water / Cumulative Pan Evaporation Ratio (IW/CPE-ratio)
iii. Evaporimeter – a. US class – A Open Pan Evaporimeter
   b. Sunken Screen Pan Evaporimeter
   c. Atmometer

Table 10. Moisture sensitive growth stages of some important crops

S.N	Crop	Moisture sensitive stages (in descending order)
1	Rice	Panicle initiation, heading, flowering, grain filling, tillering and transplanting.
2	Wheat	Crown root initiation (CRI), flowering, jointing, milking & grain filling and tillering.
3	Maize	Tasseling, silking, early grain formation and knee high
4	Sugarcane	Formative (Tillering) stage
5	Potato	Tuber initiation, tuber development and maturity.
6	Onion	Bulb formation to maturity
7	Tomato	Flowering, fruit set and development, early vegetative growth.
8	Tobacco	Early growth period
9	Sorghum	Booting, flowering and grain filling
10	Pearl millet	Heading, flowering, grain development
11	Finger millet	Panicle initiation, flowering and grain development
12	Barley	Pre-flowering and grain filling
13	Groundnut	Flowering, siliquae development and branching
14	Rapeseed & Mustered	Flowering, siliquae development and branching
15	Sunflower	Budding, flowering, seed filling, 4-6 leaf stage
16	Sesame	Flowering, seed development and branching
17	Safflower	Rosette stage, flowering and seed filling
18	Cotton	Pre-flowering and boll formation
19	All pulses	Pre-flowering and pod development
20	Soybean	Flowering and seed filling
21	Pea	Flowering and pod development
22	Cabbage	Head initiation, vegetative stage and head development
23	Citrus	Flowering, fruit setting and fruit growth and development
24	Carrot	Root enlargement
25	Bananas	Early vegetative period, flowering and fruit formation

b. *IW/CPE ratio (Irrigation water/Cumulative Pan Evaporation)*

This is one of the simplest approaches of scheduling irrigation developed by Parihar *et al.* (1978). This approach is based on the relationship between crop evapotranspiration and pan evaporation. The ratio of irrigation water to cumulative Pan evaporation for scheduling irrigation is determined experimentally for different crops and its different phenological stages under different agro-climatic conditions. The ET or the depth of irrigation or irrigation requirement can be estimated by lysimeters in field experiments.

Many field experiments have been conducted on IW/CPE ratio for scheduling irrigation of different crops (Table.13). For example, wheat should be irrigated at 0.75-1.00 IW/CPE ratio irrespective of its growth stages and 34% irrigation water can be saved compare to irrigating crop on its critical growth stages. (Parihar *et al;* 1976). The evaporimeter is used to estimate reference ET (ET_0) as given below.

Application of evaporimeter

The evaporation from free water surface in open pan may be correlated with ET requirement of the crop. Therefore, the evaporation can be measured with the help of evaporimeters like 'USWB class A Open Pan Evaporimeter (Fig .8). Sunken Open Pan Evaporimeter and Atmometer. The USWB class A Open Pan Evaporimeter is widely used for measuring the evaporation loss and consumptive use of water or ET by crop coefficient (K_{crop}) as given below.

$$ET_o = K_{pan} \times E_{Pan}$$

$$ET_{crop} = K_{crop} \times ET_o$$

Where,

ET_o = reference evapotranspiration (mm day^{-1})

E_{pan} = Pan evaporation (mm day^{-1}) from class A pan

K_{pan} = Pan Coefficient

ET_{crop} = ET of crop (mm day^{-1})

K_{crop} = crop coefficient

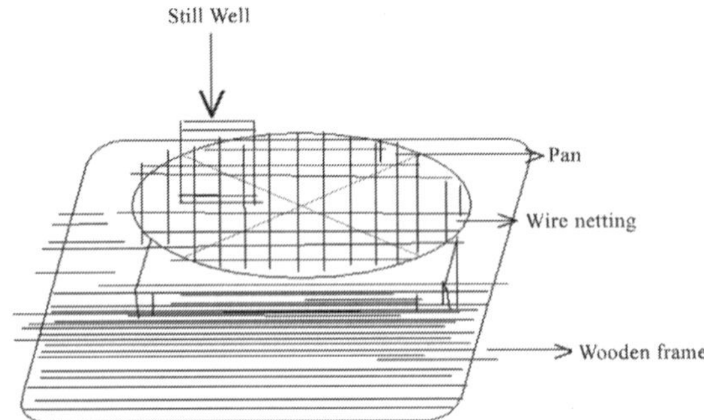

Fig. 8: USWB Open Pan Evaporimeter

The crop coefficient varies from 0.6 to 0.8 for most of crops at their different stages. The time of irrigation can be determined by deducting daily ET loss from the available soil moisture and irrigation may be applied when the ET loss approaches to allowable soil moisture depletion. The Sunken Screen Evaporimeter value can be used from the period of 25% ground coverage by crops till their period maturity (Gautam and Dastane, 1970). The Atmometers are made of porous ceramic or paper evaporating surfaces generally colored either black or white. The evaporating surface remains wet as water is supplied continuously, so linear relationship is found between water loss from atmosphere and ET. The time of irrigation is judged on the basis of daily ET loss by crops. The approximate ET values of some important crops are given in table 14. The total water requirement of crops differs with variety, soil, weather and management. The demand of irrigation water is very high for rice, sugarcane and banana while low for pulses and oilseeds (Table.15).

Table 11. Guide for judging the amount of available soil moisture range in different soil types

S.N	Available soil moisture range	Course textured (Loamy Sand)	Moderately course (Sandy Loam)	Medium texture (Loamy and Silt soil)	Fine texture (Clay Loam & Silt)
1.	Field capacity (100%)	On squeezing, no free water appears on soil, but wet outline is left on hand	Similar symptoms	Similar symptoms	Similar symptoms
2.	75-100%	Tends to stick together slightly, sometimes forms a very weak ball under pressure	Forms weak ball, breaks easily, don't stick	Forms a ball is very pliable sticks readily	Easily ribbons out between fingers has sticky feeling
3.	50-75%	Appears to be dry, don't form ball with pressure	Tends to form a ball under pressure but seldom holds together	Forms a ball some what plastic some times stick slightly with pressure	Forms a ball ribbons out between thumb and forefinger
4.	25-50%	As above, but ball is formed by squeezing very firmly	Appear to be dry, don't form a ball unless squeezed very firmly	Somewhat crumbly but holds together with pressure	Somewhat pliable forms a ball under pressure
5.	0-25%	Dry, loose single grained flows through fingers	Dry, loose flows through fingers	Powdery dry, sometimes slightly crusted out easily broken down into powdery conditions	Hard, baked cracked sometimes has loose crumbs on surface

Table 12. Allowable soil moisture depletion and suction at active root zone (0-30 cm) for scheduling irrigation of important crops.

S.N.	Crops	Allowable soil moisture depletion (%)	Soil moisture tension (bar)
1.	Rice	0-25 (1-3 days after depletion of ponded water)	Zero to 0.25
2.	Wheat	50	0.50
3.	Maize	50	0.50
4.	Sorghum	50-75	0.50-0.75
5.	Pearl millet	25-30	0.25-0.30
6.	Barley	50	0.50
7.	Soybean	60	0.60
8.	Cotton	50	0.50
9.	Sugarcane	65	0.65
10.	Berseem	25	0.25
11.	Groundnut	40	0.40
12.	Tobacco	25	0.25
13.	Potato	30	0.30
14.	Onion	65	0.65
15.	Root vegetables (radish, turnip, beet roots)	20	0.25
16.	Sugar beet	30	0.30
17.	Jute	50	0.50
18.	Other crops	?	?

Table 13. Optimum IW/CPE ratio for scheduling irrigation of different crops

S.N.	Crop	IW/CPE	S.N.	Crop	IW/CPE
1	Wheat	0.75-1.0	11	Finger millet	0.60-0.8
2	Maize	0.8-1.0	12	Pigeon pea	0.60
3	Sorghum	0.4-0.6	13	Black gram	0.8
4	Gram	0-4	14	Green gram	0.8
5	Groundnut	0.6-0.9	15	Rape seeds & Mustard	0.6-0.8
6	Sunflower	0.75	16	Soybean	0.75
7	Cotton	0.90	17	Potato	1.20-1.25
8	Sugarcane		18	Tomato	0.75
9	Grapes	0.75	19	Onion	0.70-1.0
10	Barley	0.80	20		

Table 14. Approximate values of ET of some important crops

S.N	Crop	ET(cm)	SN	Crop	ET(cm)
1	Rice	60	11	Lentil	31
2	Wheat	40	12	Sunflower	50
3	Potato	30	13	Rape	32
4	Pea	15	14	Linseed	35
5	Sorghum	50	15	Safflower	50
6	Maize	62	16	Triticale	35
7	Barley	30	17	Other crops	
8	Sugarcane	90	18	....	??
9	Pearl millet	50	19	....	
10	Berseem	90	20	....	

Table 15. Water requirement of different agricultural crops

S.N	Crops	Water requirement (mm)	SN	Crops	Water requirement (mm)
1	Rice	900-2500	13	Pea	150-200
2	Wheat	450-650	14	Tobacco	500-800
3	Sorghum	450-650	15	Potato	500-700
4	Maize	500-800	16	Tomato	600-800
5	Sugarcane	1500-2500	17	Vegetable Pea	350-500
6	Sugar beet	550-750	18	Onion	350-500
7	Groundnut	500-750	19	Bean	300-500
8	Cotton	700-1300	20	Cabbage	380-500
9	Soybean	450-750	21	Banana	1200-2200
10	Jute	400-600	22	Citrus	900-1200
11	Chickpea	250--400	23	Grape	500-100
12	Arhar	200-300	24	Pineapple	700-1000

Summary of scheduling irrigation

The irrigation schedules for different crops are summarized below based on field experimentations conducted in different agro-climatic zones of the country (Table.16).

Table 16. Criteria of scheduling irrigation and suitable irrigation methods for different crop

S.N	Crop	Irrigation criteria				Suitable irrigation method
		Critical growth stages	ASMD (%)	Soil moisture tension (Bar)	IW/CPE ratio	
1	Rice	Seedling, maximum tillering, panicle initiation, flowering and grain filling stages	Saturation stage	Saturation stage		Border check method
2	Wheat	Root crown initiation, tillering, jointing, panicle initiation, flowering and dough stages	50	0.5	0.9- 1.0	Border strip method
3	Maize	Seedling, tasseling, silking and dough stages		0.5	0.7-.9 (kharif) 1.0- 1.05 (rabi)	Furrow method Border method
4	Sorghum	Seedling, pre flowering, flowering and grain formation stages	50	0.5	0.5 – 0.6	Check method
5	Barley	Tillering and flowering stages	50	0.5	0.8-0.9	Strip method
6	Bajra	Crop establishment, pre flowering and grain filling stages	75	0.7	0.6-0.8	Border strip or check method
7	Ragi	Seedling establishment and flowering stages	50	0.5	-	Border strip or check method
8	Pigeon pea	Flowering and pod development stages	50-70	0.5-0.7	0.6	Furrow and drip
9	Green gram	Flowering and pod development stages	50	0.5	0.7 -0.8 (Summer crop)	Furrow, Border strip, Check method
10	Black gram	Branching ,Flowering and pod development stages	50	0.5	0.8	Furrow, Border strip Check method
11	Peas	Branching, flowering and pod development stages	50	0.5	0.8	Border strip Check methods

S.N	Crop	Irrigation criteria				Suitable irrigation method
		Critical growth stages	ASMD (%)	Soil moisture tension (Bar)	IW/CPE ratio	
12	Chickpea	flowering and pod development stages	-	-	0.4 (Sub tropics) and , 0.7 (Tropics)	Check method
13	Lentil	flowering and pod development stages	50	0.5	0.3-0.4	Border strip
14	Groundnut	Flowering, pegging and pod development stages	25 to 50	0.5	0.9 -1.0	Border strip method, sprinkler or drip irrigation
15	Rapeseed & Mustard	Flowering and pod development stages	50	0.4 - 0.5	0.6-0.7	Border strip furrow and Drip methods.
16	Sunflower	peak vegetative , buttoning and flowering and seed filling stages	40-60	0.4 - 0.5	0.75 -0.80	Border strip furrow and Drip methods.
17	Sesame	Branching, flowering and capsule development stages	40-50	0.4-0.5	0.35-0.45	Border strip furrow and Drip methods.
18	Castor	6-8 leaf stage, flowering and capsule development stages	-	-	-	-
19	Safflower	rosette, flowering and capsule development stages	-	-	-	-
20	Linseed	Flowering and capsule development stages	50	0.5	0.4	Border strip or corrugation methods.
21	Soybean	Flowering and pod development stages	50	0.5	0.75	Border strip and Sprinkler methods
22	Cotton		50-60	0.5	0.6-0.9	Furrow & drip
23	Sugarcane	Formative stage	50	0.4-0.6	0.6-1.0	Furrow & drip
24	Potato	Initial tuberization and stolonization	50	0.3-0.4	1.0-1.2	Furrow, Alternate furrow, Check furrow and Sprinkler
25	Tobacco	Seedling, pre flowering and flowering	50	0.5	-	Furrow

Application of remote sensing and geographical information system (RS and GIS)

The application of remote sensing and geographical information system (GIS) is gaining popularity for assessing the natural resources and so crop water management. Its application may ensure the water stressed crops and its stages accurately and so precision irrigation water management is possible. Multispectral sensors are available to provide exact and accurate spectral and temporal crops signatures that are used for assessing the water requirement and time if irrigation. The vegetation indices and ET are also estimated that will help of scheduling irrigation.

Factors affecting for efficient water management technologies

The National Agricultural Research System (NARS) have developed a plethora of technologies focusing on enhancing water use efficiency at all levels. They are briefly described below.

a. ***Land Leveling:*** Proper land leveling is one of the management options which is generally ignored. It increases the water application efficiency which leads to higher yields as well as increase in water use efficiency (Table.17). It also has a direct impact on the nutrient use efficiency.

b. **Selection of crops and varieties:** The crops and varieties are selected as per the soil, climate and water availability of the region so that production potential of crops may be realized. For example, in arid and semi arid conditions like Rajasthan and Gujarat, growing of millet, pulses and oilseeds may be more beneficial than cereals like rice and wheat.

c. **Sowing time and crop establishment method:** Crop sowing time plays crucial role for saving water. The crop must be planted as per the growing period of the region so that rain water may be utilized very effectively. Similarly the crop establishment methods like raised bed and furrow planting may not only save irrigation water but also improve cop growth and yield.

d. **Crop stages:** The crops should be irrigated at its most sensitive stages for water. Similarly, the deficit irrigation may be adopted under water scarcity and life saving irrigation should be given only to most critical stages. For example, if only one irrigation is available for wheat crop, only apply irrigation at crown root initiation (CRI) stage of wheat. If two irrigations are available, apply water at CRI and boot stages.

e. ***Methods of Irrigation***: The selection of the right method of irrigation is influenced by the soil type, land topography, crops, quality and quantity of water available for irrigation and other site-specific management variations.

Basin and border strip irrigation: Over the years many surface methods of irrigation have been developed. Among them, the check basin method of irrigation is the most popular. It is very important to employ the correct method of water application to minimize the adverse effects of irrigation. It is essential to follow the prescribed guidelines to improve productivity per unit quantity of water applied. Considerable amount of work has been done on these aspects but the guidelines for selecting the appropriate method of irrigation water application are largely ignored. Farmers go in for surface flooding which is an inefficient manner of using this precious natural resource. 10 to 15% efficiency in water use can easily be achieved by adopting the appropriate method.

Table 17. Benefits of laser leveling

Crop	Grain yield (t ha^{-1})		Water saving
	Laser leveled field	Without laser	over without laser leveled field (%)
Paddy	6.79	6.50	38
Wheat	4.75	4.55	20
Sugarcane	112.0	98.75	24
Summer mungbean	0.55	0.38	20
Potato	10.0	9.00	25
Onion	10.0	9.00	20
Sunflower	2.25	2.00	20

***Furrow irrigation*:**The furrow method of irrigation is generally used to irrigate row crops and vegetables, and is suited to soils in which the infiltration rates are between 0.5 and 2.5 cm hr^{1}. It is ideal for slopes varying from 0.2 to 0.5% and a stream size of 1-2 l s^{-1}. Many of the field crops in which water is applied through flooding, check basin or border strip methods, can easily be adapted for furrow irrigation or its modified version *i.e.* raised bed system. Table 18 shows how savings in irrigation water can be achieved by switching over to raised bed furrow irrigation systems. In areas where water for irrigation purposes is scarce, the practice of alternate or skip furrow irrigation can save considerable quantity of water without significantly affecting the yields. In areas requiring surface drainage or where the cultivated crops are sensitive to water logging, the furrows are effective in removal of excess water.

Table 18. Benefits of bed planting observed in India

Crops	Yield on beds (t ha-1)	Yield on flat (t ha-1)	Water savings (% over flat)	Yield increase (% over flat)
Cauliflower	25.9	18.9	36.4	37.0
Green peas	11.9	10.4	32.4	14.5
Okra	34.4	29.1	33.3	18.2
Carrot	36.3	28.6	31.8	26.9
Radish	34.7	26.7	29.4	30.0
Cabbage	33.0	27.8	26.8	18.7
Maize	3.3	2.4	35.5	37.4
Urd bean	1.8	1.4	26.9	33.6

Mungbean	1.6	1.3	27.9	21.8
Wheat	5.1	4.8	26.3	6.4
Pigeonpea	2.2	1.5	30.0	46.7
Gram	1.9	1.6	27.3	17.1
Average			30.3	25.7

Source: Connor *et al.* (2013)

***Surge flow irrigation*:** Excessive water intake and deep percolation losses are major limitations for irrigation through furrows and border strips. Surge flow irrigation, the intermittent application of water in a series of on and off modes of constant or variable time spans has the potential of reducing intake and percolation losses, increasing the irrigation efficiencies and conserving irrigation water. Experiments conducted at several locations have corroborated that surge flow furrow irrigation results in faster water front advance and reduces the volume of water required to complete the advance phase. The studies also confirmed that it increases water distribution efficiency, reduces infiltration rate and the cumulative infiltration depth, and consequently reduces the percolation losses.

***Pressurized irrigation systems*:** To encourage adoption of water efficient technologies, the Government has provided subsidy to popularize pressurized system of irrigation like sprinklers and drip (Trickle) which both together now occupy more than 6 Mha (Table 19).

Table 19. State wise area (ha) under micro-irrigation on March 2012

State	Drip	Sprinkler	Total
Rajasthan	55715	1098133	1153848
Maharashtra	778660	347623	1126283
Andhra Pradesh	665661	323457	989118
Karnataka	293593	385675	679268
Haryana	17772	542431	560204
Gujarat	309520	246222	555742
Madhya Pradesh	110550	176223	286774
Tamil Nadu	206756	28217	234973
West Bengal	589	150576	151165
Chattishgarh	11328	136310	147638
Orissa	12320	46090	58410
Sikkim	3460	1339	4799
Punjab	27615	11559	39174
UP	13973	17328	31301
Kerala	17301	4280	21581
Bihar	752	36628	37381
Nagaland	0	3962	3962
Jharkhand	1273	8842	10115
Goa	874	761	1635

State	Drip	Sprinkler	Total
HP	116	581	697
Arunachal Pradesh	613	0	613
Assam	116	129	245
Mizoram	72	106	178
Uttranchal	38	6	44
Manipur	30	0	30
Others	15000	30000	45000
Total	2543696	3596479	6140176

Source: NCPAH (2012)

The water savings associated with the various interventions developed are summarized below:

Proper scheduling of canals (matching supply with demand)	40 – 60 %
Precision leveling through laser levelers	15 – 20 %
Scientifically designed check basins/border strips	10 – 30 %
Zero tillage	20 – 30 %
Adoption of pressurized irrigation systems	40 – 70 %
Land configuration changes – ridge/furrow or raised / sunken beds	20 – 25 %
Use of tissue cultured eucalyptus, banana, sugarcane, papaya etc	30 %

f. Quality of irrigation water

It is projected that the share of water allocated to agriculture is likely to decrease by 10-15% in the next 20-30 years. In contrast, the demand of water for irrigation will increase for more food grain production. Therefore the non traditional water like sewage and sludge, industrial effluents, kitchen water and rain water harvesting may be taken care of irrigation purposes. The quality of sewage and sludge and industrial effluents may not be good for irrigation, so treatment may be required before irrigation.

g. Proper management

The management of crop and field including irrigation channels, sprinklers etc are required special attention for saving water and increase water use efficiency.

References

Boyer, J. S. 1982 . Plant productivity and environment. Science 218: 443–448.

Fry, Carolyn. 2008. The Impact of Climate Change: The World's Greatest Challenge in the Twenty-first Century. New Holland Publishers Ltd.

Gentine, P., Entekhabi, D., Chehbouni, A., Boulet, G. and Duchemin, B. 2007. Analysis of evaporative fraction diurnal behavior. Agricultural and Forest Meteorology 143 : 13–29

Grafton, Q. R. and Hussey, K. 2011. Water resources. New York, Cambridge University Press.

Jin, M.G., Zhang, R.Q. and Gao, Y.F. 1999. · Temporal and spatial soil water management: a case study in the Heiloonggang region, PR China. Agricultural Water Management, 42:173–187

Kang, S.Z., Gu, B.J., Du, T.S. and Zhang, J.H. 2003. Crop coefficient and ratio of transpiration to evapotranspiration of winter wheat and maize in a semi-humid region. Agricultural Water Management 59:239–254.

Kumar, P., Joshi, P.K. and Mittal, S. 2016. Demand vs Supply of Food in India - Futuristic Projection. Proc Indian Natn Sci Acad 82 (5):1579-1586.

Liu, W. Z. and Zhang, X.C. 2007. Optimizing water and fertilizer input using an elasticity index: a case study with maize in the loess plateau of china. Field Crops Research, 100 : 302–310

Liu, W.Z., Hunsaker, D.J., Li, Y.S., Xie, X.Q.and Wall, G.W.. 2002. Interrelations of yield, evapotranspiration, and water use efficiency from marginal analysis of water production functions. Agricultural Water Management, 56 : 143–151.

Liu, Yi, Li, Shiqing, Chen, Fang, Yang, Shenjio and Chen, Xinping. 2010.Soil water dynamics and water use efficiency in spring maize (*Zea mays* L.) fields subjected to different water management practices on the Loess Plateau, China. Agricultural Water Management 97 (5): 769-775.

Ministry of Agriculture & Farmers Welfare. 2015-16. State of Indian Agriculture-2015-16. Ministry of Agriculture &Farmers Welfare, Govt of India.280p.

Olesen, J.E., Mortensen, J.V., Jorgensen ,L.N. and, Andersen, M.N. 2000. Irrigation strategy, nitrogen application and fungicide control in winter wheat on a sandy soil. I. Yield, yield components and nitrogen uptake. Journal of Agricultural Science 134 : 1–11.

Parasuraman, K., Elshorbagy, A. and Carey, S.K. 2007. Modelling the dynamics of the evapotranspiration process using genetic programming. Hydrological Sciences Journal 52: 563-578.

Sandhu, K.S., Arora, V.K., and Chand, R..2002. Magnitude and economics of fertilizer nitrogen response of wheat in relation to amount and timing of water inputs. Experimental Agriculture 38 : 65–78

Suyker, A.E.and Verma, S.B. 2008. Interannual water vapor and energy exchange in an irrigated maize-based agroecosystem. Agricultural and Forest Meteorology, 148 : 417–427

Turner, N.C. 1987. Crop water deficits: a decade of progress. Advances in Agronomy, 39 : 1–51

Vaux, H.J. Jr. and Pruitt, W.O. 1983. Crop–water production functions.In: Advances in Irrigation 2 (Ed. By D. Hillel. Academic Press, New York.

Questions

a. Short and descriptive

1. Write short notes on following;
   a. Evaporation
   b. Evapotranspiration (ET)
   c. Reference ET
   d. Moisture equivalent
   e. Available water
   f. Lysimeter
   g. Consumptive use of water
   h. IW/CPE ratio.
2. Write short note on water resources of India.
3. Describe the status and significance of ground water in relation to crop production.
4. What do you mean by scheduling irrigation? Also describe the approaches of scheduling irrigation.

5. Define irrigation. Describe important irrigation methods.
6. What are the sources of irrigation and its contribution?
7. Describe the factors affecting the efficient water management.
8. What do you mean by crop water requirement? Describe different methods applied for determination of crop water requirement.

b. Objectives (MCQs)

1 Who proposed that water in 'soil plant–atmosphere continuum' was one of the most important factors affecting crop productivity?

a Boyer, 1982 b Kang *et al.*, 2002

c Jin *et al.* ((1999) d None above

2 Who reported that excessive irrigation can reduce crop WUE, while effective deficit irrigation might result into higher production and WUE?

a Jin *et al.* ((1999) b Kang *et al.*, 2002

c Boyer, 1982 d None above

3 The major source for irrigation is

a Groundwater b Tap water

c River d Pond water

4 How much WUE is to increased by 2017 as proposed by National Water Mission, one of the eight missions launched as a part of the National Action Plan on Climate Change of the Government of India?

a 20% b 25%

c 30% d 40%

5 The average annual rainfall in India is approximately

a 117 cm b 180 cm

c 150cm d 200 cm

6 How much was the net irrigated area in India during 2012-13?

a 66.1 m ha b 92.6 mha

c Both a & b d None above

7 Which of the following Indian state has the highest net irrigated area?

a Punjab b U.P.

c Delhi d Haryana

8 Which of the following is a irrigation water source in India?

a Canal b Tanks

c Tube wells d All above

9 Which of the following sequence of states leading in canal irrigation?

a U.P.> Rajasthan > Haryana > A.P.> Punjab
b U.P. >Punjab > Bihar > Rajasthan> M.P.> Haryana
c A.P.> T.N.> Maharashtra > Punjab> W.B.> Karnatika
d Rajasthan >U.P. > A.P.> Haryana >Punjab

10 Which of the following sequence of states leading in tube wells irrigation?

a U.P., >Punjab > Bihar > Rajasthan> M.P.> Haryana
b U.P.> Rajasthan > Haryana > A.P.> Punjab
c A.P.> T.N.> Maharashtra > Punjab> W.B.> Karnatika
d Rajasthan >U.P. > A.P.> Haryana >Punjab

11 Which of the following sequence of states leading in tank irrigation ?

a U.P., >Punjab > Bihar > Rajasthan> M.P.> Haryana
b U.P.> Rajasthan > Haryana > A.P.> Punjab
c A.P.> T.N.> Maharashtra > Punjab> W.B.> Karnatika
d Rajasthan >U.P. > A.P.> Haryana >Punjab

12 Which of the following sequence of states leading in wells irrigation?

a Rajasthan> M.P.> Gujrat > Maharashtra> T.N.)
b (M. P.> Karnatika > Orissa> A.P.> Kerela)
c A.P.> T.N.> Maharashtra > Punjab> W.B.> Karnatika
d Rajasthan >U.P. > A.P.> Haryana >Punjab

13 Which of the following is a soil moisture constant?

a Field capacity (FC)
b Permanent wilting point (PWP)
c Moisture Equivalent
d All above

14 Which of the following has the largest irrigation by source?

a Canal
b Tanks
c Tube wells
d All above

15 Estimated annual precipitation (including snow fall) in India is

a 4000 BCM
b 1869 BCM
c 3000 BCM
d None above

16 Average precipitation during monsoon (June-September) in India is

a 4000 BCM
b 1869 BCM
c 3000 BCM
d None above

17 What is an average annual runoff in India?

a 4000 BCM b 1869 BCM

c 3000 BCM d None above

18 Utilizable surface water resources in India is

a 690 BCM b 1121 BCM

c 431 BCM d 1869 BCM

19 Utilizable ground water resources in India is

a 690 BCM 1121 BCM

c 431 BCM 1869 BCM

20 Total utilizable water resources in India is

a 1121 BCM b 690 BCM

c 1869 BCM d 431 BCM

21 What is the ultimate irrigation potential of India?

a 58.5 mha b 64.1 mha

c 81.4 mha d 140 mha

22 How much irrigation potential has been created under major and medium irrigation projects?

a 58.5 Mha b 44.4 Mha

c 34.4 Mha d None above

23 How much irrigation potential has been created under minor projects?

a 58.5 mha b 44.4 mha

c 34.4 mha d 81.4 mha

24 How much irrigation potential has been created from ground water resources?

a 81.4 mha b 52.5 mha

c 62.8 mha d 64.1mha

25 How much irrigation potential has been created from surface water resources through minor projects?

a 17.3 mha b 52.5 mha

c 62.8 mha d 44.4 mha

26 How much was the irrigated area in India in 1951?

a 22.6 mha b 52.5 Mha

c 62.8 Mha d 140 mha

27 Permanent wilting point is

a Soil moisture level or points at which plants are wilted and cannot recoups its original state even after irrigation

b Soil moisture content that retains in soil pores after drainage of gravitational water

c Both a & b

d None above

28 At what soil water potential the plants get wilted?

a -15 bar

b 1.5 MPa

c Both a & b

d None above

29 Ultimate wilting point refers to

a A point of minimum available moisture in a soil which results a plant or a crop is just about to die

b Force at which plants get wilted

c It is the lowest limit of available water

d All above

30 Available Soil Moisture (ASM) is

a Soil water between FC and PWP

b Soil water between FC and critical available soil water

c Soil water between FC and gravitational water

d None above

31 The water holding capacity in course, fine and loamy sands is

a 6 – 10 %

b 12 – 19 %

c 10 – 14 %

d 13 – 20 %

32 The water holding capacity in sandy and fine sandy loam is

a 6 – 10 %

b 12 – 19 %

c 10 – 14 %

d 13 – 20 %

33 The water holding capacity in very fine sandy and silt loam is

a 6 – 10 %

b 12 – 19 %

c 10 – 14 %

d 13 – 20 %

34 The water holding capacity in sandy and silty clay soil is

a 6 – 10 %

b 12 – 19 %

c 10 – 14 %

d 13 – 20 %

35 The water holding capacity in clay and silty clay loam is

a 6 – 10 %

b 12 – 19 %

c 10 – 14 %

d 14 – 20 %

36 'Critical soil water level' refers to

a Water is available to plants

b Just below this level of available soil water the crop growth and development is drastically reduced

c Both a & b

d None above

37 Moisture Equivalent (ME) refers to

a Amount of water retained by an initially saturated soil sample against centrifugal force of 1000 times of gravity force for half an hour

b Moisture level at which the soil switches from water-loving to water-Repelling

c A point of minimum available moisture in a soil which results a plant or a crop is just about to die

d None above

38 Water Holding Capacity (WHC) is also referred as

a 'Saturation capacity' of the soil

b 'Maximum water holding capacity 'of the soil

c Both a & b

d None above

39 Veihmeyer and Hendrickson (1950) theory refers to

a The soil water is equally available to plants in the range of FC and PWP

b Water availability to plants is maximum at FC and decreased towards PWP and plants suffer water stress before PWP.

c The available water is divided into groups i.e. readily available water and critical level of available water.

d None above

40 Richards and Wadleigh (1952) theory of available soil water says

a The soil water is equally available to plants in the range of FC and PWP

b Water availability to plants is maximum at FC and decreased towards PWP and plants suffer water stress before PWP

c The available water is divided into groups i.e. readily available water and critical level of available water.

d None above

41 Available water holding capacities in very coarse to coarse sand is

a 18 cm

b 45 cm

c 15cm

d 57 cm

42 Which of the following soil group has the highest water holding capacity?

a Very coarse to coarse sand
b Clays and Clay Loams
c Moderately Coarse Sandy Loams
d Peats and Mucks

43 Available water holding capacity of moderately coarse sandy loam soil is

a 45cm
b 57cm
c 63cm
d 73cm

44 Available water holding capacity of medium loam soil is

a 57cm
b 45cm
c 73cm
d 63cm

45 Which of the following laboratory method is used to measure available soil water?

a Thermo gravimetric methods
b Volumetric methods
c H_2SO_4 method
d All above

46 Which of the following laboratory method is used to measure the available soil water?

a Sprit burning method
b Immersion method
c Volumetric method
d All above

47 Which of the following method is used to measure available soil water in field ?

a Tensiometer
b Electrical resistance method
c Neutron moisture meter
d All above

48 Which of the following apparatus measures available soil water in field conditions?

a Microwave radiation method (TDR)
b Pressure membrane apparatus
c Automatic moisture meter
d All above

49 Which of the following climatic condition has maximum daily consumptive use of water?

a Hot and dry weather
b Cool and dry weather
c Hot and humid weather
d Moderate humid weather

50 Peak consumptive water use under cool and humid condition will be around

a 3 mm/day
b 5 mm/day
c 4 mm/day
d 8 mm/day

51 Crop water requirement can be measured by?

a Lysimeter method
b Soil moisture depletion method
c Field experimentation
d All above

52 Which of the following climatological approach is used for measure crop water requirement?

a Blaney-criddle method (1950)
b Thornthwaite method (1948)
c Penman method
d All above

53 Lysimeter is a

a Tank or container filled of soil installed in field to measure water requirement of crops
b Apparatus to measure soil water potential
c Instrument measures soil depth
d None above

54 Lysimeters measures

a Crop ET
b Deep percolation
c Run off
d All above

55 USWB Class A Pan Evaporimeter measures

a Evaporation
b Relative humidity
c Transpiration
d All above

56 Sunken Screen Pan Evaporimeter was developed by

a Sharma (1968)
b Sharma and Dastane (1968)
c Slatyer and Mcllroy (1958)
d None above

57 Which of the following evaporimeter gives very crude measurement of evaporation?

a Pitche evaporimeter
b Pan Evaporimeter
c Sunken screen evaporimeter
d None above

58 Which of the following evaporimeter gives higher evaporation?

a USWB Class A pan evaporimeter
b Pitche evaporimeter
c Both a & b
d None above

59 Irrigation methods depends on

a Amount of water availability for irrigation
b Soil types
c Climatic conditions
d All above

60 Which of the followings irrigation method is applied for irrigating crops?

a Surface irrigation
b Drip (tickle) irrigation
c Sprinkler (over head) irrigation
d All above

61 Which of the following is a type of surface irrigation?

a Wild flooding | b Furrow method
c Check basin method | d All above

62 Furrow method is more suited to soils having infiltration rates;

a 0.5 and 2.5 cm hr^{-1} | b 0.8 and 2.5 cm hr^{-1}
c 0.5 and 3.5 cm hr^{-1} | d 0.5 and 2.8 cm hr^{-1}

63 Sub surface method is applied under conditions of

a High water table | b High infiltration rate
c Low water holding capacity of soil | d All above

64 Sprinkler irrigation is well suited in

a Arid conditions | b Undulated field conditions
c Semi arid conditions | d All above

65 Volume of irrigation water varies with

a Crop and its stage | b Soil type
c Irrigation method and irrigation efficiency | d All above

66 The depth of irrigation is influenced by

a Effective root zone | b Water retention capacity of soil
c Soil water depletion | d All above

67 Which of the following is a plant based approach for scheduling irrigation?

a Plant appearance | b Plant growth and development
c Critical growth stages | d All above

68 Which of the following is a soil based approach for scheduling irrigation?

a Soil water content | b Critical level of soil water
c Soil water potential | d All above

69 Which of the following is a climatological approach for scheduling irrigation?

a Modified Penman method | b Penman (1948) Mehtod
c Thornthwaite (1948) Method | d All above

70 Which of the following is a moisture sensitive growth stage of rice?

a Panicle initiation | b Heading
c Flowering | d All above

71 Which of the following is a moistur sensitive growth stage of wheat?

a Crown root initiation (CRF) b Jointing

c Milking & grain filling and tillering d All above

72 Which of the following is a moisture sensitive growth stage of maize?

a Tasseling b Silking

c Knee high d All above

73 Which of the following is a moisture sensitive growth stage of sorghum?

a Booting b Flowering

c Grain filling d All above

74 Which of the following is a moisture sensitive growth stage of sunflower?

a Seed filling b Flowering

c Budding d All above

75 Water requirements of rice crop is

a 900-2500mm b 700-1300

c 450-650mm d 500-800mm

76 Water requirements in wheat crop is

a 900-2500mm b 700-1300

c 450-650mm d 500-800mm

77 Water requirements in maize crop is

a 900-2500mm b 500-800mm

c 450-650mm d 450-650mm

78 Water requirements in sugarcane crop is

a 900-2500mm b 700-1300

c 450-650mm d 500-800mm

79 Water requirements in pea is

a 500-700mm b 500-800mm

c 600-800mm d 150-200mm

80 Which of the following method of irrigation is more suitable for rice?

a Border check method b Furrow method

c Check method d Strip method

81 Which of the following method of irrigation is more suitable for maize?

a Furrow method b Border method

c Both a & b d None above

82 Which of the following method of irrigation is suitable for sunflower?

a Border strip | b Drip methods
c Furrow | d All above

83 Which of the following is an efficient water management technologies ?

a Land Leveling | b Selection of crops and varieties
c Crop stages | d All above

84 Which of the following is the most popular method of irrigation?

a Check basin irrigation | b Furrow irrigation
c Surge flow irrigation | d Pressurized irrigation

85 Which of the following field intervention has the highest water saving?

a Zero tillage | b Precision leveling
c Pressurized irrigation | d Land configuration

86 How much water should be available per capita/day for normal daily consumption?

a 1560 m^3 | b 1700 m^3
c 1850 m^3 | d 1900 m^3

87 Per capita water storage in India is

a 220 m^3 | b 520 m^3
c 620 m^3 | d 820 m^3

88 What is the water scarcity level of water availability?

a $<1000\ m^3$ | b $<1200\ m^3$
c $<1300\ m^3$ | d $<1400\ m^3$

89 Which of the following formula is used to calculate potential ET?

a PET = Kc x Epan | b $ET_{crop} = K_c \times ET_0$
c $E_0 = K_p \times E_{pan}$ | d $ET_0 = C \times ET_0^*$

90 Which of the following formula is used to calculate actual crop ET?

a PET = Kc x Epan | b $ET_{crop} = K_c \times ET_0$
c $E_0 = K_p \times E_{pan}$ | d $ET_0 = C \times ET_0^*$

91 Which of the following formula is used to calculate actual evaporation under field conditions?

a PET = Kc x Epan | b $ET_{crop} = K_c \times ET_0$
c $E_0 = K_p \times E_{pan}$ | d $ET_0 = C \times ET_0^*$

92 Which of the following formula is used to calculate reference crop ET?

a PET = Kc x Epan | b $ET_{crop} = K_c \times ET_0$
c $E_0 = K_p \times E_{pan}$ | d $ET_0 = C \times ET_0^*$

93 Crop water requirement is equal to

a $ER + IR + \Delta S$

b $ER + IR + \sum_{i=1}^{n} X \frac{Msi\text{-}Mhi}{100} xASi\ X\ Di$

c Both 'a' & 'b'

d None above

94 Inflow of water is equal to

a $P + IR + \Delta S$

b $E + T + DP + R$

c $ER + IR + \Delta S$

d All above

95 Out flow of water is equal to

a $P + IR + \Delta S$

b $E + T + DP + R$

c $ER + IR + \Delta S$

d All above

96 What is the value of crop factor (K_c) of summer crops?

a 0.6

b 0.8

c 1.0

d 1.2

97 What is the value of crop factor (K_c) of winter crops?

a 0.6

b 0.8

c 1.0

d 1.2

98 What is an average value of K_c of wheat crop?

a 0.80-0.90

b 0.75-0.90

c 1.65-1.20

d 0.85-1.05

99 What is an average value of K_c of maize crop?

a 0.80-0.90

b 0.75-0.90

c 1.65-1.20

d 0.85-1.05

100 What is an average value of K_c of rice crop?

a 0.80-0.90

b 0.75-0.90

c 1.65-1.20

d 0.85-1.05

101 What is an average value of K_c of sugarcane crop?

a 0.80-0.90

b 0.75-0.90

c 1.65-1.20

d 0.85-1.05

102 What is the value of Pan evaporimeter coefficient (K_p)?

a 0.6

b 0.7

c 0.8

d 0.9

103 What is true for Veihmeyer and Hendrickson (1950) theory of available soil water to plants?

a Water is equally available in between FC to PWP
b Water availability to plants decreases from FC to PWP
c Both 'a' and 'b'
d None above

104 Which of the following soil type has the highest available water holding capacity?

a Sandy soil
b Loam soil
c Peat and muck
d Clay soils

105 Reference evapo-transpiration was proposed by

a Doorenbos and Pruitt (1973)
b Pruitt and Angus (1960)
c Penman (1948)
d M S Swaminathan (1964)

106 Which of the following factor affects the pan coefficient (K_p) values of pan evaporimeter?

a Wind velocity
b Relative humidity
c Crop type and its growth stage
d All above

107 Irrigation water of small flow can be measured by

a Current meter
b Water meter
c Float method
d All above

108 Irrigation water of high flow can be measured by

a Orifice
b Weir
c Flumes
d All above

109 Optimum IW/CPR ratio for scheduling irrigation of gram is

a 0.40
b 0.40 – 0.60
c 0.75 – 1.00
d 0.80 -1.00

110 Optimum IW/CPR ratio for scheduling irrigation of sorghum is

a 0.40
b 0.40 – 0.60
c 0.75 – 1.00
d 0.80 -1.00

111 Optimum IW/CPR ratio for scheduling irrigation of wheat is

a 0.40
b 0.40 – 0.60
c 0.75 – 1.00
d 0.80 -1.00

112 Optimum IW/CPR ratio for scheduling irrigation of maize is

a 0.40
b 0.40 – 0.60
c 0.75 – 1.00
d 0.80 -1.00

113 Optimum IW/CPR ratio for scheduling irrigation of rape & mustrad is

a 0.60 – 0.80 b 0.75

c 1.20 – 1.50 d 0.70 – 1.0

114 Optimum IW/CPR ratio for scheduling irrigation of soybean is

a 0.60 – 0.80 b 0.75

c 1.20 – 1.50 d 0.70 – 1.0

115 Optimum IW/CPR ratio for scheduling irrigation of potato is

a 0.60 - 0.80 b 0.75

c 1.20 – 1.50 d 0.70 – 1.0

116 Optimum IW/CPR ratio for scheduling irrigation of onion is

a 0.60 - 0.80 b 0.75

c 1.20 – 1.50 d 0.70 – 1.00

117 Approximate ET value of barley crop is

a 30 cm b 40 cm

c 60 cm d 90 cm

118 Approximate ET value of wheat crop is

a 30 cm b 40 cm

c 60 cm d 90 cm

119 Approximate ET value of barley rice is

a 30 cm b 40 cm

c 60 cm d 90 cm

120 Approximate ET value of sugarcane is

a 30 cm b 40 cm

c 60 cm d 90 cm

121 Approximate ET value of pea is

a 15 cm b 35 cm

c 50 cm d 90 cm

122 Approximate ET value of rape & mustard is

a 15 cm b 35 cm

c 50 cm d 90 cm

123 Approximate ET value of barley rape & mustard is

a 15 cm b 35 cm

c 50 cm d 90 cm

124 Approximate ET value of berseem is

a 15 cm
b 35 cm
c 50 cm
d 90 cm

125 Water requirement of grape is

a 500-100 mm
b 700-1000 mm
c 900-1200 mm
d 1200-2200 mm

126 Water requirement of pineapple is

a 500-100 mm
b 700-1000 mm
c 900-1200 mm
d 1200-2200 mm

127 Water requirement of citrus fruits is

a 500-100 mm
b 700-1000 mm
c 900-1200 mm
d 1200-2200 mm

128 Water requirement of banana is

a 500-100 mm
b 700-1000 mm
c 900-1200 mm
d 1200-2200 mm

129 Optimum soil moisture tension for scheduling irrigation of rapeseed & mustard is

a 0.4 – 0.5 bar
b 0.5 bar
c 0.7 – 0.8 bar
d None above

130 Optimum soil moisture tension for scheduling irrigation of wheat is

a 0.4 – 0.5 bar
b 0.5 bar
c 0.7 – 0.8 bar
d None above

131 Which of the following critical growth stages of gram are sensitive to irrigation?

a Flowering and pod development
b Crop establishment, pre flowering and grain filling
c Root crown initiation, tillering, jointing, panicle initiation, flowering and dough stages
d Seedling, tasseling, silking and dough stages

132 Which of the following critical growth stages of sorghum are sensitive to irrigation?

a Flowering and pod development
b Crop establishment, pre flowering and grain filling
c Root crown initiation, tillering, jointing, panicle initiation, flowering and dough stages
d Seedling, tasseling, silking and dough stages

133 Which of the following critical growth stages of wheat are sensitive to irrigation?

a Flowering and pod development

b Crop establishment, pre flowering and grain filling

c Root crown initiation, tillering, jointing, panicle initiation, flowering and dough stages

d Seedling, tasseling, silking and dough stages

134 Which of the following critical growth stages of maize are sensitive to irrigation?

a Flowering and pod development b

b Crop establishment, pre flowering and grain filling

c Root crown initiation, tillering, jointing, panicle initiation, flowering and dough stages

d Seedling, tasseling, silking and dough stages

135 Which of the following critical growth stages of groundnut are sensitive to irrigation?

a Flowering, pegging and pod development

b Peak vegetative , buttoning and flowering and seed filling

c 6-8 leaf stage, flowering and capsule development

d Initial tuberization and stolonization

136 Which of the following critical growth stages of sunflower are sensitive to irrigation?

a Flowering, pegging and pod development

b Peak vegetative , buttoning and flowering and seed filling

c 6-8 leaf stage, flowering and capsule development

d Initial tuberization and stolonization

137 Which of the following critical growth stages of castor are sensitive to irrigation?

a Flowering, pegging and pod development

b Peak vegetative , buttoning and flowering and seed filling

c 6-8 leaf stage, flowering and capsule development

d Initial tuberization and stolonization

138 Which of the following critical growth stages of potato are sensitive to irrigation?

a Flowering, pegging and pod development

b Peak vegetative , buttoning and flowering and seed filling

c 6-8 leaf stage, flowering and capsule development

d Initial tuberization and stolonization

139 Which of the following critical growth stages of tobacco are sensitive to irrigation?

a Seedling, pre flowering and flowering

b Rosette, flowering and capsule development

c Formative stage

d Flowering and capsule development

140 Which of the following critical growth stages of safflower are sensitive to irrigation?

a Seedling, pre flowering and flowering

b Rosette, flowering and capsule development

c Formative stage

d Flowering and capsule development

141 Which of the following critical growth stages of sugarcane are sensitive to irrigation?

a Seedling, pre flowering and flowering

b Rosette, flowering and capsule development

c Formative stage

d Flowering and capsule development

142 Which of the following critical growth stages of linseed are sensitive to irrigation?

a Seedling, pre flowering and flowering

b Rosette, flowering and capsule development

c Formative stage

d Flowering and capsule development

143 Which of the following is a most sensitive critical growth stage of wheat to irrigation?

a Root crown initiation

b Panicle initiation

c Flowering

d Dough

144 Which of the following is a most sensitive critical growth stage of maize to irrigation?

a Tasseling
b silking
c Early grain formation
d All above

145 Which of the following is a most sensitive critical growth stage of rice to irrigation?

a Panicle initiation
b Heading
c flowering
d Tillering and transplanting

146 Which of the following is a most sensitive critical growth stage of sorghum to irrigation?

a Booting
b Flowering
c Grain filling
d All above

147 Net irrigation requirement can be calculated by

a $\sum_{iri}^{n}\frac{Fci-Mbi}{100}\times Asi\times Di$
b $\sum_{iri}^{n}\frac{Msi-Mhi}{100}\times Asi\times Di$
c $\sum_{i=1}^{n}\frac{M_{1i}-M_{2i}}{100}\ xAsi\ x\ Di$
d All above

148 Soil water contribution to crop can be calculated by

a $\sum_{iri}^{n}\frac{Fci-Mbi}{100}\times Asi\times Di$
b $\sum_{iri}^{n}\frac{Msi-Mhi}{100}\times Asi\times Di$
c $\sum_{i=1}^{n}\frac{M_{1i}-M_{2i}}{100}\ xAsi\ x\ Di$
d All above

149 Seasonal consumptive use of water of a crop can be calculated by

a $\sum_{iri}^{n}\frac{Fci-Mbi}{100}\times Asi\times Di$
b $\sum_{iri}^{n}\frac{Msi-Mhi}{100}\times Asi\times Di$
c $\sum_{i=1}^{n}\frac{M_{1i}-M_{2i}}{100}\ xAsi\ x\ Di$
d All above

150 Leaf water potential can be measured by

a Pressure bomb
b TDR
c Atmometer
d Tensiometer

Answer

1. a	2. a	3. a	4. a	5. a	6. a	7. a	8. d	9. a	10. a
11. c	12. a	13. d	14. c	15. a	16. c	17. b	18. a	19. c	20. a
21. d	22. a	23. d	24. d	25. a	26. a	27. a	28. a	29. d	30. a
31. a	32. c	33. b	34. d	35. d	36. b	37. a	38. c	39. a	40. b
41. a	42. d	43. a	44. a	45. d	46. d	47. d	48. d	49. a	50. a
51. d	52. d	53. a	54. d	55. a	56. b	57. a	58. b	59. d	60. d
61. d	62. a	63. d	64. d	65. d	66. d	67. d	68. d	69. d	70. d
71. d	72. d	73. d	74. d	75. a	76. c	77. b	78. a	79. d	80. a
81. b	82. d	83. d	84. a	85. c	86. b	87. a	88. a	89. a	90. b
91. c	92. d	93. c	94. a	95. b	96. a	97. b	98. a	99. b	100. c
101. d	102. b	103. a	104. c	105. a	106. d	107. d	108. d	109. a	110. b
111. c	112. d	113. a	114. b	115. c	116. d	117. a	118. b	119. c	120. d
121. a	122. b	123. c	124. d	125. a	126. b	127. c	128. d	129. a	130. b
131. a	132. b	133. c	134. d	135. a	136. b	137. c	138. d	139. a	140. b
141. c	142. d	143. a	144. a	145. a	146. a	147. a	148. b	149. c	150. a

12

Mulching and Its Impact on Agricultural Production

Outlines

Introduction

The word mulch was coined by Wagganor in 1960. It is derived from the German word '*molsch*' means "easy to decay," and mulches have widely been used for vegetable production since ancient times (Lightfoot 1994). Therefore the term mulching refers to the practice of covering the soil surface with different organic and inorganic materials, known as mulch materials to reduce moisture loss, balance wide variations in diurnal soil temperatures especially in the root zone, control weeds, insect-pests and diseases and finally enhance crop yield (Nalayini 2007; Kader *et al*. 2019).

Mulching also known as sheet composting, has widely been used for vegetable production since ancient times (Lightfoot 1994). Mulching minimizes water runoff, improve infiltration capacity of soil, restrain weed population via shading and perform as obstacle in evapotranspiration (Rathore *et al*., 1998). Mulching also has positive environmental effects like regulation of soil temperature and plant roots, minimum nutrient losses, cut down soil erosion and compactness and improved physical conditions of soil (Ngouajio and McGiffen, 2004; Lamont, 2005). The significance of mulching was realized in the late 1930 because it can alter the surrounding conditions of agricultural lands, forest areas and horticultural lands, however some adverse effects of mulching were also reported (Bedford and Pickering, 1919). Deep mulching was practiced in 1941 for trees and shrub plantation (Pirone, 1941) for protection from drought stress, frost injury and freezing damage of crop plants. The by-products of agricultural and forestry domains have been used as mulches in the mid 1900s (Clifford and Massello, 1965). Besides, other materials i.e. trimmings of trees and shrubs, wastes of animals, stubbles and residues of crop plants were also used as mulch materials. Landscape mulches were also seen in 1957 but there was no scientific research work carried out on them (Anonymous 1957).

Mulches may be both organic and synthetic or inorganic origin. The organic mulches consist of animal and plant residues. The most commonly used organic mulches include crop residues, straws, husks, grasses and cover crops (live mulches), saw dust, compost, and manures (Rathore *et al.,* 1998), while the most frequently used inorganic mulch throughout the world is polyethylene plastic mulch. Presently, the application of plastic mulching in agriculture is very popular throughout the world and its usage is increasing day by day. In 1999, plastic mulch was used on an area of more than 22 m ha of cultivated land worldwide (Miles *et al.*, 2005). In 2002, plastic mulch was spread on an area of 15 m ha in China (Xing *et al.*, 2003). On an average, 7 m tons of plastic sheet are being used annually as mulch worldwide (Espi *et al.*, 2006) and 1.44 m tons in USA alone (Shogren, 2001). The black plastic mulch is generally used in agricultural lands globally (Schales, 1990). Utilization of black plastic mulch is mostly due to its low prices in relation to other mulches (Ngouajio and McGiffen, 2004).

Therefore mulching is practiced for various purposes:

1. To produce crops, vegetables, fruits and other high value crops in off season,
2. To protect crops from extreme weather conditions like drought, frost, snow etc.,
3. To sustain the fertility of soil,
4. To protect the soil from water and wind erosion,
5. To preserves the soil moisture,
6. To accent landscape plantings,
7. To prevent weed growth,
8. To protect the plant and their produce from attack of insect-pest and diseases,
9. To moderate the soil thermal regime throughout the cropping season,
10. To help in production of clean and quality products,
11. To provide a "finished" look to the garden, and
12. Increasing overall crop production.

Types of mulches

Mulches may be classified in two types i.e. organic and inorganic mulches as listed below:

A. Organic mulch

1. Compost/FYM/Peat
2. Crop residues
3. Saw dust
4. Peat
5. Grass clippings
6. News papers
7. Cover cropping
8. Live mulch

B. Inorganic mulch

1. Polythene mulch
2. Aluminum-coated plastic and foil
3. Gravel and stone mulch

A. Organic mulch

1. ***Compost/Manure/peat***

The compost/FYM/peat is commonly used in kitchen and home gardens to improve the soil organic matter before crop sowing. It also provides maero and micro nutrients to crop plants. These materials can also be used as mulch with 2" – 3" thick layer over soil surfaces. The compost or FYM should be well decomposed otherwise they may damage the crops. Normally the decomposed compost is free of weed seeds and cost effective.

Organic mulch

Live mulch

Plastic mulch | Bark mulch

Soil mulch | Stone mulch

Fig. 1: Types of mulching

2. Crop residue mulching

The extra production of rice residues in *Kharif* and wheat in *Rabi* season is a threat nowadays for its easy disposal. These burning of rice and wheat residues in field are a serious environmental pollution in whole North India. Therefore these crop residues can be safely utilized as a mulch @ 6t/ha either directly or after chopping. The thickness of mulching may vary from 2 to 6 inches depending upon objectives of the study and requirement of the crops. In dry region, its depth is put 2-3 inches to save soil moisture. However, these materials are prone to micro-fauna which make theme desirable for medium duration crops only (Fig.1). The grass clippings are also used as mulches, maintain 4-6 inch depth but high thickness may reduce aeration to crops. Under high moisture and heavy rainfall conditions, the rotting of grasses may smell bad. Therefore the thickness needs to be moderated as per requirement of crop conditions (Table.1).

3. Sawdust

The soft and hard wood sawdust is as mulch materials. It takes more time to decompose due to its high C:N ratio and nutrient status, so its use may immobilize the nitrogen resulting into reduced nitrogen availability in beginning. Addition of nitrogen may fasten the decomposition of sawdust. It is always better to use 2 inch thick layer of decomposed sawdust as mulch material for better results.

B. Newspaper

The newspapers may also be used as mulch. It is light and easy to handle but also easy to break and blow even in light wind velocity. Moreover the weeds may also penetrate the sheet of newspaper. Well, it is safe, free from chlorine and there is no risk of lead from ink.

4. Inorganic mulches

1. Polyethylene mulches

The use of polythene is increasing in agriculture in many forms including as mulch too. It is available in different colors and thickness. Mostly it is used in production of fruit, vegetables and high value crops. The thickness and color of polythene may suitably chosen as per requirement of the crops Normally thin film is required for better germination, weed management by soil solarization and thick is used for fruit production and saving from frost and light snowfall (Table.2). Similarly, black polythene is used in *Kharif* season, white in summer season and silver colored for insect repellent (Table.3). Polyethylene is preferred mainly due to its better permeability to long wave radiation which in turn increases the temperature around the plants during the night times.

Plastic mulch may be of following types;

i. ***Photo degradable plastic mulch*** - It is destroyed by sun in a shorter period. Examples-Plastigone, biolane etc.

ii. ***Bio degradable plastic mulch***- They are made from starch of potato, corn and wheat and easily degraded in sun light.

iii. ***Black plastic mulch*** - Normally used in winter to increase soil temperature and weed control.

iv. ***White plastic mulch*** – Normally used in summer season for establishing crops. They also repels some insects.

v. ***Reflective silver film*** - They reflect sun light and maintain root zone cooler.

vi. ***Transparent film*** - They are used for soil solarization because of raising soil temperature drastically. They are also known as clear plastic mulch.

vii. ***Selectively permeable plastic mulch*** – They are also known as reflective or colored plastic mulch. These colored films allow certain wavelength of light to pass and warm the soil.Yellow plastic mulch repels white fly.

viii. ***Photo selective plastic mulch*** – Also known as infra red transmitting (IRT) mulch. It warms soil more than black plastic mulch. They are more effective to control weeds.

2. Aluminum-coated plastic and foil

There is in general, no use of aluminum coated plastic and foil as mulch in agricultural production, however, it is used at smaller scale in vegetable production. They also reduce insect pests like aphids and viruses in vegetable crops. It is very effective in weed control but its high cost restricts its use at larger scale.

Criteria for selection of organic mulch

The selection of suitable organic mulch is pre-requisite for better result. Therefore following points should be taken care in selection of organic mulch materials;

i. It should be easily available,

ii. It should be cost effective,

iii. It should be free from the attack of insects/pests particularly termites and diseases,

iv. It should be properly decomposed if compost/FYM/Vermicompost is used because undecomposed or partially decomposed mulch material should not have a nitrogenase activity after the application of mulch,

v. It should not have any antagonistic effect on the crop, and

vi. The thickness of organic mulch should be preferred as per the requirement of crops, stage of crops and season.

Table 1. Mulching materials and specifications for use

S.N	Depth of mulch	Mulch material	Comments
1	1-2 inches	Peat Moss	Attractive, available but expensive for large scale. Moisture shall be maintained at all times.
		Bark	Ground and packaged commercially
2	2 inches	Peat	Soak well before using. Breaks down rapidly
		Sawdust	Usage of weathered sawdust if mixing with soil. Fresh sawdust causes N leaching up on break down in soils
3	2-3 inches	Decomposed litter Grass clippings	Rapid break down and formation of stable aggregates
		Maize cobs	Gives moderate acidic nature to soils. Excellent for soil structural stability
		Peanut Hulls	Improves soil structure, fairly durable. May harbor nematodes
		Shredded Tree Leaves	Cheaper, act as breeding stations for termites and ants

S.N	Depth of mulch	Mulch material	Comments
4	3-4 inches	Hay	Unattractive but repeated use adds nutrients which lasts for years. Minimizes weeds and holds moisture well.
		Pine needles	Adds acid to soil. Will not mat down. Fairly durable.
5	6 inches	Whole tree leaves	Decomposes rapidly and rich in plant nutrients
		Crop residues/ Straw	Rich in potassium but decomposition takes time and may lead to immobilization
6	¼ inches	News paper	Black ink newspapers can be used. Color dyes may be unsafe to soil micro-flora and fauna if composted and used

Source: Pedda Ghouse Peera *et al.* (2020)

Precautions for using organic mulch

Following precautions should be taken before use of organic mulches:

i. The crop should be weed free before application of mulch,
ii. Soil pH must be monitored periodically, and
iii. Rotting and fungal attack on mulches should be properly managed timely.

Table 2. Type and suitability of plastic materials

S.N		Type of plastic mulch	Suitable condition
1	***Based on thickness***		
	i	Thicker mulch	Fruit crops and plantations
	ii	Thin film	For early germination
	iii	Thin, transparent film	Soil solarization
	iv	Transparent film	Weed management through solarization
	v	Perforated mulch	Weed management in cropped land
2	***Based on color***		
	i	Black film	Monsoon season
	ii	White film	Cropped field in summer
	iii	Silver color	Film Insect repellent
	iv	Appropriate color mulch	Yield

Source: Pedda Ghouse Peera *et al.* (2020)

Table 3. Selection of mulches according to ecological conditions

S.N	Ecological situation	Type of mulch
1.	Rainy season	Perforated mulch
2.	Orchard and plantation	Thicker mulch
3.	Soil solarization	Thin transparent film
4.	Weed control through solarization	Transparent film
5.	Weed control in cropped land	Black film
6.	Sandy soil	Black film
7.	Saline water use	Black film
8.	Summer cropped land	White film
9.	Insect repellent	Silver color film
10.	Early germination	Thinner film

Source: Iqbal *et al.* (2020)

Advantages of mulching

Mulching has many beneficial effects on agriculture and environment, so it is divided into two parts;

a. Effect on agriculture.
b. Effect on environment.

a. Effect on agriculture

1. Regulation of soil temperature

The maintenance and regulation of soil temperature is a very critical factor for optimum plant growth. In extreme hot and dry conditions such as in deserts, mulches decrease the temperature of barren soils by 10°C (Martin and Poultney 1992). Mulching can keep the soil cool during very hot climatic conditions, while at normal/warm temperature in chilling days (Kader *et al.*, 2019). The soil is covered by mulches that acts as an insulating layer protecting soil by heating and maintain soil temperature. The effectiveness of mulching differs with kind and thickness of mulch materials, therefore proper selection of mulches is more important. The coarse mulches are more beneficial than the finer ones of the same category in controlling the temperature at favorable conditions (Horowitz and Thomas 1994). The soil temperature will increase more at application of plastic mulch, sand mulch, stone/gravel mulch than crop residue or straw mulch. Similarly, the dry leaves or crop residue mulch has more favorable effect than compost or FYM mulch. On other hand, heavy or thicker application of mulches is more effective in regulation of soil temperature than thin mulch application. The living mulch and organic mulch materials are far better in maintaining the favorable soil temperature compared to other mulch materials (Montague and Kjelgren, 2004). The turf mulch (living mulch) liberates more water vapors through the evapotranspiration process and decrease the soil surface temperature due to its evaporative cooling effect despite the fact that turf mulch uses more water of soil in relation to the non-living mulch materials(Montague and Kjelgren, 2004).

Among the inorganic mulches, the gravel mulch is more temperature moderating than the solid-type mulch such as concrete. The synthetic mulches (plastics, fabrics and asphalt) have been reported as inefficient in temperature regulation as these could increase the soil temperature instead of controlling at the favorable level (Kader *et al.,* 2019). Besides synthetic mulches, some of the other types of mulches like pine bark mulching increased the soil surface temperature which affected the nearest leaves to transpire more water through the leaves (Zajicek and Heilman, 1994).

Presently, the rice and wheat straw are available in plenty and is burnt during crop season making hue and cry all around in North India due to environmental pollution. It has been reported that application of 6-8 ton/ha crop residues of rice and wheat may be helpful in crops as well as vegetable for increasing water use efficiency. In one of the study in hot arid conditions of western Rajasthan demonstrated a drop of 1.1-5.6°C soil temperature during summer and 0.6-3.2°C during winter (December-January) at application of rice straw @ 6 t/ha at 20 cm depth (Awasthi *et al.*, 2006). The FYM mulch was also reported to lower the maximum temperature by 0.4 to 4.0°C (Singh *et al.*, 2011). Dawar and Pal (2021) reported an application of oat straw @ 6 t/ha improved WUE of spring sunflower at Pantnagar (Uttarakhand).

2. *Improvement in soil moisture content*

Soil and water conservation is a major issue in dryland conditions. High wind, temperature and weeds are major culprits to reduce soil moisture. Weed infestation may reduce soil moisture to the tune of 25% through ET (Harris *et al.*, 2004). Mulching helps to cover soil surface and reduce evaporation losses up to 35%. Among mulch materials, organic mulches are found more beneficial than inorganic mulches as they not only reduce evaporation losses but also act as a sponge to absorb water at rainfall and contributes to water requirement of crops and reduce irrigation requirement (Smith, 2000 and Ahmad *et al.*, 2020). The run off is also reduced up to 43% by straw mulch (Borst and Woodburn, 1942). Cover cropping also helps to conserve soil moisture but mulching was found more effective because of competition for water resources in cover cropping (Downer and Hodel, 2001). The straw mulch also improves infiltration moisture retention and surface run off leading to better root growth and crop yield. Increase in infiltration rate by 10-12% and in moisture retention (-0.01 MPa) by 6-20% was observed in the inter-row spaces of cowpea (*Vigna ungiculata*, var. FS-68) with native weeds @ 6 t ha-1 (Gupta, 1986).

a. ***Soil fertility improvement***

The organic and living mulches have beneficial impacts on soil quality including soil fertility status but the type of the material, soil characteristics and climatic conditions determine the increase, decrease or no effect on soil nutrients. Application of organic mulches was found more beneficial because they add organic carbon

mulch and hamper photosynthesis and glucose formation (Ahmad *et al.*, 2020). The crop residues and forest-produced materials are also used in uncomposted forms and cause reduction of weed species in agricultural and landscape lands (Niggli *et al.*, 1988). Mulching with polyethylene plastic sheets is more effective for weed control in cotton than organic mulches. The weed density was reduced with increasing the amount of mulch (Ahmad *et al.*, 2020).

4. *Minimizing soil compaction and erosion*

The mulching materials also protect the soil from wind and water erosion phenomenon and reduce the compaction. Mulch materials break the speed of water in hilly areas and increase the infiltration rate of soil and to maintain the slope stabilization, however engineering techniques are preferred instead of merely dependent on mulching method (Chalker-Scott, 2007). The live mulch in form of grasses, legumes and sometimes barley crop are used to reduce the run off in sloppy lands. The straw and crop residues are commonly used not only to reduce evaporation losses but also for reducing the runoff rate. In hills, pine leaves and sticks are used to reduce run off flow rate to avoid soil and water loss caused by heavy rainfall. Application of organic mulches is very effective to reduce soil compaction by heavy machinery and maintain favorable environment for both crop and soil microorganisms.

5. *Plant growth, development, and yield*

Mulching helps in improving soil aggregation and organic matter, conserving soil moisture, moderating soil temperature, reducing weed growth and wind and water erosion resulting into favorable environment for germination, plant establishment and growth and finally crop yield from 7-47% varying from crop to crop and agro-climatic conditions (Table.4). The positive effect of mulching of crop growth and yield will be higher in summer season than winter season. The organic mulches like compost, crop residues, straw, leaves, and grass clippings are better to support plant growth and development because of they are easily decomposed and add organic matter with higher water holding capacity than other types of mulches like inorganic mulches, gravel/stone and polythene mulch. Therefore the types of much materials and time of application are also important. The rapid decomposers mulch materials like compost, FYM, leaves and grass clippings are more suitable than moderate decomposers i.e. paper, straw and hay and crop residues as well as slow decomposers such as stem bark and woody chips. It is reported that sometimes the slow decomposing mulch materials like bark mulch can make nutritional problem for fast-growing tree species due to low production of plant root biomass (Davis, 1994). The root development and density was higher under organic mulches than plastic mulch (Fausett and Rom, 2001), un-mulched soil (Watson, 1988) and living mulch material (Green and Watson, 1989). Plastic mulch is used for soil solarization that in turn control weeds, conserves soil moisture and support better root development. Further, mulching effect is more in wider spaced crops than narrow spaced crops. The mulching has also been found beneficial for establishment of fruits and plantation crops. The plastic mulching is very useful in vegetables and fruits and the yield increase was recorded from 10-70% over unmulched conditions. The yield variation among different crops varies with thickness of plastic sheet, application time and management strategies (Table.5).

Table 4. Effect of mulching on yield improvement of various crops

S.N	Crop	Economic yield (t/ha)			References
		Unmulched	Mulched	% increase over unmulched	
1	Maize	2.49	4.76	47.68	Hashim *et al.* (2013)
2	Mustard	0.41	0.61	32.78	Saikia *et al.* (2014)
3	Soybean	1.32	1.57	15.92	Arora *et al.* (2011)
4	Cotton	1.67	2.22	24.77	Ahmad *et al.* (2015)
5	Lentil	0.80	0.89	10.11	Alami-Milani *et al.* (2013)
6	Mungbean	1.02	1.36	25.00	Jiang *et al.* (2012)
7	Rice	5.39	6.83	21.08	Devasinghe *et al.* (2015)
8	Potato	21.00	30.00	20.00	Dvorak *et al.* (2009)
9	Tomato	6.02	8.27	27.20	Moursy *et al.* (2015)
10	Chickpea	5.91	7.32	19.26	Fetri *et al.* (2015)
11	Barley	3.77	4.27	11.70	Malecka and Blecharczyk (2008
12	Wheat	7.24	7.79	7.06	Yang *et al.* (2006)

Source: Iqbal *et al.* (2020)

Table 5. Effect of plastic sheet thickness on yield of various vegetable crops

S.N	Name of crops)	Plastic sheet thickness (µ	Yield increased (%)
1	Brinjal	25	20-27
2	Okra	25	48-55
3	Potato	50	49-50
4	Tomato	25	65-70
5	Cucumber	25	44-52
6	Caabbage/Cauliflower	50	10-71
7	Chilli	60	60
8	Carrot	50	10-50

Source: Pedda Ghouse Peera *et al.* (2020)

It is also reported based on research findings on precision agriculture at Y S Parmar University of Horticulture and Forestry (HP) that the thickness of plastic mulch should be 20-25, 40-50 and 50-100 micron for annual-short duration, biennial-medium duration and perennial-long duration crops.

6. *Mitigation of salt stress*

The dry condition with high temperature enhance the evaporation losses and salts are coming with water from deeper soil and deposit on soil surface resulting soil saline and alkaline. Under these situations, mulching is very helpful to solve the salinity problem by reducing evaporation and more water retention in soil. Various studies exhibited that the effect of salts toxicity was reduced with mulching (Ansari *et al.*, 2001 and Yobterik and Timmer, 1994). Organic mulches again are more helpful for the reclamation of soil and induce the desalinization of soil (Dong *et al.*, 1996). Organic mulches can be decomposed by the microbes resulting in the degradation of many harmful residues and lessens the salt contamination (Gan *et al.*, 2003 and Smith and Skroch, 1995). However, the plastic mulches are not commonly used for the mitigation of salts (Sun *et al.* 1994).

7. *Reduce pest and disease infestation*

Mulching has indirect effect to reduce infestation of pest and diseases by improvement plant nutrition, better drainage, moderation of soil temperature, improved soil aggregation, and conservation of soil moisture. Consequently, mulch materials provide a healthy atmosphere for better crop growth and development with safe from pathogenic organisms (Turchetti *et al.*, 2003). Mulching mat reduce diseases infestation in following ways;

i. Mulching may protect crop plants from direct contact of pathogenic spores and other organisms coming with rainfall,
ii. Mulch materials are decomposed and produce some toxic chemicals that may degrade the pathogenic spores and reduce chances of disease occurrence (Chalker-Scott, 2007),
iii. Mulching with the aim of maximum colonization of microbes will also be helpful for the biological control of many harmful pathogens (Entry *et al.*, 2005),
iv. Use of polythene mulch sterilizes the soil and increases the soil temperature that may kill number of soil born pathogenic organisms.

However the mulching may also cause the disease infestation if the mulch materials are not good and already infested before use. The sawdust was used as a mulch material in gardens and it was noticed that sawdust caused root rot disease in many plant species (Matthews, 1948). Black polyethylene mulch caused the bacterial soft rot disease as compared to woodchips or bark mulch (Davis, 1994). Therefore, selection of the mulch material is imperative because mulches are a part of IPM plan.

8. *Soil aggregation/soil structure/aggregate stability*

Soil organic carbon (SOC) contributed to the soil on decomposition of organic mulches helps in the amelioration of soil structure. SOC is positively and strongly correlated with aggregates stability e.g. macro-water-stable aggregates are stabilized by fairly undecomposed organic binding agents.

9. *Soil biological properties*

Moisture retention and optimum soil temperature regimes and the presence of continuous substrate from mulching materials are conducive for decomposers. Organic mulch is known to attenuate the rise in soil temperature and decreased diurnal variations and also increase microbial activity. Furthermore, it was demonstrated that an organic mulch increase earthworm density by 155% over bare soil.

b. Effect on environment

1. *Remediation of heavy metals*

Some of the plant species are grown for removing the heavy metals from soil. If the dry leaves and its bark is used as mulch, it may help to reduce the toxic effect of heavy metals in soil. For examples, the leaves of arborvitae, eucalyptus, pine, and poplar may helpful for removing heavy metals from the soil solution (Salim and El-Halawa, 2002). Using woodchips and compost in forest areas can make complex with the copper metal and convert it into the form which is not toxic for proper growth of crop plants (Kiikkila *et al.*, 2002).

1. Diminishing of pesticide use

It is already clear that weed and pest-diseases infestation are reduced due to mulching. If so, there is almost no use of herbicides and pesticides to control the weeds, insect-pests and diseases. Hence mulching diminished indirectly the use of pesticides in the field. It will reduce the environmental pollutions as well as chemical residues in food products.

2. Visual enhancement by mulch

Application of mulches increases the aesthetic value of the mulched fields because of following reasons;

i. The field may looks very clean without grasses and other weeds,
ii. If the smooth and colored gravel or stone are used for mulching, the mulched area will be attractive and good looking,
iii. Mulching materials like lime grass, *Vetiveria zizanoides* etc. will not only save the soil but also spread good fragrance in the field.

Disadvantages of mulching

Mulching is not always beneficial but also has few drawbacks as listed below;

i. Large scale mulching is of course cost intensive.
ii. Sufficient availability of mulching materials is little difficult.
iii. Use of high C:N mulching materials viz., saw dust, straw results in temporary immobilization or starvation for nutrients in crops.
iv. Change in the soil reaction due to continuous use of same mulching material.
v. Application of top-dressed fertilizers is difficult.

vi. Biodegradability of plastic materials is a problematic to safe environment.

vii. Organic mulches may harbor termites, rodents and other pathogenic organisms and needs frequent irrigation and spray of termiticides and other pesticides.

viii. Some of the organic mulches have allelopathic effects on crops.

ix. Some mulches have high flammability like crop residues, polythene , bark mulches etc.

Economic importance of mulching

Mulching is a low cost and environmental friendly process as the locally produced materials like straw, crop residues, grasses, leaves and bark of tree plants, weeds, homemade compost etc. are used for mulching, so it does not cost to farmers. It also reduces the use of chemicals for controlling weeds, pests and diseases. It also decrease the cost incurred in irrigation, soil conservation and soil reclamations as mulching improves soil moisture, reduces salt deposition and also conserve soil resources from wind and water erosion. It also increases aesthetic values of the land. Mulching also help in maintaining safe and sound environment.

Conclusion

Mulching is an age old practice in agriculture mainly because of its significant role in conserving soil moisture, reducing evaporation losses and suppressing weed population. Different types of mulch materials mainly crop residues and plastic mulches are widely used in field and horticultural crops. They exhibited significant impacts on growth, yield, and quality of various crops but some adverse effects are also reported. However the mulching has more beneficial effects than few disadvantages in real field conditions. The organic mulch should be easily decomposable, capable of adding substantial amount of organic matter to the soil and should not have allelopathic effects on crops. The leguminous crop residues should be selected to poor soils as they add nitrogen in soils. The C:N ratio of organic mulches ranges between 30:1 and 80:1 and they normally do not pose too much acidic, basic or toxic effects. The selection of mulch materials, its quantity and thickness and also application time matching with crop requirement are prerequisites for larger benefits of mulching. The optimum thickness of inorganic mulches like black and white plastics is 50 to 75 micron, while for organic mulches it is 2 inch for clay and 3-4 inch for loam soils. The black polythene should be used in winter season and white in summer season. In present era of climate change and increasing weather related extreme events like drought, flood, erosion, frost and snowfall, the use of mulching will increase in agriculture for sustainable crops and horticultural crops in near future.

References

Ahmad, S., Raza, M.A.S., Saleem, M.F., Zaheer, M.S., Iqbal, R., Haider, I., Aslam, M.U., Ali, M. and Khan, I.H. 2020. Significance of partial root zone drying and mulches for water saving and weed suppression in wheat. J.Anim. Plant.Sci. 30: 154–162.

Anonymous, 1957. Handbook on Mulches. Brooklyn Botanical Garden Records 13: 1–79.

Ansari, R., Marcar, N.E., Khanzada, A.N., Shirazi, M.U. and Crawford, D.F. 2001. Mulch application improves survival but not growth of Acacia ampliceps Maslin, Acacia nilotica L. and Conocarpuslancifolius L. on a saline site in southern Pakistan. Int. J. Rev 3:158–163.

Awasthi, O.P., Singh, I.S. and Sharma, B.D. 2006. Effect of mulch on soil hydrothermal regimes, growth and fruit yield of brinjal (*Solanum melongena* L.) under arid conditions. *Indian* Journal of Horticulture, 63: 192-194.

Balasubramanian, D., Arunachalam, K., Arunachalam, A. and Das, A.K. 2013. Effect of water hyacinth (*Eichhornia crassipes*) mulch on soil microbial properties in lowland rainfed rice-based agricultural system in Northeast India. Agric. Res.2: 246–257.

Bedford, H.A.R. and Pickering, P.S.U. 1919. Science and Fruit Growing: Being an Account of the Results Obtained at the Woburn Experimental Fruit Farm Since Its Foundation in 1894. Macmillan.

Borst, H.L. and Woodburn, R. 1942. The effect of mulching and methods of cultivation on runoff and erosion from Muskingham silt loam. J. Agric. Engin. 23: 19–22.

Chalker-Scott, L. 2007. Impact of mulches on landscape plants and the environment - a review. J. Environ. *Hortic*. 25: 239–249.

Clifford, E.D. and Massello, J.W. 1965. Mulching materials for nursery seedbeds. Tree Planters' Notes 72:18–22.

Davis, J.M. 1994. Comparison of mulches for fresh-market basil production. J. Hortic. Sci. 29: 267–268.

Dong, B.B., Zhu, H.T., Zhong, Z.K., and Ye, G.F. 1996. Study on ecological effect of the forest land under-crop sowing and mulching of coastland soil by newly planted. Acta. Agric. Zheji. 8:154–157.

Downer, J. and Hodel, D. 2001. The effects of mulching on establishment of Syagrus romanzoffiana (Cham.) Becc.,Washingtonia robusta H. Wendl. and Archontophoenix cunninghamiana (H. Wendl.) H.Wendl. and Drude in the landscape. Sci. Hortic. 87: 85–92.

Entry, J.A., Strausbaugh, C.A. and Sojka, R.E. 2005. Compost amendments decrease Verticilliumdahliae infection on potato. Com. Sci. Utiliz. 13: 43–49.

Espi, E., Salmeron, A., Fontecha, A., Garcia, Y. and Real, A.I. 2006. Plastic films for agricultural applications. J. Plas. Film. Sheet. 22: 85-102.

Fausett, J.B. and Rom, C.R. 2001. The effects of transitioning a mature high-density orchard from standard herbicide ground-cover management system to organic ground-cover management systems. Arkan. Agric. Exp. Stat. Res. Ser.483:33–36.

Gan, J., Zhu, Y., Wilen, C., Pittenger, D. and Crowley, D. 2003. Effect of planting covers on herbicide persistence in landscape soils. J. Environ. Sci. Tech. 37:2775–2779.

Green, T.L. and Watson, G.W. 1989. Effects of turf grass and mulch on establishment and growth of bare root sugar maples. J. Arboric. 15: 268–272.

Harris, R.W., Clark, J.R. and Matheny, N.P. 2004. Arboriculture: integrated management of landscape trees, shrubs, and vines, 4th edition. Prentice Hall, Inc, Upper Saddle River, NJ.578 pages.

Horowitz, M., and Thomas, J.M. 1994. Couverture du sol pour la gestion des mauvaises herbes (Soil cover for weed management). In Maitrise des adventices par voie non chimique. Communications de la quatrieme conference internationale IFOAM (pp. 149–54).

ICAR. 2010. Degraded and Wastelands of India, ICAR, New Delhi.

Iqbal1, R., Raza, M. A. S., Valipour, M., Saleem, M. F., Zaheer, M. S., Ahmad, S., Toleikiene, M., Haider, I., Aslam, M. U. and Nazar, M. A. 2020. Potential agricultural and environmental benefits of mulches—a review. Bulletin of the National Research Centre 44-75. (https://doi.org/10.1186/s42269-020-00290-3).

Kader, M.A., Singha, A., Begum M.A., Jewel, A., Khan, F.H. and Khan, N.I. 2019. Mulching as water-saving technique in dry land agriculture. Bulletin of the National Research Centre 43:1–6.

Kiikkila, O., Derome, J., Brugger, T., Uhlig, C. and Fritze, H. 2002. Copper mobility and toxicity of soil percolation water to bacteria in metal polluted forest soil. J. Plant. Soil. 238: 273–280.

Lamont, W.J. 2005. Plastics: modifying the microclimate for the production of vegetable crops Hortic. Technol. 15: 477–481.

Lightfoot, D. R. 1994. Morphology and ecology of lithic-mulch agriculture. Geo. Rev. 25: 172–185.

MacLean, R.H., Litsinger, J.A., Moody, K., Watson, A.K. and Libetario, E.M. 2003. Impact of Gliricidiasepium and Cassia spectabilis hedgerows on weeds and insect pests of upland rice. J. Agric. Eco. Enviro. 94 : 275–288.

Martin, P.J. and Poultney, R. 1992. Survival and growth of clove seedlings in Zanzibar. Effects of mulching and shade crops. J. Tropic. Agric 69:365–373.

Matthews, B. 1948. Report on sawdust. New Zealand Gardener. 4: 391–394.

Miles, M. 2005. Identification, pest status, ecology and management of the green mirid, a pest of cotton in Australia. PhD thesis, University of Queensland, St Lucia, Australia.

Montague, T. and Kjelgren, R. 2004. Energy balance of six common landscape surfaces and the influence of surface properties on gas exchange of four containerized tree species. Sci. Hortic. 100: 229–249.

Nalayini, P. 2007. Poly-mulching a case study to increase cotton productivity. Senior scientist, Central Institute for Cotton Research, Regional Station, Coimbatore.

Ngouajio, M. and McGiffen, M. E. 2004. Sustainable vegetable production: effects of cropping systems on weed and insect population dynamics. Acta. Hortic. 638:77–83.

Niggli, U., Weibel. F.P., and Potter, C.A. 1988. Unkrautregulierung ineiner Dauerkulturdurch Bodenbedeckungmitorganischen Materialien (Weed control in a perennial crop using an organic mulch). Zeitschrift fur Pflanzenkrankheiten und Pflanzenschutz 11: 357–365.

Pedda Ghouse Peera., S. K., Debnath, S. and Maitra, S. 2020. Mulching: Materials, advantages and crop production. In: Protected Cultivation and Smart Agriculture edited by Sagar Maitra, Dinkar J Gaikwad and Tanmoy Shankar © New Delhi Publishers, New Delhi. pp. 55-66. (ISBN: 978-81-948993-2-7, DOI: 10.30954/NDP-PCSA.2020.6).

Pickering, J.S. and Shepherd, A. 2000. Evaluation of organic landscape mulches: composition and nutrient release characteristics. Arboric. J. 23:175–187.

Pirone, P.P. 1941. Freak weather damages trees and shrubs. New Jersey. Agric. 23: 3.

Rathore, A.L., Pal, A.R. and Sahu, K.K. 1998. Tillage and mulching effects on water use, root growth, and yield of rain-fed mustard and chickpea grown after lowland rice. J. Sci. Food. Agric. 78: 149–161.

Salim, R. and El-Halawa, R.A. 2002. Efficiency of dry plant leaves (mulch) for removal of lead, cadmium and copper from aqueous solutions. Pro.Safe. Environ. Pro. 80: 270–276.

Schales, F. 1990. Agricultural plastics use in the United States. Proceedings of the 11th International Congress of Plastics in Agriculture, 54-56.

Shogren, R.L. 2001. Biodegradable mulches from renewable resources.J. Sus. Agric. 16:33–47.

Singh, A.K. and Singh, R.B. 1999. Effect of mulches on nutrient uptake of Albiziaproceraand subsequent nutrient enrichment of coal mine overburden.J. Tropic. Sci. 11: 345–355.

Singh, I.S., Awasthi, O.P., Sharma, B.D., More, T.A. and Meena, S.R. 2011. Soil properties, root growth, water-use efficiency in brinjal (*Solanum melongena*) production and economics as affected by soil water conservation practices, Indian J. Agric. Sci., 81: 760-763.

Smith, M.W. 2000. Cultivar and mulch affect cold injury of young pecan trees. J.Amer. Pomo. Soc. 54: 29–33.

Sun, D., Dickinson, G. and Bragg, A. 1994. The establishment of Eucalyptus camaldulensis on a tropical saline site in north Queensland. Australia. J. Agric. Eco. Environ. 48: 1–8.

Tilander, Y. and Bonzi, M. 1997. Water and nutrient conservation through the use of agro-forestry mulches, and sorghum yield response. J. Plant. Soil. 197: 219–232.

Turchetti, T., Maresi, G., Nitti, D., Guidotti, A. and Miccinesi, G. 2003. Il mal dell'inchiostronelMugello (Fi): danniedapprocci di difesa (Chestnut ink disease in the Mugello area: damage and control). Monti e Boschi. 54: 22–26.

Watson, G.W. 1988. Organic mulch and grass competition influence tree root development. J. Arboric. 14: 200–203.

Wilen, C.A., Schuch, U.K. and Elmore, C.L. 1999. Mulches and sub-irrigation control weeds in container production. J. Environ. Hortic. 17: 174–180.

Xing, N.Q., Zhang, Y.Q., and Wang, L.X. 2003. The study on dry-land agriculture in North China.(In Chinese.) Chinese Agriculture Press, Beijing.

Yobterik, A.C. and Timmer, V.R. 1994. Nitrogen mineralization of agro-forestry tree mulches under saline soil conditions.In: R.B. Bryan (ed.) Adv. Geo-eco.27: 181–194.

Zajicek, J.M. and Heilman, J.L.1994. Transpiration by crape myrtle cultivars surrounded by mulch, soil, and turfgrass surfaces. J. Hortic. Sci. 26: 1207–1210.

Questions

A. Short and descriptive

1. Write short notes on following;
   a. Organic mulch
   b. Inorganic mulch
   c. Plastic mulch
   d. Sawdust
   e. Live mulch
   f. Grass clippings
2. Define mulch. Also discuss different types of mulches used in agriculture.
3. Describe the economic importance of mulching in agriculture.
4. What are the advantages of and disadvantages of mulching? Describe in detail with suitable examples.
5. Justify that mulching is not always a beneficial strategy for boosting the crop production.

B. Objective (MCQs)

1 Mulching helps in

a Improvement in soil fertility
b Moisture conservation
c Improvements soils structure
d All above

2 Positive environmental effects of mulching is

a Regulation of soil temperature
b Reduction in nutrient losses
c Cutting down soil erosion
d All above

3 Deep mulching is practiced for

a Trees
b Shrub plantation
c Protection from drought stress
d All above

4 Mulches may be

a Organic
b Synthetic
c Inorganic
d All above

5 The organic mulch consists of

a Animal residues
b Plant residues
c Both a & b
d None above

6 The most commonly used organic mulch is

a Crop residues
b Husks
c Manures
d All above

7 Mulching is practiced for

a To sustain the soil fertility
b To protect the soil from water and wind erosion
c To preserves the soil moisture
d All above

8 Mulching is useful in

a To accent landscape plantings
b To protect the plant and their produce from attack of insect-pest and diseases
c To help in production of clean and quality products
d All above

9 Mulching is practiced for

a Moderate the soil thermal regime throughout the cropping season.
b To prevent weed growth
c Increasing overall crop production
d All above

10 Compost/Manure/peat is used as a mulch in

a Kitchen garden
b Home gardens
c Both a & b
d None above

11 Which of the following organic mulch stays in filed for long time?

a Compost
b Crop residue mulching
c Sawdust
d All above

12 Which of the following organic mulch contains highest amount of C:N ratio?

a Compost
b Crop residue mulching
c Sawdust
d All above

13 Which of the following is a non-organic mulch?

a Polyethylene mulch
b Aluminum-coated plastic
c Al-Foil
d All above

14 Which of the following feature organic mulch must possess?

a It should not have any antagonistic effect on the crop
b It should be free from the attack of insects/pests particularly termites and diseases
c It should be easily decomposable
d All above

15 Which of the following irrigation method is most suitable under mulched crop.

a Drip irrigation
b Laying mulch
c Bed preparation
d All above

16 The plastic mulch is available in form of

a Rolls of 120 to 150 cm wide with length
b Rolls of 120 to 170 cm wide with length
c Rolls of 200 to 250 cm wide with length
d Rolls of 140 to 180 cm wide with length

17 Which of the following is mulch material has better permeability of light?

a PVC
b Polyethylene
c Foil
d None above

18 Which of the following is required to be conserved in dryland conditions?

a Soil conservation
b Water conservation
c Both a & b
d None above

19 The living mulch competes for

a Moisture
b Oxygen
c Nutrients
d All above

20 Which of the following is the best mulch?

a Shredded bark b Manures
c Polyethylene d PVC

21 Optimum thickness of compost mulch is

a 2– 3 inch b 2 – 6 inch
c 2 – 8 inch d 2 – 9 inch

22 Which of the following organic mulch is commonly used?

a Compost b Crop residue mulching
c Sawdust d All above

23 Which of the following mulch is mainly confined to vegetable crops?

a Aluminum-coated plastic and foil b Polyethylene mulches
c Shredded bark d None above

24 Which of the following plastic material is best suitable for fruit crops and plantations?

a Thicker polythene b Thin polythene film
c Thin transparent film d Transparent film

25 Which of the following plastic material is best suitable for weed management in cropped lands?

a Perforated mulch b Thin film
c Thin transparent film d Transparent film

26 Which of the following plastic material is best suitable for early germination?

a Thicker polythene b Thin polythene film
c Thin transparent film d Transparent film

27 Which of the following plastic material is best suitable for winter season crops?

a Back film b White film
c Silver colour film d Appropriate colour mulch

28 Which of the following plastic material is best suitable for insect repellent?

a Back film b White film
c Silver color film d Appropriate color mulch

29 Which of the following plastic material is best suitable for cropped field in summer?

a Back film b White film
c Silver color film d Appropriate color mulch

30 Which of the following plastic material is best suitable for soil solarization?

a Thicker mulch | b Thin film
c Thin transparent film | d Transparent film

31 Which of the following mulch is found to be useful in achieving early harvest and yield of muskmelon?

a Black polyethylene | b White and black polyethylene
c Red and black plastic mulches | d Thin film

32 Which of the following drawback belongs to mulching?

a Large scale mulching is cost intensive | b Unavailability of mulching materials
c Application of top-dressed fertilizers is difficult | d All above

33 Mulching may be harmful mainly due to

a Change in the soil reaction | b Release of toxic chemicals
c Source of pests and diseases | d All above

34 Mulch can act as barrier against

a Evaporation | b Beating action of rain drops
c Frost & snow | d All above

35 Mulching reduces the incidence of the diseases through

a Direct mechanism | b Indirect mechanism
c Both a & b | d None above

36 Which of the following is an indirect effect of mulching against pests and diseases?

a Improvement in nutrition of plants | b Better drainage
c Moderation of soil temperature | d All above

37 Mulch materials can decrease

a Water stress level on plants | b Weed infestation
c Erosion | d All above

38 Which of the following is the economic importance of mulching?

a Reducing soil evaporation | b Controlling soil temperature
c Reducing weed growth | d All above

39 Which of the following is used as a mulch?

a Water hyacinth | b Sawdust
c Coir pith | d All above

40 Which of the following is improved by mulching?

a Water holding capacity of soil b Infiltration rate of soil
c Moisture content of soil d All above

41 Which of the following is improved by using woody mulches?

a Nutrient level in soil b Plant foliage
c Both a & b d None above

42 Which of the following mulch reduces the soil temperature?

a White plastic film b Back plastic film
c PVC d Silver color film

43 What is the main role of mulching?

a Improves soil moisture b Prevents plants from drying
c Reduces soil erosion d All above

44 Over-Mulching can kill plants mainly due to

a Suffocation of the roots b Disease infestation
c Poor aeration d All above

45 Which of the following plastic mulch increases more soil temperature?

a Polythene mulch b Back film
c Plastic film mulching d Silver color film

46 Which color of plastic mulch is considered best?

a Brown b Red
c Black d All above

47 Which of the following type of mulching is practiced?

a Surface mulching b Polythene mulching
c Straw mulching d All above

48 Which of the following is a disadvantage of mulching?

a Cost intensive b High C:N
c Unavailability mulching materials d All above

49 Which of the following is a disadvantage of mulching?

a Change in the soil reaction b Application of top-dressed fertilizers is difficult
c Difficult biodegradation of plastic materials d All above

50 Which of the following is the most effective and suitable method of irrigation under mulched crop?

a Drip irrigation b Subsurface irrigation
c Surface irrigation d All above

51 The thickness of news paper mulch should be

a ¼ inches | b 2 inch
c 5 inch | d 7 inch

52 The thickness of hay mulch should be

a ¼ inches | b 3 inch
c 5 inch | d 7 inch

53 Polyethylene is preferred owing to its better permeability to long wave radiation which in turn

a Increase the temperature around the plants during the night times | b Decrease the temperature around the plants during the night times
c Both a & b | d None above

54 Mulching can keep the soil

a Cool during very hot climatic conditions | b Normal /warm temperature in chilling days
c Both a & b | d None above

55 The soil temperature will increase highest at application of

a Plastic mulch | b Sand mulch
c Stone /gravel mulch | d All above

56 Which of the following is a culprit to reduce soil moisture?

a High wind | b High temperature
c Weeds | d All above

57 Which of the following is not a organic mulch?

a Bark | b Dry grasses
c Paddy straw | d Stone and gravel

58 Nitrogen deficiency under mulching is caused mainly due to

a Microbial immobilization | b Reduced mineralization
c More leaching | d All above

59 Phosphorus availability will be more under mulching at

a 0-10 cm soil depth | b 10-15 cm soil depth
c 15-20 cm soil depth | d 20-25 cm soil depth

60 Which of the following plant leaves if used as mulch may reduce the level of heavy metals?

a Pine | b Eucalyptus
c Popular | d All above

after decomposition. It was reported that wood chips, straw, green manures, and bark mulches added more nutrients as compared to inorganic mulches (Ansari *et al.*, 2001, Pickering and Shepherd, 2000 and Singh *et al.*, 1991). However, application of mulches with higher ability need to be carefully managed in crops as they can damage the sensitive crops, living organisms, and water resources (Chalker-Scott, 2007). Application of uncomposted bark or straw mulches, saw dust may reduce the nitrogen content of soil without affecting the plant nutrition, resulting in reduced pollution of watershed (Niggli *et al.*, 1988). Mulches having high nitrogen content increase the yield of crop plants (Tilander and Bonzi, 1997). Husk mulch was found more effective in increasing the nutrient levels in soil than leaf or grass mulch (Singh and Singh, 1999). On contrary, it is also reported that mulching reduces N availability mainly due to microbial immobilization and less mineralization due to low temperature. Further the nitrogen deficiency may also be caused due to more possibility of leaching under mulching. In some cases, nitrogen deficiency and chlorosis was reported in maize under stubble mulching. Similarly the water hyacinth as a potential organic substrate can stimulate the growth of diversity of microbial population in agricultural soils. In general, soil respiration and microbial population were significantly greater in mulched plots compared to control. The vermicompost mulched plots were recorded higher bacterial and fungal counts at both surface and sub-surface soil layers than control (Balasubramanian *et al.*, 2013). Phosphorus accumulation was also reported higher near soil surface under stubble mulching.

3. *Reduce weed infestation*

Mulching is very helpful in reducing weed population up to great extent either by restricting germination of weed seeds (Wilen *et al.*, 1999) or smothering and killing of weeds by allelopathic effects of mulches. Besides, mulching also creates favorable environment for beneficial microbes which feed on the weed species or weed seeds (Chalker-Scott, 2007). Different type of organic and inorganic mulch materials are used for mulching but live mulch like cover cropping is also very helpful to reduce weed population mainly due to competition for space, water and light for growth and development. The black polythene mulch is better option to reduce germination of weed seeds compared to white polythene mulch. The compost mulch is full of nutrients and moisture, so not suitable for weed control rather it will act as a weed seed bank (Maclean *et al.*, 2003).

The organic and inorganic mulches are better as they make field neither compacted nor have competition with main crops. Inorganic mulches like gravel or stone mulch) when used at a 4-cm depth will avoid the colonization of weed species (Winkel *et al.*, 1995). Besides, inorganic mulches provide better soil conditions such as moisture conservation and moderation of temperature but there is no light beneath these mulches and weed seeds who require light for germination do not germinate (Chalker-Scott, 2007). Thick layer of saw dust mulching is also helpful to reduce weed seed germination by not allowing light penetration beneath the

61 Mulching must be practiced for saving soil moisture in

a Arid and semi arid areas b Humid areas

c Temperate areas d All above

62 Mulching must be practiced against frequent snow and hails in

a Arid and semi arid areas b Humid areas

c Temperate areas d All above

63 Mulching reduces infestation of diseases by act of

a Protecting from direct contact of pathogenic spores

b Producing toxic chemicals harmful to pathogenic spores

c Producing more microorgnainms killing pathogens

d All above

64 Which of the following statement related to synthetic mulch is correct?

a Photo selective plastic mulch is also known as infra red transmitting mulch

b Plastigone is an ultraviolet activated and biodegradable plastic mulch

c Biogene is a photodegradable plastic mulch and degrade into CO_2

d All above

65 Which of the following is correct for biodegradable mulch?

a It is made of starch of corn, potato and wheat b It is broken down by microbes

c It behaves like plastic mulch d All above

66 Which of the following statement is correct for soil solarization?

a It disinfects soil and reduces soil borne diseases b Clear plastic mulch is used

c It is very useful for nursery raising and early germination d All above

67 Which of the following mulch is useful for increasing soil temperature maximum?

a Crop residues b White plastic mulch

c Infra red transmitting mulch d Black plastic mulch

68 The optimum thickness of plastic mulch for annual crops is

a 20-25 micron b 40-50 micron

c 50-100 micron d None above

69 The optimum thickness of plastic mulch for biennial and medium duration crops is

a 20-25 micron b 40-50 micron

c 50-100 micron d None above

70 The optimum thickness of plastic mulch for perennial and long duration crops is

a 20-25 micron b 40-50 micron

c 50-100 micron d None above

Answer

1. **d**	2. **d**	3. **d**	4. **d**	5. **c**	6. **a**	7. **d**	8. **d**	9. **d**	10. **c**
11. **c**	12. **c**	13. **d**	14. **d**	15. **a**	16. **a**	17. **b**	18. **c**	19. **d**	20. **a**
21. **a**	22. **b**	23. **a**	24. **a**	25. **a**	26. **b**	27. **a**	28. **c**	29. **b**	30. **c**
31. **a**	32. **d**	33. **d**	34. **d**	35. **c**	36. **d**	37. **d**	38. **d**	39. **d**	40. **d**
41. **c**	42. **a**	43. **d**	44. **d**	45. **a**	46. **a**	47. **d**	48. **d**	49. **d**	50. **a**
51. **a**	52. **b**	53. **a**	54. **c**	55. **c**	56. **d**	57. **d**	58. **d**	59. **a**	60. **d**
61. **a**	62. **c**	63. **d**	64. **d**	65. **d**	66. **d**	67. **c**	68. **a**	69. **b**	70. **c**

Annexure
Important ICAR Institutes and Agricultural Universities (India)

1. **Institutes under ICAR**

a. ***Deemed Universities - 4***

1. Indian Agricultural Research Institute, New Delhi
2. National Dairy Research Institute, Karnal
3. Indian Veterinary Research Institute, Izatnagar
4. Central Institute on Fisheries Education, Mumbai

b. ***ICAR institutes - 64***

1. Central Agricultural Research Institute, Port Blair
2. Central Agricultural University, Manipur
3. Central Arid Zone Research Institute, Jodhpur
4. Central Avian Research Institute, Izatnagar
5. Central Inland Fisheries Research Institute, Barrackpore
6. Central Institute Brackish water Aquaculture, Chennai
7. Central Institute for Research on Buffaloes, Hisar
8. Central Institute for Research on Goats, Makhdoom
9. Central Institute of Agricultural Engineering, Bhopal
10. Central Institute of Arid Horticulture, Bikaner
11. Central Institute of Cotton Research, Nagpur
12. Central Institute of Fisheries Technology, Cochin
13. Central Institute of Freshwater Aquaculture, Bhubneshwar
14. Central Institute of Research on Cotton Technology, Mumbai
15. Central Institute of Sub Tropical Horticulture, Lucknow
16. Central Institute of Temperate Horticulture, Srinagar
17. Central Institute on Post harvest Engineering and Technology, Ludhiana
18. Central Marine Fisheries Research Institute, Kochi
19. Central Plantation Crops Research Institute, Kasargod

20. Central Potato Research Institute, Shimla
21. Central Research Institute for Jute and Allied Fibres, Barrackpore
22. Central Research Institute of Dryland Agriculture, Hyderabad
23. Central Sheep and Wool Research Institute, Avikanagar, Rajasthan
24. Central Soil and Water Conservation Research & Training Institute, Dehradun
25. Central Soil Salinity Research Institute, Karnal
26. Central Tobacco Research Institute, Rajahmundry
27. Central Tuber Crops Research Institute, Trivandrum
28. ICAR Research Complex for Eastern Region including Centre of Makhana, Patna
29. ICAR Research Complex for NEH Region, Barapani
30. ICAR Research Complex Goa
31. Indian Agricultural Statistical Research Institute, New Delhi
32. Indian Grassland and Fodder Research Institute, Jhansi
33. Indian Institute of Agricultural Biotechnology, Ranchi
34. Indian Institute of Horticultural Research, Bangalore
35. Indian Institute of Natural Resins and Gums, Ranchi
36. Indian Institute of Pulses Research, Kanpur
37. Indian Institute of Soil Sciences, Bhopal
38. Indian Institute of Spices Research, Calicut
39. Indian Institute of Sugarcane Research, Lucknow
40. Indian Institute of Vegetable Research, Varanasi
41. National Academy of Agricultural Research & Management, Hyderabad
42. National Biotic Stress Management Institute, Raipur
43. National Institute of Abiotic Stress Management, Malegaon, Maharashtra
44. National Institute of Animal Nutrition and Physiology, Bengaluru
45. National Rice Research Institute, Cuttack
46. National Institute of Research on Jute & Allied Fibre Technology, Kolkata
47. Sugarcane Breeding Institute, Coimbatore
48. ICAR-Vivekananda Parvatiya Krishi Anusandhan Sansthan, Almora
49. ICAR-Central Institute for Research on Cattle, Meerut, Uttar Pradesh
50. ICAR-National Institute of High Security Animal Diseases, Bhopal
51. ICAR-Indian Institute of Maize Research, New Delhi
52. ICAR- Central Agroforestry Research Institute , Jhansi

53. ICAR-National Institute of Agricultural Economics and Policy Research, New Delhi
54. ICAR- Indian Institute of Wheat and Barley Research, Karnal
55. ICAR- Indian Institute of Farming Systems Research, Modipuram
56. ICAR- Indian Institute of Millets Research, Hyderabad
57. ICAR- Indian Institute of Oilseeds Research, Hyderabad
58. ICAR- Indian Institute of Oil Palm Research, Pedavegi, West Godawari
59. ICAR- Indian Institute of Water Management, Bhubaneshwar
60. ICAR-Indian Institute of Rice Research, Hyderabad
61. ICAR- Central Institute for Women in Agriculture, Bhubaneshwar
62. ICAR-Central Citrus Research Institute, Nagpur
63. ICAR-Indian Institute of Seed Research, Mau
64. ICAR-Indian Agricultural Research Institute, Post Box No. 48, Hazaribag 825 301, Jharkhand

c. ***National Research Centres - 15***

1. National Centre for Integrated Pest Management, New Delhi
2. National Research Centre for Banana, Trichi
3. National Research Centre for Citrus, Nagpur
4. National Research Centre for Grapes, Pune
5. National Research Centre for Litchi, Muzaffarpur
6. National Research Centre for Pomegranate, Solapur
7. National Research Centre on Camel, Bikaner
8. National Research Centre on Equines, Hisar
9. National Research Centre on Meat, Hyderabad
10. National Research Centre on Mithun, Medziphema, Nagaland
11. National Research Centre on Orchids, Pakyong, Sikkim
12. National Research Centre on Pig, Guwahati
13. National Research Centre on Plant Biotechnology, New Delhi
14. National Research Centre Seed Spices, Ajmer
15. National Research Centre on Yak, West Kemang

d. **National Bureaus - 6**

1. National Bureau of Plant Genetics Resources, New Delhi
2. National Bureau of Agriculturally Important Micro-organisms, Mau, Pradesh

3. National Bureau of Agriculturally Important Insects, Bangalore
4. National Bureau of Soil Survey and Land Use Planning, Nagpur
5. National Bureau of Animal Genetic Resources, Karnal
6. National Bureau of Fish Genetic Resources, Lucknow

e. ***Directorates/Project Directorates - 13***

1. Directorate of Groundnut Research, Junagarh
2. Directorate of Soybean Research, Indore
3. Directorate of Rapeseed & Mustard Research, Bharatpur
4. Directorate of Mushroom Research, Solan
5. Directorate of Onion and Garlic Research, Pune
6. Directorate of Cashew Research, Puttur
7. Directorate of Medicinal and Aromatic Plants Research, Anand
8. Directorate of Floriculture Research, Pusa, New Delhi
9. Directorate of Weed Science Research, Jabalpur
10. Project Directorate on Foot & Mouth Disease, Mukteshwar
11. Directorate of Poultry Research, Hyderabad
12. Directorate of Knowledge Management in Agriculture (DKMA) New Delhi
13. Directorate of Cold Water Fisheries Research, Bhimtal, Nainital.

2. State Agricultural Universities

S.N	Name, email & website	Address	Telephone/Fax No.
1	Acharya NG Ranga Agricultural University Website: http://www.angrau.ac.in Email: angrau_vc@yahoo.com, raghuvardhanreddy_s@rediffmail.com	Adminstrative Office, Rajendra Nagar, Hyderabad-500030, Andhra Pradesh	040-24015035, 24013095 Fax: 040-24015031
2	Anand Agricultural University Website: http://www.aau.in Email: vc@aau.in, vc_aau@yahoo.com	Anand 388110, Gujarat	02692-261273 Fax: 02692-261520
3	Assam Agricultural University Website: http://www.aau.ac.in Email: vc@aau.ac.in, kmbujarbaruah@rediffmail.com	Jorhat 785013, Assam	0376-2340013 Fax: 0376-2340001
4	Agriculture University Website: http://aukota.org/ Email: aukota2013@gmail.com,	Borkhera, Baran Road, Kota, Rajasthan 324001	0744 232 1203

S.N	Name, email & website	Address	Telephone/Fax No.
5	Agriculture University Website: http://aujodhpur.ac.in/ Email: vcunivag@gmail.com,	Jodhpur, Mandor, Rajasthan - 342304	02912570711
6	Bidhan Chandra Krishi Viswavidyalaya Website: http://www.bckv.edu.in Email:bckvvc@gmail.com,sarojsanyal@yahoo.co.in	Mohanpur, Nadia-741252, West Bengal	033-25879772, 03473-222666 Fax:03473-222275
7	Banda University of Agriculture and Technology Website: http://buat.edu.in/ Email: vc.mskjuat@gmail.com	Banda - 210001, Uttar Pradesh	05192-221605 Fax:02426-243302
8	Bihar Agricultural University Website: http://www.bausabour.ac.in Email: vcbausabour@gmail.com	Sabour, Bhagalpur 813210, Bihar	0641-2452606 Fax:0641-2452604
9	Bihar Animal Sciences University Website: http://www.bvcpatna.org.in/ Email: principalbvc@gmail.com	Bihar Veterinary College, Patna – 800 014, Bihar	+91-612-2226644, 2222231 (O) Fax: +91-612-2222231
10	Birsa Agricultural University Website: http://www.baujharkhand.org Email: vc_bau@rediffmail.com	Kanke, Ranchi-834006, Jharkhand	0651-2450500 Fax:0651-2450850
11	Chandra Shekar Azad University of Agriculture & Technology Website:http://www.csauk.ac.in Email: vc@csauk.ac.in	Kanpur-208002, Uttar Pradesh	0512-2534155 Fax:0512-2533808
12	Chaudhary Charan Singh Haryana Agricultural University Website: https://www.hau.ac.in/ Email: vc@hau.ernet.in	Hisar-125004, Haryana	01662-231640, 284301 Fax:01662-234952
13	CSK Himachal Pradesh Krishi Vishvavidyalaya Website: http://www.hillagric.ac.in Email: vc@hillagric.ac.in	Palampur-176062, Himachal Pradesh	01894-230521 Fax:01894-230465
14	Chhattisgarh Kamdhenu Vishwavidyalaya Website: http://cgkv.ac.in	Anjora ,Durg, Chhattisgarh	
15	Dr Balasaheb Sawant Konkan Krishi Vidyapeeth Website: www.dbskkv.org Email: vcbskkv@yahoo.co.in	Dapoli Distt, Ratnagiri 415 712, Maharashtra	02358-282064 Fax:02358-282074

S.N	Name, email & website	Address	Telephone/Fax No.
16	Dr Panjabrao Deshmukh Krishi Vidyapeeth Website: http://www.pdkv.ac.in Email: vc@pdkv.ac.in	Krishinagar, Akola-444104, Maharashtra	0724-2258365 Fax:0724-2258219
17	Dr Yashwant Singh Parmar Univ of Horticulture & Forestry Website: http://www.yspuniversity.ac.in Email: vc@yspuniversity.ac.in, vcuhf@yahoo.com	Solan, Nauni-173230, Himachal Pradesh	01792-252363 Fax:01792-252242
18	Dr YSR Horticultural University Website: http://www.drysrhu.edu.in Email: vc@drysrhu.edu.in	Adminstrative office, Venkataramannagudem, PB No. 7, West Godavari Dist., Tadepalligudem- 534101 Andhra Pradesh	08818-284311 Fax: 08818-284223
19	Govind Ballabh Pant University of Agriculture & Technology Website: http://www.gbpuat.ac.in Email:vcgbpuat@gmail.com	Pantnagar-263145,Distt Udham Singh , Nagar , Uttarakhand	05944-233333,233500 Fax: 05944-233500
20	Guru Angad Dev Veterinary and Animal Science University Website: http://www.gadvasu.in Email:vijay_taneja@hotmail.com, vcgadvasu@gmail.com	Ludhiana - 141004, Punjab	0161-2553320,2553360 Fax:0161-2553340
21	Indira Gandhi Krishi Vishwavidyalaya Website: www.igau.edu.in Email: vcigkv@gmail.com	Krishak Nagar, Raipur-492006, Chhattisgarh	0771-2443419 Fax:0771-2442302, 2443121
22	Jawaharlal Nehru Krishi Viswavidyalaya Website: http://www.jnkvv.nic.in Email: gkalloo_jnkvv@yahoo.co.in	Krishi Nagar, Adhartal Jabalpur-482004, Madhya Pradesh	0761-2681706 Fax:0761-2681389
23	Junagadh Agricultural University Website: http://www.jau.in Email: vc@jau.in	Univ. Bhavan, Motibagh Junagadh-362001, Gujarat	0285-2671784 Fax:0285-2672004
24	Kamdhenu University Website: http://www.ku-guj.org/ Email: registrar.kamdhenuuniversity@gmail.com	Karmayogi Bhavan, Block-1, B1-wing, 4th Floor, Room No.414, Sector-10-A, Di: Gandhinagar-382010 Gujarat, India.	(079)65726668

S.N	Name, email & website	Address	Telephone/Fax No.
25	Karnataka Veterinary, Animal and Fisheries Sciences University Website: http://kvafsu.edu.in Email: vckvafsu@yahoo.co.in dekvafsub@yahoo.com, sskvafsu@yahoo.co.in	Nandinagar, PB No 6, BIDAR 585401, Karnataka	08482-245264 Fax:08482- 245107
26	Kerala Agricultural University Website: http://www.kau.edu Email: vc@kau.in, vicechancellorkau@gmail.com	Vellanikara, Trichur 680656, Kerala	0487-2371928, 2370034, 2438001 Fax:0487-2370019
27	Kerala University of Fisheries & Ocean Studies Website: http://www.kau.edu Email: kurup424@gmail.com	Papangad, Kochi-682506 Kerala	0487-2370117, 2703781, 2700964
28	Kerala Veterinary and Animal Sciences University Website: http://www.kcasu.ac.in Email: vc@kvasu.ac.in, vc.vetuny@gmail.com	Liason Office, Directorate of Dairy Development, Pattom, Thiruvananthapuram, 695004, Kerala	0471-2550058 Fax: 0471-2550480
29	Lala Lajpat Rai University of Veterinary & Animal Sciences Website: http://www.llruvas.edu.in/ Email: vc@llruvas.edu.in	Hisar, Haryana	01662 289332 01662- 272002
30	Nanaji Deshmukh Veterinary Science University Website: http://www.mppcvv.org Email: vcnduvs@gmail.com	South Civil Lines, Jabalpur-482001, Madhya Pradesh	0761-2678007 Fax: 0761-2620783
31	Maharana Pratap Univ. of Agriculture & Technology Website: http://www.mpuat.ac.in Email: vc@mpuat.ac.in	Udaipur, Rajasthan 313001	0294-2471101 Fax:0294-2470682
32	Maharashtra Animal Science & Fishery University Website: http://www.mafsu.in Email: vcmatsu@gmail.com, cadaba_prasad@yahoo.co.in	Seminary Hills, Nagpur-440006, Maharashtra	0712-2511088 Fax:0712-2511282
33	Mahatma Phule Krishi Vidyapeeth Website: http://mpkv.ac.in Email: vcmpkv@rediffmail.com	Rahuri-413722, Maharashtra	02426-243208 Fax:02426-243302
34	Maharana Partap Horticultural University Website: http://www.mhu.ac.in/ Email:mhuinfo.hry@gmail.com	CCS Haryana Agricultural University Campus,Hisar-125001 (Haryana)	01662-256081

S.N	Name, email & website	Address	Telephone/Fax No.
35	Vasantrao Naik Marathwada Krishi Vidyapeeth Website: http://www.vnmkv.ac.in/ Email: vcmau@rediffmail.com	Parbhani-431402, Maharashtra	02452-223002 Fax:02452-223582
36	Narendra Deva University of Agriculture & Technology Website:http://www.nduat.ernet.in Email:vc_nduat2010@yaho.co.in	Kumarganj, Faizabad -224229, Uttar Pradesh	05270-262097, 262161 Fax:05270-262097
37	Navsari Agricultural University Website: http://www.nau.in Email: vc_2004@yahoo.co.in	Navsari-396450 Gujarat	02673-283869 Fax:02673-284254
38	Orissa Univ. of Agriculture & Technology Website: http://www.ouat.ac.in Email: ouat_dproy@yahoo.co.in, bsenapati1942@yahoo.com	Bhubaneshwar-751003, Orissa	0674-2397700 Fax:0674-2397780
39	Professor Jayashankar Telangana State Agricultural University (PJTSAU) Website: www.pjtsau.ac.in Email: regrpjtsau@gmail.com	Rajendranagar, Hyderabad-500 030 Telangana, India.	EPABX: 040-24015011 to 040-24015017 040-24015161 to 040-24015163
40	P.V. Narsimha Rao Telangana Veterinary University Website: http://tsvu.nic.in/	Administrative Office, RajendraNagar, Hyderabad-500 030.	Phone : 040-24002114, Fax : 040-24002114
41	Punjab Agricultural University Website: http://www.pau.edu Email: vcpau@pau.edu	Ludhiana-141004, Punjab	0161-2401794 Fax:0161-2402483
42	Rajasthan University of Veterinary and Animal Sciences Website: http://rajuvas.org Email: vcrajuvas@gmail.com	Bijey Bhavan Place Complex (Pt Deen Dayal circle) Bikaner 334006 Rajasthan	0151-2543419 Fax: 0151-2549348
43	Rajendra Agricultural University Website: http://www.pusavarsity.org.in Email: : vcrau@sify.com	Pusa, Samastipur 848125, Bihar	06274-240226 Fax:06274-240255
44	Rajmata Vijayraje Sciendia Krishi Vishwa Vidyalaya Website: ht tp://www.rvskvv.nic.in Email: vcrvskvv@gmail.com	Race Cource Road, Gwalior 474002 Madhya Pradesh	0751-2467673 Fax: 0751-2467673
45	Sam Higginbottom University of Agriculture, Technology & Sciences, Allahabad Website:ht tp://www.shiats.edu.in/ Email: vcrvskvv@gmail.com	Rewa Road,Nani, Allahabad-211007 Uttar pradesh state	+91-532-2684781, Fax:+91 5322684346

S.N	Name, email & website	Address	Telephone/Fax No.
46	Sardar Vallabhbhai Patel University of Agriculture and Technology Website: http://www.svbpmeerut.ac.in Email:vc_agunivmeerut@yahoo.com	Modipuram, Meerut - 250110 Uttar Pradesh	0121-2888522, Fax:0121-2888505
47	Sardarkrushinagar-Dantiwada Agricultural University Website: http://www.sdau.edu.in Email: vc@sdau.edu.in	Sardar Krushinagar, Distt Banaskantha, Gujarat-385506	02748-278222 Fax:02748-278261
48	Sher-E-Kashmir Univ of Agricultural Sciences & Technology Website: http://www.skuast.org Email: vc@skuast.org	Railway Road, Jammu 18009, J&K	0191-2263714 Fax:0191-2262073
49	Sher-E-Kashmir Univ of Agricultural Sciences & Technology of Kashmir Website: http://www.skuastkashmir.ac.in Email: vc@skuastkashmir.ac.in, skuastkvc@gmail.com	Shalimar Campus, Shrinagar-191121, Jammu & Kashmir	0194-2464028, 2462159 Fax:0194-2462160, 2461543
50	Sri Karan Narendra Agriculture University Website:www.sknau.ac.in Email:info@sknau.ac.in, registrar@sknau.ac.in	Jobner,Jaipur-District,Rajasthan	9414518152 Telefax-01425-254980
51	Sri Konda Laxman Telangana State Horticultural University (SKLTSHU) Website: www.skltshu.ac.in Email:registrar@skltshu.ac.in, registrarskltshu@gmail.com	Administrative Office Rajendranagar, Hyderabad-500 030 Telangana, India.	040-24014301 (O), 24014312 (F)
52	Sri Venkateswara Veterinary University Website:http://www.svvu.edu.in Email:prabhakarvrao@yahoo.com	Admn office, Regional Library Building, Tirupati-517502	0877-2248986 Fax:0877-2249222
53	Swami Keshwanand Rajasthan Agricultural University Website: http://www.raubikaner.org Email: vcrau@raubikaner.org	Bikaner-334006,Rajasthan	0151-2250443, 2250488 Fax:0151-2250336

S.N	Name, email & website	Address	Telephone/Fax No.
54	Tamil Nadu Agricultural University Website: http://www.tnau.ac.in Email: vc@tnau.ac.in	Coimbatore-641003, Tamil Nadu	0422-2431788, 2431672 Fax:0422-2431672
55	Tamil Nadu Fisheries University Website: http://www.tnfu.org.in Email:	Nagapattinam – 611 003, Tamil Nadu	04365-253011
56	Tamil Nadu Veterinary & Animal Science University Website: http://www.tanuvas.ac.in Email:vc@tanuvas.org.in, karanmgk@gmail.com	Chennai-600051, Tamilnadu	044-25551574, 25551575 Fax:0444-225551576
57	University of Agricultural Sciences, Bangalore Website: http://www.uasbangalore.edu.in Email:vc@uasbangalore.edu.in, knarayanagowda@yahoo.co.in	GKVK, Bengaluru-560065, Karnataka	080-23332442 Fax:080-23330277
58	University of Agricultural Sciences, Dharwad Website: http://www.uasd.edu Email: vc_uasd@rediffmail.com	Dharwad-580005, Karnataka	0836-2447783, Fax:0836-2448349
59	University of Agricultural and Horticultural Sciences, Shivamogga Website: http://uahs.in/ Email:	Shimoga, Karnataka	
60	University of Horticultural Sciences Website: http://uhsbagalkot.edu.in/ Email: dandinbnm@gmail.com	Sector No 60 Navanagar Bagalkot 587102 Karnataka	08354 201310 Fax: 08354 235152
61	University of Agricultural Sciences Website: http://www.uasraichur.edu.in Email: vcuasraichur10@rediffmail.com	PB 329, Raichur – 584101 Karnataka	08532-221444 Fax: 08532- 220444

S.N	Name, email & website	Address	Telephone/Fax No.
62	UP Pandit Deen Dayal Upadhaya Pashu Chikitsa Vigyan Vishwa Vidhyalaya evam Go Anusandhan Sansthan Website: http://www.upvetuniv.edu.in Email: singhambika1945@gmail.com, duvasuvc@gmail.com	Mathura-281001 , Uttar Pradesh	0565-2470199 Fax:0565-2404819
63	Uttarakhand University of Horticulture and Forestry Website: http://www.uuhf.ac.in/ Email:	Pauri Garhwal, Uttarakhand	09412051976
64	Uttar Banga Krishi Viswavidyalaya Website:http://www.ubkv.ac.in Email: vcubkv@gmail.com, vcubkv@rediffmail.com	P.O. Pundibari, Dist. Coach Bihar-736165,West Bengal	03582-270141, 270013 Fax:03582-270249
65	West Bengal University of Animal & Fishery Sciences Website:http://www.wbuafscl.ac.in Email: wbuafs@wb.nic.in	68 KB Sarani, Kolkata-700037, West Bengal	033-25563450 Fax:033-25571986

Research Centers – CGIAR (Consultative Group for International Agricultural Research) - 15

S.N	Name of the center	Head Office	Year of Establishment
1	Africa Rice Centre (ARC) (Former name-West Africa Rice Development Association (WARDA)	Abidjan - Côte d'Ivoire (Benin)	1971
2	Center for International Forestry Research (CIFOR)	Bogor - Indonesia	1993
3	International Center for Agricultural Research in Dry Areas (ICARDA)	Beirut - Lebanon	1977
4	International Crops Research Institute for Semi-arid Tropics (ICRISAT) E-mail: icrisat@cgiar.org	Patancheru. Hyderabad (India)	1972
5	International Food Policy Research Institute (IFPRI)	Washington DC - USA	1975
6	International Institute of Tropical Agriculture (IITA)	Ibadan - Nigeria	1967
7	International Livestock Research Institute (ILRI)	Nairobi, Kenya	1994
8	International Maize and Wheat Improvement Centre (CIMMYT)	Texcoco - Mexico	1943
9	International Potato Centre (CIP)	Lima, Peru	1971

S.N	Name of the center	Head Office	Year of Establishment
10	International Rice Research Institute (IRRI)	Los Baños - Philippines	1960
11	International Water Management Institute (IWMI)	Colombo, Sri Lanka	1985
12	International Center for Tropical Agriculture (CIAT)	Rome, Italy	1967
	Biodiversity International	Rome, Italy	1974
13	International Crop Research on Agro-forestry (ICRAF): World Agroforestry	Nairobi - Kenya	1978
14	World Fish	Penang - Malaysia	1975

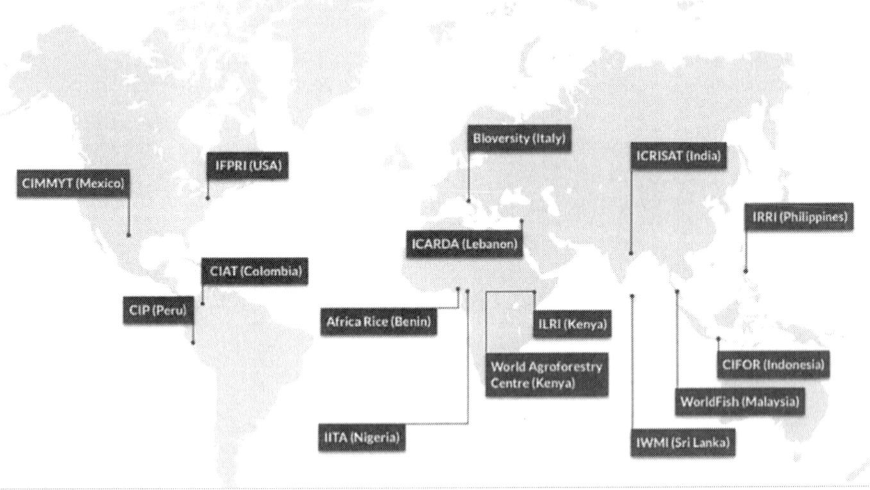

CGIAR institutes

Index

A

B

C

D

E

F

G

H

I

J

K

L

M

N

O

P

R

S

T

U

V

W

Z